Introduction to Materials Science and Engineering

Introduction to Materials Science and Engineering: A Design-Led Approach

Michael Ashby, Hugh Shercliff, and David Cebon

Department of Engineering, University of Cambridge, UK

Butterworth-Heinemann
An imprint of Elsevier

ELSEVIER

Butterworth-Heinemann is an imprint of Elsevier
The Boulevard, Langford Lane, Kidlington, Oxford OX5 1GB, United Kingdom
50 Hampshire Street, 5th Floor, Cambridge, MA 02139, United States

Notices

Knowledge and best practice in this field are constantly changing. As new research and experience broaden our understanding, changes in research methods, professional practices, or medical treatment may become necessary.

Practitioners and researchers must always rely on their own experience and knowledge in evaluating and using any information, methods, compounds, or experiments described herein. In using such information or methods they should be mindful of their own safety and the safety of others, including parties for whom they have a professional responsibility.

To the fullest extent of the law, neither the Publisher nor the authors, contributors, or editors, assume any liability for any injury and/or damage to persons or property as a matter of products liability, negligence or otherwise, or from any use or operation of any methods, products, instructions, or ideas contained in the material herein.

ISBN: 978-0-08-102399-0

For Information on all Butterworth-Heinemann publications visit our website at https://www.elsevier.com/books-and-journals

Printed and bound in Great Britain by Bell and Bain Ltd, Glasgow

Publisher: Peter Linsley
Acquisitions Editor: Stephen R. Merken
Editorial Project Manager: Shilpa Kumar
Production Project Manager: Beula Christopher
Cover Designer: Renee Duenow

Typeset by MPS Limited

Working together to grow libraries in developing countries

www.elsevier.com • www.bookaid.org

Contents

Preface

Teaching Materials Science and Engineering

Materials Science and Engineering (MSE) sits squarely at the interface between multiple under-pinning scientific disciplines and their application in all branches of engineering. The breadth of the integration of disciplines is very large – the figure shows that modern MSE spans not only branches of physics and chemistry, and the well-established branches of engineering (mechanical, electrical, civil, and so on) but increasingly biological and environmental sciences and the global engineering challenges of health care and climate change.

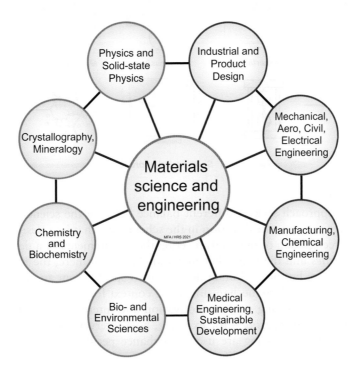

MSE is established as a bachelor's and master's discipline in its own right, but remains a minority programme within applied sciences and engineering. More commonly, Materials is a core intro-ductory course for first- and second-year university and college students in related disciplines – teaching MSE needs to meet the challenge of preparing students for an ever-increasing breadth of courses and potential careers. It is impossible, and undesirable, to try and cover everything. What is important though is to nurture a structured approach to the subject, developing thought processes and strategies for problem solving that can be translated across disciplines. Central to this philosophy is to continuously make connections between the characteristic behaviour of materials, captured by their properties, and applications of MSE – many of which involve the realisation of products through engineering design and processing.

The traditional approach to teaching Materials starts with fundamentals: atoms and electrons, atomic bonding and packing, crystallography and crystal defects, before moving up the length-scales of microstructure to alloy theory and phase transformations. This establishes the understanding of material properties, and some aspects of processing in controlling them, but gives less emphasis to the behaviour of structures and components in service, particularly the practicalities of engineering design and manufacturing. This *science-led approach* is shown in the accompanying schematic, tracking from right to left.

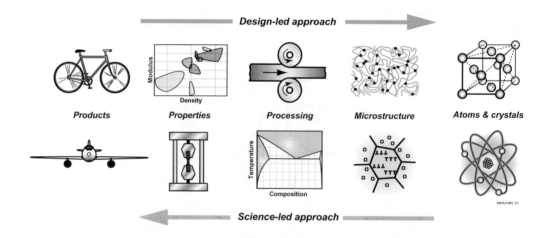

A fully integrated approach seeks a different balance between the science and engineering, by making connections in the opposite direction in the schematic – that is, a *design-led approach*. The starting point is the recognition of materials meeting the requirements of a design, to establish the central role of *properties* and the ways that materials fail. From here, the traditional science-led content naturally follows, but now 'drilling down' (so to speak) from an appreciation of the relevance of each property or failure mechanism, giving context for understanding the underpinning science (and potentially enhancing student motivation and interest). A design-led approach also gives weight to *materials engineering* – for example, how interventions in material composition, internal architecture, and processing are used to enhance performance, and the environmental impact of materials over the product life cycle. Future engineers and designers need practice in performing design calculations and in choosing suitable materials and processes – both requiring systematic handling of materials data and know-how.

The approach to teaching and learning Materials in this book

This book takes a *design-led approach* to introductory teaching and learning in MSE. It is structured, however, to meet the needs of those who prefer a *science-led* course, with design-centric sections readily being omitted. The text has *inclusivity and breadth*, with appropriate weight being given to all classes of material: metals, polymers, ceramics, 'hybrids' such as composites and foams, and natural materials. In common with most introductory Materials texts, the emphasis is on mechanical and structural properties, in line with the parallel emphasis on product design, but the same approach is applied, in less detail, to the 'functional' properties – electrical, magnetic, and optical. The core themes of properties and microstructure, and the breadth of context and applications, provide an introduction to MSE that is relevant across disciplines, extending well beyond mechanical/aerospace engineering and product design, to civil engineering, architecture, electrical engineering, and biomedical engineering.

Chapter 1 introduces material classes and their properties, while Chapters 2 and 3, respectively, provide introductions to materials in design and the fundamentals of material microstructure (including atomic and electronic scale behaviour). Chapters 4 – 9 cover core material properties and behaviours in turn – from elasticity, plasticity and fracture, to thermal and high temperature behaviour, and the surface-dominated topics of wear and corrosion. In each case the presentation follows the sequence of (i) property definition, measurement and data; (ii) 'drilling down' into the science; and (iii) applications and design calculations. Chapter 10 provides an introductory treatment of electrical, magnetic, and optical properties, emphasising the equivalent couplings between microstructure and properties, and applications. Chapters 11 and 12, respectively, focus on the role of manufacturing processes in microstructure and property control, and the sustainability of engineering materials. Practice in analytical problem solving and checking knowledge acquisition is provided in every chapter by worked examples in the text (with answers), and concluding exercises (with solutions online). Certain topics in MSE lend themselves particularly well to 'learning-by-doing,' and a number of topics are expanded separately as *Guided Learning Units 1 – 4: Crystallography, Materials Selection in Design, Process Selection in Design*, and *Phase Diagrams and Phase Transformations* (again with full solutions online).

A central feature of the book is the emphasis on *visual communication* through a unique graphical presentation of material properties as *material property charts* and numerous *schematics*. These are powerful in utilising visual memory as a learning tool, while the charts are multi-purpose: in the first instance, they support understanding of the origins of properties, their manipulation, and their fundamental limits; in addition, they provide a tool for material selection and for understanding the design trade-offs in product design.

This new text requires the level of mathematical facility appropriate to a first course in MSE and provides a balance of content between science and engineering. The design-led approach in this introductory text is derived from an equivalent higher-level text for later undergraduate or graduate-level courses[1], in which the breadth and depth are greater, notably in the applications, design case studies, and level of analysis. The design-led treatment of Materials is also developed in five further textbooks, the first relating to Mechanical Design[2],

---

[1] Ashby, M.F., Shercliff, H.R., & Cebon, D. (2019). *Materials: Engineering, science, processing and design* (4th ed.). Butterworth-Heinemann. ISBN 9780081023761. (A more advanced text developing the design-led approach to Materials in greater depth).

[2] Ashby, M.F. (2017). *Materials selection in mechanical design* (5th ed.). Butterworth-Heinemann. ISBN 9780081005996. (An advanced text developing material selection methods in detail).

the second and third to Design for the Environment[3,4], the fourth to Industrial Design[5], and the fifth to Manufacturing and Design[6].

This book and the Ansys/Granta EduPack Materials and Process Information software

Engineering design today takes place in a computer-based environment. Stress analysis (finite element method [FEM] codes, for instance), computer-aided design (CAD), design for manufacture (DFM), and product data management (PDM) tools are part of an engineering education. The Ansys/Granta EduPack Materials and Process Information software[7] for education (formerly the *Cambridge Engineering Selector [CES] EduPack*) provides a computer-based environment implementing the methodology of this textbook (e.g. for optimised materials selection and design, eco-auditing, and introducing processing, including the underlying materials science).

   This book is self-contained and does not depend on the Ansys/Granta EduPack for its use. But it is readily used with the Ansys/Granta Design software, should the instructor and student wish to do so – and the same goes for the more advanced edition[1]. Using the Ansys/Granta EduPack enhances the learning experience and provides a solid grounding in many of the domains of expertise specified by the various professional engineering accreditation bodies (analysis of components; design and manufacture in project work; economic, societal, and environmental impact; and so on). In addition, the Granta Elements database documents the fundamental properties of all 112 stable elements of the Periodic Table, offering greater depth of investigation in a science-led course.

[3] Ashby, M.F. (2021). *Materials and the environment – eco-informed material choice* (3rd ed.). Butterworth-Heinemann. ISBN 978-0-12-821521-0. (A teaching text introducing students to the concepts and underlying facts about concerns for the environment and the ways in which materials and products based on them can both help and harm it).

[4] Ashby, M.F. (2022). *Materials and sustainable development* (2nd ed.). Oxford: Butterworth-Heinemann Ltd. ISBN 9780323983617. (A text presenting a methodology for assessing whether a material, product, or even a whole technology can be regarded as sustainable).

[5] Ashby, M.F., & Johnson, K. (2013). *Materials and design, the art and science of material selection in product design* (3rd ed.). Butterworth-Heinemann. ISBN 978-0-08-098205-2. (A text that complements those above, dealing with the aesthetics, perceptions and associations of materials and their importance in product design).

[6] Tempelman, E., Shercliff, H.R., & Ninaber van Eyben, B. (2014). *Manufacturing and design – understanding the principles of how things are made*. Butterworth-Heinemann. ISBN 978-0-08-099922-7. (The source of the 'manufacturing triangle' and the structured approach to process-material-design interactions).

[7] The Ansys/Granta EduPack, Granta Design Ltd., Rustat House, 62 Clifton Court, Cambridge CB1 7EG, UK. https://www.ansys.com/products/materials/granta-selector and https://innovationspace.ansys.com/educator-hub/materials-resources/

Acknowledgements

No book of this sort is possible without advice, constructive criticism, and ideas from others. Our colleagues at Cambridge University, too numerous to mention, have been generous over many years with their time, thoughts, and feedback – as have collaborators in many other UK, North American, and European universities with the same enthusiasm for developing the teaching of this subject. Equally valuable has been the contribution of the team at Granta Design, Cambridge (now part of Ansys), responsible for the development of the Ansys/Granta EduPack software, and leaders in the evolution of Materials education through their many annual symposia.

Reviewers

We gratefully acknowledge the following reviewers of this edition, whose valuable insight helped to shape this text:

Shane Brauer (Mississippi State University)
Michael Fisch (Kent State University)
Erik Knudsen (George Mason University)
Vlado Lubarda (University of California, San Diego)

Resources that accompany this book

Resources are available to adopting instructors who register on the Elsevier textbook website, https://educate.elsevier.com/book/details/9780081023990:

Instructor's Manual of Exercises and Solutions The book includes a comprehensive set of exercises that can be completed using information, diagrams, equations, and data contained in the book itself (with a small number involving Web searches or independent student observation). Fully worked-out solutions to the exercises are freely available online to tutors and lecturers who adopt this book.

Image Bank This bank provides adopting tutors and lecturers with electronic versions of the figures from the book that may be used in lecture slides and class presentations.

PowerPoint Lecture Slides Use the available set of lecture slides in your own course as provided, or edit and reorganise them to meet your individual course needs.

Chapter 1

Introduction: materials — history, classification, and properties

Cover figure: A replica of Robert Stephenson's 'Rocket,' one of a series of early steam locomotives, built in the UK in 1829. (Image credit: Science Museum, London, UK)

https://doi.org/10.1016/B978-0-08-102399-0.00001-7

1.1 Materials science and engineering: a bit of history

Engineers *make* things – products from mobile electronics and household possessions to vehicles and civil infrastructure. These are all made from *materials*, and shaped, assembled, and finished using *manufacturing processes*. Materials support mechanical loads, they insulate or conduct heat and electricity, they channel or reject magnetic flux, transmit or reflect light, and tolerate hostile surroundings. Ideally, they do all this without damage to the environment or costing too much. So materials are central to all branches of *engineering design*: mechanical, aerospace, electrical, civil, and so on. Materials occur in nature too, showing many of the same physical behaviours, from the biomechanics and biochemistry of bones, tissues, and organs in the bodies of humans and animals, to the deformation and fracture of rocks, soils, and ice in geophysics – earthquakes and glaciers are large-scale materials problems.

Materials science and engineering is the study of the response of materials to physical and chemical stimuli – that is, their *properties*. These reflect the internal workings of materials, from the atomic and molecular scale upwards through many length-scales of *microstructure*. The field encompasses the man-made materials of engineering – metals, ceramics, polymers, fibre composites, and so on. But as noted above, the same knowledge of properties and their origins applies to naturally occurring materials, first for scientific investigation of their behaviour, but also to understand their interaction with man-made products – notably in bioengineering and civil engineering.

The parallel themes in this book are therefore materials and properties in engineering design, and the underlying science of the physical and chemical origins of these properties. A design-led approach presents materials science and engineering in familiar contexts, introducing the key properties in relation to meeting the requirements of products, including the coupling of materials to manufacturing processes.

Engineers have been making things out of materials for centuries, and successfully so – think of Isambard Kingdom Brunel, Thomas Telford, Gustave Eiffel, Henry Ford, Karl Benz and Gottlieb Daimler, the Wright brothers. But selecting materials and processes for products in the 21st century has become increasingly challenging and complex. A little history shows why. The picture on the cover page of this chapter shows Robert Stephenson's 'Rocket,' built in 1829. In Stephenson's day the number of materials available to engineers was small – a few hundred at most. There were no synthetic polymers – there are now over 45,000 of them – and no light alloys (like aluminium) or high-performance composites – now there are thousands of variants. Figure 1.1 shows this material evolution schematically, across four broad categories (elaborated in Section 1.2). The time-axis spans 10,000 years and is very non-linear – almost all the materials we use today were developed in the last 100 years. Their number is enormous: over 160,000 materials are available to today's engineer, presenting a real challenge of optimum selection. In parallel, material processing develops to suit new materials, while new techniques emerge for existing materials – for example, laser processing and additive manufacturing. And on top of this is a new imperative: the environment and climate change are driving the transition to a circular economy of recovery and re-use of materials, low carbon processing, and new materials such as biopolymers and biocomposites.

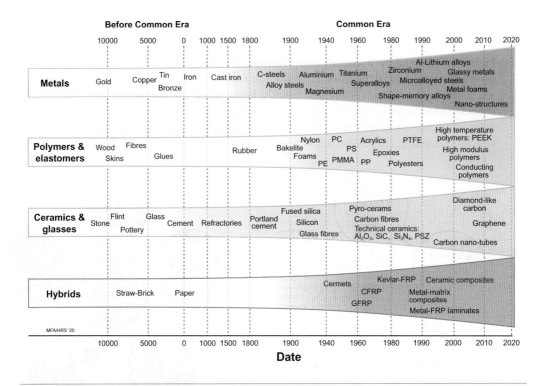

Figure 1.1 The development of materials over time. The materials of pre-history are those of nature. The development of thermo-chemistry, electro-chemistry, and polymer chemistry enabled the production of man-made materials. Three materials – stone, bronze, and iron – were so important that the era of their dominance is named after them.

The properties of engineering materials are introduced in outline in Section 1.3. Property data, and technical 'know-how' about using and processing materials, are well documented in handbooks; one such – the *ASM Materials Handbook* – runs to 22 fat volumes. How are we to deal with this vast body of information? Fortunately, another thing has changed since Robert Stephenson's day: digital information storage and manipulation. Computer-aided design (CAD) is now a standard part of an engineer's training, backed up by packages for solid modeling, finite-element analysis, optimisation, and material and process selection. Software for the last of these – the selection of materials and processes – draws on databases of the attributes of materials and processes, allowing them to be searched and displayed in ways that enable selections that best meet the requirements of a design. This text exploits novel graphical formats to display the properties of materials, in ways that are also implemented in software – examples are introduced in Section 1.4. These *material property charts* present data in a visually accessible way, revealing relationships between properties that both inform scientific understanding and enable selection in design.

1.2 Classifying materials

Figure 1.1 showed the materials under four broad categories, two of which can be sub-divided, giving the six material 'families' illustrated in Figure 1.2: metals, polymers, elastomers, ceramics, glasses, and hybrids – the last usually being made by combining two (or more) of the others. There is sense in this: the members of a family have certain characteristics in common – they have similar material properties (expanded in Section 1.3), and as a consequence, similar processing routes and applications. Here, briefly, are some of their key characteristics.

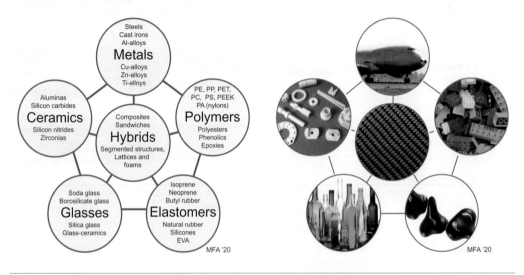

Figure 1.2 (left) The families of engineering materials: metals, ceramics, glasses, polymers, and elastomers, which can also be combined in various geometries to create hybrids. (right) Examples of each family.

Metals are crystalline solids – iron and steel, aluminium, copper, and so on are familiar in all branches of engineering. They have relatively high *stiffness* (recoverable elastic deformation). Most, when pure, are soft and easily deformed permanently – their *strength* is low. They can be made stronger by alloying and by mechanical and heat treatment, but they remain *ductile*, allowing them to be formed by deformation processes like rolling or forging (or alternatively, their melting points allow them to be melted and cast to shape in a mould). Broadly speaking, they are resistant to fracture: their *toughness* is high; and they have good *electrical* and *thermal conductivity*. But metals have weaknesses too: they are chemically reactive, and susceptible to rapid *corrosion* and *oxidation* if not protected.

Ceramics are non-metallic, inorganic solids, like porcelain and alumina. They have many attractive features: they are stiff, hard, and abrasion-resistant; they retain their strength at high temperatures; they resist corrosion well; and most are good electrical insulators. They, too, have their weaknesses: unlike metals, they are brittle (low toughness), giving ceramics a low tolerance for stress concentrations (like holes or cracks) or for high contact stresses (at clamping points, for instance). And as their melting points are high, they are much harder to shape than metals.

Glasses are non-crystalline ('amorphous') solids, a term explained more fully in Chapter 3. The commonest are silica-based glasses familiar as bottles, windows, ovenware, and optical fibres. The lack of crystal structure suppresses permanent deformation, so glasses are also hard – but thanks to a lower melting point, they are much easier to shape than ceramics. They are, of course, transparent to light and remarkably corrosion-resistant. Like ceramics, they are excellent electrical insulators – but they, too, are brittle and vulnerable to stress concentrations.

Polymers are organic solids based on long chain molecules of carbon atoms (or, in a few, silicon). Polymers are light – their densities are less than those of the lightest metals. Compared with other families they are floppy – at least 50 times less stiff than metals. But they can be strong, and because of their low density, their strength per unit weight competes with metals – particularly when drawn into *fibres*. Their properties are mostly sensitive to temperature, so a polymer that is tough and flexible at room temperature may be brittle at the −4°C of a household freezer, yet turn rubbery at the 100°C of boiling water. Few have useful strength above 150°C. But as a result they are easy to shape (which is why they are called 'plastics'). Complicated parts performing several functions can be moulded from a polymer in a single operation. Their properties are well suited for components that snap together, making assembly fast and cheap. And by accurately sizing the mould and pre-colouring the polymer, no finishing operations are needed.

Elastomers, the materials of rubber bands and running shoes, are polymers with the unique property that their stiffness is extremely low – 500 to 5000 times less than metals – and they can be stretched to many times their starting length, yet recover their initial shape when released. Despite their low stiffness they can be strong and tough – essential for tyres.

Hybrids are combinations of two (or more) materials in an attempt to get the best of both, or porous materials like foams. Glass- and carbon-fibre reinforced polymers (GFRP and CFRP) are hybrids; so, too, are sandwich structures and laminates. Almost all the materials of nature – wood, bone, skin, leaf – are hybrids. Bone, for instance, is a mix of collagen (a polymer) with hydroxyapatite (a ceramic mineral). Hybrids are often expensive, and they are relatively difficult to form, join, and recycle. So despite their attractive properties, the designer will use them only when the added performance justifies the added cost. Today's growing emphasis on high performance, light weight, and fuel efficiency provides increasing drivers for their use.

1.3 Material properties

Words like 'stiffness,' 'strength,' 'insulator' and so on describe a material characteristic, but do not quantify it for the purposes of scientific interpretation or design – for this we need material properties. Some, like density (mass per unit volume) and price (cost per unit volume or weight) are familiar enough, but others are not, and getting them straight is essential. Here they are introduced briefly; expanding on each group of properties occupies a large proportion of this book.

Mechanical Properties A steel ruler is easy to bend *elastically* – 'elastic' meaning that it springs back when released. Its elastic stiffness (here, resistance to bending) is set partly by its shape – thin strips are easier to bend than thick ones – and partly by a property of the steel itself: its *elastic modulus*, E (usually called *Young's modulus*). Materials with high E, like steel, are intrinsically stiff; those with low E, like polyethylene, are not. Figures 1.3(a,b) illustrate the consequences of inadequate stiffness.

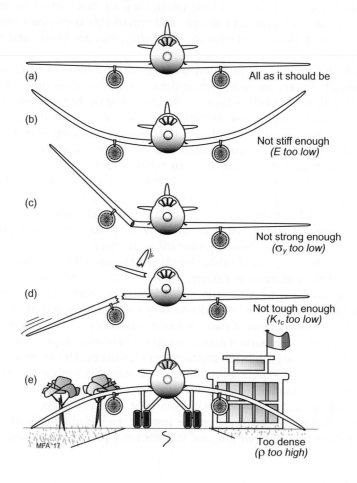

Figure 1.3 Mechanical properties.

A steel ruler bends elastically but should be hard to bend permanently – this relates to its *strength*, not stiffness. Shape again plays a role, but in combination with a different property of the steel – its *yield strength*, σ_y. Materials with large σ_y, like titanium alloys, are hard to deform permanently regardless of their stiffness, which comes from E. Those with low σ_y, like lead, can be deformed with ease. When metals deform, they generally get stronger (it is called 'work hardening'), but there is an ultimate limit, called the *tensile strength*, σ_{ts}, beyond which the material fails (in tension), while the amount it stretches before it breaks is called the *ductility*. The hardness, H, is related to the strength, σ_y. High hardness gives scratch resistance and resistance to wear. Figure 1.3(c) gives an idea of the consequences of inadequate strength.

So far so good. There is another mechanical property, which is more tricky. If the ruler were made not of steel but of PMMA (Plexiglass, Perspex), as transparent rulers are, it is likely that the ruler will fracture suddenly, before it acquires a permanent bend. Materials that break in this way are described as 'brittle'; those that deform and resist fracture are 'tough'. The resistance of materials to cracking and fracture is measured by the *fracture toughness*, K_{1c}. Steels are tough, almost always – they have a high K_{1c}. Glass epitomises brittleness – it has a very low K_{1c}. Figure 1.3(d) suggests the consequences of inadequate fracture toughness.

These are not the only mechanical properties, but they are the most important ones. We will meet them and the others in Chapters 4 – 6. A physical property that often combines with mechanical properties in design is the *density*, ρ. For a ruler it is irrelevant, but for almost anything that moves using energy from fuel, electricity, or manpower, weight carries a penalty – increasingly significant moving from bicycles to cars, trucks, and trains, but much greater for aircraft, and enormous for space vehicles. Minimising weight has much to do with clever design, including the choice of material. Aluminium has a low density, lead a high one – a lead aircraft would never get off the ground [Figure 1.3(e)]. The physical origin of density is explored in Chapter 3.

Example 1.1 Design requirements (1)

You are asked to select a material for the teeth of the scoop of a digger truck. To do so, you need to identify the design requirements and the associated material properties. What are they?

Answer. The teeth will be used in a brutal way to cut earth, scoop stones, crunch rock, often in unpleasant environments (ditches, sewers, fresh and salt water, and worse), and their maintenance will be neglected. These translate into a need for high hardness, H, to resist wear, good corrosion resistance, and high fracture toughness, K_{1c}, so they don't snap off. Does the cost of the material matter? Not much against the cost of the whole vehicle, and it is worth paying for good teeth to avoid expensive downtime.

Thermal Properties   The properties of a material change with temperature, usually for the worse. Its strength falls, it starts to "creep" (sag slowly over time), and it may oxidise, degrade, or decompose [Figure 1.4(a)]. This means that there is a limiting *maximum service temperature* T_{max} above which its use is impractical. Stainless steel has a high T_{max} – it can be used up to 800°C; most polymers have a low T_{max} and are seldom used above 150°C.

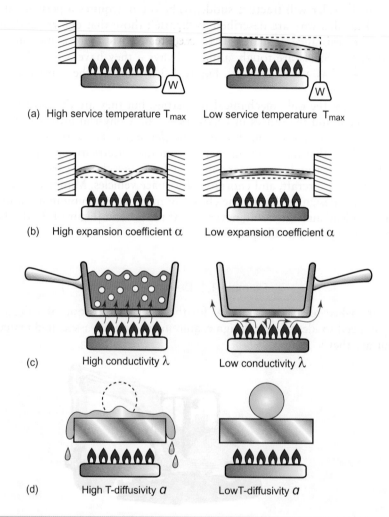

(a) High service temperature T_{max} Low service temperature T_{max}

(b) High expansion coefficient α Low expansion coefficient α

(c) High conductivity λ Low conductivity λ

(d) High T-diffusivity a LowT-diffusivity a

Figure 1.4   Thermal properties.

Most materials expand when they are heated, but by differing amounts depending on their *thermal expansion coefficient*, α. The expansion is small, but its consequences can be large. If, for instance, a rod is constrained and then heated, as in Figure 1.4(b), expansion forces the rod against the constraints, causing it to buckle. Railroad track buckles in this way if provision is not made to cope with it. Bridges have expansion joints for the same reason.

Metals feel cold; woods feel warm. This has to do with two thermal properties: *thermal conductivity* and *heat capacity*. Thermal conductivity, λ, measures the rate at which heat flows through the material when there is a temperature difference across it. Materials with high λ are what you want if you wish to conduct heat from one place to another, as in cooking pans, radiators, and heat exchangers [Figure 1.4(c)]. But low λ is useful too – for insulating homes, reducing the energy consumption of refrigerators and freezers, and enabling space vehicles to re-enter the earth's atmosphere.

These applications all relate to long-time, steady, heat flow. When we are heating things up (or cooling them), the *heat capacity*, C_p, enters the picture. It measures the amount of heat required to make the temperature of a material rise by a given amount. High heat capacity materials – copper, for instance – require a lot of heat to change their temperature; low heat capacity materials, like polymer foams, take much less. When heat is first applied, there is a more subtle thermal property that determines the rate at which the temperature changes, combining conductivity and heat capacity. Think of lighting the gas under a cold slab of material with a scoop of ice cream on top, as in Figure 1.4(d). The bottom surface heats quickly, but initially the rest is cold – it takes time until the top begins to warm up and the ice cream starts to melt. How long does this take? For a given thickness of slab, the time is inversely proportional to the *thermal diffusivity, a,* of the material, and this thermal property is proportional to λ/C_p.

Thermal properties and the effects of temperature on properties are revisited in Chapters 7 and 8.

Example 1.2 Design requirements (2)

You are asked to select a material for energy-efficient cookware. What thermal property are you looking for? And, apart from the mechanical properties discussed earlier, what other material behaviour will be important?

Answer. To be energy-efficient the pan must have a high thermal conductivity, λ, to transmit and spread the heat well. Also, it mustn't contaminate the food – that is, it must resist *corrosion* by anything that might be cooked in it – including hot salty water, dilute acids (vinegar, for example), and mild alkalis (baking soda).

Chemical properties Products often encounter hostile environments and are exposed to corrosive fluids, hot gases, or radiation. Even damp air and the sweat of your hand are corrosive. If the product is to survive for its design life, it must be made of materials – or coated with materials – that can tolerate the surroundings in which it operates. Figure 1.5 illustrates some of the commonest environmental challenges: fresh and salt water, acids and alkalis, organic solvents, oxidising flames, and ultra-violet radiation. The intrinsic resistance of a material to each of these is more difficult to quantify as a material property – a pragmatic approach based on experience is to rank material resistance to attack on a scale from 1 (very poor) to 5 (very good). Chapter 9 explains why, together with other aspects of durability (such as mechanical wear).

(a) Fresh water

(b) Salt water

(c) Acids and alkalis

(d) Organic solvents

(e) Oxidation

(f) UV radiation

Figure 1.5 Chemical properties: resistance to water, acids, alkalis, organic solvents, oxidation, and radiation.

Electrical, magnetic, and optical ('functional') properties Without electrical conduction, we would lack the easy access to light, heat, power, control, and communication that we take for granted. Metals conduct well – copper and aluminium are the best of those that are afford-able. In contrast, fuse boxes, switch casings, and the suspension supports for transmission lines all require insulators, which do not conduct electricity – that is, most plastics and glass. Here the key property is *resistivity*, ρ_e, the inverse of electrical conductivity [Figure 1.6(a)].

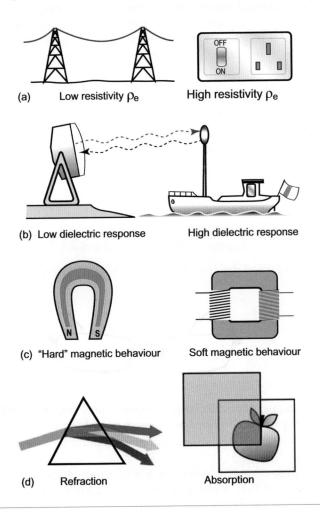

(a) Low resistivity ρ_e High resistivity ρ_e

(b) Low dielectric response   High dielectric response

(c) "Hard" magnetic behaviour   Soft magnetic behaviour

(d) Refraction   Absorption

Figure 1.6 'Functional' properties: electrical, magnetic, and optical properties.

Figure 1.6(b) suggests further electrical properties: the ability to allow the passage of microwave radiation, as in the transmitting radome, or to reflect it, as in the passive reflector of the boat. Both have to do with *dielectric* properties, particularly the *dielectric constant*, ε_D. High ε_D materials respond to an electric field, low ε_D materials do not.

Electricity and magnetism are closely linked. Electric currents induce magnetic fields; a moving magnet induces, in any nearby conductor, an electric current. The response of most materials to magnetic fields is too small to be of practical value. But ferromagnets and ferrimagnets have the capacity to trap a magnetic field permanently. These are called 'hard' magnetic materials because, once magnetised, they are hard to de-magnetise – as used in headphones, motors, and dynamos. Here the key property is the *remanence*, a measure of the intensity of the retained magnetism. In contrast, 'soft' magnetic materials are easy to magnetise and de-magnetise – they are the materials of transformer cores. They have the capacity to 'conduct' a magnetic field but not to retain it permanently [Figure 1.6(c)]. For these, a key property is the *saturation magnetisation*, which measures how large a field the material can conduct.

Materials respond to light as well as to electricity and magnetism – hardly surprising, since light itself is an electro-magnetic wave. Materials that are opaque *reflect* light; those that are transparent *refract* it, changing the light path in a way controlled by the *refractive index*. Some are opaque to all wavelengths (colours), others *absorb* specific wavelengths while allowing others to pass freely [Figure 1.6(d)].

Electrical, magnetic, and optical properties and materials are explored in more depth in Chapter 10.

Example 1.3 Design requirements (3)

What are the essential and the desirable requirements of materials for eyeglass (spectacle) lenses?

Answer. Essentials: Optical clarity is paramount, together with the requirement that it can be shaped and polished with great precision, to meet the required optical prescription. And it must resist sweat and be sufficiently scratch-resistant to cope with normal handling.

Desirables: A high refractive index and a low density allow thinner, and thus lighter, lenses. And a material that is cheap allows either a lower cost for the consumer or a greater profit margin for the maker.

Environmental ('eco') properties　Extracting materials and processing them into products consumes nearly one-third of global energy demand. The associated emissions are a cause for international concern over climate change, and demand for materials is forecast to double in the next 40 years. It is essential, therefore, to understand the properties of materials linked to their environmental impact, and to seek ways to use them more sustainably. Chapter 12 introduces the key ideas of material life-cycle analysis, material efficiency, and sustainability.

Three quantities describe the environmental character of a material (Figure 1.7):

- *Embodied energy*: the energy required to produce 1 kg of the material, including that used to mine and refine the feedstocks, and to convert these into usable material stock.
- *Carbon footprint*: the mass of carbon released due to the production of 1 kg of the material (or, strictly, the carbon-equivalent release – a way of including the global warming effect of all the gaseous emissions).
- *Water demand*: the net volume of water that is used during the production of 1 kg of material.

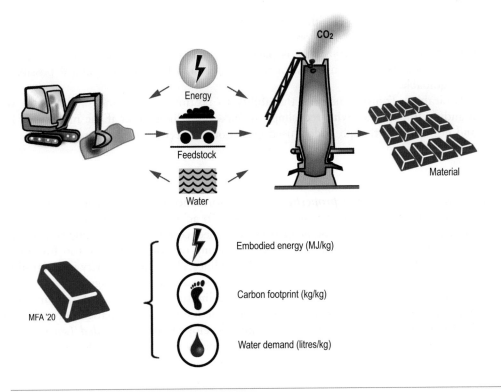

Figure 1.7 The environmental (eco) properties of materials.

Example 1.4 Design requirements (4)

What are the essential and the desirable requirements of materials for a single-use (disposable) water bottle that does minimal environmental harm?

Answer. Essentials: Health aspects come first: the material of the bottle must be non-toxic and able to be processed in a way that leaves no contaminants. As the contents are what the consumer is paying for, the container is regarded as disposable and must be cheap. The material of the bottle should be recyclable and, if possible, be biodegradable.

Desirables: Transparency (or the ability to be tinted easily and safely) helps with branding (though not with recycling). It makes handling easier if the material is light and not brittle, although widely used glass bottles meet neither criterion.

Summary: design-limiting properties   The performance of a component is limited by the properties of the materials of which it is made. To achieve a desired level of performance, the values of the design-limiting properties must meet certain targets. In Figure 1.3, stiffness, strength, and toughness are design-limiting – if any one of them were too low the plane wouldn't fly. In the design of power transmission lines, electrical resistivity is design-limiting; in the design of a camera lens, the optical qualities constrain the design.

Materials are chosen by identifying the design-limiting properties and applying limits to them, screening out materials that do not meet the limits – a systematic approach to material selection is introduced in Chapter 2, and developed later in *Guided Learning Unit 2*. Processes, too, have characteristics that can be equally design-limiting, leading to a parallel scheme for choosing viable processes (Chapter 11 and *Guided Learning Unit 3*).

1.4 Material property charts

Data sheets for materials list their properties, but they give no perspective or comparison between materials. The way to achieve these is graphically, through *material property charts*. The simplest is a *bar chart*, showing one property for all materials (or a subset of them). Figure 1.8 shows an example, for Young's modulus E (characterising stiffness). The largest is more than 1 million times greater than the smallest. Many other properties have similarly large ranges, so it makes sense to plot them on *logarithmic*[1] (or 'log') scales, not linear. The length of each bar shows the range of the property for each material, here segregated by material family. The differences between families now become apparent: metals and ceramics are stiff, polymers much less so – by a factor of about 50 – while the moduli of elastomers are some 500 times smaller still.

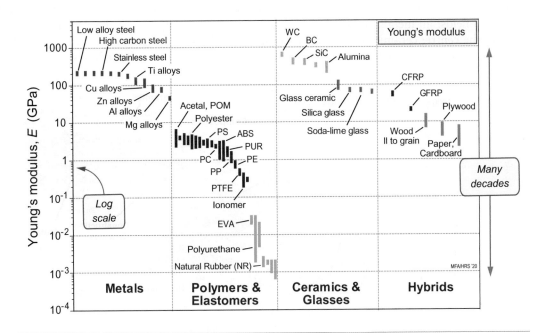

Figure 1.8 A bar chart of Young's modulus E. It reveals the difference in stiffness between material families.

---

[1] *Logarithmic* means that the scale goes up in constant multiples, usually of 10. We live in a logarithmic world – our senses, for instance, all respond in this way.

Example 1.5 Use of property bar charts

In a cost-cutting exercise, a designer suggests that certain die-cast zinc parts could be replaced by cheaper moulded polyethylene (PE) parts with the same shape. A team member expresses concern that the PE replacement might be too flexible. By what factor will the stiffness of the part change if the substitution is made?

Answer. The stiffness of two parts with the same shape, but made of different materials, will differ by the same factor as the moduli of the materials. The bar chart of Figure 1.8 shows that the modulus of PE is less, by a factor of about 100, than that of zinc alloys. The concern is a real one.

Much more information is packed into the picture if two properties are plotted against one another to give a *property chart*, as in Figure 1.9, here showing Young's modulus E and density ρ (again using log scales). Now the material families are more distinctly separated into fields: all metals lie in the reddish zone, top right; polymers in the dark blue envelope in the centre; ceramics (with construction materials shown separately) in the yellow fields, and so on. Within each field, individual materials appear as smaller ellipses.

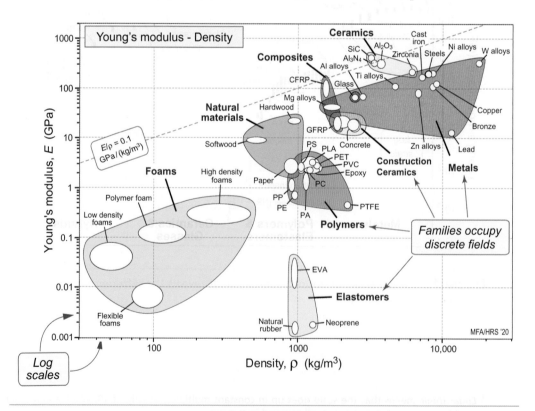

Figure 1.9 A property chart of Young's modulus and density. Families occupy discrete areas of the chart.

Example 1.6 Use of property charts

Steel is stiff (high modulus E) but heavy (high density ρ). Aluminium alloys are less stiff but also less dense. One criterion for lightweight design is a high value of the ratio E/ρ, defining materials that have a high stiffness per unit weight. Can you think of a graphical technique to answer the following: does aluminium have a higher value of E/ρ than steel? And what about CFRP?

Answer. The $E - \rho$ chart of Figure 1.9 can answer the question in seconds: materials with equal values of E/ρ lie along a line of slope 1 – one such line is shown. Materials with high E/ρ lie towards the top left, those with low values towards the bottom right. If a line with the same slope is drawn through Al alloys on the chart, it passes almost exactly through steels: the two materials have almost the *same* value of E/ρ, a surprise when you think that aircraft are made of aluminium, not steel (the reason will become clear in *Guided Learning Unit 2*). CFRP, by contrast, has a much higher E/ρ than either aluminium or steel.

Material property charts like these are a core tool used throughout this book:

- They give an overview of the physical, mechanical, and functional properties of materials, presenting the information about them in a compact way.
- They give perspective on the physical origins of properties – a help in understanding the underlying science.
- They are a tool for optimised selection of materials to meet given design requirements, and they help us understand the use of materials in existing products.

1.5 Summary and conclusions

This chapter introduced the broad families of materials: metals, ceramics, glasses, polymers, and elastomers, plus hybrids that combine the properties of two or more of the others. Material properties lie at the centre of materials science and engineering. They characterise how materials respond to a wide range of physical and chemical stimuli, enabling scientific understanding of their origins and stimulating the development of new materials. Engineering design depends on selecting materials that meet the requirements of a product, including how it will be made – here we think of properties as *design-limiting*. The same core knowledge of properties is the starting point for understanding materials in the natural world, from the biomechanics of human bodies to earthquakes and the flow of glaciers.

Here we introduced the main properties of engineering and natural materials: *physical properties* like density; *mechanical properties*, including Young's modulus and yield strength, governing deformation and failure; *functional properties*, describing the thermal, electrical, magnetic, and optical behaviour; *chemical properties*, determining how the material reacts with its environment; and *eco-properties* that relate to the impact that production of the material has on the world and its climate.

Understanding the science of properties and selecting materials requires a clear perspective on the quantitative data: *material property charts* are a unique and powerful way of achieving this. The concept was introduced in this chapter, and these charts will be one of the central features of this book.

1.6 Further reading

Askeland, D. R., Phulé, P. P., & Wright, W. J. (2010). *The science and engineering of materials* (6th ed.). Thomson. ISBN 9780495296027. (A well-established materials text that deals well with the science of engineering materials).

Callister, W. D., & Rethwisch, D. G. (2018). *Materials science and engineering: An introduction* (10th ed.). John Wiley & Sons. ISBN 978-1-119-40549-8. (A well-established text taking a science-led approach to materials).

Delmonte, J. (1985). *Origins of materials and processes.* Technomic Publishing Company. ISBN 87762-420-8. (A compendium of information about materials in engineering, documenting the history).

Hummel, R. (2004). *Understanding materials science: History, properties, applications* (2nd ed.). Springer Verlag. ISBN 0-387-20939-5. (An introduction to the ages of materials and their properties with a historical slant).

Miodownik, M. (2014). *Stuff matters.* Penguin. ISBN 978-0-241-95518-5. (An engaging exposition of the diversity of uses of materials, covering modern innovations and their historical evolution).

Shackelford, J. F. (2014). *Introduction to materials science for engineers* (8th ed.). Prentice Hall. ISBN 978-0133789713. (A well-established materials text with a design slant).

1.7 Exercises

Exercise E1.1	Use Google to research the history and uses of one of the following materials: • Tin • Glass • Cement • Titanium • Carbon fibre
Exercise E1.2	What is meant by the *design-limiting properties* of a material in an application?
Exercise E1.3	There have been many attempts to manufacture and market plastic bicycles. All have been too flexible. Which design-limiting property is insufficiently large?
Exercise E1.4	What are the design-limiting properties of the material for the blade of a knife that will be used to gut fish?
Exercise E1.5	What are the design-limiting properties of the material for an oven glove?

Exercise E1.6   What are the design-limiting properties of the materials for light bulbs? Consider a variety of bulb types.

Exercise E1.7   A material is needed for a tube to carry fuel from the fuel tank to the carburettor of a motor mower. The design requires that the tube can bend and that the fuel be visible. List what you would think to be the design-limiting properties.

Exercise E1.8   A material is required as the magnet for a magnetic soap holder. Soap is mildly alkaline. List what you would judge to be the design-limiting properties.

Exercise E1.9   List three applications that need high stiffness and low weight.

Exercise E1.10   List three applications that need optical quality glass.

Exercise E1.11   List three applications that require high thermal conductivity.

Exercise E1.12   List three applications that require low thermal expansion. Think of things that lose accuracy or won't work if they distort.

Exercise E1.13   Aluminium has a high carbon footprint – its value is 13.1 kg/kg. What does this mean?

Exercise E1.14   List the six main families of engineering materials. Use your own experience to rank them approximately:

(a) by stiffness (Young's modulus E). A sheet of a material that has a high modulus is hard to bend when in the form of sheet. A sheet of material with a low modulus is floppy.
(b) by thermal conductivity (λ). Materials with high conductivity feel cold when you pick them up on a cold day. Materials with low conductivity may not feel warm, but they don't freeze your hands as quickly as those with high conductivity.

Exercise E1.15   Examine the Young's modulus chart of Figure 1.9. By what factor, on average, are polymers less stiff than metals? By what factor is Neoprene less stiff than Polypropylene (PP)?

Exercise E1.16   Wood is a natural polymeric material, largely made up of cellulose and lignin, and they, like engineering polymers, are almost entirely hydrogen, carbon, oxygen, and a little nitrogen. Has nature devised a cocktail of these elements that has a higher modulus than any bulk man-made polymer? Engineering polymers and woods appear on the charts of Figures 1.8 and 1.9. Use either one to answer the question.

Exercise E1.17   Windows can be made of glass (house windows), polycarbonate, PC (conservatory roofs), or polymethyl-methacrylate, PMMA (aircraft cockpit canopies). Would a glass window be more flexible than a replacement of the same thickness made of polycarbonate? Would it be heavier? Use the property chart of Figure 1.9 to find out.

Exercise E1.18   Do zinc alloys have a higher *specific stiffness E/ρ* than PP? Use the property chart of Figure 1.9 to find out (E is the modulus, ρ is the density). Contours of constant E/ρ are diagonal lines on the log axes of the figure – one is shown at $E/\rho = 0.1$ GPa/(kg/m^3) – with higher values towards the top left of the chart.

Chapter 2
Materials, processes, and design

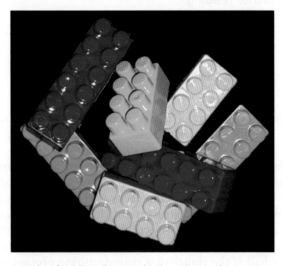

Lego bricks are made of ABS, a thermoplastic polymer. (Image courtesy of A-Best Fixture Co., 424 West Exchange Street, Akron, Ohio, 44302, USA.)

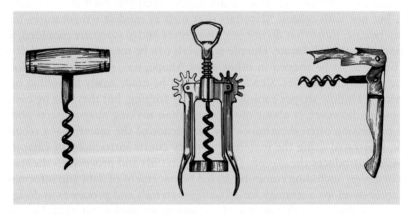

Embodiment designs for corkscrews using a direct pull concept (istock.com)

2.1 Introduction and synopsis

A successful product – one that performs well, gives pleasure to the user, and is good value for money – uses the best materials for the job and fully exploits their potential and characteristics.

The families of materials – metals, polymers, ceramics, and so forth – were introduced in Chapter 1. But it is not, in the end, a *material* that we seek in design, it is a certain *profile of properties* – the one that best meets the needs of the design. The members of each family can be organised into a hierarchical structure of classes and sub-classes based on character- istics they have in common, enabling selection to proceed progressively from a broad level of a material class down to a specific variant in that class. Choosing a material is only half the story. The other half is the choice of a manufacturing process route to shape, join, and finish it. Process families, too, can be organised into a family tree, based on the underlying physics of how they work, with a corresponding set of processing attributes.

The choice of material and process are tightly coupled: each process works with some materials but not with others. Materials that melt at modest temperatures to low-viscosity liquids can be cast; those that do not (e.g., polymers and glasses) are moulded, while ceramics are compacted in the solid state. Ductile materials can be forged, rolled, and drawn, but those that are brittle must be shaped in other ways. Shape, too, influences the choice of process, due to the underlying physics. Slender shapes can be made easily by rolling or drawing, but not by casting. Hollow shapes cannot be made by forging, but they can by casting or mould- ing. Processing, however, is about much more than making shaped parts and joining them together: processing often determines key properties of the material – a recurring theme in this book, illustrated using the material property charts introduced in Chapter 1.

Section 2.2 introduces the classifications of materials and processes – the starting point for selection in design (including computer-based management of data and information). The rest of the chapter develops a strategy for selecting materials and processes in design, introducing some of the vocabulary of design, and the stages in its implementation. Design starts with a *market need*, which is analysed to produce a set of *design requirements*. Ways to meet these requirements (*Concepts*) are sought, developed (*Embodied*), and refined (*Detailed*) to give a *product specification*. The choice of material and process evolves in parallel with this process.

The selection strategy involves four steps: *translation, screening, ranking,* and *technical evaluation.* These steps are outlined for material selection, with an example of the first translation step; the rest of the selection process for materials, and the parallel treatment of processes, are covered in later chapters and *Guided Learning Units 2 and 3.*

2.2 Classification of materials and processes

Classifying materials   As introduced in Chapter 1, engineering materials fall into six broad families: metals, polymers, elastomers, ceramics, glasses, and hybrids. These are chosen to reflect the broad similarities in properties between members of each family, directly reflecting the underlying physics and chemistry of atomic bonding and microstructure. Figure 2.1 illustrates these families expanded in a tree structure of classes, sub-classes, and members. As an example, the Materials Universe contains the family *Metals*, which in turn contains the class *Aluminium alloys*, which contains the sub-class *6000 series*, within which we find the particular member *Al 6061*. Every member of the Materials Universe is characterised by a set of *attributes*: the properties mentioned in Chapter 1, but also its processing characteristics and typical applications.

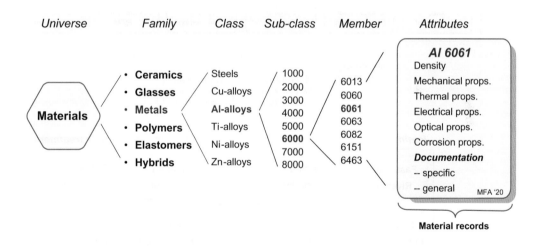

Figure 2.1   The taxonomy of the Universe of Materials and their attributes. Computer-based selection software stores data in a hierarchical structure like this.

Classifying processes   A *process* is a method of shaping, joining, or finishing a material. *Casting, injection moulding, fusion welding,* and *electro-polishing* are all processes, and there are hundreds of them to choose from. It is important to select the right process route at an early stage in the design before the cost penalty of making changes becomes large. The choice for a given component depends on the material of which it is to be made, and the target outcome in terms of properties; on its shape, dimensions, and precision; and on how many are to be made – in short, on the *design requirements.*

Manufacturing processes are organised under the headings shown in Figure 2.2. *Primary processes* create shapes – the first row lists six primary shaping processes: casting, moulding, deformation, powder methods, methods for forming composites, and special methods including additive manufacture. *Secondary processes* modify shapes or properties: here they are shown as *machining*, which adds features to an already shaped body, and *heat treatment*, which enhances bulk properties. Below these come *joining* and *surface treatment* (many of which also involve heat). Figure 2.3 illustrates some of these, organised in the same way, representing the progression through a multi-stage manufacturing route. It should not be treated too literally: the order of the post-shaping steps will vary to suit the needs of the design and the optimisation of production and performance.

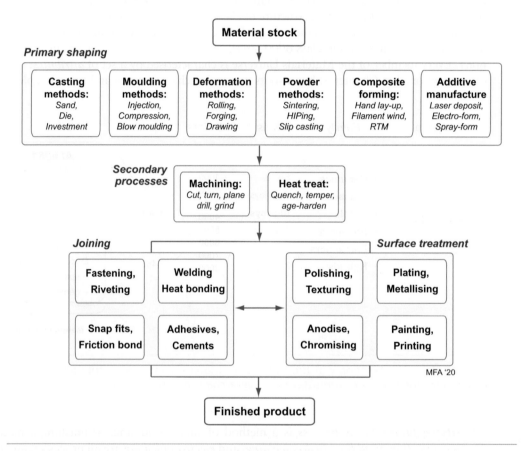

Figure 2.2   The classes of process. The first row contains the primary shaping processes; below lie the secondary processes of machining and heat treatment, followed by the families of joining and surface treatment processes.

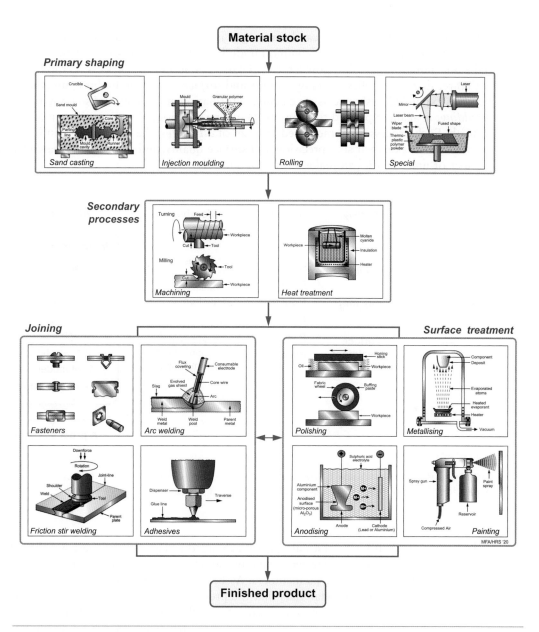

Figure 2.3 Examples of the families and classes of manufacturing processes. The arrangement follows the general pattern of Figure 2.2.

Information about processes can be arranged in a hierarchical classification like that used for materials, but now based on common underlying process physics, particularly the material state (liquid, bulk solid, powder, etc). Figure 2.4 shows the process hierarchy, starting with three distinct families: *shaping, joining,* and *surface treatment.* In this figure, the *shaping* family is expanded to show classes: casting, deformation, moulding, and so on. One of these, *moulding,* is again expanded to show its members: rotation, blow, injection moulding, and so forth. Each process is characterised by a set of *attributes*: the materials it can handle, the shapes it can make, their size, precision and surface quality, and an economic batch size (the number of units that it can make most economically).

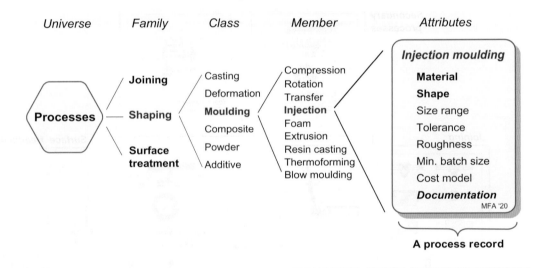

Figure 2.4 The taxonomy of the universe of processes with part of the shaping family expanded. Each member is characterised by a set of attributes.

The other two process families are partly expanded in Figure 2.5. There are three broad classes of *joining* process: adhesives, welding, and fasteners. In this figure one of them, *welding,* is expanded to show its members. As before, each member has attributes, including (as for all processes) the materials that the process can join. After that the attribute list differs from that for shaping – now the geometry of the joint and the way it will be loaded are important, plus other joint-specific requirements such as whether it can easily be disassembled. The lower part of the figure expands the family of *surface treatment* processes, expanded to show some of its classes and members. The dominant attributes distinguishing between processes are the materials that can be treated by each process, and the design purpose in applying it.

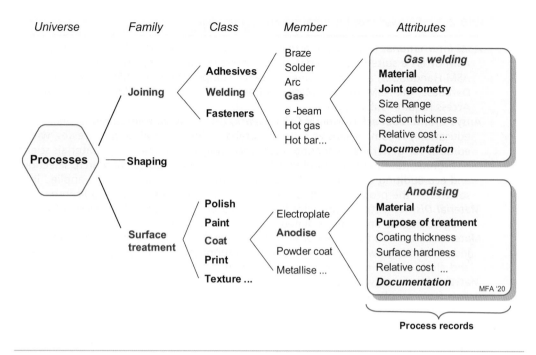

| Universe | Family | Class | Member | Attributes |

Braze
Solder
Arc
Gas
e-beam
Hot gas
Hot bar...

Gas welding
Material
Joint geometry
Size Range
Section thickness
Relative cost ...
Documentation

Adhesives
Welding
Fasteners

Joining

Processes —**Shaping**

Surface treatment

Polish
Paint
Coat
Print
Texture ...

Electroplate
Anodise
Powder coat
Metallise ...

Anodising
Material
Purpose of treatment
Coating thickness
Surface hardness
Relative cost ...
Documentation
MFA '20

Process records

Figure 2.5   The taxonomy of the process universe again, with the families of joining and surface treatment partly expanded.

Process-property interaction   Processing can change properties. If you deform a metal it gets harder (*work hardening*); if you then heat it up it gets softer again (*annealing*). If polyethylene – the stuff of plastic bags – is drawn to a fibre, its strength is increased by a factor of 5 because the polymer chains are drawn into alignment. Soft, stretchy rubber is made hard and brittle by vulcanising. Glass can be chemically and thermally treated to give it the impact resistance to withstand a projectile ('bullet-proof glass'). And composites like carbon-fibre reinforced epoxy have no useful properties at all until processed – prior to processing they are just a soup of resin and a sheaf of fibres. Joining, too, changes properties. Fusion welding involves the local melting and re-solidifying of the faces of the parts to be joined, also producing an adjacent 'heat-affected zone.' The weld region has properties that differ from those of the base material – usually worse. Surface treatments, by contrast, are generally chosen to improve properties: electroplating to improve corrosion resistance, or carburising to improve wear. We return to process-property interactions in depth in Chapter 11; but many examples of the effect of processing on properties will be encountered along the way.

Data and information sources for materials and processes   Software is now available to manage data and information about materials and processes, making it easier to find data and to manipulate it. They vary greatly in their breadth of coverage and the audience at which they are aimed. Table 2.1 lists a selection.

Table 2.1 Software for information about materials

ASM International (http://www.asminternational.org/online-databases-journals) The ASM offers a suite of online databases, accessed via this link. They include the ASM Handbooks, a multi-volume reference compendium; the ASM Failure Analysis Database, the ASM Medical Materials Database, and the ASM Micrograph Database. Access requires subscription.

Ansys Granta EduPack (formerly *CES EduPack*) (http://www.grantadesign.com/education/) A comprehensive suite of databases for materials and processes, with editions for general engineering, aerospace, polymer engineering, materials science, environmental design, and industrial design. It includes powerful search, selection, and eco-auditing tools. (The selection methods in this book can all be applied in this software – the property charts were made using it.) Access requires a license.

Material District (https://materialdistrict.com/material/) A materials news-sheet with an image database and articles aimed at industrial design.

Material Connexion (www.materialconnexion.com) A materials library emphasising industrial design with records, images, and supplier, backed up by an advisory service and alerts as new materials enter the market.

MatWeb (www.matweb.com) A large database of the engineering properties of materials, drawn from suppliers' data sheets.

2.3 The design process, and selection of material and process

Original design starts from a new concept and develops the information necessary to implement it. *Evolutionary design* (or *redesign*) starts with an existing product and seeks to change it in ways that increase its performance or ease of use, reduce its cost, or both.

Original design   New ideas or working principles lead to products being developed from scratch (the vinyl disk, the audio tape, the compact disc, the MP3 player were all, in their day, completely new). Original design can be stimulated by new materials – Figure 1.1 made the historical connection between materials and industrial evolution through the ages. Most recently, high-purity silicon enabled the transistor; high-purity glass, the optical fibre; high coercive-force magnets, the miniature earphone and now the electric car; solid-state lasers, the barcode. Sometimes a new material suggests the new product. In other cases, new endeavours demand the development of new or significantly improved materials: nuclear technology drove the development of new zirconium alloys and stainless steels; space technology stimulated growth in beryllium alloys and lightweight composites; turbine technology today drives improvement in high-temperature alloys and ceramics; concern for the environment drives the development of bio-polymers.

The central column of Figure 2.6 shows the design process. The starting point is a *market need* or a *new idea*; the end point is the full *product specification* for a product that fills the need or embodies the idea. It is important to define the need precisely and to formulate a clear set of *design requirements*. The iterative design process to take the need through to product specification is captured in three stages in Figure 2.6: *conceptual design, embodiment design*, and *detailed design*.

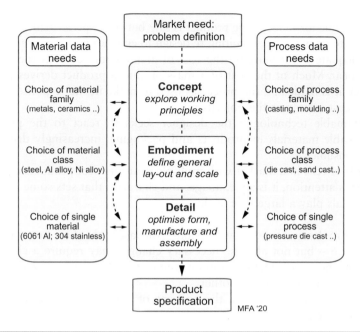

Figure 2.6 The design flow chart, showing how material and process selection enter. Information about materials is needed at each stage, but at very different levels of breadth and precision. The broken lines suggest the iterative nature of original design and the path followed in redesign.

At the conceptual design stage, all options are open: the designer considers alternative concepts and how they work. The next stage, embodiment, takes the more promising concepts and analyses their operation, requiring approximate quantification of loading, component sizing, performance, and indicative costs. The embodiment stage ends with a feasible layout to pass to the detailed design stage, where full specifications for the components and assembly are drawn up. Critical components may be subjected to precise mechanical or thermal analysis to finalise dimensions, and optimisation conducted to enhance performance and minimise cost. Throughout this process, materials and processes are considered and chosen – ideally starting from the widest possible range of options, then steadily refining the specification of both to a final material choice and production route. Developing methodologies for selection in design is a core theme of this book, discussed later in the chapter.

Redesign   Most design is not 'original' in the sense of starting from a totally new idea. It starts with an existing product and corrects its shortcomings, enhances its performance, or reduces its cost or environmental impact, making it more competitive. Here are some scenarios that call for redesign:

- 'Product recall': if a product is released and fails to meet safety standards, urgent redesign is required. This is often a material failure (sometimes due to a negative impact of processing) so alternative materials or processing are needed to overcome the problem.

- 'Poor value for money': the product is safe but offers performance that, at its price, is perceived to be mediocre, requiring re-design to enhance performance and/or reduce the cost.
- 'Inadequate profit margin': the cost of manufacture exceeds the price that the market will bear. Much of the cost of a mass-produced product derives from the materials of which it is made and the processes used to make it; the response is to re-examine both with cost-cutting as the objective.
- 'Sustainable technology': the designer seeks to react to the profligate use of non-renewable materials in products and packaging, increasingly driven by demand from the consumer.
- 'Apple effect': in a market with multiple near-identical products competing for the consumer's attention, it is style, image, and character that sets some products above others. Materials play a large role in this.

Much of re-design has to do with detail – the last of the three boxes in the central window of Figure 2.6 – but not all. The necessary changes may require a change of configuration and layout (the embodiment phase) or even of the basic concept, replacing one of the ways of performing a function by another.

Described in abstract terms, the concept of stages in design is not easy to grasp; an example will help.

Devices to open corked bottles   Bottled wine, traditionally, is sealed with a cork, either made from the natural material or often nowadays a plastic substitute – and we also now have a completely different design solution: the screwcap. Corked wine bottles are unlikely to disappear any time soon, so there remains a market need, in summary: 'a device to allow convenient access to wine in a corked bottle, at modest cost, and without contaminating the wine.'

Three concepts for doing this are shown in Figure 2.7, removing the cork by axial traction (= pulling), by shear traction, or by pushing it out with internal air pressure.

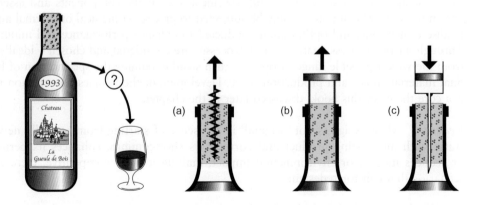

Figure 2.7   Three concepts to meet a market need: gaining access to wine in corked bottles.

In the first, a screw is threaded into the cork to which an axial pull is applied; in the second, slender elastic blades inserted down the sides of the cork apply shear traction when pulled; in the third, the cork is pierced by a hollow needle through which a gas is pumped. The cover page of this chapter shows bottle openers based on the first of these concepts.

Figure 2.8(a–d) shows four embodiment sketches for devices based on the first concept of axial traction. One uses a direct pull; the other three use some sort of mechanical advantage – levered pull, geared pull, and spring-assisted pull. The embodiments suggest layouts, mechanisms, and scale, and approximate loads will be known – enough for an initial selection of material options. In the final, detailed stage of design, the components are dimensioned to carry the working loads safely, their precision and surface finish are defined, and final choices of material and manufacturing route are made (Figure 2.8, right). We now consider selection in design.

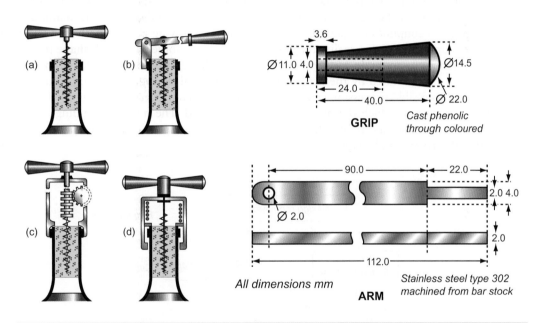

Figure 2.8 Embodiment sketches for the first concept: direct pull, levered pull, geared pull, and spring-assisted pull. Each system is made up of components that perform a sub-function. Detailed design drawings for the lever of embodiment (b) are shown on the right.

Material and process selection in design   A common approach may be taken to material selection at each stage of the design process (see Figure 2.6, left-hand column), but what differs is the level of precision and breadth in the data. At the concept stage, the designer requires approximate property values, easily accessed for the widest possible range of materials. All options are open: a polymer may be the best choice for one concept, a metal for another. At the embodiment stage for a particular concept, the materials options narrow – say to 'Al alloys' or 'fibre composites' – but a higher level of precision is needed, requiring more specialised handbooks and databases. The final stage of detailed design requires manufacturers' data for the specific materials chosen. A given material can have a wide range of property values, depending on the ways different producers make it. And however good the data, some prototyping and testing are always required, particularly for safety critical components, using the actual material that will be supplied. Much can also be learned from products that have been in service – manufacturers commonly analyse failures or worn-out products, looking for unanticipated problems, which re-design or re-selection may eliminate.

The selection of a material cannot be separated from that of process and of shape. To make a shape, a material undergoes a chain of processes that involve shaping, joining, and finishing (see Figures 2.2 and 2.3). The selection of processes follows a parallel route to that of materials (see Figure 2.6, right-hand column). Again, the starting point is a broad catalogue of all processes, which is then narrowed by eliminating those that cannot make the desired shape in the target material. And as the requirements expand and are refined, the processing information needs become more detailed, moving rapidly to dialogue with manufacturers.

Figure 2.6 simplifies the design process somewhat, showing parallel paths for selection of material and process – in reality they interact with one another, and with shape (Figure 2.9). The interactions are multi-faceted: specification of shape restricts the choice of material and process – often limited by the properties of the material *during* processing, combined with limits to accessible processing conditions. And it is often processing that is used to develop the properties on which the material has been selected. The more demanding and sophisticated the design, the tighter the specifications and the greater the interactions between material, process, and shape.

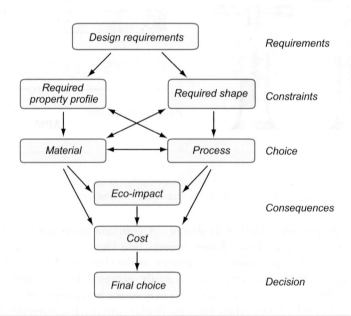

Figure 2.9 The interaction between design requirements, material, shape, and process.

2.4 The selection strategy: translation, screening, ranking, and technical evaluation

Choice of material and process is central to the design process: seeking the best match between the design requirements and the attribute profiles of the materials and processes. Given the enormous breadth in options for materials (and processes), we need a strategy, as shown in Figure 2.10 (applied to materials). The first task is *translation*: converting the design requirements into a profile of key properties and other needs the material must satisfy. This proceeds by identifying the functional and geometric *constraints* that must be met, and the *objectives* that we seek to optimise in the design. These become the filters for *screening*: eliminating the materials that cannot meet one or more of the constraints. This is followed by *ranking*: ordering the remaining shortlist by their ability to meet a criterion of excellence, such as minimising mass or cost. The final task is to explore the most promising candidates in depth, examining how they are used at present, case histories of failures, and how best to design and manufacture with them – in other words, a *technical evaluation*.

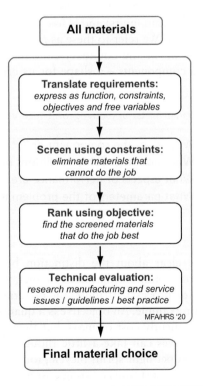

Figure 2.10   The selection strategy, applied to materials. A similar strategy is later adapted for processes. There are four steps: translation, screening, ranking, and technical evaluation.

Process selection follows a parallel route in many respects. In this case translation means identifying the processing constraints that must be met – material compatibility, shape, dimensions, precision, quality, and target properties. Some of these may be used to screen out processes that cannot meet the constraints, while in this context ranking is almost always a matter of cost. But for processing, many aspects require early technical evaluation, seeking expertise and best practice guides to assess the multiple interactions between design, material, and process, while refining design details and the choice of material, and identifying optimal processing conditions. We return to this in Chapter 11 (and the associated *Guided Learning Unit 3*). For now, we stick to materials and look at the selection steps more closely.

Translation   Any engineering component has multiple *functional constraints*: to support a load or contain a pressure without failure or deflecting too much, and/or to transmit heat or electricity efficiently, while functioning for a required lifetime in the design environment, and at an acceptable price, weight, and impact on the environment. Then there are clearly *geometric constraints* on the dimensions of the component, depending on its function and how it works. Out of all of these constraints, the designer will have one or more *objectives*: to make it as cheap as possible, or as light, or as environmentally benign – or some combination of these. Certain parameters can be adjusted to optimise the objective – the designer is free to vary some dimensions (within limits) and is free to choose the material for the component and the processes to make it. We refer to these as *free variables*. Constraints, objectives, and free variables (Table 2.2) define the boundary conditions for selecting a material and – in the case of load-bearing components – a shape for its cross section.

Table 2.2   Function, constraints, objectives, and free variables

Function	• What does the component do?
Constraints	• What non-negotiable conditions must the material meet?
Objectives	• What aspects of performance are to be maximised or minimised?
Free variables	• What parameters of the problem is the designer free to change?

It is important to be clear about the distinction between constraints and objectives. A constraint is an essential condition that must be met, which may sometimes be expressed as a limit on a material (or process) attribute: the material must cost less than $4/kg and be an electrical insulator, for example. An objective is a quantity for which a maximum or minimum value is sought, frequently cost or mass, but there are others (Table 2.3). Identifying the constraints and objectives can take a little thought, but it is an essential first step. In choosing materials for a super-light sprint bicycle, for example, the objective is to minimise mass, with an upper limit on cost, thus treating cost as a constraint. But in choosing materials for a cheap 'shopping' bike, the two are reversed: now the objective is to minimise cost with an upper limit on mass, treating it as a constraint.

Table 2.3   Common constraints and objectives

Common constraints*	Common objectives
Meet a target value of	**Minimise**
● Stiffness	● Cost
● Strength	● Mass
● Toughness	● Volume
● Thermal conductivity	● Impact on the environment
● Electrical resistivity	● Heat loss
● Magnetic remanence	
● Optical transparency	**Maximise**
● Cost	● Energy storage
● Mass	● Heat flow

*All these constraints, and products that rely on them, appear in later chapters.

The outcome of the translation step is therefore a clear statement of the function of the component, a list of constraints, and the overall objectives and free variables. As we saw in Chapter 1, it is then usually straightforward to identify the relevant *design-limiting properties*. Here we consider the translation step for the corkscrew.

Example 2.1 The corkscrew lever

Figure 2.8 shows the lever for one of the corkscrews in the design case study. In use, it is loaded in bending and must not deflect too much (stiffness), bend permanently (strength), or snap off (toughness). Finally, it must not corrode in wine or water. The length of the lever is specified, but the cross section is not – we can vary the section to bear the use-loads. The lever should be as cheap as possible. Formulate the translation, and list the design-limiting properties.

Answer: Translation for the corkscrew lever:

Function	● Lever (beam loaded in bending)	
Constraints	● Stiff enough	
	● Strong enough	
	● Tough enough	*Functional constraints*
	● Resist corrosion in water and wine	
	● Length L specified	*Geometric constraint*
Objective	● Minimise the material cost of the beam	
Free variables	● Cross-section area of beam, A	
	● Choice of material	

The design-limiting properties are those directly related to the objective and the constraints: cost/kg C_m (and density ρ, to convert to cost/volume), Young's modulus E, strength σ_y, fracture toughness K_{1c}, and corrosion resistance.

Screening Screening (see Figure 2.10) eliminates candidates that cannot do the job at all, because one or more of their attributes lies outside the limits set by the constraints. For example, the requirement that 'the component must function in boiling water' or that 'the component must be transparent' imposes obvious limits on the attributes of *maximum service temperature* and *optical transparency* that successful candidates must meet.

Ranking To rank the materials that survive the screening step we need a criterion of excellence. For this we use *material indices*, which measure how well a candidate delivers the objective set by the design (see Figure 2.10). Performance is sometimes limited by a single property: the best materials for buoyancy are those with the lowest density, ρ; the best materials for thermal insulation have the smallest values of the thermal conductivity, λ – provided, of course, that in each case they also meet all other constraints imposed by the design. Here, maximising or minimising a single property maximises performance. Often, though, it is a trade-off between a combination of properties that is relevant. Thus, the best materials for a light, stiff tie-rod are those with the greatest value of *specific stiffness*, E/ρ, combining Young's modulus E and density ρ. Good materials for springs have high values of σ_y^2/E, where σ_y is the yield strength. The property or property group that maximises performance for a given design is called its *material index*. There are many such indices, each associated with maximising some aspect of performance given particular functional and geometric constraints. *Guided Learning Unit 2* shows how material indices are derived from careful consideration of the objectives, constraints, and free variables, and in particular how to select materials using property charts. Subsequent chapters use indices frequently to compare material performance.

Technical evaluation The outcome of the steps so far is a ranked short-list of candidates that meet the constraints and that maximise or minimise the criterion of excellence, whichever is required. You could just choose the top-ranked candidate, but what hidden weaknesses might it have and is it commonly used? Can it be manufactured successfully into the target product? To proceed further we need a *technical evaluation* (see Figure 2.10, bottom). Technical documentation on materials (and processes) is largely descriptive, graphical, or pictorial – a 'knowledge base' of expertise and best practice, as much as a quantitative database. For example, we may refer to case studies of previous applications (and failures) of the material, details of its corrosion behaviour in particular environments, data for its availability and pricing, and warnings of its environmental impact or toxicity. Such information is found in handbooks, suppliers' data and websites, failure analyses, and case studies. And at an early stage, designers must engage with manufacturers to understand processing limitations, including the sensitivity of its properties to the way it is processed. Technical evaluation narrows the short-list to a final choice, allowing a definitive match to be made between design requirements and material attributes, including manufacturing know-how.

Why are all these steps necessary? Without systematic screening, ranking, and evaluation, the available material-process combinations are enormous, and the volume of technical documentation is overwhelming. Diving in and hoping to stumble on a solution gets you nowhere. But identifying a small number of strong candidates means that the search for detailed information and expertise is targeted, and the task becomes viable.

2.5 Summary and conclusions

Classification is the first step in creating an information management system for materials and processes. In it the records for the members of each Universe are organised in tree-like hierarchies, based on their common characteristics. There are six broad families of materials for mechanical design: metals, ceramics, glasses, polymers, elastomers, and hybrids that combine the properties of two (or more) of the others. Processes, similarly, can be grouped into families: those that shape and join components, or modify surface properties. This structure forms the basis of computer-based materials-selection systems, now widely available.

The starting point of a design is a *market need* captured in a set of *design requirements*. *Concepts* for a product that meet the need are devised and explored with rough estimates of design parameters. Viable concepts proceed to the *embodiment* stage: refinement of working principles, size and layout, and estimates of performance and cost. If the outcome is successful, the designer proceeds to the *detailed design* stage: optimisation of performance, full analysis of critical components, and preparation of detailed production drawings using CAD, showing dimensions, specifying precision and quality, and identifying the material and manufacturing path. Design is not a linear process – some routes lead to a dead end, requiring iteration of earlier steps. And, frequently, the task is one of redesign, requiring that objectives and constraints are re-assessed.

The selection of material and process run parallel to this set of stages, starting with the widest possible range of options for both. As the design requirements are formulated in increasing detail, constraints emerge that must be met, and one or more objectives is formulated. The constraints narrow the options, and the objective(s) allow ranking of the materials that remain. Identifying the constraints, the objectives, and the free variables (the translation step) comes first in the selection strategy. The quantitative steps in material selection – screening and ranking – are covered in *Guided Learning Unit 2*; the resulting measures of performance – material indices – are applied frequently in the rest of the book. When the task is the choice of process, a similar strategy applies for translation, screening, ranking, and technical evaluation of the options – we return to this in Chapter 11 and *Guided Learning Unit 3*.

2.6 Further reading

Ashby, M. F. (2017). *Materials selection in mechanical design* (5th ed.). Butterworth Heinemann. ISBN 978-0081-00599-6. (An advanced text developing material selection methods in detail).

Budinski, K. G., & Budinski, M. K. (2010). *Engineering materials, properties and selection* (9th ed.). Prentice Hall. ISBN 978-0-13-712842-6. (A well-established materials text that deals well with both material properties and processes).

Thompson, R. (2007). *Manufacturing processes for design professionals*. Thames and Hudson. ISBN 978-0-500-51375-0. (A beautifully produced and lavishly illustrated introduction to materials and to forming, joining, and finishing processes).

Ulrich, K. T. (2011). *Design: Creation of artefacts in society*. University of Pennsylvania Press. ISBN 978-0-9836487-0-3. (An excellent short introduction to the kind of structured reasoning that lies behind good product design. The text is available on: http://www.ulrichbook.org/).

2.7 Exercises

Classification exercises A good classification looks simple – think, for instance, of the Periodic Table of the Elements. Creating it in the first place, however, is not so simple. This chapter introduced two hierarchical classifications (for materials and processes) that work well, meaning that every member of the classification has a unique place in it, and any new member can be inserted into its proper position without disrupting the whole. Exercises E2.1 – E2.4 ask you to create hierarchical classifications following the same pattern of Universe – Family – Class – Member. Make sure that each level of the hierarchy properly contains all those below it. There may be more than one way to do this, but one is usually better than the others. Test it by thinking how you would use it to find the information you want.

Exercise E2.1 Classify the different ways in which sheets of paper can be attached to each other, temporarily or permanently. The Universe is 'Paper attachment.' Create the Family level by listing the distinctly different mechanisms for attachment. Then add the Class level – the distinct subsets of each family.

Exercise E2.2 In how many ways can wood be treated to change its surface appearance and feel? The Universe is 'Wood surface treatment.' Develop a classification into distinct techniques at the Family level, and subsets of these at the Class level, based on your own observation of wood products.

Exercise E2.3 You run a bike shop that stocks bikes of many types (children's bikes, street bikes, mountain bikes, racing bikes, folding bikes, etc.). You need a classification system to allow customers to look up your bikes on the internet. The Universe is 'Bicycles.' Develop the Family and Class levels of the classification, and note the further levels of hierarchy that would be needed to uniquely classify a given bike.

Exercise E2.4 You are asked to organise the inventory of fasteners in your company. Fasteners come in various sizes and materials, and there is more than one type. What would be a logical hierarchical classification for the Universe of 'Fasteners'?

How was it made? Examining product images (or, better, real products) offers clues to how things were shaped, joined, and finished, as in the next three exercises. We return to this in Chapter 11 and *Guided Learning Unit 3*.

Exercise E2.5 The four products shown were produced using the following *shaping* processes:

    A. Additive manufacturing
    B. Casting
    C. Forging and milling
    D. Turning

By thinking about characteristics of the shape being made, and doing a bit of research, match each product to a process.

1   2   3   4

Exercise E2.6   The four products shown illustrate the four *joining* methods:

A. Adhesives
B. Arc welding
C. Brazing
D. Threaded fasteners

Match each product to a process, identifying features that confirm your choice.

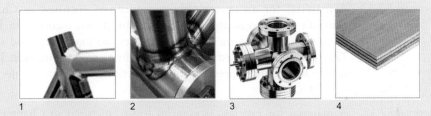

1   2   3   4

Exercise E2.7   The four products shown illustrate the four *surface treatment* methods:

A. Anodising
B. Chrome plating
C. Enameling
D. Powder coating

Match each product to a process, using the appearance of the product.

1   2   3   4

Design process, and translation of design requirements The next three questions revise key terminology and concepts in the processes of design and selection. There are then nine examples of the translation step in the material selection strategy. We can learn a lot about materials and design by examining the same product made in more than one material – we finish with two examples of these.

Exercise E2.8   Summarise the steps in developing an original design.

Exercise E2.9   What is meant by a *constraint* and what by an *objective* in the requirements for a design? How do they differ?

Exercise E2.10   Outline the purposes of each of the key steps in the selection strategy for materials described in this chapter: translation, screening, ranking, and technical evaluation.

Exercise E2.11   The teeth of a scoop for a digger truck, pictured in Chapter 1, must cut earth, scoop stones, crunch rock, often in the presence of fresh or salt water and worse. Translate these requirements into a prescription of function, constraints, objectives, and free variables.

Exercise E2.12   A material for an energy-efficient saucepan, pictured in Chapter 1, must transmit and spread the heat well, it must resist corrosion by foods, and must withstand the mechanical and thermal loads expected in normal use. The product itself must be competitive in a crowded market. Translate these requirements into a prescription of function, constraints, objectives, and free variables.

Exercise E2.13   A material for eyeglass (spectacle) lenses, pictured in Chapter 1, must have optical-quality transparency. The lens may be moulded or ground with precision to the required prescription. It must resist sweat and be sufficiently scratch-resistant to cope with normal handling. The mass-market end of the eyeglass business is very competitive, so price is an issue. Translate these requirements into a prescription of function, constraints, objectives, and free variables.

Exercise E2.14   Bikes come in many forms, each aimed at a particular sector of the market:

- Sprint bikes
- Touring bikes
- Mountain bikes
- Shopping bikes
- Children's bikes
- Folding bikes

Use your judgement to identify the primary objective for each of these.

Exercise E2.15   A material is required for the windings of an electric air-furnace capable of temperatures up to 1000°C. Think out what attributes a material must have if it is to be made into windings and function properly in a furnace. List the function and the constraints; set the objective to 'minimise material price' and the free variables to 'choice of material.'

Exercise E2.16 A material is required to manufacture office scissors. Paper is an abrasive material, and scissors sometimes encounter hard obstacles like staples. List function and constraints; set the objective to 'minimise material price' and the free variables to 'choice of material.'

Exercise E2.17 A material is required for a heat exchanger to extract heat from geothermally heated, saline water at 120°C (and thus under pressure). List function and constraints; set the objective to 'minimise material price' and the free variables to 'choice of material.'

Exercise E2.18 A material is required for a disposable fork for a fast-food chain. List the objective and the constraints that you would see as important in this application.

Exercise E2.19 Formulate the constraints and objective you would associate with the choice of material to make the forks of a racing bicycle.

Exercise E2.20 Cheap coat hangers used to be made of wood – now it is only expensive ones that use this material. Most coat hangers are now metal or plastic, and both differ in shape from the wooden ones, and from each other. Examine wood, metal, and plastic coat hangers, comparing the designs, and comment on the ways in which the choice of material has influenced them.

Exercise E2.21 Cyclists carry water in bottles that slot into bottle cages on their bicycles. Examine metal and plastic bottle holders, comparing the designs, and comment on the ways in which the choice of material has influenced them.

Chapter 3
Material properties and microstructure — overview and atom-scale fundamentals

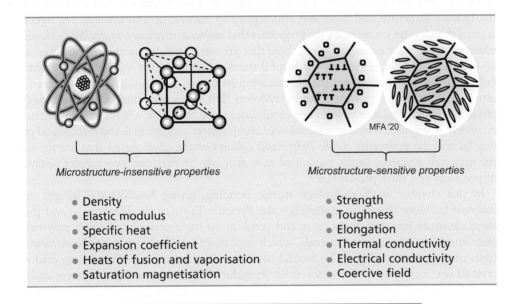

MFA '20

Microstructure-insensitive properties

- Density
- Elastic modulus
- Specific heat
- Expansion coefficient
- Heats of fusion and vaporisation
- Saturation magnetisation

Microstructure-sensitive properties

- Strength
- Toughness
- Elongation
- Thermal conductivity
- Electrical conductivity
- Coercive field

Chapter contents

https://doi.org/10.1016/B978-0-08-102399-0.00003-0

3.1 Introduction and synopsis

The material properties used by engineers to design products from mobile electronics to suspension bridges are a consequence of the physics of atoms and molecules, and a whole array of internal features we call *microstructure*. Atoms are around 10^{-10} m in diameter; man-made products start with tiny devices at a fraction of a millimetre, up to civil engineering construction at the kilometre scale. Materials include quantifiable microstructural features from clusters of a few atoms right up to the scale of the products themselves – a rather bewildering range of length-scale spanning at least six orders of magnitude. The conceptual challenge therefore is to drill down to the relevant features and length-scales to identify the underlying physics that govern the properties.

We start with a high-level distinction that has wide-ranging consequences. It is between properties that depend directly on atomic bonding and atom packing, and are largely insensitive to microstructure, and those that also depend on microstructural features. The figure at the start of this chapter summarises the properties that are *microstructure-insensitive* – elastic modulus and density are examples – and those that are *microstructure-sensitive* – such as strength and electrical conductivity. This is a helpful starting point for two reasons. First, atomic-scale behaviour broadly determines the classification of materials introduced in Chapters 1 and 2 into metals, ceramics, polymers, and elastomers. But more importantly, atomic behaviour is physically proscribed and beyond our capabilities to be modified, whereas microstructure may be widely manipulated by a combination of composition and thermal and mechanical processing. So it is the properties in the right-hand column of the cover figure that are the focus of the materials engineer, in the pursuit of new property profiles and ever-greater optimisation of products and their performance.

In this chapter we first explore atomic bonding, giving fundamental insight into the simplest building blocks of materials – the Periodic Table of the Elements – and the ways these elements interact: some react and combine to form ceramics, glasses, polymers, and elastomers; the majority are metals, which mix more freely across multiple elements – the basis of engineering *alloys*. The second fundamental aspect is atomic packing into regular crystals (or occasionally into amorphous irregular structures) and (for polymers and elastomers) into long chain molecules, which then interact with one another. Crystallography is a substantial field of endeavour in its own right. The best way to comprehend the essential geometry is by doing worked examples – so this is provided in the separate *Guided Learning Unit 1 – Simple ideas of crystallography*.

Equipped with a knowledge of atomic bonding and packing, we can interpret the microstructure-insensitive properties – for example, this chapter and *Guided Learning Unit 1* explore why the density of every solid is so well determined by the mass and packing of the atoms it contains. Discussion of another atomically determined property – elastic modulus – follows in Chapter 4; later chapters introduce the complexities of microstructure and the properties that depend on it. But at many points we will return to the atomic scale to capture some essential physics – for example, diffusion in solids, the electrochemistry of corrosion, and the processing of metallic alloys. First we set out the big picture of microstructure, properties, and length-scales in the major material classes, by way of an overview of the central chapters of this book and introduce some of the technical terminology of the subject.

3.2 Material properties and length-scales

Metals Figure 3.1 summarises the main microstructural features in metals. Starting at the bottom with atoms, we have crystalline packing (with the exception of the unusual amorphous metals) – responsible for density and elastic moduli (Chapters 3 and 4). Crystals contain atom-scale defects – vacancies (missing atoms) and atoms in solution – and this smallest scale of microstructure directly influences thermal, electrical, optical, and magnetic behaviour (Chapters 7 and 10). Yield strength, toughness, fatigue, friction, and wear depend firmly on microstructural features spanning a wide range of length-scales, observed and quantified with all kinds of microscopy and diffraction techniques. This spans crystal defects called 'dislocations' and various obstacles that impede their motion (e.g. precipitates), through to crystal 'grains,' surface roughness, porosity, and cracks. Dislocation motion governs the strength of metals and alloys (Chapter 5), while crack initiation and growth from surface roughness and internal defects control fracture and fatigue (Chapter 6), and surfaces naturally control friction and wear (Chapter 9). Diffusion is fundamentally an atomic-scale phenomenon, but its effects are governed by the larger length scales over which it operates – those of precipitation, dislocation motion at temperature, grains, porosity, and electrochemical reactions, i.e. the microstructural phenomena behind a host of service and processing behaviours: corrosion, oxidation, creep, sintering, and heat treatments (Chapters 8, 9, and 11). Metal products themselves span a huge range: kitchen foil is around $10\,\mu m$ thick, automotive panels a millimetre or so, ship propellers are several metres in diameter, whereas bridges and buildings reach the kilometre scale.

Ceramics and glasses Ceramics are crystalline, glasses are amorphous. Figure 3.2 shows both structures near the bottom of our materials length-scale. Many of the properties directly reflect the atomic layout and the intrinsically strong nature of the covalent or ionic bonding – from elastic modulus and density to electrical insulation and semiconduction. Atomic scales also govern viscosity in molten glass and diffusion in compacted ceramic powders – though again diffusion operates at the scale of porosity during processing (Chapter 11). Cracks in ceramics are closely related to grain size, and it is cracks and surface finish that dominate brittle fracture in both ceramics and glasses (Chapter 6).

Polymers and elastomers Polymers and elastomers are inherently molecular rather than atomic. Figure 3.3 shows the wide diversity at this scale: depending on the chemistry, polymer molecules arrange into amorphous, crystalline, or cross-linked structures, and some can be aligned by drawing. Density, modulus, and the functional properties again reflect behaviour at this scale (Chapters 3, 4, and 10); strength and toughness bring in larger features such as crazes and cracks (Chapters 5 and 6).

Hybrids: foams, composites, and natural materials We noted earlier that microstructure-insensitive properties such as modulus and density are proscribed by the physics of atomic or molecular bonding and packing. But there is a way to get around this by making *hybrid materials* – mixing together two (or more) solids, as in fibre composites, or making the material porous, as in foams (both analysed with respect to elastic modulus in Chapter 4). These hybrids have their own characteristic length-scales, which we will refer to as their

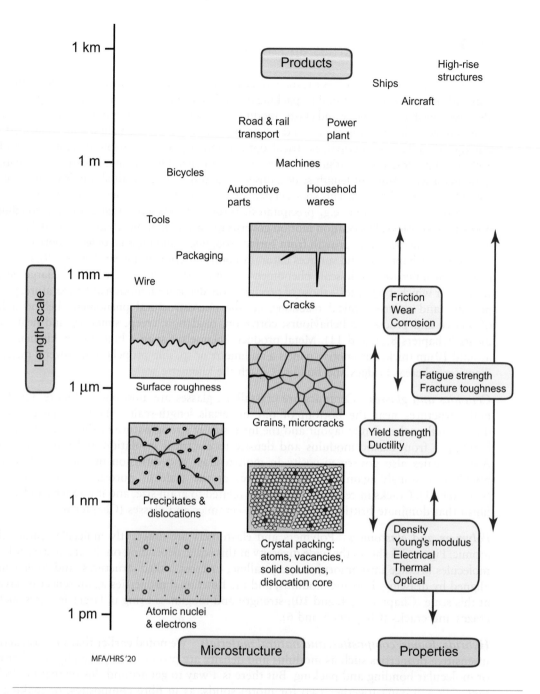

Figure 3.1   Microstructural features in metals, showing their typical length-scale and the properties they determine. Each interval on the length-scale axis is a factor of 1000.

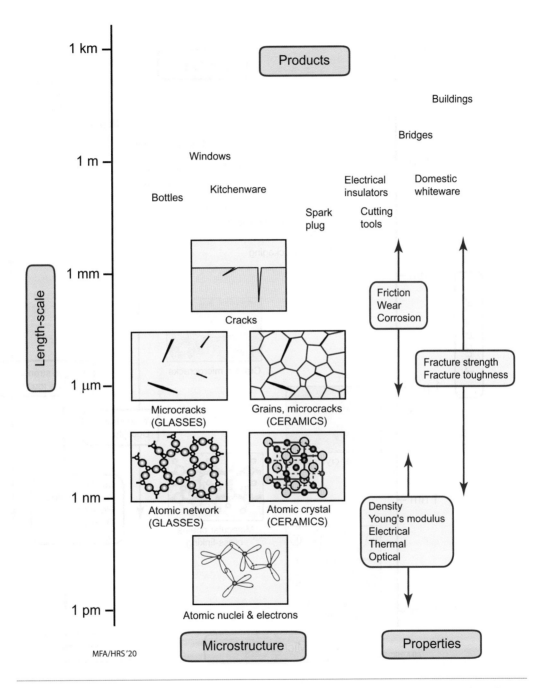

Figure 3.2 Microstructural features in ceramics and glasses, showing their length-scale and the properties they determine. Each interval on the length-scale axis is a factor of 1000.

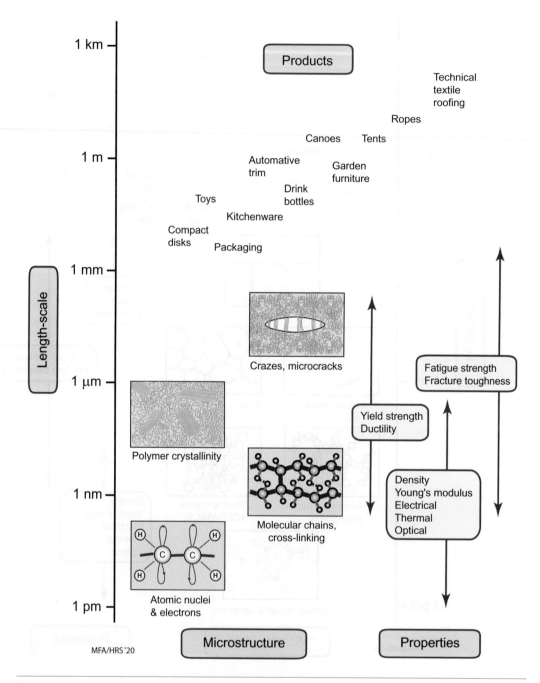

Figure 3.3   Microstructural features in polymers and elastomers, showing their length-scale and the properties they determine. Each interval on the length-scale axis is a factor of 1000.

'architecture' (since the constituent solids still have their own internal microstructure) – for example, the cell size in a foam, and the fibre diameter and ply thickness in laminated fibre composites. Natural materials such as wood (or those made from natural materials, such as paper) are commonly hybrids. Wood proves to have layer upon layer of fascinating internal structure from cellulose microfibrils embedded in lignin (natural polymers), making up the multi-layered walls of hollow tubular cells, varying in size and density within the tree (including its annual growth rings). The mechanical performance is so good in fact, relative to their weight, that this has inspired the whole field of 'biomimetics' – artificially replicating aspects of natural architectures in man-made materials. Figure 3.4 shows a range of hybrids, all falling within a central range of our length-scale diagram, including

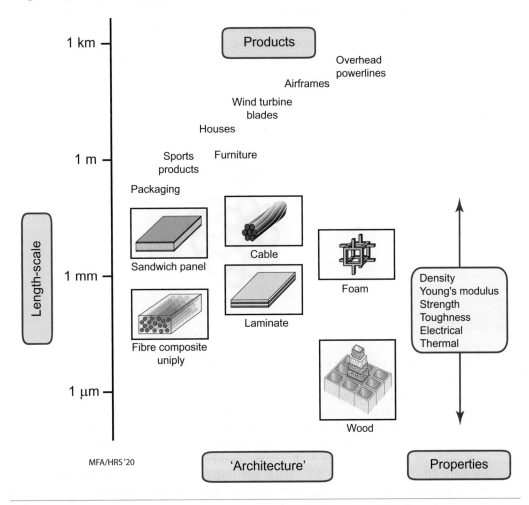

Figure 3.4 Material 'architecture' in hybrids: man-made (foams, composites, sandwich panels, cables) and natural (wood), showing their length-scales. Hybrid properties all depend on the length-scale of the constituents. Each interval on the length-scale axis is a factor of 1000.

other examples where the boundary between material and product becomes a little blurred, such as multi-stranded cables and sandwich panels. The properties of the solids within a hybrid come, of course, from the bulk materials (as in Figures 3.1, 3.2, and 3.3). Given these properties, however, hybrid properties all depend directly on the length-scales and volume fractions of their constituents – examples will be found later in this book.

3.3 Atomic structure and inter-atomic bonding

The structure of the atom   The core of an atom is its *nucleus* (Figure 3.5). The nucleus accounts for almost all the atomic mass and almost none of its volume. It is made up of *protons* (positive charge) and *neutrons* (no charge), known collectively as *nucleons*. The nucleus of a given element contains protons equal in number to its *atomic number*, but can exist as *isotopes* with differing numbers of neutrons and thus atomic weights. Isotopes have identical chemical properties but differ in their nuclear properties, particularly in their nuclear stability.

Nucleons:
neutrons and
protons

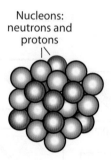

Figure 3.5   The atomic nucleus.

The most stable nuclei are those with the greatest *binding energy* per nucleon – they cluster around iron, atomic number 26 (Figure 3.6). Nuclei above iron in atomic number can release energy by fission; those below iron can do so by fusion. Fission releases, at most, a few hundred keV per event; fusion can release much more.

The most stable elements are found to be the most abundant in the earth's crust: silicon, aluminium, iron, calcium, sodium, magnesium, and oxygen; combined they account for over 80% of its content. The high atomic number platinum group, by contrast, account for only 10^{-7}% by mass (or 10^{-3} parts per million [ppm]). Natural abundance fractions are increasingly important to understand in relation to sustainable development (Chapter 12) – key technologies for high field magnets, electronic devices, and vehicle batteries depend on elements such as lithium and 'rare earths' with natural abundance fractions of 1 to 100 ppm.

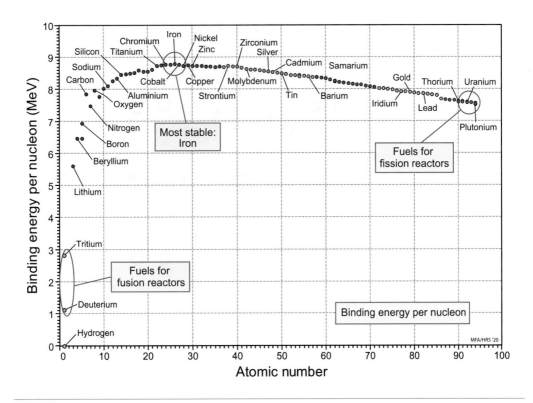

Figure 3.6 The binding energy per nucleon, a measure of nuclear stability.

Electronic structure of atoms Electrons (negative charge) orbit the nucleus in discrete, quantised energy levels (Figure 3.7). Their behaviour requires quantum mechanics for explanation. This involves accepting the idea that electrons behave both as particles and as waves, a more diffuse existence in which the probability of their presence in wave-mode (shown in green on this and subsequent figures) has meaning, although their precise position in particle mode (shown as blue spheres) does not. The wavelengths of electrons around an atomic core, like the vibrational modes of a taut string, have discrete values characterised by a principal quantum number $n = 1, 2, 3, 4$, or 5. The principal shells split into sub-shells corresponding to a second quantum number taking the values 0, 1, 2, or 3, designated by the letters *s, p, d*, and *f* [1]. Figure 3.7 suggests the first five energy levels (there are more) and the maximum number of electrons each can contain. The filling of a given orbit is described by the two quantum numbers with a superscript indicating the number of electrons in it – a number that cannot exceed the values indicated on the figure.

---

[1] *s, p, d*, and *f* stand for 'sharp,' 'principal,' 'diffuse,' and 'fundamental,' describing the spectral lines of an atom.

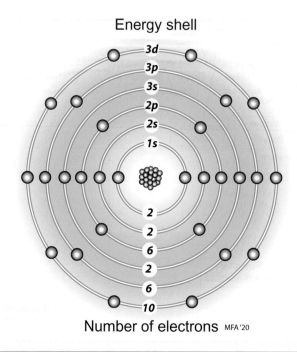

Energy shell

Number of electrons MFA '20

Figure 3.7   Electron orbitals and the number of electrons they can contain.

Thus, if the first, lowermost orbit ($n = 1$, subshell s) contains two electrons – the maximum allowed – it is described as $1s^2$. If the level $n = 3$, subshell d has 6 electrons in it (it can hold 10), and it becomes $3d^6$.

Filled principal shells impart maximum stability. The inert gases helium ($1s^2$), neon ($1s^2 – 2s^2\ 2p^6$), and argon ($1s^2 – 2s^2\ 2p^6 – 3s^2\ 3p^6\ 3d^{10}$) are so stable that almost nothing reacts with them. Atoms with unfilled shells like sodium, with one $3s$ electron, can revert to the electron-count of neon by giving up an electron. Chlorine, with five $3p$ electrons, can acquire the electron-count of argon by capturing one. Bringing one sodium atom and one chlorine atom together and transferring one electron from the sodium (making it a positive ion) to the chlorine (making it negative) allows both to acquire 'inert-gas' electron shells (Figure 3.8).

As the atomic number of an atom increases, the electrons fill the energy levels, starting with the lowest, the $1s$. The electrons in the outermost shell, known as *valence electrons*, largely determine the chemical character of an atom. This leads to a periodic pattern of properties as the shells fill, moving from inert-gas behaviour when a set of shells is filled completely, to a reactive, electropositive (electron-donating) character as the first electrons enter the next level, to an electronegative (electron-attracting) response when the shell is nearly full, and reverting to inert-gas behaviour when the next level is completely full. The cycle is reflected in the Periodic Table of the Elements (Figure 3.9).

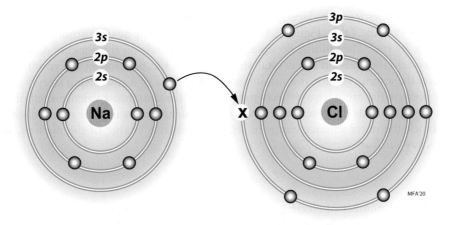

Figure 3.8 Electron transfer from Na to Cl, creating positive and negative ions with more stable electron configurations (*1s* shell not shown).

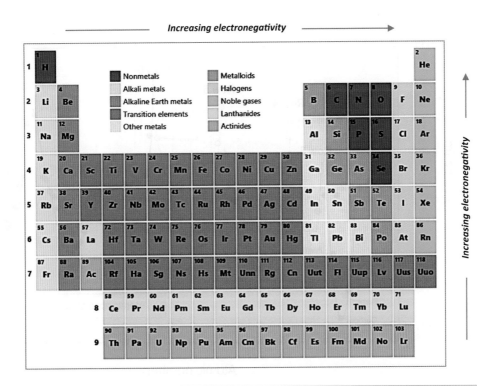

Figure 3.9 The Periodic Table of the Elements.

Material properties that depend on electronegativity change in a cyclic way across the rows of the Periodic Table. Figure 3.10 shows atomic volume against atomic number, colour-coded row by row. The volume occupied per atom depends on the packing density of atoms in the crystal (and thus on the crystal structure), but the position in the row has a greater influence: the atomic volume is a minimum in the middle of a row, rising steeply at the ends. This is because the atom 'size' reflects the radius of the outermost valence shell of electrons. Explaining this needs the concept of *screening*, the degree to which the inner (filled) electron shells block the outer valence electrons from the positive charge of the nucleus. We define the *effective nuclear charge*, Z_{eff}, as

$$Z_{eff} = Z - S \tag{3.1}$$

Here S is the number of electrons in the inner filled shells, equal to the atomic number of the noble gas that ends the row above the element in the Periodic Table. As an example, the S-value for chlorine is 10, the atomic number of neon. The quantity Z is the total number of electrons, equal to the atomic number of the element itself, 17 for chlorine. The 10 inner electrons (S) of chlorine screen out the positive charge of 10 protons, leaving an effective nuclear charge Z_{eff} of $(17 - 10) = +7$ that pulls the outer electrons closer to the nucleus, reducing the atomic radius. The imperfect screening is a maximum in the middle of the

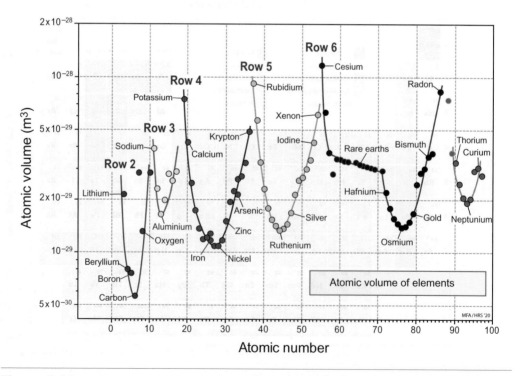

Figure 3.10   The cyclic change of atomic volume across the Periodic Table of Elements.

row, leading to the minimum. The effective nuclear charge increases, and the atomic volume initially decreases from left to right across a row in the Periodic Table. Moving down a column of the table, the number of filled electron shells increases. The effective nuclear charge remains the same, but now the orbitals are farther from the nucleus, which exerts less pull on the outer electrons, increasing the atomic radius.

Figure 3.10 shows atomic volume, not radius. Screening explains the plunging volume from the start of a row toward the middle (lithium to carbon in row 2, for instance), and the slower increase in volume in moving down a column of the Periodic Table (the sequence lithium – potassium – rubidium – cesium, for example). Superimposed on this is the effect of atom packing: elements at the start of a row are metallic and mostly close-packed. Elements toward the end of a row are non-metallic and have crystal structures with lower packing density.

Example 3.1

What is the effective nuclear charge Z_{eff} of aluminium? Would you expect it to have a larger or smaller atomic radius than calcium?

Answer.   Referring to the Periodic Table (Figure 3.9), the atomic number of aluminium is $Z = 13$, and that of the noble gas one row above it in the table, neon, is $S = 10$. The effective nuclear charge [Equation (3.1)] of aluminium is $Z_{eff} = 13 - 10 = 3$. The atomic number of calcium is $Z = 20$, and that of the noble gas above it in the table, argon, is $S = 18$. The effective nuclear charge of calcium is $Z_{eff} = 20 - 18 = 2$. We expect aluminium to be a smaller atom than calcium for two reasons: it has a greater effective nuclear charge, pulling the valence electrons more strongly toward the core, and it lies in row 3 of the Periodic Table while calcium is in row 4, so the calcium has an extra filled inner shell screening its valence electrons from the charge on the core.

Microstructure-insensitive properties (such as the elastic moduli – Chapter 4) show similar progressions. This is partly explained by the cyclic atomic volume, but there is a second effect superimposed (discussed next): the change in *cohesive energy* across a row, associated with the number of free electrons per ion (in a metal), and the number of shared electrons (in a covalently bonded solid).

Inter-atomic bonding and cohesive energy   Many aspects of inter-atomic bonding can be understood in terms of *electronegativity*, the propensity of an atom to attract one or more electrons to become a negative ion. Electronegativity is measured on an arbitrary, dimensionless scale proposed by Pauling[2] that ranges from 0.7 (francium, the least electronegative) to 4.0 (fluorine, the most), and it too varies in a cyclic way across the rows of the Periodic Table (Figure 3.11). We will use this plot in a moment as a tool for estimating bond type.

---

[2] Linus Carl Pauling (1901 –1994), American chemist and peace activist, twice Nobel laureate, originator of the exclusion principle and author of *How to Live Longer and Feel Better* advocating massive doses of vitamin C to ward off colds.

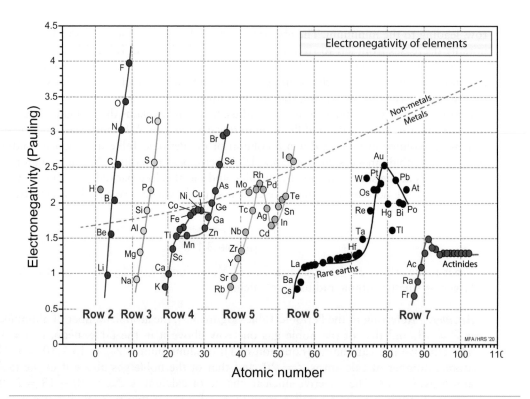

Figure 3.11 Electronegativity variation with atomic number.

As neutral atoms are brought together from infinity, their electrons begin to interact, tending to form a bond. Some bonds are so weak that, at room temperature, thermal energy is enough to disrupt them. Others are so strong that, once formed, they are stable to over 3000°C. As the bond forms and the atoms are drawn together, the energy at first falls; but when filled electron shells start to overlap the energy rises again because, while unfilled shells can share electrons, the Pauling exclusion principle prohibits filled shells from doing the same. This generates a balancing repulsive force, causing the atoms to sit at an equilibrium spacing a_o, where the energy is a minimum (Figure 3.12). The *cohesive energy*, H_c, is the energy per mol (a mol is 6.022×10^{23} atoms) required to separate the atoms of a solid completely, giving neutral atoms at infinity. The greater the cohesive energy, the stronger are the bonds between the atoms.

We anticipate therefore that the cohesive energy will correlate closely with properties that are governed directly by atomic bonding, and this proves to be the case – for example, for elastic moduli (Chapter 4). There are several characteristic types of bonding in solids, depending on how the valence electrons interact – here we examine the four that dominate material properties.

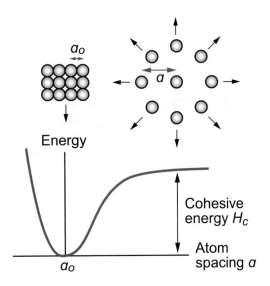

Figure 3.12 Cohesive energy.

Metallic bonding About 80% of the elements of the Periodic Table are metals – the box lists some examples, showing the wide range of cohesive energy. Valence electrons in metallic solids detach themselves from their parent atoms (leaving them as positive ions) and form a negatively charged swarm of almost-freely moving charge carriers (Figure 3.13). The swarm screens the array of positive ions from their mutual repulsion, shutting each off, so to speak, from its repulsive neighbours. The Coulombic attraction between the swarm and the array of positive ions creates the bond. It is not directional, so metallic solids are usually close-packed because this minimises the average distance between the positively charged ions and the negatively charged electron swarm. Electronegative atoms attract rather than donate electrons, so they are generally non-metallic. Those that are electropositive do the opposite, so that a boundary (blue line) can be sketched onto Figure 3.11, dividing the elements on it, very roughly, into those with metallic character and those with non-metallic.

A key aspect of metals is their ability to freely mix with one another, and with non-metallic elements, to form *alloys*. Impurity elements 'dissolve' as single dispersed atoms up to some concentration, substituting for an atom of the host element; in higher concentrations, it is usually preferable for dissimilar elements to combine together within the metal, forming regions around particular atomic proportions. Much more on this later – for now we note that it is the diffuse nature of the metallic bond that enables this freedom to mix metallic elements, and that the essential characteristics of metallic bonding (and thus microstructure-insensitive properties) persist in alloys.

> **Metallic-bonded materials**
> - H_c = 60 – 850 kJ/mol
> - Copper
> - Aluminium
> - Lead
> - Tungsten

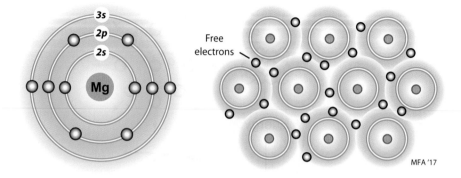

Figure 3.13   Metallic bonding in magnesium. The two **3s** valence electrons detach and move freely through the array of positively charged ions.

Ionic bonding If a strongly electro-positive atom approaches one that is strongly electro-negative, there is a tendency for electrons to leave the first and join the second, giving both filled outer shells in the way illustrated in Figure 3.8. The first atom acquires a positive charge, the second a charge that is negative, and the two are drawn together by Coulombic attraction. The resulting ionic bond is non-directional, attracting equally in all directions. Ionically bonded compounds tend to form when the difference in electronegativity of the atoms (plotted in Figure 3.11) is greater than 1.7. Crystal structures held together by ionic bonding must meet two further conditions: that of overall charge neutrality and that of ensuring every positive ion has only negative ions as its nearest neighbours and vice versa. The box lists some common ionically bonded materials. The cohesive energy is generally higher than for metallic bonding.

> **Ionically bonded materials**
> - H_c = 600 – 1600 kJ/mol
> - Sodium chloride, NaCl
> - Potassium chloride, KCl
> - Calcium chloride, $CaCl_2$
> - Magnesium oxide, MgO

Covalent (or molecular) bonding Covalent bonding is so called because it arises from the sharing of valence electrons between atoms in such a way that each acquires an inert-gas electron structure. It operates in a diverse range of elements and compounds, including carbon crystals and molecules, with cohesive energies spanning a wide range overlapping both metallic and ionic bonding (see box).

> **Covalently bonded materials**
> - H_c = 100 – 1200 kJ/mol
> - Diamond
> - Silicon
> - Polyethylene, along chain
> - Quartz

To understand covalent bonding, however, we start with the water molecule (Figure 3.14). Here the *1s* electrons of the two hydrogens are shared with the almost-full *2p* shell of the oxygen giving it the electron structure of neon. It in turn shares an electron with each hydrogen giving them the helium configuration.

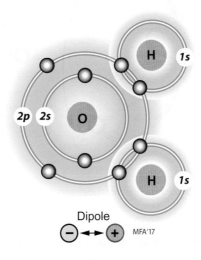

Figure 3.14 Covalently bonded water molecule.

Covalent bonds tend to form in molecules when the electronegativity difference between the two atoms is less than 1.7. For water this difference is 1.4, large enough that the electrons tend to spend more time on the oxygen than on the two hydrogens, giving a *polar* bond (i.e. the bond is directional, with a permanent *dipole moment*). When the difference is less than about 0.4, the electrons are more equally shared and the bond is non-polar.

Carbon, particularly, forms strong covalent bonds both with itself and with other atoms. To understand these, we need the idea of orbital mixing or *hybridisation*. Carbon, $1s^2 - 2s^2\ 2p^2$, can mix *2s* with *2p* electrons to create new orbitals that have directionality (Figure 3.15). Mixing one *2s* electron with the two *2p* electrons gives three sp^2 orbits lying in a plane and directed at 120° to each other. Mixing all four *2s* with *2p* electrons gives four new sp^3 orbits directed at the four corners of a tetrahedron. Each hybrid orbit can hold two electrons but has only one until it shares it with another atom, creating a covalent bond.

These new orbits have fundamental significance for the compounds of carbon. Graphite is a hexagonal net of sp^2-bonded carbon atoms, each of the three orbits sharing its one electron with another identical atom to form covalently bonded sheets. Diamond is a

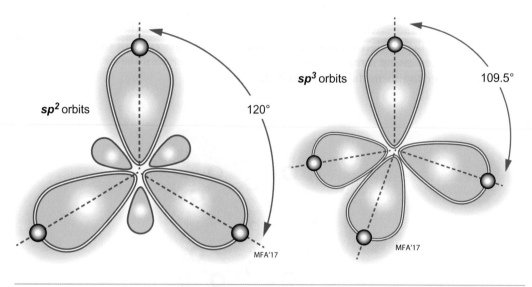

Figure 3.15 *sp²* and *sp³* orbitals of carbon atoms.

three-dimensional network of *sp³*-bonded carbon atoms, similarly bonded to other identically hybridised C atoms. Many polymers and many organic molecules are based on *sp³*-bonded carbon: two of the four orbits are used to make long chain molecules, with other atoms (such as hydrogen) or molecular groups being attached to the other two. The bonding within the polymer chain is covalent, but we also need to consider the bonding between them – which comes next.

Dipolar (or van der Waals) bonding Inert gases become liquid or solid at sufficiently low temperatures. They already have full electron shells, so what is binding the atoms together? The bond has its origins in dipole-dipole attraction (Figure 3.16). The charge distribution on an atom fluctuates as the electrons move; at any instant the charge is unevenly distributed, giving the atom a temporary *dipole moment*. The dipole of one atom induces a dipole in an adjacent atom and the pair of dipoles attract, creating a weak bond.

> **Dipolar-bonded materials**
> * $H_c = 7 - 50$ kJ/mol
> * Ice
> * Dry ice (solid CO_2)
> * Paraffin wax
> * Polyethylene, between chains

Dipolar bonding can arise in another way. As explained earlier, polar covalent bonds like that of water (see Figure 3.14) carry a permanent dipole moment. It is this that binds water molecules together to form ice. The same dipolar bonding glues one polymer chain to its neighbours, such as in polyethylene and polypropylene. The box shows that the cohesive energy associated with dipolar bonding is much lower than metallic, ionic, or covalent.

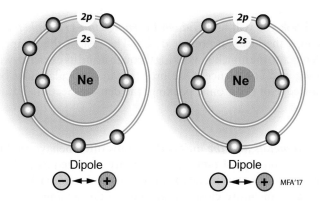

Figure 3.16 Dipolar attraction between neutral atoms.

Example 3.2

Boron sulphide is used in high-tech glasses. Would you anticipate that it has metallic, ionic, or covalent bonding? Use the electronegativity chart (Figure 3.11) to reach a conclusion.

Answer. Both boron and sulphur are non-metals. The difference in their electronegativity is about 0.5 on the Pauling scale, far below the value of 1.7 at which ionic bonding becomes dominant. We anticipate that the bonding is covalent.

3.4 Atomic and molecular packing in solids

Introduction to Guided Learning Unit 1: Simple ideas of crystallography

The properties of materials depend on the way the atoms or molecules within them are packed. In *glasses* the arrangement is a disordered one with no regularity or alignment. *Polymers* are largely made up of tangled long-chain molecules. *Metals* and *ceramics*, by contrast, are crystalline, with a regularly repeating pattern of atomic packing in structural units. Crystals are described using the language of *crystallography*, which provides a framework for understanding their three-dimensional geometry. The key ideas of crystallography are introduced briefly in the following pages.

Later in the book you will find *Guided Learning Unit 1: Simple ideas of crystallography*, which develops the ideas in greater depth. The rest of this chapter will be intelligible without reference to the unit, but working through it and its exercises will give more complete understanding and confidence. One property in particular may be calculated directly from a knowledge of the atomic packing: the density (Section 3.5 and *Guided Learning Unit 1*).

Atom packing in metals and the unit cell   Atoms often behave as if they were hard, spherical balls. The balls on a pool table, when set, are arranged as a close-packed layer [Figure 3.17(a)]. The atoms of many metals form extensive layers packed in this way. There is no way to pack the atoms more closely than this, so this arrangement is called 'close-packed.' These two-dimensional layers of atoms can then be close-packed in three dimensions. Surprisingly, there are two ways to do this. Where three atoms meet in the first layer, they form natural locations for the atoms of a second close-packed layer – but only half can be occupied by layer 2 – essentially there are six of these locations around and above each atom, but only space for three of them to be occupied. Now when a third layer is added, there are two options. In one, its atoms can be placed exactly above those in the first layer, and the sequence repeated to give a crystal with ABABAB... stacking [Figure 3.17(b)]; it is called *close-packed hexagonal (CPH)*, for reasons explained in a moment. But in the alternative stacking sequence, the third layer is offset from both the first and second layers, giving (on repeating) ABCABCABC... [Figure 3.17(c)]; it is called *face-centred cubic (FCC)*. Many metals (copper, silver, aluminium, and nickel) have the FCC structure; many others (magnesium, zinc, and titanium) have the CPH structure. The two structures have the same packing fraction, 0.74, meaning that the spheres occupy 74% of the available space. The difference appears negligible – you have to look very closely at Figure 3.17 to see how they differ. But this small difference in layout significantly influences some properties, particularly those having to do with plastic deformation (Chapter 5).

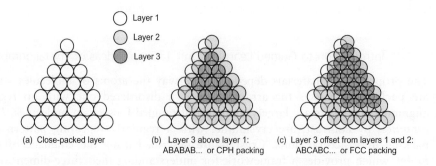

(a) Close-packed layer   (b) Layer 3 above layer 1: ABABAB... or CPH packing   (c) Layer 3 offset from layers 1 and 2: ABCABC... or FCC packing

Figure 3.17 Two alternative options for close-packing of spheres.

Not all structures are close-packed. Figure 3.18 shows one of these, made by stacking square-packed layers. It has a lower packing density than the hexagonal layers of the FCC and CPH structures. An ABABAB ... stacking of these layers builds the *body-centred cubic (BCC)* structure, with a packing fraction of 0.68. Iron and most steels have this structure. There are many other crystal structures, but for now these three are enough, covering most important metals.

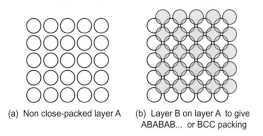

(a) Non close-packed layer A   (b) Layer B on layer A to give ABABAB... or BCC packing

Figure 3.18 Square-packing of spheres.

A regular packing of atoms that repeats itself is called a *crystal*. It is possible to pack atoms in a non-crystallographic way to give what is called an *amorphous* structure, sketched as two-dimensional layers in Figure 3.19 – fully amorphous structures are random in three dimensions. This is a less efficient way to fill space with spheres: the packing fraction is 0.64 at best.

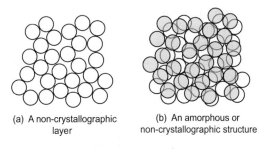

(a) A non-crystallographic layer   (b) An amorphous or non-crystallographic structure

Figure 3.19 Amorphous structures.

The characterising unit of a crystal structure is its *unit cell*. Figure 3.20 shows three; the red lines define the cell. The atoms have been shrunk to reveal the cell more clearly; in reality, they touch in close-packed directions. In the first [Figure 3.20(a)], the cell is a hexagonal prism. The atoms in the top, bottom, and central planes form close-packed layers like that of Figure 3.17(b), with ABAB... stacking – that's why this structure is called 'close-packed

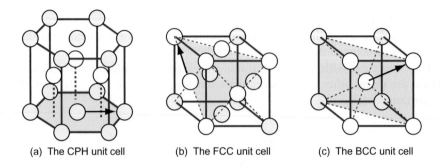

(a) The CPH unit cell   (b) The FCC unit cell   (c) The BCC unit cell

Figure 3.20 The close-packed hexagonal (CPH), face-centred cubic (FCC), and body-centred cubic (BCC) unit cells. All atoms are the same but are shaded differently to emphasise their positions.

hexagonal' (CPH) or sometimes 'hexagonal close-packed' (HCP). The second [Figure 3.20(b)] is also made up of close-packed layers, though this is harder to see: the shaded triangular plane is one of them. If we think of this as an A plane, atoms in the plane above it nest in the B position, and those in the plane below it, in the C position, giving ABCABC... stacking [see Figure 3.17(c)]. The unit cell itself is a cube with an atom at each corner and one at the centre of each face – hence, 'face-centred cubic' (FCC). The final cell [Figure 3.20(c)], is the characterising unit of the square-layer structure of Figure 3.18; it is a cube with an atom at each corner and one in the middle, i.e. 'body-centred cubic' (BCC).

Unit cells pack to fill space (Figure 3.21): the resulting array is called the *crystal lattice*; the points at which cell edges meet are called *lattice points*. The crystal itself is generated by attaching one or a group of atoms to each lattice point so that they form a regular, three-dimensional, repeating pattern. The cubic and hexagonal cells are among the simplest; there are many others with edges of differing lengths meeting at differing angles. The one thing they have in common is their ability to stack with identical cells (without rotation) to completely fill space.

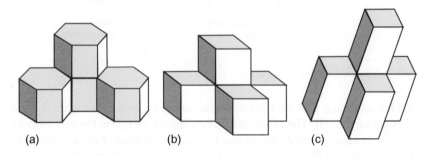

(a)   (b)   (c)

Figure 3.21 Unit cells stacked to fill space: (a) a hexagonal cell; (b) a cubic cell; (c) a cell with edges of differing length that do not meet at right angles.

Atom packing in ceramics   Most ceramics are compounds, made up of two or more atom types, ionically or covalently bonded. They, too, have characteristic unit cells. Figure 3.22 shows those of two hard ceramics used for tooling and abrasives: tungsten carbide (WC) and silicon carbide (SiC). The cell of the first is hexagonal, that of the second is cubic, but now a pair of different atoms is associated with each lattice point: a W-C pair in the first structure and a Si-C pair in the second. More examples are investigated in *Guided Learning Unit 1*.

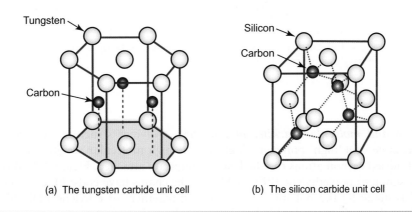

(a) The tungsten carbide unit cell   (b) The silicon carbide unit cell

Figure 3.22 Unit cells of ceramic compounds.

Atom packing in glasses   The crystalline state is the lowest energy state for elements and compounds. Melting disrupts the crystallinity, scrambling the atoms and destroying the regular order. The atoms in a molten metal look like the amorphous structure of Figure 3.19. On cooling through the melting point most metals crystallise easily, though by cooling them very quickly it is sometimes possible to trap the molten structure to give an amorphous metallic 'glass.' With compounds it is easier to do this, and with one in particular, silica (SiO_2), crystallisation is so sluggish that its usual state is the amorphous one. Figure 3.23 shows, on the left, the atom arrangement in crystalline silica: identical hexagonal Si–O rings, regularly arranged. On the right is the more usual amorphous state. Now some rings have seven sides, some have six, some five, and there is no order – the next ring could be any one of these. Amorphous silica is the base structure of almost all glasses, mixed with other oxides to make soda glass (windows, bottles) or borosilicate glasses (Pyrex). This is why the structure itself is called 'glassy,' a term used interchangeably with 'amorphous.'

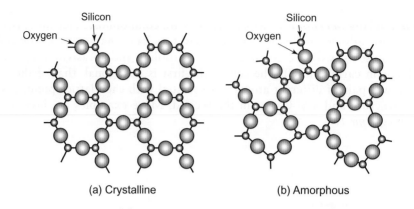

Figure 3.23 Two alternative structures for silica, the basis of most glasses. (a) Crystalline silica; (b) glassy or amorphous silica.

Atom and molecular packing in polymers   Polymer structures are quite different. The backbone of a 'high' polymer ('high' means high molecular weight) is a long chain of sp^3 hybridised carbon atoms to which side groups are attached. Figure 3.24 shows a segment of the simplest: polyethylene (PE) $(-CH_2-)_n$. The chains have ends: PE ends with a CH_3 group. PE is made by the polymerisation (linking together) of ethylene molecules, $CH_2 = CH_2$, where the = sign is a double bond, broken by polymerisation to give covalent sp^3 links to carbon neighbours along the chain. Figure 3.25 shows the chain structure of four more of the most widely used linear polymers.

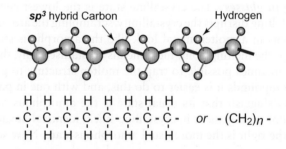

Figure 3.24 Polymer chains have a carbon-carbon backbone with hydrogen or other side groups. The figure shows three alternative representations of the polyethylene molecule.

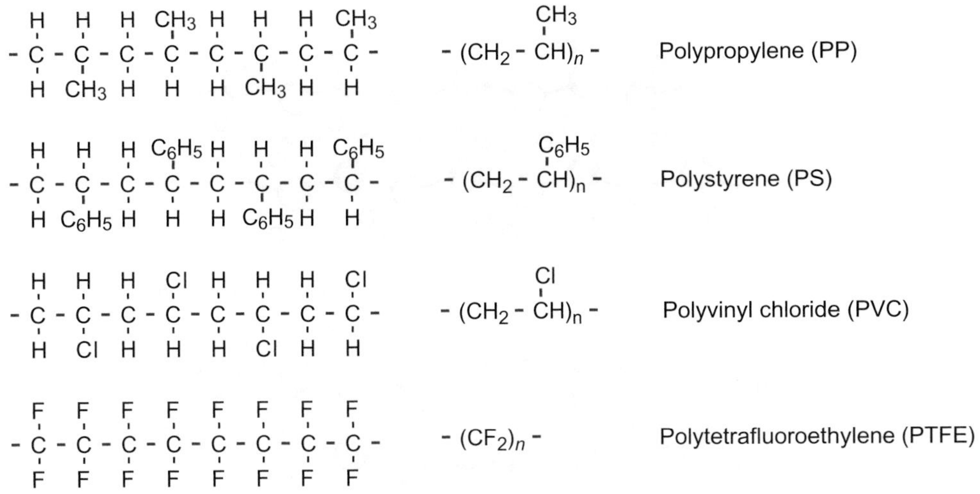

$$- \overset{\overset{\displaystyle H}{|}}{\underset{\underset{\displaystyle H}{|}}{C}} - \overset{\overset{\displaystyle H}{|}}{\underset{\underset{\displaystyle CH_3}{|}}{C}} - \overset{\overset{\displaystyle H}{|}}{\underset{\underset{\displaystyle H}{|}}{C}} - \overset{\overset{\displaystyle CH_3}{|}}{\underset{\underset{\displaystyle H}{|}}{C}} - \overset{\overset{\displaystyle H}{|}}{\underset{\underset{\displaystyle H}{|}}{C}} - \overset{\overset{\displaystyle H}{|}}{\underset{\underset{\displaystyle CH_3}{|}}{C}} - \overset{\overset{\displaystyle H}{|}}{\underset{\underset{\displaystyle H}{|}}{C}} - \overset{\overset{\displaystyle CH_3}{|}}{\underset{\underset{\displaystyle H}{|}}{C}} -$$

$- (CH_2 - \overset{\overset{\displaystyle CH_3}{|}}{CH})_n -$ Polypropylene (PP)

$$- \overset{\overset{\displaystyle H}{|}}{\underset{\underset{\displaystyle H}{|}}{C}} - \overset{\overset{\displaystyle H}{|}}{\underset{\underset{\displaystyle C_6H_5}{|}}{C}} - \overset{\overset{\displaystyle H}{|}}{\underset{\underset{\displaystyle H}{|}}{C}} - \overset{\overset{\displaystyle C_6H_5}{|}}{\underset{\underset{\displaystyle H}{|}}{C}} - \overset{\overset{\displaystyle H}{|}}{\underset{\underset{\displaystyle H}{|}}{C}} - \overset{\overset{\displaystyle H}{|}}{\underset{\underset{\displaystyle C_6H_5}{|}}{C}} - \overset{\overset{\displaystyle H}{|}}{\underset{\underset{\displaystyle H}{|}}{C}} - \overset{\overset{\displaystyle C_6H_5}{|}}{\underset{\underset{\displaystyle H}{|}}{C}} -$$

$- (CH_2 - \overset{\overset{\displaystyle C_6H_5}{|}}{CH})_n -$ Polystyrene (PS)

$$- \overset{\overset{\displaystyle H}{|}}{\underset{\underset{\displaystyle H}{|}}{C}} - \overset{\overset{\displaystyle H}{|}}{\underset{\underset{\displaystyle Cl}{|}}{C}} - \overset{\overset{\displaystyle H}{|}}{\underset{\underset{\displaystyle H}{|}}{C}} - \overset{\overset{\displaystyle Cl}{|}}{\underset{\underset{\displaystyle H}{|}}{C}} - \overset{\overset{\displaystyle H}{|}}{\underset{\underset{\displaystyle H}{|}}{C}} - \overset{\overset{\displaystyle H}{|}}{\underset{\underset{\displaystyle Cl}{|}}{C}} - \overset{\overset{\displaystyle H}{|}}{\underset{\underset{\displaystyle H}{|}}{C}} - \overset{\overset{\displaystyle Cl}{|}}{\underset{\underset{\displaystyle H}{|}}{C}} -$$

$- (CH_2 - \overset{\overset{\displaystyle Cl}{|}}{CH})_n -$ Polyvinyl chloride (PVC)

$$- \overset{\overset{\displaystyle F}{|}}{\underset{\underset{\displaystyle F}{|}}{C}} - \overset{\overset{\displaystyle F}{|}}{\underset{\underset{\displaystyle F}{|}}{C}} - \overset{\overset{\displaystyle F}{|}}{\underset{\underset{\displaystyle F}{|}}{C}} - \overset{\overset{\displaystyle F}{|}}{\underset{\underset{\displaystyle F}{|}}{C}} - \overset{\overset{\displaystyle F}{|}}{\underset{\underset{\displaystyle F}{|}}{C}} - \overset{\overset{\displaystyle F}{|}}{\underset{\underset{\displaystyle F}{|}}{C}} - \overset{\overset{\displaystyle F}{|}}{\underset{\underset{\displaystyle F}{|}}{C}} - \overset{\overset{\displaystyle F}{|}}{\underset{\underset{\displaystyle F}{|}}{C}} -$$

$- (CF_2)_n -$ Polytetrafluoroethylene (PTFE)

Figure 3.25 Four more common polymers, showing the chemical make-up. The strong carbon-carbon bonds are shown in red.

Polymer molecules bond together to form solids. The carbon-carbon chains that form the backbone of a linear polymer are covalently bonded and strong. The molecules attract each other, but only weakly [Figure 3.26(a)]; the weak van der Waals (dipole) bonds are easily stretched or rearranged. The resulting structure is sketched in Figure 3.27(a): a dense spaghetti-like tangle of molecules with no order or crystallinity. This is the structure of polymers like those of Figure 3.25; the weak bonds melt easily, allowing the polymer to be moulded when hot – hence, they are called *thermoplastics*.

The weak bonds of thermoplastics enable segments of the molecules to line up, at least in polymers with the simpler molecules, which can arrange into small crystalline regions [Figure 3.27(b)]. These crystallites are small – typically 1 to 10 microns across – just the right size to scatter light. So amorphous polymers with no crystallites can be transparent: polycarbonate (PC), polymethyl-methacrylate (PMMA, Plexiglas) and polystyrene (PS) are examples. Those that are semi-crystalline, like PE and nylon (PA), scatter light and are translucent.

The real change comes when chains are *cross-linked* by replacing some of the weak van der Waals bonds by strong covalent C-C bonds [Figure 3.26(b)], making the whole array into one huge multiply connected network. *Elastomers* (rubbery polymers) have relatively few cross-links [Figure 3.27(c)]. *Thermosets* like epoxies and phenolics have many cross-links [Figure 3.27(d)], making them stiffer and stronger than thermoplastics. Cross-links are not broken by heating, so once the links have formed, elastomers and thermosets cannot be thermally moulded, welded, or (importantly) recycled.

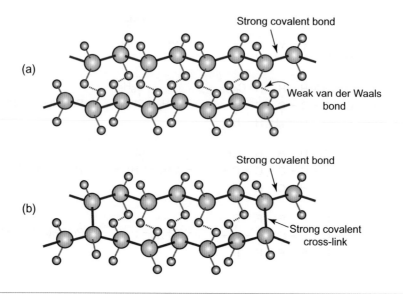

Figure 3.26 The bonding between polymer chains: weak dipolar van der Waals bonds, and strong covalent cross-links.

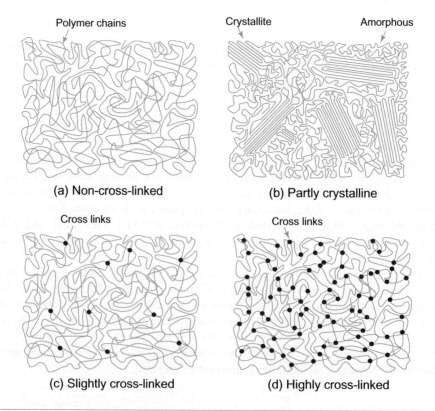

Figure 3.27 Polymer structures: (a) amorphous thermoplastic; (b) semi-crystalline thermoplastic; (c) elastomer; (d) thermoset.

3.5 The physical origin of density

Many applications (e.g. aerospace components, automotive bodies, sports equipment) require stiffness and strength at low weight, so an important factor in design is the *density* ρ of the materials used (i.e. the mass per unit volume, measured in kg/m^3).

Measurement of density The density of samples with regular shapes can be determined using a precision mass balance and accurate measurements of the dimensions (to give the volume), but this is not the best way. Better is the 'double weighing' method: the sample is first weighed in air, and then when fully immersed in a liquid of known density. When immersed, the sample feels an upward force equal to the weight of liquid it displaces (Archimedes' principle[3]). The density is then calculated as shown in Figure 3.28, avoiding the difficult measurement of volume for irregular shapes – care is only needed to expel all air from the immersed sample.

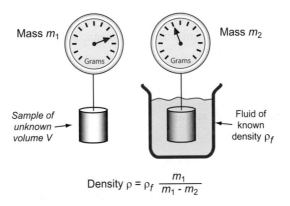

Figure 3.28 Measuring density.

Physical origin of density As discussed earlier, atoms differ greatly in weight but relatively little in size. Among solids, the heaviest stable atom, uranium (atomic weight 238), is about 35 times heavier than the lightest, lithium (atomic weight 6.9), yet when packed to form solids their atomic diameters are almost exactly the same (0.32 nm). The largest atom, cesium, is only 2.5 times larger than the smallest, beryllium. So density is mainly determined by the atomic weight and only to a lesser degree by the atom size and the way they are packed. Metals are dense because they are made of heavy atoms, packed closely together.

---

[3] Archimedes (287–212 BC), Greek mathematician, engineer, astronomer and philosopher, designer of war machines, the Archimedean screw for lifting water, evaluator of π (as 3 + 1/7) and conceiver, while taking a bath, of the principle that bears his name.

Ceramics, for the most part, have lower densities than metals because they contain light Si, O, N, or C atoms. Polymers have low densities because they are largely made of light carbon (atomic weight 12) and hydrogen (atomic weight 1) in low packing density amorphous or semi-crystalline configurations. The lightest atoms, packed in the most open way, give solids with a density of around 1000 kg/m^3 – the same as that of water. Materials with lower densities than this are foams (and woods), made up of cells containing a large fraction of porosity. For pure crystalline elements and compounds, the density can be calculated directly from the atomic weight and radius – as shown in *Guided Learning Unit 1*, the atomic radius and packing configuration (FCC, BCC, etc.) determine the number of atoms (of known mass) per unit volume, and hence the density – a microstructure-insensitive property.

Density of metallic alloys   Most metallic alloys are not pure but contain two or more different elements. Often they dissolve in each other, like sugar in tea, but as the material is solid we call it a *solid solution*: examples are brass (a solution of Zn in Cu) and stainless steel (a solution of Ni and Cr in Fe). As we shall see in later chapters, some material properties are changed a great deal by making solid solutions; density is not. As a rule the density $\tilde{\rho}$ of a solid solution lies between the densities ρ_A and ρ_B of the materials that make it up, following a linear *rule of mixtures* (an arithmetic mean, weighted by atomic fraction) known, in this instance, as Vegard's law:

$$\tilde{\rho} = f\rho_A + (1-f)\rho_B \qquad (3.2)$$

where f is the fraction of A atoms.

3.6 Summary and conclusions

Material properties have their origins in the characteristics of the bonds between atoms and in their packing in solids. Microstructure-insensitive properties (like elastic modulus and density) are those for which the atomic length-scale governs the behaviour directly. Other properties (e.g. strength, electrical resistivity) are microstructure-sensitive, involving internal microstructural features between the atomic scale and that of the product itself. Interpreting the material physics across this wide range of length-scales, shown schematically at the start of this chapter, provides the basis for understanding the properties of materials and their distinctive classes – metals, ceramics, and polymers, as well as man-made (or natural) hybrids.

   Understanding the properties of engineering materials starts from a grasp of the Periodic Table of Elements and the electronic origins of electronegativity and atomic volume, which cycle systematically along each row of the Periodic Table. Bonding between atoms can be understood in terms of the electronegativity of atoms: their propensity to capture one or more electrons to become negative ions, which leads to several characteristic forms of bonding. Metallic, covalent, and ionic bonds are stiff and strong; dipolar van der Waals bonds are much less so.

The second atomic-scale behaviour that underpins properties is the atomic or molecular packing – regular crystalline structures in metals and ceramics, to random amorphous glasses, polymers, and elastomers (with a bit of organised packing in semi-crystalline thermoplastics). The geometry and interpretation of crystal structure is expanded in the separate *Guided Learning Unit 1*. Together, the differences in bonding and packing explain the significant variation in many key microstructure-insensitive properties between the material classes. For example, the density of a material is the mass of its atoms divided by the volume they occupy. Atoms differ relatively little in size and packing density (the ranges cover a factor of about 2), whereas they differ a great deal in mass. So density is principally set by the atomic weight; in general, the further we go down the Periodic Table the greater the density.

Elastic modulus, too, is microstructure-insensitive and dominated by the type of bonding – see Chapter 4. Later chapters explore many important microstructure-sensitive properties – strength, toughness, and thermal and electrical conductivity. While microstructural features at many length-scales then play a lead role, we will also need to draw on the atomic scale behaviour to interpret both the differences between material classes and the ways in which the properties of metals and polymers can be manipulated by composition and processing.

3.7 Further reading

Askeland, D. R. Phulé, P. P., & Wright, W. J. (2010). *The science and engineering of materials* (6th ed.). Thomson. ISBN 9780495296027. (A well-established materials text that deals with the science of engineering materials).

Callister, W. D., & Rethwisch, D. G. (2014). *Materials science and engineering: An introduction* (9th ed.). John Wiley & Sons. ISBN 978-1-118-54689-5. (A well-established text taking a science-led approach to the presentation of materials teaching).

Shackelford, J. F. (2014). *Introduction to materials science for engineers* (8th ed.). Prentice Hall. ISBN 978-0133789713. (A well-established materials text with a materials and a design slant).

3.8 Exercises

Exercise E3.1   Bone is a natural hybrid material made of ceramic and polymer constituents (and aqueous solutions), with a complex multi-scale architecture. Conduct a search online to explore some of the different types of bone architecture in the human body, identifying their key constituents and architectural length-scales.

Exercise E3.2   On the Pauling scale, the electronegativity of germanium is 1.8, that of bromine is 2.8. Will they form an ionic bond?

Exercise E3.3   Assess what sort of bonding you would expect in:

(a) strontium (Sr, atomic number 38)?
(b) magnesium iodide (MgI_2)?
(c) aluminium phosphide (AlP)?

Use the electronegativity chart (Figure 3.11) as a basis for your reasoning.

Exercise E3.4   Rank the following bonds in order from the most covalent to the most ionic: (a) Na-Cl; (b) Li-H; (c) H-C; (d) H-F; (e) Rb-O, given the following electronegativities:

Element	Na	Cl	Li	H	C	F	Rb	O
Electronegativity	0.9	3.0	1.0	2.1	2.5	4.0	0.8	3.5

Exercise E3.5   With reference to key microstructural details, explain briefly what is meant by the following:

• crystalline solid
• amorphous solid
• thermoplastic
• thermoset
• elastomer

Exercise E3.6   The table shows sample density data for two brasses (Cu-Zn alloys) and for pure Cu and Zn. The notation Cu-10Zn means 10 atomic % Zn (and 90% Cu). Vegard's law [Equation (3.2)] suggests that density is expected to follow a simple linear rule of mixtures with atomic % for these properties – density is microstructure-insensitive. Explain why this is the case, and investigate the fit to Vegard's law for these brasses.

Alloy	Density ρ (kg/m³)
Pure Cu	8960
Cu-10Zn	8710
Cu-30Zn	8450
Pure Zn	7130

Chapter 4
Elastic stiffness and stiffness-limited applications

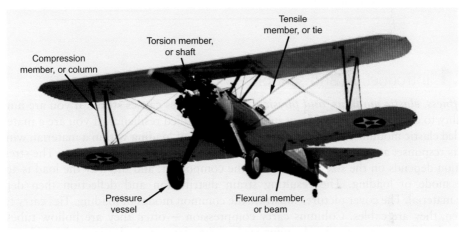

Tensile member, or tie

Torsion member, or shaft

Compression member, or column

Pressure vessel

Flexural member, or beam

Tensile member, or tie

Compression member, or column

Flexural member, or beam

Modes of loading.
(istock.com)

https://doi.org/10.1016/B978-0-08-102399-0.00004-2

4.1 Introduction and synopsis

Stiffness, elastic modulus, and physical origins   Stress causes strain. If you are human, the ability to cope with stress without undue strain is called resilience. If you are a material, it is called elastic modulus. Stress measures the intensity of loading within a material, while strain is its response: a component that is strained will change shape and deflect. The stress distribution depends on the size and shape of the component, and the way the load is applied – the mode of loading. The resulting strain distribution and deflection then depend on the material. The cover pictures illustrate the common modes of loading. Ties carry tension – often, they are cables. Columns carry compression – often they are hollow tubes. Beams carry bending moments, like the wing spar of the plane or the horizontal roof beams of the airport. Shafts carry torsion, as in the propeller shaft of the plane. Pressure vessels contain a pressure, as in the tyres of the plane.

Stiffness is the resistance to *elastic* shape-change, meaning that the material returns to its original shape when the stress is removed. *Strength* (see Chapter 5) is its resistance to permanent distortion or total failure. Stress and strain are not material properties; they describe a stimulus and a response. Stiffness and strength are central to mechanical design, often in combination with the density, ρ. This chapter introduces stress, strain, and the elastic moduli that relate them. These properties are neatly summarised in a *material property chart* – the modulus-density chart – the first of many in this book.

The elastic modulus derives directly from the atomic spacing and the nature of the bonds between the atoms – it is a microstructure-insensitive property (see Chapter 3). There is not much you can do to change the packing and bonding of atoms – the range of modulus for a given composition is narrow for crystalline solids but wider for polymers and elastomers (due to greater variety in their molecular structure and bonding). At a higher length scale, however, there are two ways in which modulus (and density) can be manipulated: by combining two materials together to make a *composite* or by making them porous, as in *foams*. Property charts are a good way to show how this works.

Solid mechanics and stiffness in design   A few years back, with the millennium approaching, countries and cities around the world turned their minds to iconic building projects. In Britain there were several. One was – and still is – a new pedestrian bridge spanning the River Thames, linking St Paul's Cathedral to the Museum of Modern Art facing it across the river. The design was daring: a suspension bridge with suspension cables that barely rise above the level of the deck instead of the usual great upward sweep. The result was visually striking: a sleek, slender span like a 'shaft of light' (the architect's words). Just one problem: it wasn't stiff enough. The bridge opened but when people walked on it, it swayed and wobbled sideways, so alarmingly that it was promptly closed. A year and $5 million later it reopened, much modified with vibration dampers attached in many places, and now it is fine.

The first thing you tend to think of with big structures such as buildings is *strength*: they must not fall down. *Stiffness*, often, is automatically achieved by a strength-limited analysis, and taken for granted. But even buildings deflect elastically, often due to wind loading, and in extreme cases even by the loading imposed by people, such as with London's Millennium Bridge. Any product of engineering carrying a load, however, is a type of 'structure', from aircraft, ships, and cars, to bicycles, sports products, and furniture, to children's toys, mobile phones, and medical implants. A quick reflection shows that for all these products they only function properly if they have the right stiffness.

To design a product, we therefore need methods to evaluate stress, strain, and deflection for the common modes of loading: the discipline known as *solid mechanics*. In this chapter we introduce the mechanics of elastic stiffness and deflection; in Chapters 5 and 6 the problems will be strength and failure, by plastic yielding or by fracture. The resulting toolbox of standard solutions may be used to perform simple design calculations, and provides the basis for *material selection in design*, presented in *Guided Learning Unit 2*.

4.2 Stress, strain, and elastic moduli

Modes of loading   Most engineering components carry loads, under one or more of the modes of loading introduced earlier: tension, bending, and so on. Usually one mode dominates, and the component can be idealised as one of the simply loaded cases in Figure 4.1. *Ties* carry axial tension, shown in (a); *columns* do the same in simple compression, as in (b). Bending of a *beam*, case (c), creates axial tension on one side of the neutral axis (the centre line, for a beam with a symmetric cross section) and compression on the other. *Shafts* carry twisting or torsion, as in (d), which generates shear rather than axial load. Pressure differences are supported by *shell* structures, like the cylindrical tube shown in (e), generating biaxial tension if the internal pressure is greater than that outside (or compression if the difference is reversed).

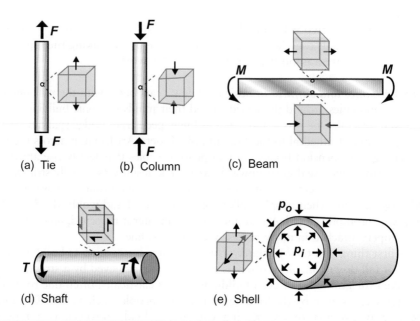

Figure 4.1   Modes of loading and states of stress.

Stress   To visualise the stress and strain in a component, we consider a representative element of material – usually a cube, as in Figure 4.2. If a force F is carried through the material normal to the face of the element, as on the left of Figure 4.2(a), equilibrium requires that equal and opposite forces apply to the top and bottom faces, and indeed every plane normal to F carries this force. If the area of such a plane is A, the *tensile stress* σ in the element (neglecting its own weight) is

$$\sigma = \frac{F}{A} \tag{4.1}$$

If the sign of F is reversed, the stress is compressive and given a negative sign. Forces[1] are measured in *newtons* (N), so stress has units of N/m², also called *pascals* (Pa) – the same unit as for fluid pressure[2] (though note that stress in a solid is quite different – it has direction associated with it). But a stress of 1 N/m² (= 1 Pa) is tiny – atmospheric pressure is 10^5 N/m² – so the usual unit is MN/m² (10^6 N/m²), or *megapascals* (MPa).

---

[1] Isaac Newton (1642–1727), scientific genius and alchemist, formulator of the laws of motion, the inverse-square law of gravity (though there is some controversy about this), laws of optics, differential calculus, and much more.

[2] Blaise Pascal (1623–1662), philosopher, mathematician, and scientist, who took a certain pleasure in publishing his results without explaining how he reached them. Almost all, however, proved to be correct.

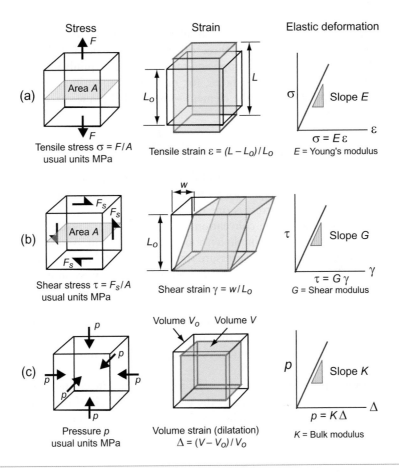

Figure 4.2 The definitions of stress, strain, and elastic moduli, for different loading on a cubic material element: (a) uniaxial normal force; (b) shear force; (c) hydrostatic pressure.

Example 4.1

A brick column is 50 m tall. The bricks have a density of $\rho = 1800$ kg/m^3. What is the axial compressive stress at its base? Does the shape of the cross section matter?

Answer. Let the cross-section area of the column be A and its height be h. The force exerted on the column at its base is equal to its weight, $F = \rho A h g$, where g is the acceleration due to gravity, 9.8 m/s^2. Using Equation (4.1), the axial compressive stress at the base is $\sigma = -F/A = -\rho g h$, where the '$-$' indicates compression. Note that the area A cancels out – uniaxial loading gives a stress that is independent of the shape or size of the cross section.

Hence $\sigma = -1800 \times 9.8 \times 50 = -8.8 \times 10^5$ N/m$^2 = -0.88$ MPa.

(Note that the same formula gives the static pressure at depth h in a fluid of density ρ – but fluid pressure acts equally in all directions; the stress in a solid is directional).

If the force lies *parallel* to the face of the element, this is called a *shear force*, F_s. To maintain static equilibrium of the element, three other forces are needed [left of Figure 4.2(b)]: an equal and opposite force on the opposing face, for force equilibrium, and a complementary pair of forces on the other two faces, for moment equilibrium, preventing rotation. This creates a state of *pure shear* in the element – the shaded plane carries a *shear stress* τ, again with units of MPa, where

$$\tau = \frac{F_s}{A} \tag{4.2}$$

One further state of stress is useful in defining the elastic response of materials: that produced by applying equal forces to all six faces of the cubic element [as on the left of Figure 4.2(c)]. The state of stress is one of *hydrostatic pressure*, symbol p, again with units of MPa. There is an unfortunate convention here. Pressures are positive when they push – the reverse of the convention for simple tension and compression. And be careful that you understand the difference between this pressure state inside a solid and the gas pressures loading the shell in Figure 4.1(e), which lead to biaxial tension or compression in the solid walls.

Engineering components can have complex shapes and can be loaded in different ways, creating complex distributions of stress. But no matter how complex, the stresses in any small element within the component can always be described by a combination of tension, compression, and shear acting on the cube faces. Commonly the simple cases of Figure 4.2 suffice, using superposition of two cases to capture, for example, bending plus compression.

Strain   Strain is the response of materials to stress (middle column of Figure 4.2). A tensile stress σ applied to an element causes the element to stretch. If the element in Figure 4.2(a), originally of side length L_o, stretches by $\delta L = L - L_o$, the nominal *tensile strain* ε is

$$\varepsilon = \frac{\delta L}{L_o} \tag{4.3}$$

A compressive stress shortens the element; the nominal *compressive strain*, defined in the same way, is negative. Since strain is the ratio of two lengths, it is dimensionless and has no units.

A shear stress τ causes a *shear strain* γ [middle of Figure 4.2(b)]. If the element shears by a distance w, the *shear strain* γ is given by

$$\tan \gamma = \frac{w}{L_o} = \gamma \tag{4.4}$$

In practice $\tan \gamma \approx \gamma$ because elastic strains are almost always small ($<<1$).

Finally, a hydrostatic pressure p causes an element of volume V to change in volume by δV. The volumetric strain, or *dilatation* Δ [middle of Figure 4.2(c)], is

$$\Delta = \frac{\delta V}{V} \tag{4.5}$$

Stress–strain curves and moduli   Figure 4.3 shows typical tensile stress–strain curves for a ceramic, a metal, and a polymer, each taken all the way to failure. The initial part, up to the

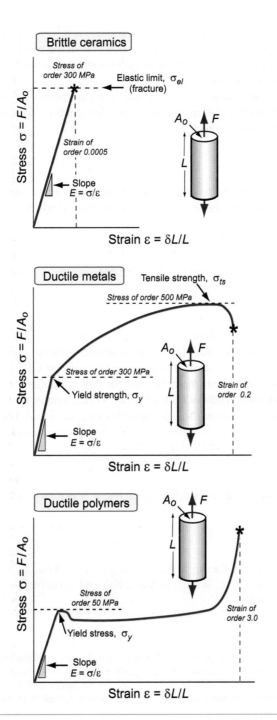

Figure 4.3 Tensile stress–strain curves for ceramics, metals, and polymers.

elastic limit σ_{el}, is approximately linear (Hooke's law[3]), where elastic means that the strain is recoverable – the material returns to its original shape when the stress is removed. Stresses above the elastic limit cause permanent deformation (ductile behaviour) or brittle fracture – more on these in Chapters 5 and 6.

Within the linear elastic regime, strain is proportional to stress – the right column of Figure 4.2, each with a characteristic slope. For uniaxial loading, the tensile strain is proportional to the tensile stress:

$$\sigma = E\varepsilon \tag{4.6}$$

and the same applies in compression. The constant of proportionality, E, is *Young's modulus*.[4] Similarly, the shear strain γ is proportional to the shear stress τ:

$$\tau = G\gamma \tag{4.7}$$

and the dilatation Δ is proportional to the pressure p:

$$p = K\Delta \tag{4.8}$$

where G is the *shear modulus* and K the *bulk modulus*. All three moduli have the same dimensions and units as stress (as strain is dimensionless): N/m^2 or Pa. As with stress it is convenient to use a larger unit, this time *gigapascals*, or GPa (10^9 N/m^2).

Young's modulus, the shear modulus, and the bulk modulus are related, but to relate them we need one more quantity, *Poisson's ratio*.[5] When loaded axially, materials generally contract in the transverse directions, as shown for the element of Figure 4.2(a). Poisson's ratio, ν, is the negative of the ratio of the lateral or transverse strain ε_t, to the axial strain ε, in uniaxial tensile loading:

$$\nu = -\frac{\varepsilon_t}{\varepsilon} \tag{4.9}$$

Since the transverse strain itself is negative, ν is positive.

Example 4.2

The brick column in Example 4.1 has bricks with Young's modulus $E = 25$ GPa, and Poisson's ratio $\nu = 0.2$. What are the axial and transverse strains of the bricks at the bottom of the column?

---

[3] Robert Hooke (1635–1703), able but miserable man, inventor of the microscope, and perhaps, too, of the idea of the inverse-square law of gravity. He didn't get along with Newton.

[4] Thomas Young (1773–1829), English scientist, expert on optics and deciphering ancient Egyptian hiero-glyphs (among them, the Rosetta stone). It seems a little unfair that the modulus carries his name, not that of Hooke.

[5] Siméon Denis Poisson (1781–1840), French mathematician, known both for his constant and his distribution. He was famously uncoordinated, failed geometry at university because he could not draw, and had to abandon experimentation because of the disasters resulting from his clumsiness.

Answer. Using the result of Example 4.1 with Equation (4.6), the axial strain is $\varepsilon = \sigma/E = -8.8 \times 10^5/25 \times 10^9 = -3.5 \times 10^{-5}$ (often written as -35 *microstrain* [$\mu\varepsilon$]). From Equation (4.9), the transverse strain is $\varepsilon_t = -\nu\,\varepsilon = 7.0\ \mu\varepsilon$. This time the strain is positive, indicating a small expansion at the base of the column.

In an isotropic material (one for which the moduli do not depend on the direction in which the load is applied) the moduli are related in the following ways:

$$G = \frac{E}{2(1+\nu)}; \quad K = \frac{E}{3(1-2\nu)} \tag{4.10}$$

For metals and other crystalline materials, $\nu \approx 1/3$, so that

$$G = \frac{3}{8}E \quad \text{and} \quad K = E \tag{4.11a}$$

Elastomers are exceptional. For these $\nu \approx 1/2$, so that

$$G = \frac{1}{3}E \quad \text{and} \quad K \gg E \tag{4.11b}$$

This means that rubber (an elastomer) is easy to stretch in tension (very low E), but if constrained from changing shape, or loaded hydrostatically, it is very stiff (large K) – a feature for which designers of shoes must allow.

Hooke's law in three dimensions The simple loading states in Figure 4.1 lead to stresses in one or two perpendicular directions. But, as we have now seen, even under uniaxial load the strain is inherently three dimensional, thanks to Poisson's ratio. So it is helpful to relate stress and strain in a more general way that allows for loading acting in all three perpendicular directions at once. Figure 4.4 shows the cubic element of material, now subjected to three unequal stresses σ_1, σ_2, and σ_3, resulting in strains ε_1, ε_2, and ε_3.

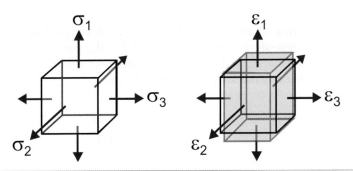

Figure 4.4 Multi-axial states of stress and strain.

Now to relate stress to strain we use the *principle of superposition* – applying each stress in turn, finding the strains, and then summing these to find the overall strain when all three stresses act together. First applying a stress σ_1, the resulting strains [from Equations (4.6) and (4.9)] are

$$\varepsilon_1 = \frac{\sigma_1}{E}, \quad \varepsilon_2 = \varepsilon_3 = -v\varepsilon_1 = -\frac{v\sigma_1}{E} \tag{4.12}$$

Repeating for stresses σ_2 and σ_3, and summing the strains gives us *Hooke's law in three dimensions*:

$$\varepsilon_1 = \frac{1}{E}\left(\sigma_1 - v\sigma_2 - v\sigma_3\right)$$

$$\varepsilon_2 = \frac{1}{E}\left(-v\sigma_1 + \sigma_2 - v\sigma_3\right) \tag{4.13}$$

$$\varepsilon_3 = \frac{1}{E}\left(-v\sigma_1 - v\sigma_2 + \sigma_3\right)$$

These results are particularly helpful in design problems in which strain is constrained.

Elastic energy   If you stretch an elastic band, energy is stored in it. The energy can be considerable: catapults can be lethal; the elastic deflection of truck springs can cushion the truck and its contents. The super-weapon of the Roman arsenal at one time was a wind-up mechanism that stored enough elastic energy to hurl a 10 kg stone projectile 100 metres.

How do you calculate this energy? A force F acting through a small displacement dL does work $F \, dL$. Dividing by area and length, respectively, gives us a stress $\sigma = F/A$ acting through a strain increment $d\varepsilon = dL/L$, so this becomes work done per unit volume, dW, with units of J/m^3:

$$dW = \frac{F \, dL}{AL} = \sigma \, d\varepsilon \tag{4.14}$$

If the stress is acting on an elastic material, this work is stored as elastic energy. The elastic part of all three stress–strain curves of Figure 4.3 – the part of the curve before the elastic limit – is linear, with $\sigma = E\varepsilon$. The work done per unit volume as the stress is raised from zero to a final value σ is the area of the triangle under the stress–strain curve:

$$W = \frac{1}{2}\sigma \times \varepsilon = \frac{1}{2}\frac{\sigma^2}{E} \tag{4.15}$$

This is the energy that is stored, per unit volume, in an elastically strained material. The energy is released when the stress is relaxed – the basis of using springs as an energy store.

Example 4.3

A steel rod has length $L_o = 10$ m and a diameter $d = 10$ mm. The steel has Young's modulus $E = 200$ GPa, elastic limit (the highest stress at which it is still elastic and has not yielded) $\sigma_y = 500$ MPa, and density $\rho = 7800$ kg/m^3. Calculate the force in the rod, its extension, and the elastic energy stored per unit volume, when the stress just reaches the elastic limit. Compare the elastic energy per unit mass with the chemical energy stored in gasoline, which has a heat of combustion (calorific value) of 43,000 kJ/kg.

Answer. The cross-section area of the rod is $A = \pi d^2/4 = 7.85 \times 10^{-5}$ m^2.

At the elastic limit, from Equation (4.1), the force in the rod is $F = \sigma_y A = 39.2$ kN (3.92 tonnes).

From Equation (4.6), the corresponding strain is: $\varepsilon = \sigma/E = 500 \times 10^6/200 \times 10^9 = 0.0025$ (i.e. 0.25%).

The extension of the rod from Equation (4.3) is $\delta L = \varepsilon\, L_o = 0.0025 \times 10 = 25$ mm.

The elastic energy per unit volume, from Equation (4.15), is $W = \sigma^2/2E = (500 \times 10^6)^2/(2 \times 200 \times 10^9) = 625$ kJ/m^3, giving an elastic energy per kg of $W/\rho = 80$ J/kg. This is much less than that of gasoline, by a factor of more than 500,000.

Measurement of Young's modulus You might think that the way to measure the elastic modulus of a material would be to apply a stress in the linear elastic regime, measure the strain, and divide one by the other. In reality, moduli measured as slopes of stress–strain curves are rather inaccurate because strains are very small, and difficult to distinguish from apparent contributions to the strain from deflection of the test machine. Accurate moduli are in fact measured dynamically: by measuring the frequency of natural vibrations of a small beam, or (surprisingly) by measuring the velocity of sound waves in the material. Both depend on $\sqrt{E/\rho}$, so if you know the density ρ you can calculate E.

Stress-free strain Stress is not the only stimulus that causes strain. Some materials respond to a magnetic field by shrinking – an effect known as *magnetostriction*. Others respond to an electrostatic field in the same way – they are known as *piezo-electric* materials. In each case a material property relates the magnitude of the strain to the intensity of the stimulus (Figure 4.5). The strains are small, but the stimulus can be controlled with great accuracy, and the material response is extremely fast. Magnetostrictive and piezo-electric strain can be driven at high frequency. This is exploited in precision devices and actuators – applications we return to in Chapter 10.

A more familiar effect is that of thermal expansion: strain caused by change of temperature. The thermal strain ε_T is linearly related to the temperature change ΔT by the *expansion coefficient*, α:

$$\varepsilon_T = \alpha \Delta T \qquad (4.16)$$

where the subscript T is a reminder that the strain is caused by temperature change, not by stress (more on this in Chapter 7).

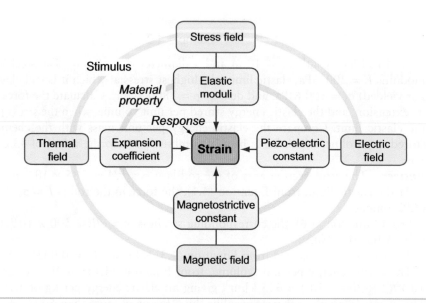

Figure 4.5 Stimuli and material properties that give a strain response.

The term 'stress-free strain' is a little misleading. It correctly conveys the idea that the strain is not *caused* by stress but by something else. But these strains can nonetheless give rise to stresses — for example, if the body experiencing the stimulus is constrained from straining. Thermal stress, arising from constrained thermal expansion, can be a particular problem, causing mechanisms to jam and railway tracks to buckle.

4.3 The big picture: material property charts

The modulus–density chart The moduli and densities of common engineering materials are plotted in Figure 4.6; material families are enclosed in coloured envelopes. The modulus E spans seven decades[6], from 0.0001 to nearly 1000 GPa; the density ρ spans a factor of 2000, from less than 10 to 20,000 kg/m^3. Ceramics (yellow envelopes) and metals (red envelope) have high moduli and densities; none has a modulus less than 10 GPa or a density less than 1700 kg/m^3. Polymers, by contrast, all have moduli below 10 GPa and densities that are almost all lower than any metal or ceramic — most are close to 1000 kg/m^3 (the density of water). Elastomers have roughly the same density as other polymers, but their moduli are lower by a further factor of 100 or more. Materials with even lower densities are porous: man-made foams and natural cellular structures like wood and cork. The property chart gives a visual overview of where material families and their members lie in

---

[6] Very low-density foams and gels (which can be thought of as molecular-scale, fluid-filled foams) can have lower moduli than this. As an example, gelatine (as in Jell-O) has a modulus of about 10^{-5} GPa.

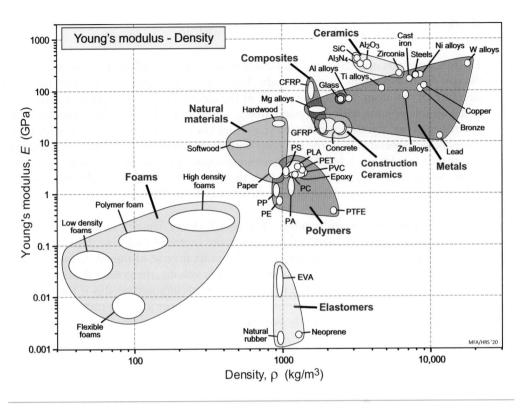

Figure 4.6 The Young's modulus–density chart.

E–ρ space, helpful in understanding the physical origin of the properties (see next section). We also use it as a design tool in *Guided Learning Unit 2*, to select materials for stiffness-limited applications in which weight must be minimised.

Anisotropy Glasses have disordered structures with properties that are *isotropic*, meaning they are the same no matter in which direction they are measured. Crystals, by contrast, are ordered and because of this their properties depend on the direction in which they are measured – they are *anisotropic*. Most real metals and ceramics are *poly-crystalline*, made up of many tiny, randomly oriented crystals. This averages out the directionality in properties, so the aggregate behaves as if it was isotropic. Sometimes, though, anisotropy is important. Single crystals, drawn polymers, and natural materials like wood are *anisotropic*; their properties depend on the direction in the material in which they are measured. The property bubbles for woods in Figure 4.6 are for the stiff direction, loading parallel to the grain. Fibre composites can have a modulus parallel to the fibres that is larger by a factor of 20 than that perpendicular to them (we explore why in Section 4.5). Anisotropy must therefore be considered when wood and composite materials are selected in design.

4.4 The physical origin of elastic moduli

The fundamentals of atomic and molecular bonding were explained in Chapter 3, revealing the row-by-row patterns in the Periodic Table of the Elements governed by the electronic structure of the atom. One consequence of the electron shell structure is a cyclic variation in the atomic volume with increasing atomic number (see Figure 3.10). Another is that the electronegativity of the elements also cycles from low to high along each row, with the highest values being associated with the top rows with the lower atomic numbers, characterising the non-metals (see Figure 3.11). Microstructure-insensitive properties are those that depend on atomic scale, electronic behaviour, and are therefore expected to track the progressions with atomic number. Let's see if this is the case for elastic modulus.

Atomic bonding and elastic moduli Figure 4.7 shows Young's modulus of the elements against atomic number. The modulus peaks at the centre of each row, falling sharply at the end of one row and the start of the next – the exact inverse of the cyclic minima in atomic volume (see Figure 3.10). So the smaller the atomic volume, the greater the number of bonds per unit volume, and the larger the modulus. But superimposed on this is the effect of *cohesive energy*, associated with the number of free electrons per ion (in a metal), and the number of shared electrons (in a covalently bonded solid).

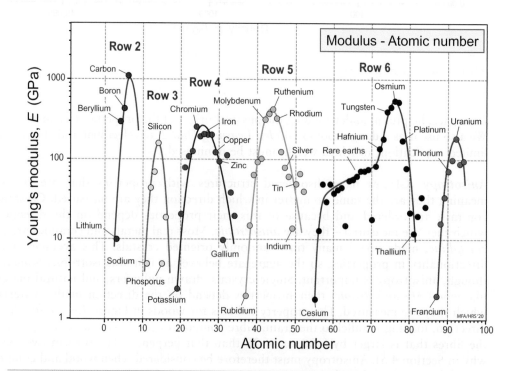

Figure 4.7 The cyclic variation of elastic modulus across the Periodic Table of Elements.

Recall from Section 3.3 that the *cohesive energy* H_c is the energy per mol required to separate the atoms of a solid completely, giving neutral atoms at infinity (see Figure 3.12). The minimum energy state is when the atoms sit at their equilibrium spacing, a_o. When atoms are displaced by a small amount from this spacing, the energy will rise in tension or compression, and the restoring force back towards equilbrium is the gradient of the energy curve (Figure 4.8). It is therefore the shape of this energy minimum that produces the linear elastic force-displacement response. So we can think of the atomic bonds as little springs linking the atom centres, as in Figure 4.8. As the atoms are pulled apart or pushed together, elastic energy is stored – shown as the area under the F-δ response. The gradient of this response is the stiffness of the atomic spring, $S = F/\delta$. In general, a high cohesive energy gives a deeper energy minimum – this in turn leads to a steeper F-δ response and higher bond stiffness.

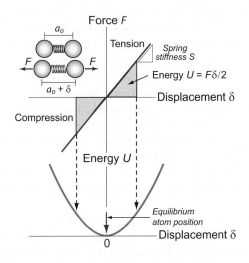

Figure 4.8   Bond stiffness around the equilibrium atomic spacing.

Each atomic bond applies a force F to an atom of diameter a_o. Crystal packing arranges atoms efficiently in planes (see Chapter 3), so each bond is associated with the area occupied per atom in a plane perpendicular to F, which will be of the order of a_o^2 (imagine a plane with atoms packed in a square array). So this corresponds to an atomic-scale stress $\sigma \approx F/a_o^2$. A displacement of δ between two atoms initially separated by a distance a_o corresponds to an atomic-scale strain $\varepsilon = \delta/a_o$. So the elastic modulus E can be derived directly from the atomic-scale ratio of stress and strain:

$$E = \frac{\sigma}{\varepsilon} \approx \frac{F/a_o^2}{\delta/a_o} = \frac{S}{a_o} \tag{4.17}$$

Table 4.1 lists the typical ranges, for each of the main types of bonding, of the cohesive energy, bond stiffness S, and corresponding Young's modulus, E. The broad correlation between elastic modulus and atomic bond stiffness is clear, remembering that this is superimposed on

Table 4.1   Bond energies, stiffnesses, and Young's moduli

Bond type	Example	Cohesive energy H_c (kJ/mol)	Bond stiffness S (N/m)	Young's modulus E (GPa)
Covalent	Diamond (carbon)	100–1200	20–200	100–1000
Metallic	Engineering metals	60–850	8–60	20–400
Ionic	Sodium chloride	600–1600	4–100	30–400
Dipolar (van der Waals)	Polyethylene	7–50	0.5–5	0.5–5

the effect of atomic spacing. The covalent bond is particularly stiff – diamond has a very high modulus because the carbon atom is small (giving a small spacing and high bond density), and its atoms are linked by very stiff springs ($S = 200$ N/m).

For most engineering metals, the metallic bond is somewhat less stiff ($S = 8$–60 N/m). But metals are close-packed, or nearly so, giving them high moduli – though below that of diamond. Alloys contain a mixture of metallic bonding, between similar and dissimilar atoms, but these are of comparable stiffness. So when two elements are mixed, the modulus generally lies between those of the pure metals. Ionic bonds, found in many ceramics, have stiffnesses comparable with those of metals, giving them high moduli, too. As noted earlier, the stiffness of a single crystal metal or ceramic will vary with direction relative to the atomic packing – they are *anisotropic* at the crystal scale. Most solids are *polycrystalline*, containing very many *grains* stuck together, each with their own lattice orientation (more on this in Chapter 5). This averages out the directionality in the crystal stiffness, giving *isotropic* elastic moduli.

Polymers contain both strong, covalent bonds along the polymer chain and weak van der Waals bonds ($S = 0.5$–5 N/m) between the chains; it is the weak bonds that stretch when the polymer is deformed, giving them low moduli. The random molecular structure is automatically isotropic. However, if the polymer is heavily drawn the molecules align enough for the covalent bonds to be stretched, giving a much higher modulus (see the right-hand end of the polymer stress-strain curve of Figure 4.3). This is exploited in the production of polymer fibres, giving properties that are anisotropic and far superior along the fibre compared to a bulk polymer (see Chapter 11).

Table 4.1 allows an estimate to be made of the lower limit for Young's modulus for a true solid. The largest atoms ($a_o = 4 \times 10^{-10}$ m) bonded with the weakest bonds ($S = 0.5$ N/m) will have a modulus of roughly

$$E = \frac{0.5}{4 \times 10^{-10}} \approx 1\,\text{GPa} \tag{4.18}$$

Many polymers have moduli of about this value, but as the E/ρ chart (Figure 4.6) shows, solid materials exist that have moduli that are much lower than this limit – the *elastomers*. (Foams have low moduli because of their porosity – we return to this in Section 4.5). The origin of the moduli of elastomers takes a little more explaining.

The elastic moduli of elastomers An elastomer is a tangle of long-chain molecules with occasional cross-links, as in Figure 4.9(a). The bonds between the molecules, apart from the cross-links, are weak – so weak that, at room temperature, they have melted, meaning that segments are free to slide over each other. Were it not for the cross-links, the material would have no stiffness at all; it would be a viscous liquid. This tangled structure is random with no order or structure – expressed in the language of thermodynamics, it has high *entropy*. Stretching it, as in Figure 4.9(b), partly aligns the molecules, resembling the crystallites in polymers (see Figure 3.27). Crystals are ordered: their entropy is low. So here there is a resistance to stretching – a stiffness – that has nothing to do with bond stretching at all, but with a thermodynamic resistance to the molecular ordering caused by strain. The cross-links give the elastomer a 'memory' of the disordered shape it had to start with, allowing large strains that are elastic and fully recoverable. A full theory is complicated – it involves the statistical mechanics of long-chain tangles – so it is not easy to estimate the value of the modulus. The main thing to know is that the moduli of elastomers are uniquely low because of this strange thermodynamic origin.

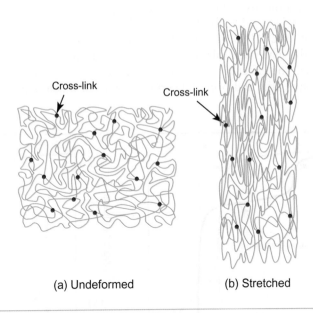

(a) Undeformed   (b) Stretched

Figure 4.9 Elastomer extension induces order, reducing entropy.

Temperature-dependence of polymer moduli: the glass transition temperature Crystalline solids have well-defined, sharp melting points. Amorphous polymers behave in a different way – the bonding is weaker and more diffuse, with the inter-chain bonding spread over a spectrum of atomic spacing because of the tangled molecular structure. The amorphous regions gradually change properties from solid to liquid over a range centred on the *glass transition temperature*, T_g, as the weak inter-chain bonds progressively melt on heating. In semi-crystalline thermoplastics, however, the crystallites have their own sharp melting temperature, typically about $1.5 \times T_g$, due to the closer, regular packing of the chains. Elastomers and thermosets have a glass transition but do not melt on heating – because of the cross-linking, they degrade and burn instead.

How does the glass transition affect the Young's modulus? The answer is dependent on the type of polymer, since even if the weak bonding has gone, there can still be elastic behaviour above T_g. Figure 4.10 shows schematics of the variation of modulus with temperature. Below T_g, polymers are referred to as glassy, with the modulus controlled by the van der Waals bonding. Above T_g, amorphous thermoplastics (left-hand figure) show a tiny residual elasticity that stems from entanglements in the molecular structure, blurring into viscous flow. Semi-crystalline thermoplastics also show a drop in modulus across their glass transition, on the order of 10 times if the crystallinity is high, to a broad plateau in modulus where the response is referred to, somewhat misleadingly, as 'rubbery.' The degree of crystallinity determines how stiff the material remains above T_g, but eventually all thermoplastics melt and flow. Note the position of room temperature (RT) relative to T_g – in amorphous thermoplastics, T_g effectively sets the upper limit of usefulness of the polymer; in semi-crystalline it sets the lower limit (due to brittleness in the glassy state). Cross-linked behaviour is shown in Figure 4.10(b). Elastomers display a dramatic drop in modulus through T_g, by a factor of order 1000: the very low modulus of rubbers reflects the fact that at room temperature the material is already above T_g and is governed by thermodynamics (as discussed earlier). The effect is well illustrated by immersing a rubber tube in liquid nitrogen, when it becomes stiff and brittle! Thermosets have a much higher degree of cross-linking,

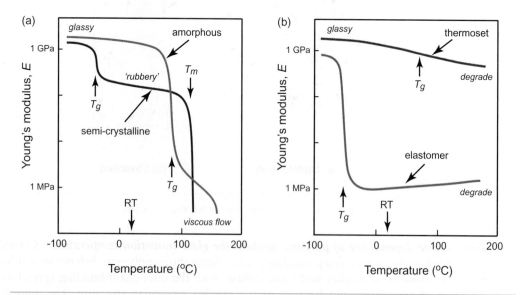

Figure 4.10 Schematic temperature-dependence of Young's modulus for (a) thermoplastics (amorphous and semi-crystalline); (b) elastomers and thermosets. The glass transition temperatures T_g are where the weak inter-chain bonding is lost. *RT* indicates room temperature.

retaining a higher modulus as temperature falls, with heavily cross-linked thermosets barely showing a glass transition effect at all.

Polymer deformation shows other unusual characteristics. Since stretching the material is accompanied by some sliding of the molecules past one another, the stiffness is particularly sensitive to the rate of loading. Rapid loading does not give time for chain sliding; slow loading enables it, giving a quite different response. The glass transition temperature is therefore sensitive to the deformation rate. So at the same temperature, a polymer can switch from being floppy and resistant to fracture, to being much stiffer and glass-like, by pulling on it rapidly. Design to cope with the effect of temperature on polymers is discussed further in Chapter 8.

4.5 Manipulating the modulus and density

The property chart in Figure 4.6 shows that each ceramic (e.g. SiC, Al_2O_3) or class of metallic alloys (e.g. steels, Ni alloys) occupies a small, well-defined bubble. Polymers show a little more variation, but as we have seen it is apparent that the modulus and density of solids reflect some inherent physical characteristics – the *atomic bonding* and *packing*. So there's not much you can do to change these properties in a given solid material, but there are other ways to manipulate modulus and density, to occupy new spaces on the chart. *Hybrid materials* are mixtures of solid materials (or a solid with air) to give internal structure at a length scale greater than the atomic. Polymers and metals can be reinforced with stiff ceramic particles or fibres, to make *composites*. And almost all materials can be made into porous *foams* or some other cellular structure. Materials such as woods are in fact natural hybrids of mixed polymers, exploiting both foam and composite characteristics. Man-made foams and composites either occupy blank spaces on the chart, or overlay existing solids, creating new opportunities and competition in design. A bit of simple modelling shows how they work.

Composites   The development of high-performance composites is one of the great material advances of the last 50 years. The concept is shown in Figure 4.11, with fibres or particles being embedded in a matrix made of a polymer (polymer matrix composites [PMCs]), a metal (MMCs), or a ceramic (CMCs). The aim is to produce a hybrid material that exploits the attractive properties of the components. Fibres have exceptional stiffness and strength, so PMCs have high stiffness and strength per unit weight but now in a usable, mouldable form, while MMCs and CMCs offer improved high-temperature performance.

When a volume fraction f of a reinforcement r (density ρ_r) is mixed with a volume fraction $(1 - f)$ of a matrix m (density ρ_m) to form a composite with no residual porosity, conservation of mass dictates that the composite density $\tilde{\rho}$ is given exactly by the *rule of mixtures*:

$$\tilde{\rho} = f\rho_r + (1 - f)\rho_m \tag{4.19}$$

The geometry or shape of the reinforcement does not matter except in determining the maximum packing fraction of reinforcement and thus the upper limit for f (typically 50%).

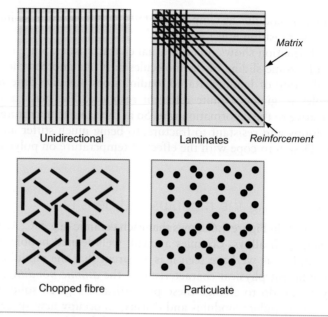

Figure 4.11 Composite structures.

The modulus of a composite is bracketed by two *bounds* – limits between which the modulus must lie. The *upper bound*, $\tilde{E}_u$, is found by assuming that, on loading, the two components strain by the same amount, like springs in parallel. The stress is then the volume fraction weighted average of the stresses in the matrix and the reinforcement, giving, once more, a rule of mixtures:

$$\tilde{E}_u = f E_r + (1-f)E_m \tag{4.20}$$

where E_r is the Young's modulus of the reinforcement and E_m that of the matrix. To calculate the *lower bound*, $\tilde{E}_l$, we assume that the two components carry the same stress, like springs in series. The strain is the volume fraction weighted average of the strains in each component, and the composite modulus is

$$\tilde{E}_l = \left[\frac{f}{E_r} + \frac{(1-f)}{E_m} \right]^{-1} = \frac{E_m E_r}{f E_m + (1-f)E_r} \tag{4.21}$$

Figure 4.12 shows the range of modulus and density that can, in principle, be obtained by mixing two materials together. The boundaries are calculated from Equations (4.19), (4.20), and (4.21). There are practical limits, of course: the matrix and reinforcement must be chemically compatible and available in the right form, and processing the mixture must be

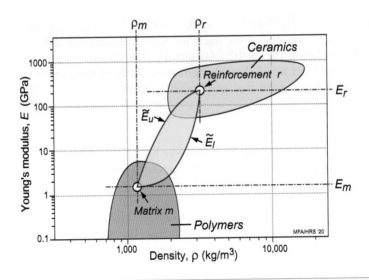

Figure 4.12 Bounds for the modulus and density of composites.

achievable at a sensible cost. The 'Composites' bubble in Figure 4.6 shows two successful examples: glass fibre-reinforced polymer (GFRP) and carbon fibre-reinforced polymer (CFRP). Note that the bubble lies between 'Polymers' and 'Ceramics,' as expected. They are as stiff as metals but lighter – a desirable combination.

Example 4.4

A composite material has a matrix of polypropylene and contains 10% glass reinforcement. Estimate the Young's modulus and density of the composite if (i) the glass is in the form of long parallel fibres (use the upper bound for loading parallel to the fibres), (ii) the glass is in the form of small particles (corresponding to the lower bound). Compare the *specific stiffness* (E/ρ) of the constituent materials and the two composites.

Use the following property data:

	E (GPa)	ρ (kg/m^3)	E/ρ (GPa.m^3/kg)
Polypropylene (*m*)	1	900	1.1×10^{-3}
Glass (*r*)	70	2200	31.8×10^{-3}

Answer. The density of both composites is the same: from Equation (4.19) with $f = 0.1$, $\rho = 1030$ kg/m^3. Using the upper bound [Equation (4.20)] for Young's modulus of the fibre-reinforced composite (parallel to the fibres): $E_u = 7.9$ GPa, with a specific stiffness $E_u/\rho = 7.9/1030 = 7.7 \times 10^{-3}$ GPa.m³/kg. Using the lower bound [Equation (4.21)] for the modulus of the particle-reinforced composite: $E_l = 1.1$ GPa, with a specific stiffness $E_l/\rho = 1.07 \times 10^{-3}$ GPa.m³/kg. The specific stiffness of the fibre-reinforced composite is seven times that of the matrix. That of the particle-reinforced composite is marginally lower than that of the matrix alone.

Foams   Foams are made much as you make bread: by mixing a matrix material (the dough) with a foaming agent (the yeast) and controlling what then happens in such a way as to trap the bubbles. Polymer foams are familiar for insulation and flotation, and as the filler in cushions and packaging. You can, however, make foams from other materials: metals, ceramics, and even glass. They are light (they are mostly space), and they have low moduli. This might sound like a combination of little use, but that is mistaken: if you want to cushion or to protect a delicate object (such as yourself), what you need is a material with a low, controlled stiffness – foams provide this.

   Figure 4.13 shows an idealised cell of a low-density foam. It consists of hollow cubic cells edged with solid of thickness t, with the length of the cells being L from the mid-thickness of the edges. Cellular solids are characterised by their *relative density*, the fraction of the foam occupied by the solid. For the structure shown here (with $t << L$) it is

$$\frac{\tilde{\rho}}{\rho_s} = 3\left(\frac{t}{L}\right)^2 \tag{4.22}$$

where $\tilde{\rho}$ is the density of the foam and ρ_s is the density of the solid of which it is made.

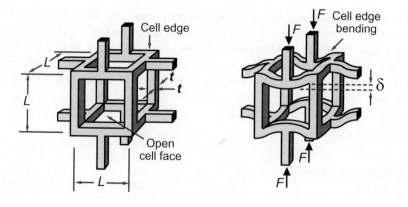

Figure 4.13   Idealised foam cell, and its elastic response to load.

When the foam is loaded, the cell walls bend, as shown on the right of the figure. This behaviour can be modelled (using the solid mechanics in the next section), giving the foam modulus $\tilde{E}$

$$\frac{\tilde{E}}{E_s} = \left(\frac{\tilde{\rho}}{\rho_s} \right)^2 \tag{4.23}$$

where E_s is the modulus of the solid from which the foam is made. This gives us a second way of manipulating the modulus – in this case, decreasing it. At a relative density of 0.1 (meaning that 90% of the material is empty space), the modulus of the foam is only 1% of that of the material in the cell wall, as sketched in Figure 4.14. The range of modulus and density for real polymer foams is illustrated in the chart of Figure 4.6 – as expected, they fall below and to the left of the polymers of which they are made.

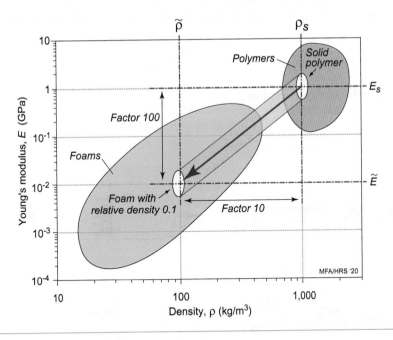

Figure 4.14   Estimates for modulus and density of foams.

4.6 Introductory solid mechanics – elasticity and stiffness

Modelling is a key part of design. Analytical models are sufficient in the early stages of design to establish whether a concept will work at all and to guide the choice of materials that optimise performance. Concluding the design will involve computational modelling to give precise values of key design parameters – but reliable use of numerical methods requires an understanding of the fundamentals on which the computation is based. For components carrying loads, engineers have produced analytical models for many standard geometries under idealised loading – treating real loads as if they act at a point for example. Standard solutions exist for stresses, strains, displacements, and types of failure under loading in tension, bending, torsion, pressure loading, and so on. A selection of key results are presented here, without full details of their derivation – there is no shortage of books on structural mechanics for this. The important thing in design is to know that the results exist, how they work, and when to use them.

Elastic extension or compression Figure 4.15(a) shows a tensile stress $\sigma = F/A$ applied uniaxially to a tie of length L_o and constant cross-section area A, leading to an extension δ expressed as a strain $\varepsilon = \delta/L_o = \sigma/E$. Thus the relation between the load F and deflection δ is

$$\delta = \frac{L_o F}{AE} \tag{4.24}$$

and the stiffness S is defined as

$$S = \frac{F}{\delta} = \frac{AE}{L_o} \tag{4.25}$$

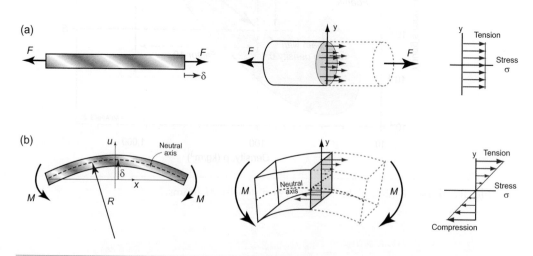

Figure 4.15 (a) A tie loaded in tension with a force F has a stiffness $S = F/\delta$, and carries an axial tensile stress. (b) A beam of rectangular cross section loaded in bending with a moment M, giving a radius of curvature R, with stress σ varying linearly from tension to compression either side of the neutral axis.

In uniaxial tension, the *shape* of the cross-section area does not matter because the stress is uniform over the section. In a strut carrying compression, the stress and strain are both negative, but the stiffness is the same. If the strut is slender, however, it will not carry axial stress for long but will *buckle* – we come to this later.

Elastic bending of beams   When a beam is loaded by a *bending moment*, M, its initially straight axis is deformed into a curve of radius R [Figure 4.15(b)]. Provided the transverse deflection of the beam, u(x), is reasonably small compared to its length (as is the case for most materials in the elastic regime), the curvature κ (= 1/R) can be written as

$$\kappa = \frac{\mathrm{d}^2 u}{\mathrm{d}x^2} = \frac{1}{R}$$

This curvature generates a linear variation of axial strain ε across the section, with tension on one side and compression on the other – the position of zero strain being the *neutral axis*. As stress is proportional to strain in uniaxial tension and compression, the bending stress varies linearly with distance y from the neutral axis [Figure 4.15(b)]. Material is more effective at resisting bending the farther it is from that axis, so the shape of the cross section is now important. Elastic beam theory (see Further Reading) gives the stress σ caused by a moment M in a beam made of material of Young's modulus E, as

$$\frac{\sigma}{y} = \frac{M}{I} = E\kappa = E\frac{\mathrm{d}^2 u}{\mathrm{d}x^2} \tag{4.26}$$

where I is the *second moment of area*, defined as

$$I = \int_{\text{section}} y^2 \mathrm{d}A = \int_{\text{section}} y^2 b(y)\mathrm{d}y \tag{4.27}$$

The distance y to each element dA of the cross-sectional area is measured perpendicular to the neutral axis, and b(y) is the width of the section at y. The second moment I characterises the resistance of the section to bending – it includes the effect of both size and shape. For standard shapes, the maths has already been completed for us – expressions for the cross-section area A and the second moment of area I are shown for four common sections in Figure 4.16.

Section shape	Area, A (m²)	Second moment of area, I (m⁴)
	bh	$\dfrac{bh^3}{12}$
	πr^2	$\dfrac{\pi}{4}r^4$
	$\pi(r_o^2 - r_i^2)$ $\approx 2\pi r_o t$ $(t \ll r_o)$	$\dfrac{\pi}{4}(r_o^4 - r_i^4)$ $\approx \pi r_o^3 t$ $(t \ll r_o)$
	$2t(h+b)$ $(t \ll h,b)$	$\dfrac{1}{6}h^3 t\left(1 + 3\dfrac{b}{h}\right)$ $(t \ll h,b)$

Figure 4.16   Key structural parameters for four section shapes: cross-section area (A), and second moment of area for *bending* (I).

Example 4.5

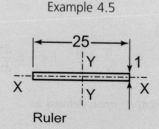

Ruler

A steel ruler has a width $w = 25$ mm and a thickness $t = 1$ mm. Calculate the second moment of area for bending about the X-X axis I_{XX} and about the perpendicular Y-Y axis, I_{YY}, comparing the two values.

For bending about the (flexible) X-X axis:

$$I_{XX} = \frac{wt^3}{12} = \frac{25 \times 1^3}{12} = 2.1 \, \text{mm}^4$$

For bending about the (stiff) Y-Y axis:

$$I_{YY} = \frac{tw^3}{12} = \frac{1 \times 25^3}{12} = 1300 \, \text{mm}^4$$

I_{YY} is more than 600 times greater than I_{XX} because the material of the ruler is located close to the X-X axis but (on average) much farther away from the Y-Y axis.

Example 4.6

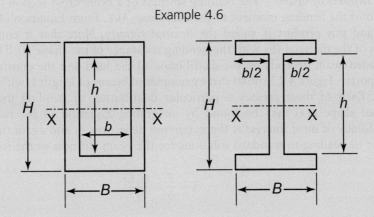

Moments of inertia for more complex sections can be calculated by *superposition* of the results for simple shapes like the rectangle. Consider the rectangular hollow section shown in the figure: integrating Equation (4.27) over the whole area $(B \times H)$ led to $I_{XX} = BH^3/12$, but this includes the area of the hollow central region. The contribution of this area to the integral is simply I_{XX} for the small rectangle, $(b \times h)$, that is, $I_{XX} = bh^3/12$. Hence the second moment of area for the hollow rectangular section is:

$$I_{XX} = \frac{BH^3}{12} - \frac{bh^3}{12}$$

Use the same reasoning to find the second moment of area I_{XX} for the I-beam shown in the figure.

Answer. The external dimensions of the I-beam are again $(B \times H)$, so the enclosing large rectangle has second moment of area $I_{XX} = BH^3/12$. Two smaller rectangles $(b/2 \times h)$ are included in this result and need to be subtracted to give the second moment for the I-beam:

$$I_{XX} = \frac{BH^3}{12} - 2 \times \frac{(b/2)h^3}{12} = \frac{BH^3}{12} - \frac{bh^3}{12}$$

The result is exactly the same as for the hollow section.

[*Note*: This method works only if the two rectangles share a common X-X axis – each area element dA has the same value of y in the integration behind both expressions for I_{XX}. If this is *not* the case, the analysis is more complex, requiring the use of the 'parallel axis theorem' – as for example to find I_{YY} for the I-section; see any standard text on structural mechanics for more details. Second moments of area for many common shapes are available in textbooks and handbooks, or online, as are tables of values for commercially available structural sections in steels, aluminium, wood, and GFRP.]

Elastic stiffness of beams   The bending stiffness of a beam cross section is characterised by the ratio of the bending moment to the curvature, M/κ. From Equation (4.26), this is equal to EI, and this product is called the *flexural rigidity*. Note that it combines the elastic stiffness of the material (E) with the bending resistance of the shape (I). The deflected shape of a loaded beam depends on the distribution of the load and the constraints imposed by the supports – Figure 4.17 shows three examples of beams of length L with a total transverse load F. Each of these creates a particular distribution of bending moment $M(x)$. The deformed shape $u(x)$ may be found by integrating Equation (4.26) twice starting from $M(x)$. Usually of most interest is the *maximum* deflection, δ, and again the maths has been done for us, leading to standard solutions for the beam stiffness of the form:

$$S = \frac{F}{\delta} = \frac{C_1 EI}{L^3} \tag{4.28}$$

Note that the form of the result is the same for all simple load distributions – the effects of the support arrangements and load distribution are captured in the value of the constant C_1 (shown in Figure 4.17). We show in *Guided Learning Unit 2* that the best choice of material is independent of the value of C_1 with the convenient outcome that the optimum solution for a complex distribution of loads is the same as that for a simple one.

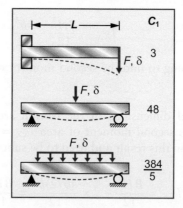

Figure 4.17   Elastic deflection of beams. The maximum deflection δ of a span L under a load F depends on the flexural rigidity (EI) of the cross section, and the way the beam is supported and the load is distributed.

Example 4.7

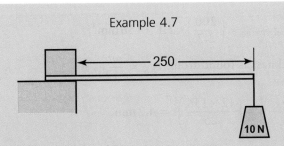

The ruler in Example 4.5 is made of stainless steel with Young's modulus of 200 GPa. A student supports the ruler as a horizontal cantilever with 250 mm protruding from the edge of a table, and hangs a weight of 10 N on the free end. Choose the appropriate solution from Figure 4.17, and calculate the vertical deflection of the free end if the ruler is mounted with the X-X axis horizontal. How much stiffer would the ruler be if it was clamped with the Y-Y axis horizontal instead?

(You may neglect the deflection due to the self-weight of the ruler).

Answer. Using Equation (4.28) with $C_1 = 3$ (from Figure 4.17), with I_{XX} from Example 4.5 and $E = 200$ GPa:

$$\delta = \frac{FL^3}{C_1\, EI_{XX}} = \frac{10\,(0.25^3)}{3\,(200 \times 10^9)\,(2.1 \times 10^{-12})} = 124 \text{ mm}$$

From Example 4.5, with Y-Y horizontal, the second moment of area (and thus the flexural rigidity EI and beam stiffness S) increased by a factor of 1300/2.1. The ruler would only deflect by $124 \times (2.1/1300) = 0.02$ mm. Try bending a ruler each way yourself to feel this difference. This shows the dramatic benefit of locating the material of a beam as far as possible from the bending axis, so as to maximise the second moment of area, and thus why I-sections are often used for structural girders.

Example 4.8

A new ruler is to be made from polystyrene, with Young's modulus of 2.7 GPa. Its length and width are to be the same as the stainless steel ruler, but its thickness can be changed by the designer, in order to match the bending stiffness of the steel ruler in Example 4.7. Calculate the thickness needed to achieve this. Compare the masses of the two rulers. (The densities of stainless steel and polystyrene are 7700 kg/m^3 and 1000 kg/m^3, respectively).

Answer. To provide the same stiffness, F/δ, we need the same flexural rigidity:

$$(EI)_{\text{polystyrene}} = (EI)_{\text{stainless steel}}$$

Hence: $(I)_{\text{polystyrene}} = \left(\dfrac{200}{2.7}\right) \times 2.1 = 156\,\text{mm}^4$

So the thickness is found from: $I_{XX} = \dfrac{wt^3}{12}$, so

$$t = \left(\frac{12\,I_{XX}}{w}\right)^{\frac{1}{3}} = \left(\frac{12 \times 156}{25}\right)^{\frac{1}{3}} = 4.2\,\text{mm}.$$

The masses are $m_{SS} = 7700 \times 0.3 \times 0.025 \times 0.001 = 58$ g, and $m_{PS} = 1000 \times 0.3 \times 0.025 \times 0.0042 = 32$ g. Despite its much lower density, the polystyrene ruler is only 50% lighter than the stainless steel one, because of the greater thickness needed for stiffness.

Buckling of columns and plates   If it is sufficiently slender, a strut, column, or plate loaded axially in compression will deflect sideways, and this sets the limiting load for failure in design. This *elastic buckling* occurs at a critical load, F_{crit}, which may be derived using the same beam theory used for the deflection of beams subjected to transverse loads. The critical buckling load again depends on the length L and the flexural rigidity, EI:

$$F_{crit} = \frac{n^2\,\pi^2\,EI}{L^2} \tag{4.29}$$

where n is a constant that depends on the end constraints: clamped, or free to rotate, or free to rotate and translate (Figure 4.18). The value of n is effectively the number of 'half-wavelengths' of the buckled shape (the half-wavelength is the distance between points of zero bending moment – i.e. simple supports that can rotate – or inflection points). Some of the great historical engineering disasters have been caused by buckling. It can occur without warning, and only slight misalignment is enough to significantly reduce the load at which it happens. When dealing with compressive loads on columns, or in-plane compression of plates, it is advisable to check that you are well away from the buckling load.

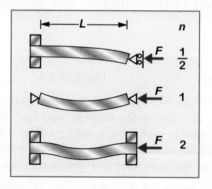

Figure 4.18   The buckling load of a column of length *L* depends on the flexural rigidity *EI* and on the end constraints; three are shown here, together with the value of *n*.

Example 4.9

The stainless steel ruler in Example 4.7 is placed upright on a table. If the length of the ruler is 300 mm, what is the axial compressive force that needs to be applied to cause it to buckle? Assume that both ends are free to rotate but remained aligned (the middle case in Figure 4.18).

Answer. Using Equation (4.29) with $n = 1$ (from Figure 4.18), with $I_{XX} = 2.1$ mm^4 and $E = 200$ GPa (from Example 4.5), and $L = 300$ mm:

$$F_{crit} = \frac{n^2 \pi^2 EI}{L^2} = \frac{1 \times \pi^2 \times (200 \times 10^3) \times 2.1}{300^2} = 46.1\,\text{N}.$$

This is a modest load – roughly the weight of a mass of 4.6 kg.

4.7 Introduction to material selection

This chapter has explored the elasticity of materials, first defining the key properties, their physical origins, and the limited options for modifying them. The analysis of the elastic stiffness of simple components in design was then introduced. With this knowledge, we can now turn to the problem of material selection.

Chapter 2 introduced the systematic pathway to choosing the best materials (and processes) for a product. Recall that consideration of the design requirements identifies the *constraints* on a mechanical or structural design – for example, the working or failure loads, the maximum deflection that is acceptable, limits on the component dimensions, and considerations of cost and sustainability. These are assessed against the properties of candidate materials, screening out those that do not meet one or more criteria, and ranking the remaining short-list against one particular measure of performance – the *objective* – for example, minimum mass, cost, or environmental impact.

Introduction to Guided Learning Unit 2: Material selection in design

Material selection is a topic that is best learnt by doing it – this is the purpose of *Guided Learning Unit 2: Material selection in design*. This works through the methodology first for stiffness-limited problems, and then for designs limited by material strength (the topic of Chapter 5). This *Unit* may therefore be worked through in two halves, after each of Chapters 4 and 5, or completed in its entirety after studying Chapter 5. A key aspect of the selection method is the use of material property charts – for example, the modulus–density chart naturally captures the property trade-offs needed to optimise a light, stiff design. *Guided Learning Unit 2* focusses on stiffness and strength in design, but the methodology extends to problems in fracture, high temperature or thermally-limited design, electrical materials, and so on (examples will be found in later chapters).

4.8 Summary and conclusions

When a solid is loaded, it initially deforms elastically. 'Elastic' means that, when the load is removed, the solid springs back to its original shape. The material property that measures stiffness is the elastic modulus. Three related moduli capture the different ways in which solids can be loaded:

- *Young's modulus, E,* measuring resistance to stretching and compressing, axially or in bending
- *Shear modulus, G,* measuring resistance to twisting (torsion)
- *Bulk modulus, K,* measuring resistance to hydrostatic compression

The modulus–density (E–ρ) property chart helps to visualise the relative values of these two properties for materials in the different material classes, rationalising the underlying materials science. Elastic moduli have their origins in the stiffness of the bonds between atoms in a solid and in the packing of the atoms, and thus the number of bonds per unit area. Since the atomic packing does not vary much from one solid to another, the moduli mainly reflect the stiffness of the bonds. Bonding can take several forms, depending on how the electrons of the atoms interact. Metallic, covalent, and ionic bonds are stiff; dipolar (van der Waals) bonds are much less so, which is why steel has high moduli and polyethylene has low.

There is very little that can be done to change the bond stiffness or atomic weight and packing of a solid, so at first sight we are stuck with the moduli and densities of the materials we already have. But there are two ways to manipulate them via a higher length scale: by mixing two materials to make composites, or by mixing a material with space to make foams. Both are powerful ways of creating 'new' materials that occupy regions of the E–ρ map that were previously empty.

Adequate stiffness is central to the design of structures that are *deflection-limited*. Stiffness is influenced by the size and shape of the cross-section, the way it is loaded (tension, compression, or bending), and the material of which it is made (via the elastic moduli). Solid mechanics is the domain for the analysis of the loading on a component and the resulting stresses, strains, and deflections. Equipped with these standard solutions for elasticity and the appropriate material property charts, we can choose the optimum materials for design to meet stiffness targets, while minimising weight, for instance – the methodology explored via *Guided Learning Unit 2*.

4.9 Further reading

Ashby, M. F. (2017). *Materials Selection in Mechanical Design* (5th ed.). Butterworth Heinemann. ISBN 978-0-08-100599-6. (A text that takes the solid mechanics analysis of stiffness in this chapter to a much higher level).

Askeland, D. R., Phulé, P. P., & Wright, W. J. (2010). *The Science and Engineering of Materials* (6th ed.). Toronto. ISBN 9780495296027. (A well-established materials text that deals well with the science of engineering materials).

Budinski, K. G., & Budinski, M. K. (2010). *Engineering Materials, Properties and Selection* (9th ed.). Prentice-Hall. ISBN 978-0-13–712842-6. (Like Askeland, this is a well-established materials text that deals well with both material properties and processes).

Callister, W. D., & Rethwisch, D. G. (2014). *Materials Science and Engineering: An Introduction* (9th ed.). John Wiley & Sons. ISBN 978-1-118–54689-5. (A well-established text taking a science-led approach to materials teaching).

Clyne, T. W., & Hull, D. (2019). *An Introduction to Composite Materials* (2nd ed.). Cambridge, Cambridge University Press. ISBN 9781–139050586. (A concise and readable introduction to composites that takes an approach that minimises the mathematics and maximises the physical understanding).

Shackelford, J. F. (2014). *Introduction to Materials Science for Engineers* (8th ed.). Prentice-Hall. ISBN 978–0133789713. (A well-established materials text with a materials and a design slant).

Young, W. C. (2012). *Roark's Formulas for Stress and Strain* (8th ed.)McGraw-Hill. ISBN 078–0071742474. (This is the 'Yellow Pages' of formulae for elastic problems – if the solution is not here, it probably doesn't exist).

4.10 Exercises

Exercise E4.1 Identify which of the five modes of loading (Figure 4.1) is dominant in the following components:

- Fizzy drinks container
- Overhead electric cable
- Shoe soles
- Wind turbine blade
- Climbing rope
- Bicycle forks
- Aircraft fuselage
- Aircraft wing spar

Can you think of another example for each mode of loading?

Exercise E4.2 The cable of a hoist has a cross-sectional area of 80 mm^2. The hoist is used to lift a crate weighing 250 kg. What is the stress in the cable? The free length of the cable is 3 m. How much will it extend if it is made of steel (modulus 200 GPa)? How much if it is made of polypropylene, PP (modulus 1.2 GPa)?

Exercise E4.3 Figure 4.4 shows a cubic element of material with Young's modulus E and Poisson's ratio v that is subjected to normal stresses σ_1, σ_2, and σ_3, resulting in strains ε_1, ε_2, and ε_3 [Equation (4.13)].

(a) Find an expression for the dilatation Δ (fractional volume change) when a unit cube experiences strains of ε_1, ε_2, and ε_3, assuming the strains are elastic and therefore small ($<< 1$).

(b) Find the dilatation Δ for an element subjected to a uniaxial tensile stress $\sigma_1 = \sigma$ (i.e. $\sigma_2 = \sigma_3 = 0$). For what value of Poisson's ratio v is volume conserved? For a Poisson's ratio $v = 0.33$, what is the % change in material density while subjected to a tensile stress $\sigma = 0.001\ E$ (i.e. a typical stress at the elastic limit)?

(c) Find the dilatation Δ for an element subjected to equal pressure p in all directions (i.e. $\sigma_1 = \sigma_2 = \sigma_3 = -p$. Hence derive the formula relating the bulk modulus, K, to Young's modulus, E, and Poisson's ratio, ν [Equation (4.10)].

Exercise E4.4   The figure shows a cube of material located in a rigid slot of the same width as the cube. A vertical downwards stress $\sigma_1 = -\sigma$ is applied to the top face of the cube. The cube wishes to expand laterally (due to Poisson's ratio), and in the 2-direction it can do so, but in the 3-direction it is constrained and the strain $\varepsilon_3 = 0$. Use Hooke's law in three dimensions [Equation (4.13)] to find the stress in the 3-direction. Hence find the 'effective modulus,' given by σ_1/ε_1, and comment on its magnitude for typical values of Poisson's ratio.

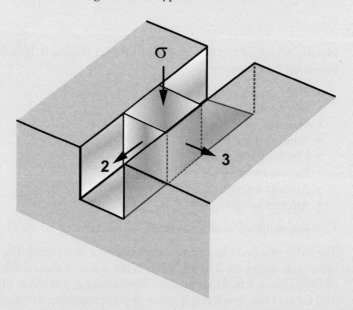

Exercise E4.5   A mounting block made of butyl rubber of modulus $E = 0.0015$ GPa is designed to cushion a sensitive device against shock loading. It consists of a cube of the rubber of side-length $L = 40$ mm located in a slot in a rigid plate, exactly as shown in Exercise E4.4. The slot and cube have the same cross-sectional dimensions, and the surrounding material is much stiffer, so the strain in one direction is constrained to be zero. A vertical compressive load F is applied, and the downward deflection of the top of the block is δ. The maximum expected value of F is 50 N. Use the result of Exercise E4.4 to find the stiffness of the cube (F/δ) in this constrained condition, assuming $\nu \approx 0.5$. Hence find the maximum deflection δ of the top face of the block. What would this deflection be without any constraint?

Exercise E4.6   A catapult has two rubber arms, each with a square cross section with a width 4 mm and length 300 mm. In use its arms are stretched to three times their original length before release. Assume the modulus of rubber is 10^{-3} GPa and that it does not change when the rubber is stretched. How much energy is stored in the catapult just before release?

Exercise E4.7   Use the modulus–density chart of Figure 4.6 to find, from among the materials that appear on it:

     (a) the material with the highest density.
     (b) the metal with the lowest modulus.
     (c) the polymer with the highest density.
     (d) the approximate range of modulus of elastomers.
     (e) the range of modulus of CFRP (carbon-fibre reinforced polymer).

Exercise E4.8   Use the modulus–density chart of Figure 4.6 to identify the polymers with the lowest density, and then use Appendix A to state some of their applications. Comment on whether you think the density was a factor in the choice of polymer for each application.

Exercise E4.9   Both magnesium alloys and CFRP have excellent stiffness-to-weight characteristics as measured by the ratio E/ρ. Use the typical property values in Appendix A to identify which is the better of the two.

Exercise E4.10   The speed of longitudinal waves in a material is equal to $\sqrt{E/\rho}$. The density ρ of carbon steel is 7500 kg/m^3 and its Young's modulus E is 210 GPa. What is the velocity of longitudinal sound waves in steel? Watch the units!

Exercise E4.11   The stiffness S of an atomic bond in a material is 50 N/m and its centre-to-centre atom spacing is 0.3 nm. What, approximately, is its elastic modulus?

Exercise E4.12   The table shows Young's modulus data for two copper alloy systems (Cu-Zn and Cu-Ni alloys), and for pure Cu, Zn, and Ni. Here the notation Cu-10Zn means 10 weight % Zn (and 90% Cu). The atomic packing and metallic bonding in alloys suggest that modulus is a microstructure-insensitive property, and expected to follow a simple linear rule of mixtures (in atomic %). Investigate graphically if this is the case for these Cu alloys (neglecting the difference between atomic and weight %, as Cu, Zn, and Ni are sufficiently close in atomic weight).

Alloy	Young's modulus E (GPa)
Pure Cu	120
Cu-10Zn	118
Cu-30Zn	112
Pure Zn	95
Cu-10Ni	125
Cu-30Ni	145
Cu-70Ni	165
Pure Ni	190

Exercise E4.13   Polypropylene (PP) may be drawn into fibres to enhance Young's modulus. A PP fibre used for marine ropes and fishing lines has a Young's modulus $E = 15$ GPa and a density $\rho = 910$ kg/m^3. Why are these properties important for these applications? Locate the property values on the E–ρ chart and compare them with the values for bulk PP. Can you suggest a microstructural explanation for the properties of PP in the form of fibres?

Exercise E4.14   Derive the upper and lower bounds for the modulus of a composite quoted as Equations (4.20) and (4.21) of the text. Regardless of the type of composite (fibre, particle, laminate), both matrix and reinforcement may be aggregated into a sandwich of two layers, as in the figures, with thicknesses determined by the volume fraction of reinforcement, f.

To derive the upper bound for the modulus, assume that the matrix and reinforcement behave like two springs in *parallel* (the left-hand figure) – i.e. each must *strain* by the same amount, and the total load F is the sum of the load in each material; in this orientation, note that f corresponds to an *area* fraction of reinforcement.

To derive the lower bound, assume that the matrix and reinforcement behave like two springs in *series* (the right-hand figure) – i.e. they carry the same *stress*, and the total extension δ is the sum of the extension of each material; in this orientation, f corresponds to the through-thickness *length* fraction of reinforcement.

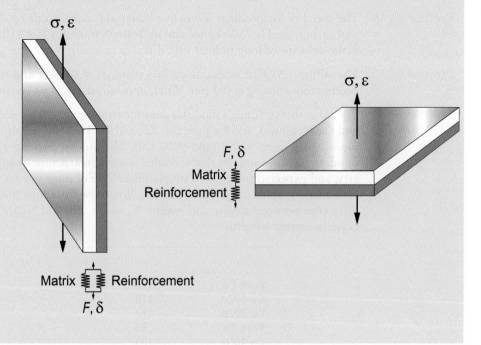

Exercise E4.15   Laminated panels are made by gluing together alternate layers of two different materials, as shown in the figure. Consider a laminate made of isotropic materials of Young's modulus E_1 and E_2, with a (thickness) fraction f of material 1. Determine whether the through-thickness and in-plane Young's moduli, respectively, will follow the upper bound or the lower bound. Hence find the in-plane modulus for a laminate consisting of 50% Al sheet and 50% chopped glass-fibre reinforced polyester, for which the Young's moduli are 70 and 17 GPa, respectively.

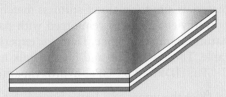

Exercise E4.16   (a) A sheet of unidirectional fibre-reinforced material contains 50% of fibres with Young's modulus E_f in a matrix of Young's modulus E_m, with $E_f / E_m = 100$. Find an expression for the Young's modulus parallel and perpendicular to the fibres, in terms of E_f.

(b) A composite panel is manufactured from an even number of layers of this unidirectional material stacked and stuck together. Fibres in alternate layers are oriented at 90° to one another. Derive an expression for the in-plane Young's modulus in either of the two directions parallel to sets of fibres. (Note that you may use the expression derived in Exercise E4.15 for the in-plane modulus, even though the layers are now anisotropic, as the number of layers of the two materials are equal).

(c) The layered composite panel in part (b) is manufactured using drawn polymer fibres ($E_f = 120$ GPa) in a thermoset matrix ($E_m = 1.2$ GPa). Calculate the Young's modulus in either of the two directions parallel to sets of fibres. The density of the composite is $\rho = 950$ kg/m³. Use the $E - \rho$ property chart in Figure 4.6 to identify the nearest material that would compete directly with this new composite.

Exercise E4.17   A volume fraction $f = 0.2$ of silicon carbide (SiC) particles is combined with an aluminium matrix to make a metal matrix composite. The modulus and density of the two materials are listed in the table. Assume the Young's modulus of particle-reinforced composites lies close to the lower bound [Equation (4.21)]. Calculate the density and approximate modulus of the composite. Is the specific modulus, E/ρ, of the composite greater than that of unreinforced aluminium? How much larger is the specific modulus if the same volume fraction of SiC is used in the form of parallel continuous fibres instead? For continuous fibres the modulus lies very close to the upper bound [Equation (4.20)].

	Density, kg/m³	Modulus, GPa
Aluminium	2700	70
Silicon carbide	3150	420

Exercise E4.18   Medical prosthetic implants such as hip replacements are at present made of metals such as stainless steel or titanium. These metals are much stiffer than the bone, giving poor transfer of load into the bone. New composite materials aim for a closer stiffness match. One uses a matrix of high-density polyethylene (HDPE) reinforced with particulate hydroxyapatite (HA), the natural mineral in bone. Data for experimental composites are listed in the table. Those for bone and for the bulk materials in the composite are bone, 7 GPa; HDPE, 0.65 GPa; hydroxyapatite, 80 GPa.

(a) Plot the upper and lower bounds for the Young's modulus of HDPE-HA composites against the volume fraction of reinforcement, f (from 0 to 1), together with the experimental data.

(b) Use the lower bound as a guide to extrapolate the experimental data to estimate the volume fraction of hydroxyapatite needed to match the modulus of bone. Is this practical?

Volume fraction of HA	Young's modulus (GPa)
0	0.65
0.1	0.98
0.2	1.6
0.3	2.73
0.4	4.29

Exercise E4.19   Find the typical value of Young's modulus E for polypropylene (PP) from Appendix A. Use the foam model in the text [Equation (4.23)] to estimate the modulus of a PP foam with a relative density $\tilde{\rho}/\rho_s$ of 0.2.

Exercise E4.20   Find the typical values of density and Young's modulus E for Mg alloys and Ni alloys from Appendix A. Use the model relationship in Equation (4.23) to estimate the ranges of Young's modulus and density achievable by foaming these metals to relative densities $\tilde{\rho}/\rho_s$ in the range 0.1 to 0.3. Locate approximate 'property bubbles' for these new metallic foams on the E–ρ property chart in Figure 4.6. With which existing materials would these foams compete, in terms of modulus and density? Which other properties of these metal foams might differ significantly from the competition?

Exercise E4.21   A material is required for a packaging foam 9 cm thick. The maximum compression allowed is 3 mm, when a load of 100 kg is applied over an area of 10×10 cm.

(a) Find the required Young's modulus of the foam.

(b) Use the theoretical relationship for the modulus of a foam in Equation (4.23) to select a suitable relative density, and actual density ρ, if the foam is manufactured out of: (i) polypropylene; (ii) polyurethane elastomer. (Use values for Young's modulus and density from Appendix A). How do the properties of the proposed materials compare with the foams shown on the $E - \rho$ property chart in Figure 4.6?

Exercise E4.22 The second moment of an area I_{XX} for an I-beam with the dimensions shown in the figure was found in Example 4.6. By treating the shape as three separate rectangles, find the corresponding second moment of area about a perpendicular axis, I_{YY}. Simplify the expression for the special case of uniform thickness, case $B - b = (H - h)/2 = t$.

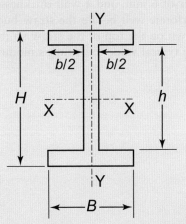

Exercise E4.23 A beam is to be made by gluing and screwing together four wooden planks of length 2 m, thickness 20 mm, and width 120 mm. The Young's modulus of the wood is 10 GPa.

(a) Determine the second moment of area I for the three configurations shown in cross section in the accompanying figure.
(b) Calculate the mid-span deflection δ of each of the beams when they are simply supported at their ends and loaded at mid-span with a force F of 100 N.
(c) Determine the stiffness F/δ for each of the beam configurations.

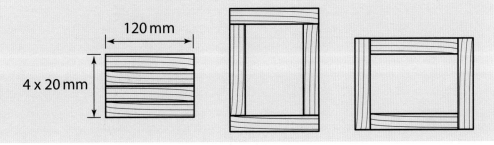

Exercise E4.24   Consider a solid cylindrical rod of length L and diameter d, loaded axially in tension with a force F, giving an axial extension δ. The same rod may also be loaded as a cantilever with a tip load of F, giving a transverse deflection δ. Find an expression for the ratio of the stiffnesses (F/δ) in tension and bending. Evaluate this ratio for the case of a rod of length 1 m and diameter 4 mm, and explain why it does not depend on the material.

Exercise E4.25   The lid of a take-away drinking cup is designed to be perforated using a straw, with the required force being no greater than 5 N. To combat concerns about plastic waste, a new straw is being developed using the biodegradable polymer PLA. The drinking straws have a length of 25 cm, an outer diameter of 6 mm, and a wall thickness of 0.2 mm. Check whether the lid will perforate well before the straw buckles elastically. Assume the user pushes down on the top of the straw so that both ends are free to rotate, and use the typical value for Young's modulus for PLA in Appendix A.

Chapter 5
Plasticity, yielding and ductility, and strength-limited applications

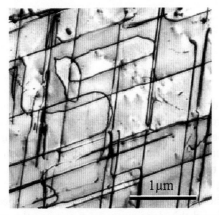

Dislocations in the intermetallic compound, Ni_3Al.
(Image courtesy of C. Rentenberger and H. P. Karnthaler, Institute of Materials Physics, University of Vienna, Austria.)

Strength-limited design can be elastic – ensuring that the cabin of the car does not deform in a crash – or plastic – absorbing the impact energy in the front of the car (left). Plasticity allows metals to be formed (right).
(Image of Crash Testing: Courtesy of Euro NCAP, Test Results – Honda Jazz 2004. Image of Forging Press: iStock.com.)

5.1 Introduction and synopsis

The verb 'to yield,' in everyday speech, can mean to submit to external demands or to produce something useful, like crops or a profit. The *yield strength*, when speaking of a material, is the stress beyond which it becomes plastic. The term is well-chosen: yielding and plastic flow are a type of failure – the material may strain too much or in an uncontrollable way; but they are also useful, allowing metals to be shaped and structures to safely absorb energy. This chapter is about strength, yielding, and plasticity mainly (but not wholly) about metals: the plasticity of iron and steel enabled the Industrial Revolution and the engineering achievements of the likes of Telford[1] and Brunel,[2] and this dominance continues to the present day.

In contrast to stiffness (Chapter 4), strength is a microstructure-sensitive property. Metals in particular can be manipulated via their composition and process history to deliver a wide range of yield strengths (in combination with other key properties, such as toughness, discussed in later chapters). The key to understanding this is a type of crystal defect – the dislocation – and the way it moves through a lattice under the action of stresses above the yield strength, enabling permanent strain. Increasing the strength therefore involves the introduction of obstacles to prevent that motion. Property charts are again a valuable tool for illustrating the comparative strength of different material classes, and the effects of alloying and processing in metals.

---

[1] Thomas Telford (1757–1834), Scottish engineer, brilliant proponent of the suspension bridge at a time when its safety was a matter of debate. Telford may himself have had doubts – he was given to lengthy prayer on the days that the suspension chains were scheduled to take the weight of the bridge. Most of his bridges, however, still stand.

[2] Isambard Kingdom Brunel (1806–1859), perhaps the greatest engineer of the Industrial Revolution (c. 1760–1860) in terms of design ability, personality, power of execution, and sheer willingness to take risks – the *Great Eastern*, for example, was five times larger than any previous ship ever built. He took the view that 'great things are not done by those who simply count the cost.' Brunel was a short man and self-conscious about his height; he favoured tall top hats to make himself look taller.

Stiffness-limited design (Chapter 4) is design to avoid excessive elastic deflection. *Strength-limited design*, our concern here, is design to avoid permanent deformation or plastic collapse. So it is still *elastic design*, but now to ensure that the stresses remain below the yield strength (or in some cases constrained to yield locally, avoiding overall collapse). That is only part of the strength story: *plastic design* aims to permit controlled extensive plasticity. The safety of modern cars relies on the front of the car absorbing the kinetic energy in a collision by plastic deformation. And the manufacturing processes of metal rolling, forging, extrusion, and drawing use plastic flow. Here, the strains are large and the focus is on the forces and work necessary to achieve a prescribed change of shape.

Plasticity problems are solved using solid mechanics. The analysis of standard geometries and modes of loading in Chapter 4 – tension, compression, bending, and so on – are extended here to calculate the stresses for simple design scenarios and to lay the foundation for material selection to avoid failure in design, presented in *Guided Learning Unit 2*.

5.2 Strength, ductility, plastic work, and hardness: definition and measurement

Yield properties and ductility are measured using the standard tensile tests introduced in Chapter 4, with the materials taken to failure. Figure 5.1 shows the types of stress–strain behaviour observed in different material classes. The *yield stress* or *yield strength* σ_y (or *elastic limit* σ_{el}) – units: MPa (i.e. MN/m^2) – requires careful definition. For metals, the onset of plasticity is sometimes marked by a distinct kink in the stress–strain curve; if not, we identify σ_y with the *0.2% proof stress* – that is, the stress at which the stress–strain curve for axial loading deviates by a permanent strain of 0.2% [this is the case shown in Figure 5.1(a) – note that the material continues to carry an elastic strain after yielding, so this proof stress is the stress from which we unload elastically to an offset strain of 0.2%, following the slope of Young's modulus]. The elastic limit or yield strength is the same in tension and compression. When strained beyond the yield point, most metals *work harden*, causing the rising part of the curve. Figure 5.1(a) shows that during work hardening the sample extends stably, with a corresponding uniform reduction in the cross-section area, conserving volume (for reasons we will see later). In tension, a maximum stress – the *tensile strength*, σ_{ts} – is reached. After this peak, the deformation localises to a particular region of the sample, with all subsequent plastic strain confined to this region, leading to accelerated narrowing of the section – this is called *necking* – with eventually a ductile failure as voids open up and coalesce within the material.

For polymers, σ_y is identified as the stress at which the stress–strain curve becomes markedly non-linear: typically, a strain of 1% [Figure 5.1(b)]. The behaviour beyond yield depends on the temperature relative to the glass transition temperature T_g (defined in Chapter 4). Well below T_g most polymers are brittle. As T_g is approached, plasticity becomes possible until, at about T_g, thermoplastics exhibit *cold drawing*: large plastic extension of a stable necking region at almost constant stress, during which the molecules are pulled into alignment with the direction of straining, eventually leading to a steep rise in stress to fracture when alignment is complete. At still higher temperatures, thermoplastics become viscous and can be moulded; thermosets soften a little but finally decompose. The yield strength σ_y of a polymer–matrix

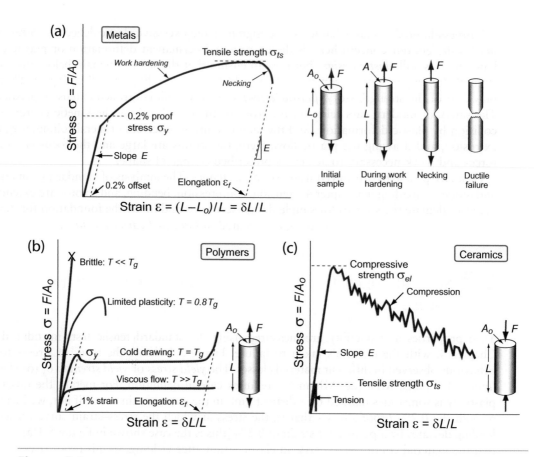

Figure 5.1 Stress–strain responses to uniaxial loading: (a) ductile metals in tension, also showing the changes in sample geometry in the work hardening and necking regions; (b) polymers in tension, at different temperatures relative to the glass transition; (c) ceramics in compression.

composite is similarly defined by a set deviation from linear elastic behaviour, typically 0.5%. Composites that contain aligned fibres (including natural composites like wood) are a little weaker (up to 30%) in compression than in tension because the fibres buckle on a small scale.

For ductile metals and polymers, the *plastic strain*, ε_{pl}, is the permanent strain resulting from plasticity; thus it is the total strain ε_{tot} minus the recoverable, elastic, part:

$$\varepsilon_{pl} = \varepsilon_{tot} - \frac{\sigma}{E} \tag{5.1}$$

The *ductility* is a measure of how much plastic strain a material can tolerate. It is measured in standard tensile tests by the *elongation* ε_f (the plastic tensile strain at fracture) expressed as a percentage [Figure 5.1(a,b)]. Strictly speaking, ε_f is not a material property because it depends on the sample dimensions – the values that are listed in handbooks and databases are for a standard test geometry – but it remains useful as an indicator of the ability of a material to be deformed.

Ceramics and glasses are brittle at room temperature [Figure 5.1(c)]. They have yield strengths, but these are so enormously high that, in tension, they are never reached; the materials fracture first. Even in compression, ceramics and glasses crush before they yield – more on this in Chapter 6. For now, it is useful to have a practical measure of the strength of ceramics to allow their comparison with other materials. The measure used here is the *compressive crushing strength*, and since it is not a yield point but is still the end of the elastic part of the stress–strain curve, we call it the *elastic limit*, σ_{el}.

In Chapter 4, the area under the elastic part of the stress–strain curve was identified as the elastic energy stored per unit volume ($\sigma_y^2/2E$). Beyond the elastic limit, *plastic work* is done in deforming a material permanently by yield or crushing. The increment of plastic work done for a small permanent extension or compression dL under a force F, per unit volume $V = A_o L_o$, is

$$dW_{pl} = \frac{F\,dL}{V} = \frac{F}{A_o}\frac{dL}{L_o} = \sigma\,d\varepsilon_{pl}$$

The total plastic work done per unit volume at fracture is, as for the elastic case, the area under the curve – but as this is not now a simple triangle, we would need to integrate to find the area under the stress–strain curve (remembering to subtract the elastic triangle at the end):

$$W_{pl} = \int_0^{\varepsilon_f} \sigma\,d\varepsilon_{pl} \tag{5.2}$$

This aspect of plasticity is important in energy-absorbing applications, such as the crash boxes in the front of cars or roadside crash barriers.

Example 5.1

A steel rod with a yield strength $\sigma_y = 500$ MPa is uniformly stretched to a strain of 0.1 (10%). Approximately how much plastic work is absorbed per unit volume by the rod in this process? Assume that the material is 'elastic perfectly-plastic,' that is, from the yield point it deforms plastically at a constant stress σ_y. Compare this plastic work with the elastic energy per unit volume calculated in Example 4.3.

Answer. Assuming that the material is 'elastic perfectly-plastic,' that is, the stress equals the yield strength up to failure, and neglecting the elastic contribution, the plastic energy per unit volume at failure is approximately:

$$W_{pl} = \sigma_y\,\varepsilon_f = 500 \times 10^6 \times 0.1 = 50\ MJ/m^3$$

(strictly minus the elastic stored energy, which is still there before unloading). This plastic energy is 80 times the elastic strain energy of 625 kJ/m^3 in Example 4.3 (but is still only 0.015% of the chemical energy stored in the same mass of gasoline).

Tensile and compression tests are not always convenient: the component or sample is destroyed, or the material may not be available in the size and shape needed. The hardness test (Figure 5.2) is a *non-destructive test*, which avoids this problem. A pyramidal diamond or a hardened steel ball is pressed into the surface of the material. This leaves a tiny shallow indent, the width of which is measured with a microscope, and converted to a projected area A perpendicular to the load. The indent means that plasticity has occurred, and the hardness H measures the resistance to plastic indentation. By the simplest definition, hardness is defined as:

$$H = \frac{F}{A} \qquad\qquad (5.3)$$

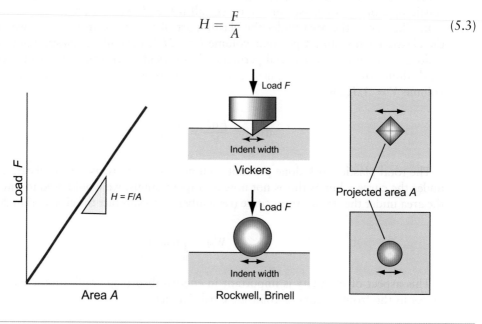

Figure 5.2 The hardness test. The Vickers test uses a diamond pyramid; the Rockwell and Brinell tests use a steel sphere, producing the indent shapes shown on the right.

In practice, small corrections may also be made to convert the projected area A to the actual area of contact between indenter and material. The plastic region under the indenter is surrounded by elastic material that has not deformed, and this constrains sideways plastic deformation, so that H is larger than the (uniaxial) yield strength σ_y – in practice it is about $3\sigma_y$. The plastic strain under the indenter also means that work hardening gives a strength somewhere between the yield and tensile strengths. Strength, as we have seen, is measured in units of MPa, and since H is also a force per unit area it would be logical to measure it in MPa, too, but historical convention dictates otherwise: for example, the *Vickers hardness*, H_v, has units of kg/mm^2. But this is equivalent to H in MPa if we multiply H_v by g (= $9.81\,m/s^2$, the acceleration due to gravity), with the approximate result that

$$H_v \approx \frac{3\,\sigma_y}{g} \approx \frac{\sigma_y}{3} \qquad\qquad (5.4)$$

(remembering that this will be a higher 'yield strength' than the true yield stress due to the hardening caused by the test itself). Each of the hardness scales (Vickers, Rockwell, Brinell) slightly differ in regard to strength – but these are simply tabulated and evaluated in handbooks and online. The hardness test is less accurate than tensile testing but has the advantage of being non-destructive, so strength can be measured for a component in service, or the test may be used for rapid, low-cost quality assurance purposes in production.

True stress and true strain The graphs in Figure 5.1 strictly show the *nominal stress–nominal strain* responses (i.e. the area and length used are the original values at the start of the test). But once materials yield, their dimensions change [Figure 5.1(a)]. The *true stress*, σ_t, takes account of the current dimensions, so in tension for example [Equation (4.1)]

$$\sigma_t = \frac{F}{A} \qquad (5.5)$$

To relate this to the *nominal stress* σ_n, we note that in plastic deformation the volume of the sample is conserved, for reasons discussed later in this chapter, that is

$$Volume = A_o L_o = AL \qquad (5.6)$$

where A and L are the current area and length of the test sample, initially A_o and L_o. Hence the nominal stress is

$$\sigma_n = \frac{F}{A_o} = \frac{F}{A}\left(\frac{L_o}{L}\right) \qquad (5.7)$$

Returning to the definition of nominal strain, ε_n [Equation (4.3)]

$$\varepsilon_n = \frac{\delta L}{L_o} = \left(\frac{L - L_o}{L_o}\right) = \left(\frac{L}{L_o}\right) - 1 \qquad (5.8)$$

Combining with Equations (5.5) and (5.7) gives the true stress as

$$\sigma_t = \sigma_n \left(1 + \varepsilon_n\right) \qquad (5.9)$$

True strain, ε_t, is a bit more involved. An incremental change in length, dL, gives an incremental strain relative to the current length L. To find the cumulative true strain, we therefore integrate these increments over the full extension from L_o to L, giving

$$\varepsilon_t = \int_{L_o}^{L} \frac{dL}{L} = \ln\left(\frac{L}{L_o}\right) = \ln\left(1 + \varepsilon_n\right) \qquad (5.10)$$

To see how different the true and nominal quantities are in practice, let's do an example.

Example 5.2

Find the true strain for the following values of the nominal strain during a tensile test on a metal: (a) the 0.2% offset strain, used to define the proof stress (Figure 5.1); (b) a strain of 20%, typical for the onset of failure at maximum stress. What is the ratio of true to nominal stress at these points?

Answer. The true strains are: (a) $\ln(1 + 0.002) = 0.002$ (to three decimal places); (b) $\ln(1.2) = 0.182$.
The ratio of true to nominal stress is $(1 + \varepsilon_n)$, giving: (a) 1; (b) 1.2.

The example shows that for the elastic regime and small plastic strains, the difference between nominal and true stresses and strains is negligible. This explains the effectiveness of using nominal values to define key properties such as the yield stress. By the end of the tensile test in a metal, the discrepancy between the two is typically 20%.

What about compressive true stress and strain? Now the cross-sectional area increases with plastic deformation (to conserve volume). Equations (5.5) to (5.10) all hold, so long as care is taken with the sign of the applied force F and the strain increments dL, which are negative. Figure 5.3 shows nominal and true stress–strain curves superimposed, for both compression testing and for the stable part of a tensile test up to the tensile strength. As observed in Example 5.2, in tension the true stress is higher than nominal, whereas for strain

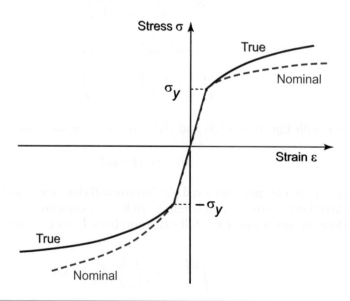

Figure 5.3 Nominal and true stress–strain curves for a ductile metal, in tension and compression.

the reverse is true. In compression, nominal stress exceeds true, and again the reverse for strain. Figure 5.3 shows one benefit of true stress-strain: the curves are identical in shape for tension and compression. The compressive side also extends to larger values of stress and strain, as the instability of necking in tension is avoided. The true stress–strain behaviour is most useful for *metal forming*, where plastic strains of the order of 50% or more are common. And some forming processes use multiple passes, with several plastic deformations one after another. In the Exercises at the end of the chapter you can show that the total strain in this case can simply be found by adding up the true strains, but this is not the case for nominal strains.

5.3 The big picture: charts for yield strength

Strength can be displayed on material property charts. Two are particularly useful: strength–density and modulus–strength charts.

The strength–density chart   Figure 5.4 shows the yield strength σ_y or elastic limit σ_{el} plotted against density ρ. The range of strength for engineering materials, like that of the modulus,

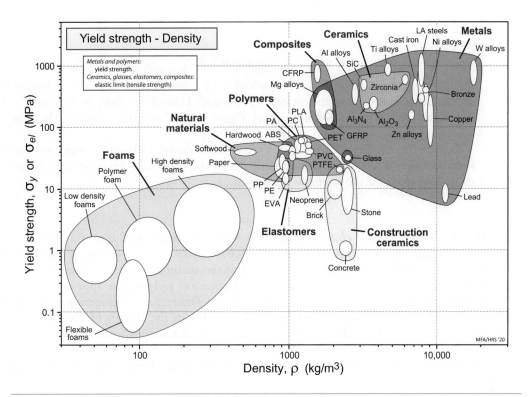

Figure 5.4   The strength–density property chart.

spans about 6 decades: from less than 0.01 MPa for foams, used in packaging and energy-absorbing systems, to 10^4 MPa for diamond (not shown in the figure), exploited in diamond tooling for machining and as the indenter of the Vickers hardness test. Members of each family again cluster together and can be enclosed in envelopes, each of which occupies a characteristic part of the chart.

Comparison with the modulus–density chart (Figure 4.6) reveals some marked differences. The modulus and density of a solid are both well-defined quantities with a narrow range of values. The strength is not: it is microstructure-sensitive. The strength range for a given class of metals, such as stainless steels, can span a factor of 10 or more, while the spread in stiffness or density is at most 10%. So the strength bubbles on Figure 5.4 for metals are elongated. The wide ranges for metals reflect the underlying physics of yielding and present designers with an opportunity for manipulation of the strength by varying composition and process history – discussed later in this chapter (and in Chapter 11).

Polymers cluster together with strengths between 10 and 100 MPa, and the modulus and strength vary by a similar amount for a given polymer. The composites CFRP and GFRP have strengths that lie between those of polymers and ceramics, as we might expect since they are mixtures of the two. The analysis of the strength of composites is not as straightforward as for modulus in Chapter 4, though the same bounds (with strength replacing modulus) generally give realistic estimates.

Example 5.3

An unidentified metal from a competitor's product is measured to have a Vickers hardness of 500 and a density of approximately 7900 kg/m^3. Estimate its yield strength and locate its position on the strength–density chart (Figure 5.4). What material is it likely to be?

Answer.   From Equation (5.4) a Vickers hardness of 500 gives a yield strength of approximately 1500 MPa. Plotted with the density, this falls near the top of the low alloy (LA) steels on the strength–density chart.

The modulus–strength chart   Figure 5.5 shows Young's modulus, E, plotted against yield strength, σ_y or elastic limit, σ_{el}. This chart allows us to examine a useful material characteristic, the *yield strain*, σ_y/E, meaning the strain at which the material ceases to be linearly elastic. On log axes, contours of constant yield strain appear as a family of straight parallel lines (of slope = 1), as shown in Figure 5.5. Engineering polymers have large yield strains, between 0.01 and 0.1 – useful in the design of elastic hinges; the values for metals are at least a factor of 10 smaller. Composites and woods lie on the 0.01 contour, as good as the best metals. Elastomers, because of their exceptionally low moduli, have values of σ_y/E in the range 1 to 10, much larger than any other class of

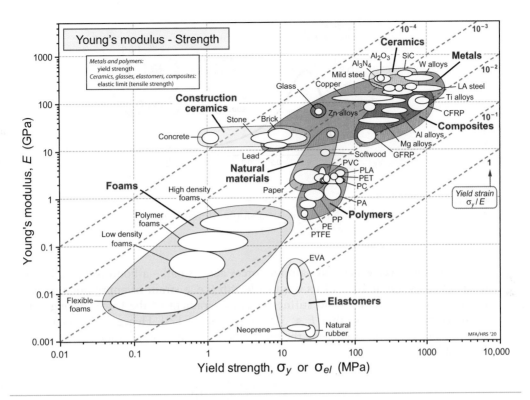

Figure 5.5 The Young's modulus–strength chart. The contours show the strain at the elastic limit, σ_y/E.

material. This chart is also used in material selection, for example for efficient springs (see *Guided Learning Unit 2*).

5.4 The physical origins of strength and ductility

The ideal strength The bonds between atoms are elastic for small changes in atomic spacing, but ultimately have a breaking point. In Chapter 4 we treated the elastic force-displacement response of the atomic spring as an equivalent stress–strain response for a single bond (assuming each atom occupies a cube of side length a_o). Figure 5.6 continues the stress–strain curve for a single bond to much larger strains: the stress goes through a maximum and falls to zero as the atoms lose their interaction altogether. The peak is the bond strength – the stress you need to break it and separate the atoms altogether.

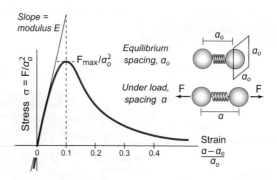

Figure 5.6 The stress–strain curve for a single atomic bond (it is assumed that each atom occupies a cube of side length a_o).

The distance over which inter-atomic forces act is small – a bond is typically broken if it is stretched by about 10% of its original length (i.e. a strain of approximately 0.1). Recall from Chapter 4 that the bond stiffness S is the gradient of the force-spacing curve at the equilibrium spacing, and that $E \approx S/a_o$ [Equation (4.17)]. Hence a notional linear-elastic stress at a strain of 0.1, following the black line on Figure 5.6, of slope equal to the modulus E, would be

$$\sigma = 0.1\frac{S}{a_o} = \frac{E}{10} \tag{5.11}$$

This over-estimates the peak stress, F_{max}/a_o^2, due to the curvature of the atomic stress–strain curve. More refined calculations that account for the shape of the curve show that the ratio of the peak stress to the modulus E is around 1/15. We can therefore define the *ideal strength* of a solid as

$$\sigma_{ideal} = \frac{S}{15a_o} = \frac{E}{15} \tag{5.12}$$

Figure 5.7 shows σ_y/E for metals, polymers, and ceramics. None achieves the ideal value of 1/15; most don't even come close. To understand why, we need to look at the microstructure and imperfections in each material class.

Strength of polymers   Figure 5.1(b) showed a range of stress–strain responses in polymers. Recall first from Chapter 4 that the glass transition temperature T_g is the temperature at which the inter-molecular hydrogen bonds melt. This is usually associated with a marked drop in the material stiffness, depending on whether the polymer is amorphous, semi-crystalline, or cross-linked. Not surprisingly, the strength of a polymer also depends strongly on whether the material is above or below the glass transition, and a number of different failure mechanisms can occur.

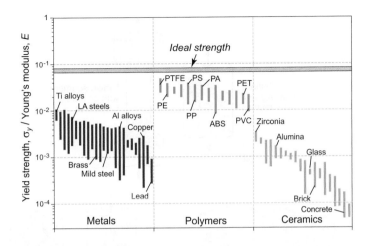

Figure 5.7 The ideal strength is predicted to be about $E/15$, where E is Young's modulus. The figure shows σ_y/E with a shaded band at the ideal strength.

At low temperatures $(T < 0.75\ T_g)$, polymers are elastic-brittle, like ceramics, and fail by propagation of a dominant flaw (see Chapter 6). Above this temperature, they become plastic. When pulled in tension, the chains slide over each other, unraveling, so that they become aligned with the direction of stretch, by the spreading of a stable necked region, as in Figure 5.8(a) – a process called *drawing*. The drawn material is stronger and stiffer than before, by a factor of about 8, giving drawn polymers exceptional properties, but limited in shape to fibre filaments.

Many polymers, including PE, PP, and nylon, draw at room temperature. Others with higher glass temperatures, such as PMMA, do not – at room temperature they *craze*. Small crack-shaped regions open up within the polymer, but bridged by drawn-out ligaments of aligned molecules, as in Figure 5.8(b). Crazes scatter light, so their presence causes whitening, easily visible when cheap plastic articles are bent. If stretching is continued, one or more crazes develop into proper cracks, and the sample fractures.

Crazing gives limited ductility in tension, but large plastic strains may still be possible in compression, by *shear banding* [Figure 5.8(c)]. These bands allow blocks of material on either side of the band to shear with respect to one another, without the material coming apart. Progressive formation of new shear bands leads to further strain.

Across all of these mechanisms, polymers have a strength that is not very far below their ideal value (Figure 5.7). But we shouldn't get too excited – their moduli are much lower than metals to begin with, but it does mean that polymers are not so far below metals when compared on strength (observe this for yourself on the modulus – strength property chart of Figure 5.5).

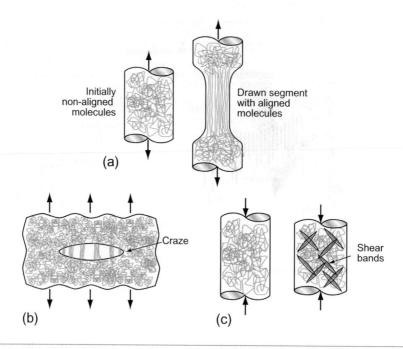

Figure 5.8 (a) Cold drawing – one of the mechanisms of deformation of thermoplastics. (b) Crazing – local drawing across a crack. (c) Shear banding.

Crystalline imperfection: defects in metals and ceramics Crystals contain imperfections of several kinds (Figure 5.9). A missing atom is a point defect called a vacancy [Figure 5.9(a)]. These play a key role in diffusion and creep (Chapter 8) but do not influence strength directly.

No crystal is made up of a single pure element. Other elements are always present at some concentration – either as impurities inherited from the process by which the material was made or as deliberate additions, creating *alloys* (a material in which two or more elements are mixed). If the foreign elements 'dissolve' and are spread out atomically in the crystal – like salt in water – then we have a *solid solution*. Figure 5.9(b) shows both a *substitutional solid solution* (the dissolved atoms, shown in red, replace those of the host) and an *interstitial solid solution* (the dissolved atoms, shown in black, are small enough to squeeze into the spaces or 'interstices' between the host atoms). Substitutional solutes rarely have the same size as those of the host material, so they distort the surrounding lattice – as do those small atoms that dissolve *interstitially*. So these impurities are useful – the lattice distortion is one of the reasons that alloys are stronger than pure materials, as we shall see in a moment.

Now to the key player, portrayed in Figure 5.9(c): the *dislocation*. These are responsible for making metals soft, well below their ideal strength, while making them ductile, enabling extensive shape change by plastic flow, with the material remaining intact. The upper part

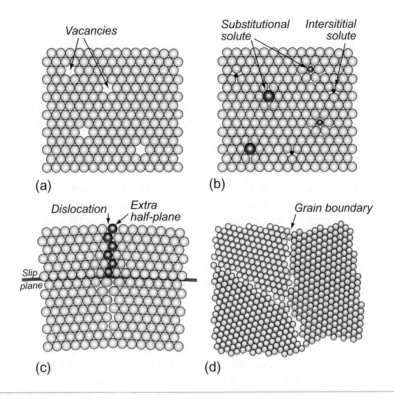

Figure 5.9   Defects in crystals. (a) Vacancies – missing atoms. (b) Foreign (solute) atoms in interstitial and substitutional sites. (c) A dislocation – an extra half-plane of atoms. (d) Grain boundaries.

of the crystal in Figure 5.9(c) has an extra 'half-plane' of atoms compared to the lower part – the region around the bottom of this half-plane is the dislocation. They are actually *line defects*, extending perpendicular to the view shown in the figure. Dislocations distort the lattice – here the green atoms (the *dislocation core*) are the most distorted – and because of this they have elastic energy associated with them. So why are they there? To grow a perfect crystal just 1 cm³ in volume from a liquid or vapour, about 10^{23} atoms have to find their proper sites on the perfect lattice, and the chance of this happening is just too small – all crystals contain vacancies and dislocations.

Most crystalline solids are polycrystalline, made up of very many tiny crystals, or *grains*. Figure 5.9(d) shows three perfect grains stuck together – here shown with different colours, though they are all the same atoms. Due to the different orientations in each grain, the interfaces between them have irregular packing with a bit of extra space between the atoms. These planar defects are called *grain boundaries*. Grain boundaries form in pure materials (as shown here) but also in alloys (when the mixture of atoms in one grain may differ in chemical composition from those of the next).

So metals are full of defects of the types shown in Figure 5.9, from the atomic scale upward. Between them they explain diffusion of atoms in solids, and the *microstructure-sensitive properties*: strength, ductility, electrical resistance, thermal conductivity, and much more. At the heart of metal alloying and processing is the direct manipulation of these defects to control properties – in this chapter we learn how strength depends on microstructure; in Chapter 11 and *Guided Learning Unit 4* we add the context of processing. For the rest of this section we focus on dislocations, as these control the plastic behaviour of metals. First we explore why the strength of metals is below ideal (Figure 5.7), and then address how strength is increased by forcing dislocations to interact with obstacles made of defects.

Ceramics can also deform plastically – but only if they are very hot (and generally they also need to be in compression). But their tensile strength shown in Figure 5.7 is also well below yield. To explain this we need to consider a different type of defect: *cracks*, and their interaction with stress, analysed by *fracture mechanics* and tackled in Chapter 6.

Dislocations and plastic flow   Recall that the strength of a perfect crystal computed from inter-atomic forces gives an 'ideal strength' around $E/15$ (where E is the modulus). The strengths of engineering alloys are much lower – typically 1 to 10 % of it (Figure 5.7). This was a mystery until the 1930s when Englishman G. I. Taylor[3] and Hungarian Egon Orowan[4] realised that a 'dislocated' crystal could deform at stresses far below the ideal. So what is a dislocation, and how does it enable deformation?

Figure 5.10(a) shows conceptually how to visualise a dislocation. Imagine that the crystal is cut along a horizontal atomic plane up to the line shown as ⊥—⊥, and the top part above the cut is then slipped across the bottom part by one full atom spacing. Most of the atoms either side of the cut re-align across the slip plane, except at the very end of the cut where there is a misfit in the atom configuration, as shown in Figure 5.10(b) for a simple cubic lattice packing. The problem is that there is now effectively an extra 'half-plane' of atoms in the top part of the crystal, with its lower edge along the ⊥—⊥ line. This is the *dislocation line* – the line separating the part of the plane that has slipped from the part that has not. This particular configuration is called an *edge dislocation* because it is formed by the edge of the extra half-plane, represented by the symbol ⊥.

Dislocations control plastic strain, since if a dislocation moves across its slip plane, then it makes the material above the plane slide one atomic spacing relative to that below. Figure 5.11 shows how this happens. At the top is a perfect crystal. In the central row a shear stress τ is acting parallel to the slip planes, and this drives an edge dislocation in from the left, it sweeps through the crystal, and exits on the right. Note that no individual atom

[3] Geoffrey (G. I.) Taylor (1886–1975), known for his many fundamental contributions to aerodynamics, hydrodynamics, and the structure and plasticity of metals; it was he, with Egon Orowan, who realised that the ductility of metals implied the presence of dislocations. One of the greatest of contributors to theoretical mechanics and hydrodynamics of the 20th century, he was also a supremely practical man – a sailor himself, he invented (among other things) the anchor used by the Royal Navy.

[4] Egon Orowan (1901–1989), Hungarian/US physicist and metallurgist, who, with G. I. Taylor, realised that the plasticity of crystals could be understood as the motion of dislocations. In his later years he sought to apply these ideas to the movement of fault lines during earthquakes.

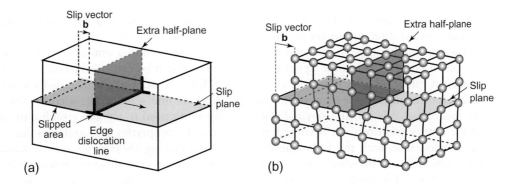

Figure 5.10   (a) Visualising a dislocation by cutting, slipping, and rejoining bonds across a slip plane. The slip vector **b** is perpendicular to the edge dislocation line ⊥–⊥. (b) The atom configuration around an edge dislocation in a simple cubic crystal.

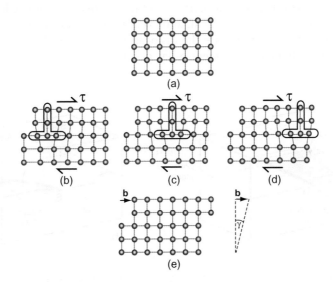

Figure 5.11   An initially perfect crystal is shown at (a). A shear stress τ drives the passage of an edge dislocation across the slip plane, shown in the sequence (b, c, d), shearing the upper part of the crystal over the lower part by the slip vector **b**, leading at (e) to a shear strain γ.

moves very far as the dislocation moves – the bonds around the dislocation are stretched to and fro to allow the pattern of misfitting atoms to move along the slip plane. By the end of the process the upper part has slipped relative to the lower part by a slip step **b**, known as the Burger's vector, equal (or close) to one atomic spacing. The result is the shear strain γ shown at the bottom. Note that in this configuration the shear stress is perpendicular to the edge dislocation, which moves in the direction of the stress, and the resulting slip step **b** is also in the direction of the shear stress.

There is another type of dislocation that may be visualised in a similar way. After making the cut in Figure 5.10(a), the upper part of the crystal can be displaced *parallel* to the edge of the cut rather than *normal* to it, as in Figure 5.12(a). That, too, creates a dislocation, with a different configuration of misfitting atoms along its line – more like a corkscrew – and for this reason it is called a *screw dislocation*. We don't need to try and visualise the details of the atomic positions – it is enough to know that its properties are similar to those of an edge dislocation, with the following distinction: the driving shear stress is now *parallel* to the screw dislocation, which sweeps sideways through the crystal, again moving normal to itself, but with the resulting slip step **b** remaining in the direction of the applied shear stress. In fact, dislocations also form curves as they move over a slip plane, in which case they are called *mixed dislocations*: part edge and part screw. The 'pure' edge and screw configurations are then found at two positions on the dislocation at 90° to one another [Figure 5.12(b)]. Once we've grasped this, it is easiest to forget about where the atoms are and just to think of dislocations as flexible line defects gliding over the slip planes under the action of an applied shear stress. It does not matter whether the dislocation is edge, screw, or mixed: it always moves perpendicular to itself, and every part of it produces the same slip vector **b** in the direction of the shear stress – the dislocation line being the boundary on the plane up to which point a fixed displacement **b** has occurred.

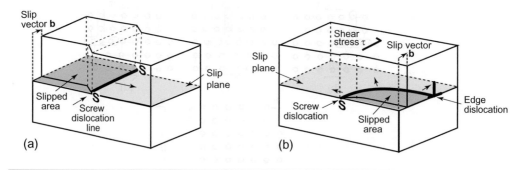

Figure 5.12   (a) A screw dislocation, with the slip vector **b** parallel to the dislocation line S – S; (b) a curved, mixed dislocation, connecting a pure edge to a pure screw dislocation.

The dislocation is key to understanding why metals have a strength that is so far below their ideal strength: it is far easier to move a dislocation through a crystal, stretching and remaking bonds only locally along its line as it moves, than it is to simultaneously break *all* the bonds in the plane before remaking them. It is like moving a heavy carpet by shuffling a fold across it rather than sliding the whole thing at one go. In real crystals it is easier to make and move dislocations on some planes than on others. The preferred planes are called *slip planes* and the preferred directions of slip in these planes are called *slip directions*. We have seen these previously in the CPH, FCC, and BCC unit cells of Figure 3.21 – dislocations prefer planes and directions that are close-packed, or nearly so.

Single slip steps are tiny – one dislocation produces a displacement of about 10^{-10} m. But if large numbers of dislocations traverse a crystal, moving on many different planes, the shape of a material changes at the macroscopic length-scale. Figure 5.13 shows just two dislocations traversing a sample loaded in tension, and then another two on a different pair of slip planes. The slip steps (here very exaggerated) cause the sample to get a bit thinner and longer. Repeating this millions of times on many planes gives the large plastic extensions observed in practice. Since none of this changes the average atomic spacing, this explains why the volume remains unchanged. So this is how materials deform plastically at stresses well below the ideal strength – but note the crucial advantages: the material remains fully intact without any loss of strength as its shape changes significantly. This ductility is essential for safe design – we can live with a bit of unintended strain, but not with everything being brittle – and it also enables us to shape components from lumps of metal in the solid state.

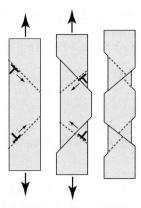

Figure 5.13 Dislocation motion leads to strain, but conserves volume.

Why does a shear stress make a dislocation move?   A shear stress is needed to move dislocations, because crystals resist their motion with a friction-like resistance f per unit length. For yielding to take place, the shear stress must overcome the resistance f. Imagine that one dislocation moves right across a slip plane, as in Figure 5.14. In travelling the distance L_2, it shifts the upper half of the crystal by a distance b (the magnitude of the Burger's vector) relative to the lower half. The shear stress τ acts on an area $L_1 L_2$, giving a shear force $F_s = \tau L_1 L_2$ on the surface of the block. As the displacement of the block in the direction of the force is b, the force does work

$$W = \tau L_1 L_2 b \tag{5.13}$$

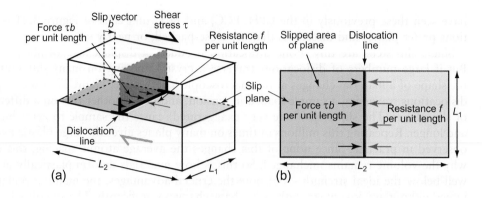

Figure 5.14 The force on a dislocation. (a) Perspective view; (b) plan view of slip plane.

This work is done against the resistance force f per unit length, or $f L_1$ on the length L_1, and it does so over a displacement L_2 (because the dislocation line moves this far against f), giving a total work against f of $f L_1 L_2$. Equating this to the work W done by the applied stress τ gives

$$\tau b = f \tag{5.14}$$

This result holds for any dislocation – edge, screw, or mixed. So, provided the shear stress τ exceeds the value f/b it will make dislocations move and cause the crystal to shear.

Line tension The atoms near the 'core' of a dislocation are displaced from their proper positions, as shown by green atoms back in Figure 5.9(c), and thus they have higher potential energy. To keep the potential energy of the crystal as low as possible, the dislocation tries to be as short as possible – it behaves as if it has a *line tension*, T, like an elastic band. The tension can be calculated by atom-scale elasticity theory (see Further Reading for the analysis), giving the simple answer that the line tension T, which is strictly an energy per unit length (just as a surface tension is an energy per unit area), is

$$T = \frac{1}{2} G b^2 \tag{5.15}$$

where G is the shear modulus (since dislocations are associated with local lattice straining in shear), and b is the magnitude of the Burger's vector. Recall that for metals G scales with Young's modulus E, with $G \approx 3/8\ E$ (Chapter 4). The line tension has an important bearing on the way in which dislocations interact with obstacles, as we shall see in a moment. At first it is perhaps surprising that we explain *plastic* deformation via dislocation motion, which turns out to depend on an *elastic* property – the shear modulus, G. But it derives from the fact that dislocations strain the lattice elastically by small (elastic) changes in the atomic spacing. This local elastic distortion enables the dislocations to move large distances, leading to large plastic strains.

The lattice resistance   Where does the resistance to slip, f, come from? There are several contributions. Consider first the *lattice resistance*, f_i: the intrinsic resistance of the crystal structure to plastic shear. Pure metals are soft because the non-localised metallic bond does little to obstruct dislocation motion, whereas ceramics are hard because their more localised covalent and ionic bonds (which must be stretched and reformed when the lattice is sheared) lock the dislocations in place. When the lattice resistance is high, as in ceramics, further hardening is superfluous – the problem becomes that of suppressing fracture. On the other hand, when the lattice resistance f_i is low, as in metals, the material needs to be strengthened by introducing obstacles to slip. This is done by adding alloying elements to give *solid solution hardening*, and using precipitates or dispersed particles to give *precipitation hardening*, while other dislocations give what is called *work hardening*, and polycrystals also have *grain boundary hardening*. These techniques for manipulating strength are expanded next, and are central to alloy design.

5.5 Manipulating strength

Strengthening metals: dislocation pinning   The way to make crystalline materials stronger is to make it harder for dislocations to move. In a pure crystal they move when the force τb (per unit length) exceeds the lattice resistance f_i. There is little we can do to change this – it is an intrinsic property like the modulus E. So obstacles to dislocation motion must be added: Figure 5.15 shows views of a slip plane containing obstacles from the perspective of

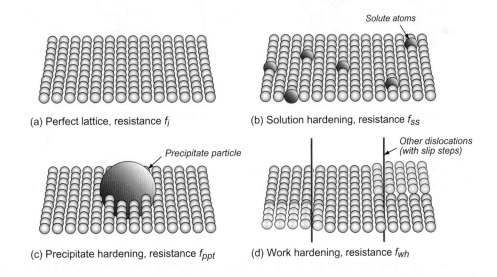

(a) Perfect lattice, resistance f_i

(b) Solution hardening, resistance f_{ss}

Solute atoms

Precipitate particle

(c) Precipitate hardening, resistance f_{ppt}

(d) Work hardening, resistance f_{wh}

Other dislocations (with slip steps)

Figure 5.15   A 'dislocation-eye' view of the slip plane across which it must move.

an advancing dislocation. In the perfect lattice shown in (a) the only resistance is the intrinsic strength of the crystal; solution hardening, shown in (b), introduces atom-size obstacles to motion; precipitation hardening, shown in (c), presents larger obstacles; and in work hardening, shown in (d), the slip plane becomes stepped where it is intersected by dislocations in other orientations.

Figure 5.16(a) shows what happens when dislocations encounter obstacles, which act as *pinning points*. From Equation (5.14), the force on the dislocation between two pinning points of spacing L is $F = \tau bL$. From the figure, there is one obstacle for each length L of dislocation, so the force on each obstacle is also $F = \tau bL$. To keep the dislocation moving, an increase in the shear stress is needed to 'bow out' the dislocation through the gap. The *bowing angle* θ and applied shear stress τ increase until the force F can overcome the *pinning force* p that the obstacle can apply to the dislocation. Figure 5.16(b,c) show two ways in which dislocations escape from pinning points: 'weak' obstacles are passed when the bowing angle θ lies between $0°$ and $90°$; for 'strong' obstacles, the bowing angle reaches the limiting value $\theta = 90°$, and the dislocation escapes by other means – the arms of the dislocations link up and the dislocation advances by 'bypassing' the obstacles, leaving a dislocation loop around the obstacle [Figure 5.16(c)]. A helpful way to think about dislocation pinning is to recall that dislocations have a 'line tension' trying to make the dislocation as short as possible. As the dislocation bows out, the line tension 'pulls' against the obstacle until it gives way, or until the maximum pull is exerted (when $\theta = 90°$), and it bypasses it instead.

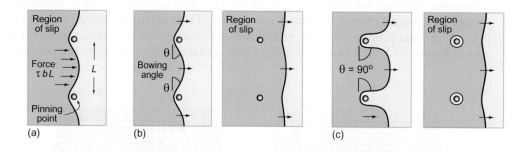

Figure 5.16 Dislocation pinning: (a) force on dislocation between pinning points; (b, c) dislocation bowing and bypassing from weak and strong obstacles, respectively.

In both weak and strong cases in Figure 5.16, the dislocation escapes when $F = p$. Hence the additional contribution to the shear stress τ needed to overcome the pinning points is

$$\Delta\tau = \frac{p}{bL} \tag{5.16}$$

Dislocation pinning is an elastic effect, locally distorting the atoms around the dislocation. As a result, the pinning force p, for each type of obstacle in a given material, scales with the dislocation energy per unit length (or line tension), $Gb^2/2$ [Equation (5.15)]. Combining this with Equation (5.16), the shear stress τ needed to force the dislocation through a field of obstacles therefore has the form

$$\tau = \alpha \frac{Gb}{L} \tag{5.17}$$

where α is a dimensionless constant characterising the obstacle strength. Take a close look at this equation: G and b are material constants, so in order to manipulate the strength, the parameters that we can control are: α (the obstacle strength) and L (the obstacle spacing). Armed with Equation (5.17), we can now explain strengthening mechanisms.

Solid solution hardening   Solid solution hardening is strengthening by alloying with other elements that disperse atomically into the host lattice [Figure 5.17(a)]. They are added when the alloy is molten, and are then trapped in the solid lattice as it solidifies. Adding zinc to copper makes the alloy *brass* – copper dissolves up to 30% zinc. The zinc atoms replace copper atoms to form a random *substitutional solid solution*. The zinc atoms are bigger than those of copper and, in squeezing into the copper lattice, they distort it. This 'roughens' the slip plane, so to speak, making it harder for dislocations to move – they are pinned by the solute atoms and bow out, thereby adding an additional resistance f_{ss} to dislocation motion.

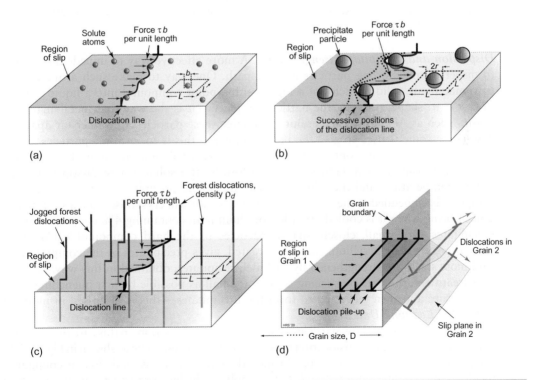

Figure 5.17   (a) Solid solution hardening. (b) Precipitation or dispersion hardening. (c) Forest hardening (work hardening). (d) Grain boundary hardening.

The figure illustrates that the concentration of solute, c, expressed as an atom fraction within a given plane, is on average:

$$c = \frac{b^2}{l^2}$$

where L is the spacing of obstacles in the slip plane and b is the atom size. Thus,

$$L = \frac{b}{c^{1/2}}$$

Plugging this into Equation (5.17) relates the concentration of the solid solution to the additional shear stress τ_{ss} required to move the dislocation:

$$\tau_{ss} = \alpha\, G c^{1/2} \tag{5.18}$$

Brass, bronze, stainless steels, and many other metallic alloys derive their strength in this way. They differ only in the extent to which the solute distorts the host crystal, described by the constant α. The bowing angle θ for solute is small – much less than 90° – so solute provides weak pinning.

Dispersion and precipitate strengthening   A more effective way to impede dislocations is to disperse small, strong particles in their path. One way to make such a microstructure is to mix small, solid particles of a high melting point compound into a liquid metal, and to cast it to shape, trapping the particles in place – it is the way that metal–matrix composites such as Al–SiC are made. But in practice these particles are rather far apart, so an alternative route is used to give the best outcomes – this is to form the particles *in situ* by a *precipitation* process, controlled by *heat treatment*. Solute atoms often form solid solutions more easily at high temperature than low. So if a solute is first dissolved in a metal at high temperature, and the alloy is cooled, the solute precipitates out as tiny crystals of compounds embedded in the host lattice – much as salt will crystallise from a saturated solution of brine when it is cooled. An alloy of aluminium containing 4% copper, for instance, precipitates very small, closely spaced particles of the hard compound $CuAl_2$. Copper alloyed with a little beryllium, similarly treated, gives precipitates of the compound CuBe. And most steels are strengthened by carbide precipitates, obtained in this way. In practical heat treatment, the particle spacing is kept as small as possible by first quenching to room temperature and then reheating to form the precipitates – this is discussed in more depth in *Guided Learning Unit 4*.

Figure 5.17(b) shows how particles obstruct dislocation motion. In this case the particles are 'strong' obstacles, and the dislocation bows out to $\theta = 90°$, escaping as shown in Figure 5.16(c), leaving a loop of dislocation wrapped round the particle. In this semi-circular configuration, the dislocation of line tension $T = \frac{1}{2} Gb^2$ is applying the biggest force it can on each obstacle, equal to $2T$, so this is the maximum pinning force p. From Equations (5.15) and (5.16), the strength contribution from precipitates is therefore

$$\tau_{ppt} = \frac{2T}{bL} = \frac{Gb}{L} \tag{5.19}$$

Comparison with Equation (5.17) shows that the maximum value of the constant α is therefore 1. Precipitation hardening is the most effective way to increase strength: precipitation-hardened aluminium alloys can be 15 times stronger than pure aluminium.

Work hardening   The rising part of the stress–strain curve of Figure 5.1 is known as *work hardening*: the plastic deformation causes dislocations to accumulate. The *dislocation density*, ρ_d, is defined as the length of dislocation line per unit volume (m/m^3). Even in an annealed soft metal, the dislocation density is around 10^{10} m/m^3, meaning that a 1 cm cube (the size of a cube of sugar) contains about 10 km of dislocation line. When metals are deformed, dislocations multiply, causing their density to grow to as much as 10^{17} m/m^3 or more – 100 million km/cm^3! The reason that the yield stress rises is because a moving dislocation now finds that its slip plane is penetrated by a 'forest' of intersecting dislocations with a steadily decreasing average spacing $L = \rho_d^{-1/2}$ (since ρ_d is a number per unit area). Figure 5.17(c) shows the idea schematically – a gliding dislocation intersecting an array of forest dislocations in other orientations. As a moving dislocation advances, it bows out between pinning points at dislocation intersections – a relatively weak obstacle ($\alpha < 1$). As the pinning point gives way, material above the slip plane slips relative to that below, and creates a little step called a *jog* in each forest dislocation. These jogs have potential energy – they are tiny segments of dislocation of length b – and this is the source of the pinning force p, with the result that p again scales with $Gb^2/2$. Assembling these results into Equation (5.17) gives the work hardening contribution to the shear stress:

$$\tau_{wh} = \alpha \frac{Gb}{L} = \alpha Gb\sqrt{\rho_d} \tag{5.20}$$

The greater the density of dislocations, the smaller the spacing between them, and so the greater their contribution to τ_{wh}.

Virtually all metals work harden. It can be a nuisance: if you want to roll thin sheet, work hardening quickly raises the yield strength so much that you have to stop and *anneal* the metal (i.e. heat it up to remove the accumulated dislocations) before you can go on – a trick known to blacksmiths for centuries. But it is also useful: it is a potent strengthening method in manufacturing, utilising processes such as rolling and forging to simultaneously shape and harden alloys – particularly for those that cannot be heat-treated to give precipitation hardening.

Grain boundary hardening   Grain boundary hardening is a little different to the other mechanisms. Almost all metals are polycrystalline, made up of tiny, randomly oriented grains, meeting at grain boundaries like those of Figure 5.9(d). The grain size, D, is typically 10 to 100 μm. These boundaries obstruct dislocation motion, as shown in Figure 5.17(d). Dislocations in Grain 1 can't just slide into the next one (Grain 2) because the slip planes don't line up. Instead, new dislocations have to nucleate in Grain 2 with combined slip vectors that match the displacement caused by dislocations in Grain 1. The mechanism is as follows: dislocations 'pile up' at the boundary, creating the stress needed in the next grain to nucleate the matching dislocations. The number of dislocations in a pile-up scales with the size of the grain; and the bigger the pile-up, the lower the extra shear stress needed to enable slip to continue into the next grain. This gives another contribution to the shear stress, τ_{gb}, that scales as $D^{-1/2}$, giving

$$\tau_{gb} = \frac{k_p}{\sqrt{D}} \tag{5.21}$$

where k_p is called the Hall-Petch constant, after the scientists who first measured it. For normal grain sizes τ_{gb} is small and not a significant source of strength, but for materials that are microcrystalline ($D < 1$ μm) or nanocrystalline (D approaching 1 nm) it becomes significant. Controlling the grain size is routine in conventional processing – in the initial casting of a liquid to make a solid shape, and then by deformation and heat treatments that lead to 'recrystallisation' in the solid – more on this in Chapter 11. More exotic processing is needed to produce really fine grain sizes.

Relationship between dislocation strength and yield strength   To a first approximation the strengthening mechanisms add up, giving a dislocation shear yield strength, τ_y, of

$$\tau_y = \tau_i + \tau_{ss} + \tau_{ppt} + \tau_{wh} + \tau_{gb} \tag{5.22}$$

Strong materials either have a high intrinsic strength, τ_i (like diamond), or they rely on the superposition of solid solution strengthening τ_{ss}, precipitation hardening τ_{ppt}, and work hardening τ_{wh} (like high-tensile steels). Nanocrystalline solids exploit, in addition, the contribution of τ_{gb}. But before we can use this information, one problem remains: we have calculated the shear yield strength of one crystal, loaded in shear *parallel to a slip plane*. What we want is the yield strength of a *polycrystalline material* in *tension*. To link them there are two simple steps.

First, when a remote shear stress is applied to an aggregate of crystals, some crystals will have their slip planes oriented favorably with respect to the shear stress, others will not. This randomness of orientation means that the shear stress applied to a polycrystal that is needed to make *all* the crystals yield is higher than τ_y, by a factor of about 1.5 (called the Taylor factor, after the same G. I. Taylor who postulated the dislocation mechanism, with Orowan). This remote shear stress is referred to as the *yield strength in shear, k*.

Second, yield strength is measured in uniaxial tension, so how does this axial stress produce the shear stress k that leads to yield of the polycrystal? The answer is that a uniform tensile stress σ creates shear stress on all planes that lie at an angle to the tensile axis. Figure 5.18 shows a tensile force F acting on a rod of cross-section A. An inclined plane is shown with the normal to the plane at an angle θ to the loading direction. It carries components of force $F\sin\theta$ and $F\cos\theta$, parallel and perpendicular to the plane. The area of this plane is $A/\cos\theta$, so the shear stress parallel to the plane is

$$\tau = \frac{F\sin\theta}{A/\cos\theta} = \sigma\sin\theta\,\cos\theta$$

where $\sigma = F/A$ is the axial tensile stress. The variation of τ is plotted against σ in the figure – the maximum shear stress is found on planes at an angle of $45°$, when $\tau = \sigma/2$. The yield strength in shear k is therefore equal to $\sigma_y/2$, where σ_y is the tensile yield stress.

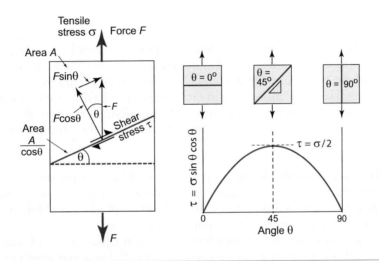

Figure 5.18 The variation of stress with angle of the reference axes. A tensile stress σ gives a maximum shear stress $\tau = \sigma/2$ on a plane at $45°$ to the tensile axis.

Combining these two factors, the tensile stress to cause yielding of a polycrystalline sample is approximately three times the shear strength of a single crystal:

$$\sigma_y = 2k \approx 3\tau_y \tag{5.23}$$

As a result, the superposition of strengthening mechanisms in Equation (5.22) applies equally to the yield strength, σ_y.

Example 5.4

A polycrystalline aluminium alloy contains a dispersion of hard particles of diameter 10^{-8} m and an average centre-to-centre spacing of 6×10^{-8} m, measured in the slip planes. The shear modulus of aluminium is $G = 26$ GPa and $b = 0.286$ nm. Estimate the contribution of the hard particles to the yield strength of the alloy.

Answer. For the alloy, $L = (6 - 1) \times 10^{-8}$ m $= 5 \times 10^{-8}$ m, giving

$$\tau_{ppt} = \frac{Gb}{L} = \frac{27 \times 10^9 \times 0.286 \times 10^{-9}}{5 \times 10^{-8}} = 154 \text{ MPa}$$

Since $\sigma_y = 3\tau_y$ for polycrystals [Equation (5.23)], the contribution to the yield stress $\sigma_y = 462$ MPa.

Example 5.5

Two cubes of aluminium of side length 1 cm are produced in annealed (soft) and cold worked (hard) conditions. The dislocation densities in the two samples are 10^5 mm^{-2} and 10^9 mm^{-2}, respectively. Assuming a square array of parallel dislocations, estimate: (a) the total length of dislocation in each sample; (b) the distance between dislocations in the cold worked sample; (c) the ratio of the contribution made by dislocation hardening to the strength of the cold worked sample, to that of the annealed sample.

Answer.
(a) The dislocation density ρ_d is the length of dislocation per unit volume. For a 1 cm cube, the volume = 1000 mm^3. For a dislocation density $\rho_d = 10^5$ mm^{-2} (mm/mm^3), the total length of dislocation $= 10^5 \times 1000 = 10^8$ mm = 100 km; for $\rho_d = 10^9$ mm^{-2} (mm/mm^3), the length is 10^{12} mm = 1 million km!
(b) Assume that the dislocations form a uniform square array of side d. In this case, each dislocation sits at the centre of an area d^2. For a unit length of dislocation, the volume of material per dislocation is $1 \times d^2$, so the dislocation density is $\rho_d = 1/d^2$, hence for the cold worked sample

$$d = \frac{1}{\sqrt{\rho_d}} = \frac{1}{\sqrt{10^9}} = 32 \text{ nm}$$

(c) From Equation (5.20), $\tau_{wh} \propto \sqrt{\rho_d}$
Hence the ratio of the strength contributions $= \sqrt{10^9}/\sqrt{10^5} = 100$.

Strength and ductility of alloys Of all the properties that materials scientists and engineers have sought to manipulate, the strength of metals and alloys is probably the most explored. It is easy to see why: Table 5.1 gives a small selection of the applications of metals and their alloys – their importance in engineering design is clearly enormous. The hardening mechanisms are often used in combination. This is illustrated graphically for copper alloys in Figure 5.19. Good things, however, have to be paid for. Here the payment for increased strength is, almost always, loss of ductility, so the elongation ε_f is reduced. The material is stronger but it cannot be deformed as much without fracture. This general trend is evident

Table 5.1 Metal alloys with typical applications, indicating the strengthening mechanisms used

Alloy	Typical uses	Solution hardening	Precipitation hardening	Work hardening
Pure Al	Kitchen foil			✓✓✓
Pure Cu	Wire			✓✓✓
Cast Al, Mg	Automotive parts	✓✓✓	✓	
Bronze (Cu–Sn), Brass (Cu–Zn)	Marine components	✓✓✓	✓	✓✓
Non-heat-treatable wrought Al	Ships, cans, structures	✓✓✓	✓✓✓	
Heat-treatable wrought Al	Aircraft, structures	✓	✓✓✓	✓
Low carbon steels	Car bodies, structures, ships, cans	✓✓✓	✓✓✓	
Low alloy steels	Automotive parts, tools	✓	✓✓✓	✓
Stainless steels	Pressure vessels	✓✓✓	✓	✓✓✓
Cast Ni alloys	Jet engine turbines	✓✓✓	✓✓✓	

Symbols: ✓✓✓ = Routinely used. ✓ = Sometimes used.

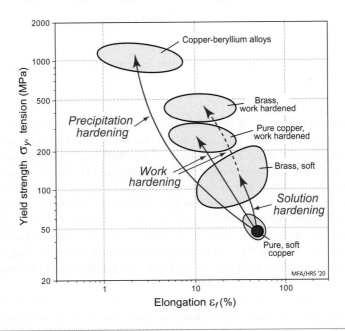

Figure 5.19 Strengthening and ductility (elongation to fracture) for copper alloys. The hardening mechanisms are frequently combined, but in general the greater the strength, the lower the ductility.

in Figure 5.20, which shows the nominal stress–strain curves for a selection of engineering alloys. Chapter 11 and *Guided Learning Unit 4* will explore the practicalities of manipulating the strength and ductility of alloys in the context of processing.

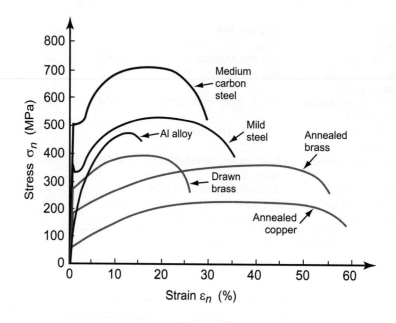

Figure 5.20 Nominal stress–strain curves for a selection of engineering alloys. Stronger alloys tend to have lower ductility.

Strengthening polymers In non-crystalline solids the dislocation is not a helpful concept. We think instead of some unit step of the flow process: the relative slippage of two segments of a polymer chain or the shear of a small molecular cluster in a glass network. Their strength has the same origin as that underlying the lattice resistance: if the unit step involves breaking strong bonds (as in an inorganic glass), the materials will be strong and brittle, as ceramics are. If it only involves the rupture of weak bonds (the van der Waals bonds in polymers, for example), it will have low strength. Polymers, too, must therefore be strengthened by impeding the slippage of segments of their molecular chains. This is achieved by *blending*, by *drawing*, by *cross-linking*, and by *reinforcement* – again, we manipulate the microstructure-sensitive property of strength by a combination of composition and processing (see Chapter 11 for more details).

A blend is a mixture of two polymers, stirred together in a sort of industrial food-mixer. The strength and modulus of a blend are just the average of those of the components, weighted by volume fraction (a rule of mixtures again).

Drawing is the deliberate use of the molecule-aligning effect of stretching, like that sketched in Figure 5.8(a), to greatly increase stiffness and strength in the direction of stretch. Fishing line is drawn nylon, Mylar film is a polyester with molecules aligned in the plane of the film. Fabrics are woven or felted sheet materials made from fibres – an industrial example is geotextiles made from drawn polypropylene and used to constrain earthworks.

Cross-linking (sketched in Figures 3.27 and 3.28) creates strong bonds between molecules that were previously linked by weak van der Waals forces. Vulcanised rubber is rubber that has been cross-linked, and the superior strength of epoxies derives from cross-linking.

Reinforcement is possible with particles of cheap fillers – sand, talc, or wood dust. Far more effective is reinforcement with fibres – usually glass or carbon – either continuous or chopped, as explained in Chapter 4.

5.6 Introductory solid mechanics – plasticity and strength

In Chapter 4, analytical modelling in design was introduced via the solid mechanics of elastic stiffness. Recall that solutions are available for standard geometries and modes of loading – tensile ties, bending beams, and so on. In the previous chapter, the key results were related to displacement and stiffness of components in the elastic regime. Now we extend those solutions to the analysis of the stress distribution, with a view to predicting the failure loading. So the scenarios are mostly those of *strength-limited design*, governed by avoiding the onset of plasticity at the yield stress – note therefore that this is still elastic design. Design for full plasticity is less common but also important – for vehicle crashworthiness and for many manufacturing forming processes. Finally we introduce the key idea of a *stress concentration* – the local magnification of stress associated with a change in cross-section or hole – which has proved to be the origin of countless failures, particularly in combination with cyclic loading (or fatigue – developed in Chapter 6).

Yielding of ties and columns   Recall that a tie is a rod loaded in tension, a column is a rod loaded in compression [see Figure 4.1(a,b)]. The state of stress within them is uniform, as shown in Figure 4.15(a). If this stress, σ, is below the yield strength σ_y the component remains elastic; if it exceeds σ_y, it yields. Yield in compression is only an issue for short, squat columns. Slender columns and panels in compression are more likely to buckle elastically first (Chapter 4).

Yielding of beams and panels   The stress state in bending was also introduced in Chapter 4, in relation to the strain that leads to curvature. A bending moment M generates a linear variation of longitudinal stress σ across the section [Figure 5.21(a)] defined by

$$\frac{\sigma}{y} = \frac{M}{I} = E\kappa \tag{5.24}$$

where y is the distance from the neutral axis, κ is the beam curvature, and the influence of the cross-section shape is captured by I, the second moment of area. For elastic deflection, we were interested in the last term in the equation: that containing the curvature. For yielding, it is the first term. The maximum longitudinal stress σ_{max} occurs at the surface [Figure 5.21(a)], at the greatest distance y_m from the neutral axis

$$\sigma_{max} = \frac{M\,y_m}{I} = \frac{M}{Z_e} \tag{5.25}$$

The quantity $Z_e = I/y_m$ is a function of the cross-sectional shape called the *elastic section modulus* (not to be confused with the elastic modulus of the material, E). If the applied moment is increased above ($Z_e \, \sigma_y$), the linear profile in stress is truncated at σ_y and plasticity spreads inwards from the surface, as in Figure 5.21(b). If unloaded, the beam has some permanent curvature, so it is damaged but has not failed completely. It is important to recognise that although the plastic zone has yielded, it still carries the full yield load – yielding is not a loss of strength, and here the strain is limited because the interior remains elastic. But at a higher moment the plastic zones penetrate through the section of the beam and link up to form a *plastic hinge* [Figure 5.21(c)]. This is the maximum collapse moment of the beam.

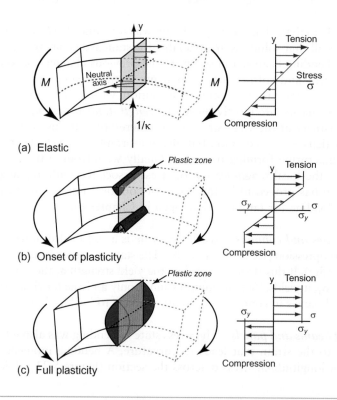

Figure 5.21 A beam loaded in bending. The stress state is shown on the right for: (a) purely elastic loading, (b) the onset of plasticity, and (c) full plasticity.

Figure 5.22 shows three simple configurations of beams loaded in bending, with the magnitude of the maximum moment and its position indicated in red. This is where the yield stress is first exceeded at the moment given by Equation (5.25), and also, when the moment reaches the collapse value, where plastic hinges will form. This collapse moment, M_f, is found

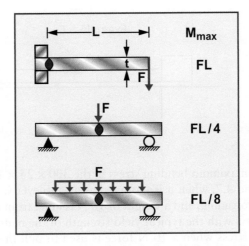

Figure 5.22 Standard configurations for beam bending, showing the maximum moment in terms of the load and span, and its location (in red), where first yield and plastic collapse occur.

by integrating the moment caused by the constant stress distribution over the section [as in Figure 5.21(c)], with compression on one side, tension on the other:

$$M_f = \int_{section} b(y)|y|\, \sigma_y\, dy = Z_p\, \sigma_y \qquad (5.26)$$

where Z_p is the *plastic section modulus*, a second function of the section shape. For first yielding and full plasticity, the moment required is simply $Z\,\sigma_y$, using Z_e or Z_p, respectively. Example expressions for both are shown in Figure 5.23. The ratio Z_p/Z_e is always greater than 1 and is a measure of the safety margin between initial yield and collapse. For efficient shapes, like tubes and I-beams, the ratio is quite close to 1 – yield spreads quickly from the surface to the neutral axis. Rectangular sections have a greater safety margin (see Example 5.7).

Section shape	Area A (m²)	Elastic section modulus Z_e (m³)	Plastic section modulus Z_p (m³)
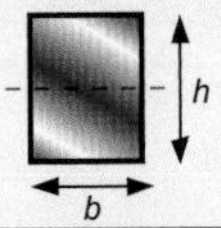 rectangle, $b \times h$	$b\,h$	$\dfrac{bh^2}{6}$	$\dfrac{bh^2}{4}$
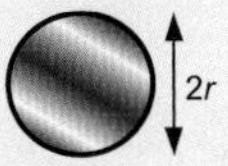 circle, $2r$	πr^2	$\dfrac{\pi}{4} r^3$	$\dfrac{\pi}{3} r^3$
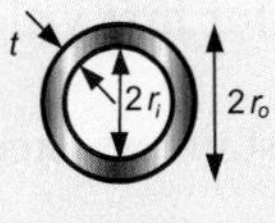 tube, $2r_i$, $2r_o$, t	$\pi(r_o^2 - r_i^2)$ $\approx 2\pi rt$ $(t \ll r)$	$\dfrac{\pi}{4r_o}(r_o^4 - r_i^4)$ $\approx \pi r^2 t$ $(t \ll r)$	$\dfrac{\pi}{3}(r_o^3 - r_i^3)$ $\approx \pi r^2 t$ $(t \ll r)$

Figure 5.23 The area A, elastic section modulus Z_e, and fully plastic section modulus Z_p for three simple shapes.

Example 5.6

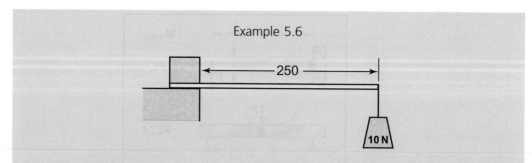

Calculate the maximum bending stress in the $300 \times 25 \times 1$ mm stainless steel ruler in Examples 4.5 and 4.7, when it is mounted as a cantilever with the X-X axis horizontal with 250 mm protruding, and a weight of 10 N is hung from its end, as shown. Compare this bending stress with the typical yield strength of the material from Appendix A, and with the tensile stress when a 10 N force is used to pull the ruler axially.

Answer. The maximum bending moment M in the cantilever (at its base – adjacent to the support) is $M = WL = 0.25 \times 10 = 2.5$ Nm. Using Equation (5.25) with $I_{XX} = 2.1$ mm^4 (from Example 4.5), and $y_m = 0.5$ mm at the extreme fibre of the beam:

$$\sigma = \frac{M y_m}{I_{XX}} = \frac{2.5 \times 0.0005}{2.1 \times 10^{-12}} = 595 \, \text{MPa}$$

From Appendix A, the yield strength of stainless steel is 540 MPa. So the ruler would not support this load without yielding.

When loaded by the same force in tension, the stress would be

$$\sigma = \frac{F}{wt} = \frac{10}{25 \times 1} = 0.4 \, \text{MPa}$$

This stress is *far* less than the bending stress for the same load, and well below the strength of the material.

Example 5.7

For a rectangular cross section loaded in bending, find the ratio of the plastic and elastic section moduli. Hence find the ratio of the moments for first yield and full collapse for this shape (i.e. the safety margin).

What load would cause plastic collapse of the stainless steel ruler, loaded as in Example 5.6?

Answer. From Figure 5.26, the ratio of the plastic and elastic section moduli for a rectangular section is:

$$\frac{Z_p}{Z_e} = \frac{bh^2/4}{bh^2/6} = 1.5$$

Since $M_f = Z_p \sigma_y$, this is also the ratio of the failure moments. It takes 50% more load to collapse a rectangular section than to cause the onset of yielding: a substantial safety margin.

For the ruler, $Z_p = bh^2/4 = 25 \times 1^2/4 = 6.25\,\text{mm}^3$.

Hence the collapse moment $M_f = Z_p \sigma_y = 6.25 \times 10^{-9} \times 540 \times 10^6 = 3.375\,\text{Nm}$, and the collapse load is $W = M_f/L = 3.375/0.25 = 13.5\,\text{N}$.

Yielding of springs Helical springs are a special case of elastic loading, in which the coiled wire is loaded in *torsion* (twisting) (Figure 5.24): when the spring is loaded axially, the individual turns twist. For a spring with n turns of wire of diameter d and shear modulus G, wound to give a spring of radius R, the spring stiffness S is

$$S = \frac{F}{u} = \frac{Gd^4}{64nR^3} \tag{5.27}$$

where F is the axial force applied to the spring and u is its extension. Springs may fail by yielding, fracture, or (in compression) by buckling. The elastic limit to extension due to plastic yielding occurs at the force

$$F_{crit} = \frac{\pi}{32}\frac{d^3 \sigma_y}{R} \tag{5.28}$$

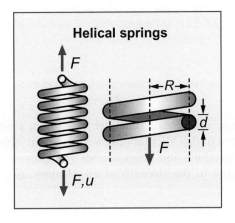

Helical springs

Figure 5.24   Helical springs are a special case of elastic loading.

Stresses in pressure vessels   Many products are hollow structures designed to contain internal pressure (or withstand external pressure) – from aircraft fuselages and submarines, to pressure vessels for liquified gas, to cans and bottles for fizzy drinks, and party balloons. Most are cylindrical or spherical in shape, with a wall that is thin compared to their radius and length. In this idealised geometry, there are simple solutions for the magnitude of the stress in the wall of the structure in terms of the pressure difference across the wall. The stress state is *biaxial*, with stresses acting in two perpendicular directions in the wall; the through-thickness stress turns out to be negligibly small (being equal to the applied pressure or less).

To analyse the stresses, we consider the equilibrium of the vessel cut into two, balancing the forces due to the pressure and the stresses in the wall. Figure 5.25 shows a cylindrical pressure vessel of radius R and wall thickness t, with an internal pressure p relative to the outside. Note that this thin-walled analysis assumes $t \ll R$ (typically $R/t = 10$ or higher). Two sections are shown: a cut straight across the cylinder, to find the longitudinal stress σ_l; and a longitudinal cut along an arbitrary length L, to find the circumferential or 'hoop' stress, σ_h. The analysis is reserved for the Exercises; the results are:

$$\sigma_l = \frac{pR}{2t} \quad \text{and} \quad \sigma_h = \frac{pR}{t} \tag{5.29}$$

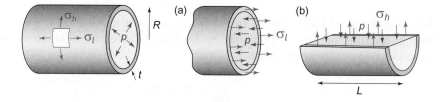

Figure 5.25   Finding the biaxial stresses σ_l and σ_h in a thin-walled cylindrical pressure vessel: (a) a transverse section; (b) a longitudinal section, of length L.

Example 5.8

A small can of fizzy drink has a diameter of 50 mm, a wall thickness of 0.1 mm, and an internal pressure of 0.2 MPa (about two times atmospheric pressure). Find the magnitude of the longitudinal and hoop stresses and comment on their magnitude relative to the pressure, and to the strength of an Al alloy used for cans (about 300 MPa).

If a thin-walled cylinder is pressurised until the wall splits, in which direction will the failure run?

Answer. From Equation (5.29), the stresses are:

$$\text{Longitudinal stress, } \sigma_l = \frac{pR}{2t} = \frac{0.2 \times \frac{50}{2}}{2 \times 0.1} = 25\text{MPa};$$

$$\text{hoop stress, } \sigma_h = \frac{pR}{t} = \frac{0.2 \times \frac{50}{2}}{0.1} = 50\,\text{MPa}.$$

Both are much greater than the pressure causing the stresses, but well below the failure strength of the Al alloy.

The hoop stress is twice the longitudinal stress, so when the pressure is increased to failure, the cylinder will split longitudinally, perpendicular to the greater stress. (Meat-eaters may have observed that sausages always split lengthwise when cooked, never into two halves: this is why!)

Stress concentrations   Holes, slots, threads, and changes in section concentrate stress locally (Figure 5.26). Yielding will therefore start at these places, though as the bulk of the component is still elastic this initial yielding is not usually catastrophic. The same cannot be said for *fatigue* (cyclic loading – Chapter 6), where stress concentrations are often implicated as the origins of failure.

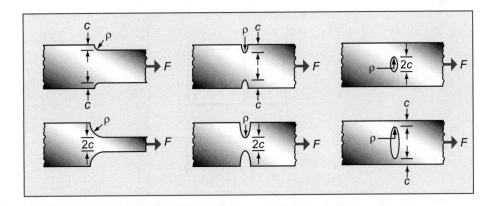

Figure 5.26   Stress concentrations. The change of section concentrates stress most strongly where the radius of curvature ρ of the feature is smallest.

First we use standard analysis to find the *nominal stress* in the component σ_{nom}, ignoring features that cause the stress concentration. Then the *maximum local stress* σ_{max} is found approximately by multiplying the nominal stress σ_{nom} by a *stress concentration factor Ksc*:

$$K_{sc} = \frac{\sigma_{max}}{\sigma_{nom}} = 1 + \alpha \left(\frac{c}{\rho_{sc}} \right)^{1/2} \tag{5.30}$$

Here ρ_{sc} is the minimum radius of curvature of the stress-concentrating feature and c is a characteristic dimension associated with it: either the half-thickness of the remaining ligament, the half-length of a contained notch, the length of an edge notch, or the height of a shoulder, whichever is least (Figure 5.26). The factor α is roughly 2 for tension, but is nearer 1/2 for torsion and bending. Though inexact, the equation is an adequate working approximation for many design problems. More accurate stress concentration factors are tabulated in design handbooks.

Example 5.9

The skin of an aircraft is made of aluminium alloy with a yield strength of 300 MPa and a thickness of 8 mm. The fuselage is a circular cylinder of radius 2 m, with an internal pressure that is $\Delta p = 0.5$ bar (5×10^{-2} MPa) greater than the outside pressure. A window in the side of the aircraft is nominally square (160 mm × 160 mm). It has corners with a radius of curvature of $\rho = 5$ mm.

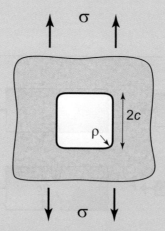

Calculate the nominal hoop stress σ_h in the aircraft wall due to pressure and estimate the peak stress near the corner of the window. Ignore the effects of longitudinal stress in the wall. To avoid the risk of *fatigue failure*, due to repeated pressurisation cycles (Chapter 6), the maximum allowable stress is one-third of the yield strength. Is 5 mm a suitable radius for the corners of the window?

Answer. The nominal hoop stress is:

$$\sigma_h = \frac{\Delta p R}{t} = \frac{5 \times 10^{-2} \times 2}{0.008} = 12.5 \, \text{MPa}$$

The stress concentration factor for the window geometry is found from the top RH standard case in Figure 5.26. Taking c as half the width of the window (80 mm) and $\rho_{sc} = 5$ mm, the stress concentration factor is:

$$K_{sc} \approx 1 + \alpha \left(\frac{c}{\rho_{sc}} \right)^{1/2} = 1 + 2 \left(\frac{80}{5} \right)^{1/2} = 9$$

Consequently the peak stress near the corner of the window is approximately $9 \times 12.5 = 112$ MPa, which is greater than the allowable value, risking fatigue failure after a small number of pressurisations. Such was the fate of the de Havilland Comet – the world's first pressurised commercial jet airliner, which first flew in 1949. Hence the large corner radii of aircraft windows today.

5.7 Strength in design and manufacturing

Strength-limited material selection   This chapter has extended the analysis of simple geometries and loading states to the problem of finding the maximum stress, which controls failure governed by material strength. 'Avoiding failure' is an alternative constraint to 'allowable elastic deflection' in the systematic choice of materials when designing of a component. The second part of *Guided Learning Unit 2: Material selection in design* extends the methodology using this functional constraint, for the same objectives of minimum mass, cost, or environmental impact, for standard loading cases and geometries with similar constraints as are applied in stiffness-limited design. Property charts including strength, as in Figures 5.4 and 5.5, are again central to implementing the method.

Full plasticity: metal forming   Large strain plasticity is unusual in design but is central to manufacturing. Solid-state metal forming is an ancient practice. Once metals could be extracted and cast, it did not take long to discover that they could be shaped by heating, beating, and bending them, and that their properties improved, too. Almost any metal part that is not cast directly to shape will undergo some plastic forming in the solid state, either hot or cold: standard sizes of plate, sheet, tube, and I-sections, as well as other prismatic shapes like railway track, are shaped by rolling or extrusion; three-dimensional shapes like tooling, gears, and cranks are forged; and rolled sheet is shaped further by stamping (car body panels) and deep drawing (beverage cans). Figure 5.27 shows schematics of some of these processes, characterised in most cases by compressive loading (to avoid early failure in tension) and very large plastic strains. We revisit metal-forming processes in Chapters 8, 11, and *Guided Learning Unit 3*, in relation to microstructure evolution during shaping and in the choice of process in design.

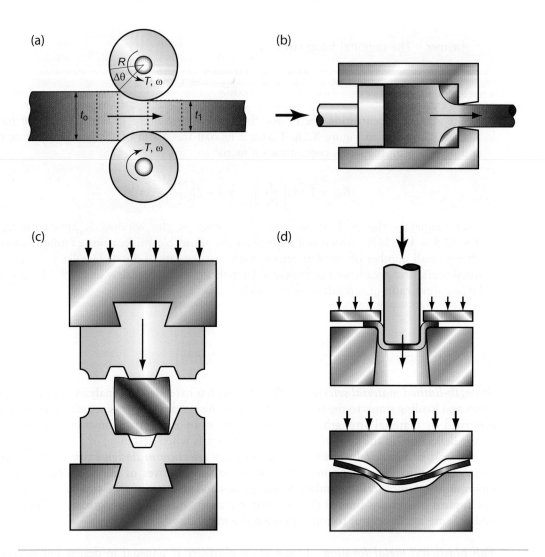

Figure 5.27 Metal forming processes: (a) rolling; (b) extrusion; (c) forging (closed die); (d) sheet forming (deep drawing, stamping).

5.8 Summary and conclusions

Load-bearing structures and components require materials with reliable strength, for which there is more than one measure. *Elastic design* requires that the stresses must nowhere exceed the *yield strength*, σ_y (for ductile materials) or the *elastic limit* σ_{el} (for those that are not ductile). *Plastic design* and *deformation processing*, by contrast, require some or all of the product to deform plastically, either to absorb energy in crash-resistant applications or to enable the shape to be changed (as in metal forming). Then two further properties become relevant: the *ductility*, ε_f, and the *tensile strength*, σ_{ts}, which are the maximum strain and the maximum stress the material can tolerate (in tension) before fracture. The tensile strength is larger than the yield strength because of *work hardening*.

Charts plotting strength again show that material families occupy different areas of material property space, depending on the strengthening mechanisms on which they rely. Strength for metals is a particularly microstructure-sensitive property. Crystal defects – particularly *dislocations* – are central to the understanding of alloy hardening. Dislocation motion gives plastic flow in crystalline solids, though the intrinsic lattice resistance is weak in pure metals, which are unexpectedly low strength. Deliberately obstructing dislocation motion with solute, precipitates, more dislocations, and grain boundaries is needed to give useful strength. The strongest materials utilise more than one mechanism.

Non-crystalline solids – particularly polymers – deform in a less organised way by sliding of the tangled polymer chains, sometimes aligning them with the direction of deformation to give fibres of high stiffness and strength. Polymers have useful intrinsic strength relative to metals – better than their elastic properties. Strength can be enhanced further by blending, cross-linking, and reinforcement with particles or fibres.

Elastic design is design to avoid yield. That means ensuring the maximum stress in a loaded component is less than the yield strength, σ_y (or the elastic limit σ_{el} in low ductility materials). Account must sometimes be taken of changes in section that lead to stress concentrations. Standard solid mechanics solutions give the maximum stress for ties, beams, and many other components in terms of their geometry and the loads applied to them. These are also used for material selection in strength-limited design, as shown in *Guided Learning Unit 2*.

5.9 Further reading

Hosford, W.F. (2015). *Mechanical behavior of materials* (2nd ed.). Cambridge University Press. ISBN 978-0-521-19569-0. (A text that nicely links stress–strain behaviour to the micromechanics of materials).

Hull, D., & Bacon, D.J. (2001). *Introduction to dislocations* (4th ed.). Butterworth-Heinemann. ISBN 0-750-064681-0. (An introduction to dislocation mechanics).

Jenkins, C.H.M., & Khanna, S.K. (2006). *Mechanics of materials*. Elsevier Academic. ISBN 0-12-383852-5. (A simple introduction to mechanics, emphasising design).

Young, W.C. (2012). Roark's formulas for stress and strain (8th ed.). McGraw-Hill. ISBN 078-0-07-174247-4. (This is the 'Yellow Pages' of formulae for elastic problems – if the solution is not here, it probably doesn't exist).

5.10 Exercises

Note that to complete some of these Exercises you will need to draw on solid mechanics analysis and formulae from Chapters 4 and 5.

Exercise E5.1   Sketch a stress–strain curve for a typical metal. Mark on it the yield strength, σ_y, the tensile strength σ_{ts}, and the ductility ε_f. Indicate on it the work done per unit volume in deforming the material up to a strain of $\varepsilon < \varepsilon_f$ (pick your own strain ε).

Exercise E5.2   On a sketch of a typical stress–strain curve for a ductile metal, distinguish between the maximum elastic energy stored per unit volume and the total energy dissipated per unit volume in plastic deformation. Give examples of engineering applications in which a key material characteristic is (i) maximum stored elastic energy per unit volume; (ii) maximum plastic energy dissipated per unit volume.

Exercise E5.3   The figure shows four tensile nominal stress–strain curves for Materials A, B, C, and D. Identify which of these corresponds to metal, polymer, or ceramic, giving reasons for your choice in terms of the relative magnitudes of key properties and the underlying deformation and failure mechanisms.

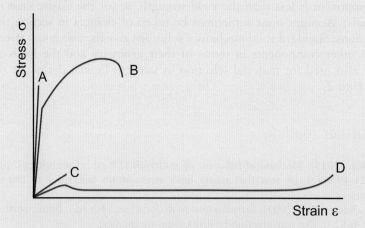

Exercise E5.4   In Exercise E4.2, the extension of the cable of a hoist was considered when lifting a mass of 250 kg. Find the stress in the cable, of cross-sectional area 80 mm^2, and evaluate the factor of safety between the applied stress and the yield stress, for steel (yield stress 300 MPa) and polypropylene (yield stress 32 MPa). What mass would lead to failure in each material?

Exercise E5.5   (a) Write down expressions for the nominal strain and the true strain for the following situations:

        (i) in a tensile test where the length of the specimen increases from L_o to L;

        (ii) in a compression test where the height of the specimen decreases from h_o to h.

    (b) In what circumstances can the true strain be approximated by the nominal strain?

Exercise E5.6   In a two-stage elongation of a specimen, the length is first increased from l_o to l_1 and then from l_1 to l_2. Write down the true strains $\varepsilon 1$ and $\varepsilon 2$, respectively, for each of the two elongation processes considered separately. Show that the true strain for the overall elongation l_o to l_2 is given by the sum of the true strains for each of the two separate processes. Is the same true for the nominal strains?

Exercise E5.7   In a 'tandem' rolling process, a wide metal strip passes continuously between four sets of rolls, each of which apply a reduction in thickness. The strip elongates without getting wider, and therefore accelerates: why is this? For an incoming thickness $h_{in} = 25$ mm moving at speed $v_{in} = 5$ cm.s^{-1}, and an exit thickness $h_{out} = 3.2$ mm, calculate: (i) the overall true strain applied to the material; (ii) the exit speed v_{out}. State any assumptions made.

Exercise E5.8   Explain briefly why the material property bubbles for metals are very elongated on the Young's modulus – strength chart (Figure 5.5). Comment on whether the same is true for the polymers. Why is this?

Exercise E5.9   Use the yield strength–density chart or the yield strength–modulus chart (Figures 5.4 and 5.5) to find the following:

- The metal with the lowest strength
- The approximate range of strength of the composite GFRP
- Whether there are any polymers that are stronger than wood
- How the strength of GFRP compares with that of wood
- Whether elastomers, that have moduli that are far lower than polymers, are also far lower in strength

Exercise E5.10 The table shows the ranges of the densities, the Young's moduli, and the yield strengths of the ferrous alloys (alloys based on iron, usually with carbon, and commonly other alloying elements). What do you observe about the spread of values for each property, for all classes of ferrous alloy? Hence, suggest which of the properties is *microstructure-sensitive*. Choose one class of ferrous alloy and compare the typical values of these properties from Appendix A with the ranges given. Why is more caution advisable in using 'typical' values for some properties than for others?

Material	Density (kg/m³)	Young's modulus (GPa)	Yield strength (MPa)
Cast iron (ductile)	7000–7200	170–180	250–630
High carbon steel	7800–7900	200–220	430–930
Medium carbon steel	7800–7900	200–220	370–900
Low alloy steel	7800–7900	200–210	470–1600
Low carbon steel	7800–7900	200–220	250–360
Stainless steel	7600–7900	190–210	250–1100

Exercise E5.11 CFRP, Al, and Ti alloys have excellent strength-to-weight characteristics as measured by the ratio σ_y/ρ. Use the typical property values in Appendix A to rank them in order of performance according to this measure.

Exercise E5.12 Using a suitable material property chart and the tables in Appendix A, determine which polymers have the highest yield strength and state some of their applications.

Exercise E5.13 Abrasives have high hardness, H. Assuming that Vickers hardness and yield strength are related by $H_v = \sigma_y/3$ [Equation (5.4)], use the data in Appendix A to identify materials that would be suitable for use as abrasives.

Exercise E5.14 What is meant by the ideal strength of a solid? Which material class most closely approaches it?

Exercise E5.15 (a) Explain briefly the four main microstructural mechanisms by which metals are hardened. For each of the following, identify which mechanisms account for the following increases in yield stress (NB there may be more than one):

(i) Pure annealed aluminium: 25 MPa; cold rolled non-heat-treatable Al-Mn-Mg alloy: 200 MPa

(ii) Pure annealed copper: 20 MPa; cast 60-40 brass (60% Cu, 40% Zn): 105 MPa

(iii) Pure annealed iron: 140 MPa; quenched and tempered medium carbon steel: 600 MPa

(iv) Pure annealed copper: 20 MPa; nanocrystalline pure Cu: 500 MPa

(b) For each of mechanisms (i) – (iii) in part (a), give other examples of important engineering alloys that exploit these mechanisms.

Exercise E5.16 An aluminium alloy used for making cans is cold rolled into a strip of thickness 0.3 mm and width 1 m. It is coiled around a drum of diameter 15 cm, and the outer diameter of the coil is 1 m. In the cold rolled condition, the dislocation density is approximately $10^{15}\,m^{-2}$. Estimate:

(i) the mass of aluminium on the coil;

(ii) the total length of strip on the coil;

(iii) the total length of dislocation in the coiled strip.

Exercise E5.17 The table shows yield stress data for two copper alloy systems that are solid solution strengthened (Cu-Zn and Cu-Ni alloys), and for pure Cu, Zn, and Ni, all in the annealed condition (i.e. with no work hardening). Here the notation Cu-10Zn means 10 weight % Zn (and 90% Cu). Dislocation theory suggests that strength is a microstructure-sensitive property. Investigate graphically how much the data for these Cu alloys deviate from a simple linear rule of mixtures (in weight %).

Alloy	Yield stress σ_y (MPa)
Pure Cu	50
Cu-10Zn	95
Cu-30Zn	115
Pure Zn	100
Cu-10Ni	135
Cu-30Ni	170
Cu-65Ni	220
Pure Ni	140

Exercise E5.18 (a) Explain briefly how impeding the motion of dislocations with dispersed precipitates in a metal can raise the yield stress.

(b) A straight segment of dislocation line of length L is pinned at two particles in a crystal. The application of a shear stress τ causes the dislocation to bow outwards as shown in the figure, decreasing its radius of curvature as it does so. The force per unit length acting on the dislocation is τbL, where b is the Burger's vector. Given that the dislocation line tension $T = Gb^2/2$, show that the stress just large enough to fully bow out the dislocation segment into a semi-circle is given by $\tau = Gb/L$.

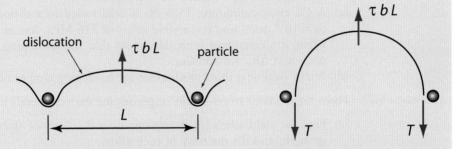

(c) Estimate the tensile yield stress of aluminium containing an array of equally spaced second phase precipitates, where the volume fraction of the second phase is 5% and the precipitate diameter is 0.1 μm. The Burger's vector for aluminium is 0.29 nm, and the shear modulus $G = 27$ GPa.

(d) If incorrect heat treatment of the aluminium alloy results in a non-uniform distribution of precipitates, explain the effect this would have on the yield stress.

Exercise E5.19   A metal matrix composite consists of aluminium containing hard particles of silicon carbide SiC with a mean spacing of 1 μm. The yield strength of the composite is found to be 8 MPa greater than the unreinforced matrix alloy. If a new grade of the composite with a particle spacing of 0.2 μm were developed, what would you expect the strength contribution of the particles to be?

Exercise E5.20   Nano-crystalline materials have grain sizes in a range 0.01 to 0.1 μm. If the contribution of grain boundary strengthening in an alloy with grains of 0.1 μm is 20 MPa, what would you expect it to be if the grain size were reduced to 0.01 μm?

Exercise E5.21   The yield strength σ_y of low carbon steel is moderately dependent on the grain size D and the relation can be described by the equation $\sigma_y = \sigma_o + k \sqrt{(1/D)}$, where σ_o and k are material constants. The yield strength is 322 MPa for a grain size of 180 μm and 363 MPa for a grain size of 22 μm.

   (a) Calculate the yield strength of the steel for a grain size of 11 μm.
   (b) Explain briefly the physical significance of the constant σ_o.

Exercise E5.22   (a) The work hardening contribution to the yield stress $\Delta\sigma_{wh}$ can be assumed to be proportional to ρ_d, where ρ_d is the dislocation density. A hot rolled sheet of pure copper (Cu) has a yield stress of 90 MPa and $\rho_d = 10^{13}$ m/m^3. A cold rolled sheet of the same material has a yield stress of 360 MPa and $\rho_d = 10^{15}$ m/m^3. Find the yield stress of fully annealed pure Cu.
   (b) The solid solution hardening contribution to the yield stress $\Delta\sigma_{ss}$ can be assumed to be proportional to $(C_{ss})^n$, where C_{ss} is the concentration in solid solution. An annealed Cu alloy containing $C_{ss} = 10\%$ Ni has a yield stress of 110 MPa, while an alloy containing $C_{ss} = 20\%$ Ni has a yield stress of 131 MPa. Find the exponent n.
   (c) A Cu alloy containing 15% Ni is cold rolled to a dislocation density $\rho_d = 10^{14}$ m/m^3 and has a yield stress of 216 MPa. Assess whether these values are consistent with the theory that the hardening contributions $\Delta\sigma_{wh}$ and $\Delta\sigma_{ss}$ are additive.
   (d) Briefly outline a third mechanism for hardening copper alloys.

Exercise E5.23   From the uniaxial stress–strain responses for various metals in Figure 5.20:

   (a) Find the yield stress (or if appropriate a 0.2% proof stress), the tensile strength, and the ductility of each alloy.
   (b) Account for the differences in strength, comparing annealed copper with each of the other alloys in turn.

Exercise E5.24   The figure shows typical data for a selection of Al alloys on a strength-ductility property chart. Compare this with the corresponding data for Cu alloys (Figure 5.19). Do the same overall trends apply with regard to:

(i) the trade-off between strength and ductility?

(ii) the ranking of the hardening mechanisms in order of increasing effectiveness?

(iii) Estimate the ratio of the highest strength alloy to that of the pure, annealed base metal, for Al and Cu.

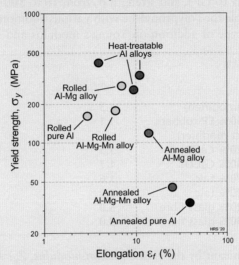

Exercise E5.25 From the uniaxial stress–strain responses for various polymers in the figure:

(a) Estimate the tensile strengths and ductility of each material.
(b) Account for the main differences between the shapes of the graphs for PMMA, ABS, and Polypropylene.

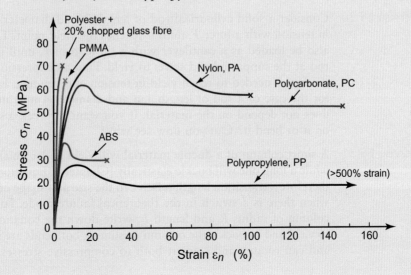

Exercise E5.26   Polycarbonate, PC (yield strength 70 MPa) is blended with polyester PET (yield strength 50 MPa) in the ratio 30%/70%. If the strength of blends follows a rule of mixtures, what would you expect the yield strength of this blend to be?

Exercise E5.27   Sketch a Young's modulus – strength chart using log scales with ranges of E from 1 to 10 GPa, and strength σ_y from 10 to 100 MPa. Plot the data given in the table for Polypropylene with various additions. Comment on the effect of each type of addition on Young's modulus and strength.

Material	Young's modulus, E (GPa)	Strength, σ_y (MPa)
PP (unmodified)	1.3	27
PP + 10% EP rubber (impact modifier)	1.1	20
PP + 20% CaCO$_3$	1.3	21
PP + 40% CaCO$_3$	1.7	19
PP + 20% talc	1.6	25
PP + 40% talc	3.0	24
PP + 20% glass fibre	3.0	47
PP + 40% glass fibre	8.0	90

Exercise E5.28   What is meant by the *elastic section modulus*, Z_e? A beam carries a bending moment M. In terms of Z_e, what is the maximum value that M can take without initiating plasticity in the beam?

Exercise E5.29   Calculate the elastic bending moment M_e that causes first yield in a beam of square cross section $b \times b$ with tensile yield strength σ_y. Calculate the fully plastic moment M_p in the same beam. What is the ratio of M_p/M_e?

Exercise E5.30   Consider a solid cylindrical rod of length L and diameter d, loaded axially in tension with a force F, until it reaches the yield point. The same rod may also be loaded as a cantilever with a tip load of F, until the surface of the rod at the supported end starts to yield. Find an expression for the ratio of the forces needed to reach yield in tension and bending. Evaluate this ratio for the case of a rod of length 1 m and diameter 4 mm, and explain why it does not depend on the material. If you want to break a stick, do you pull on it or bend it? Can you now see why?

Exercise E5.31   A short column of a ductile material will yield under uniaxial compression, while a long one will buckle elastically [Chapter 4, Equation (4.29)]. There is therefore a transition length relative to the size and shape of the cross section when there is a switch in the theoretical failure mode. For a solid circular column of radius R and length L, write down the equations for the forces to yield and to buckle the column (assuming both ends are simply supported and can rotate), and convert both to compressive stresses. Sketch how the stresses for yield and for buckling vary with the 'slenderness ratio' (L/R), and find an expression for the transition value of (L/R), when the applied stress is equal for both mechanisms. Use the modulus–strength property chart (Figure 5.5) to determine whether the transition slenderness is generally higher for polymers or for metals.

Exercise E5.32   An elastic-plastic wire of diameter d, with modulus E and yield strength σ_y, is made into a helical spring with five turns of radius $R = 4d$.

(a) What is the ratio of the stiffness of the spring to the axial stiffness of the same length of wire, loaded in tension? Assume that $G = 3/8\ E$.
(b) What is the ratio of the plastic failure load of the spring to that of the wire loaded axially in tension?
(c) How much greater is the extension of the spring at failure compared to that of the wire?

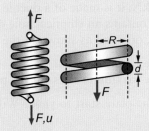

Exercise E5.33   Figure 5.25 shows the biaxial stress state in a closed cylindrical pressure vessel with internal pressure p, radius R, and wall thickness t $(\ll R)$. Two sections are shown: (a) transverse; (b) longitudinal, over an arbitrary length L. The cut surface of the solid material carries the relevant stress (longitudinal and hoop, respectively) normal to the cut, so to find the force due to the stress we need to find the area of the solid carrying the stress. The pressure acts normal to the conceptual 'cut' through the gas, so similarly we need the area on which the pressure acts to find the force due to the pressure. For equilibrium, these forces balance and are equal. Follow these steps to find, in terms of p, R and t:

(a) the longitudinal stress, σ_l;
(b) the hoop (circumferential) stress, σ_h.

Exercise E5.34   A thin-walled spherical pressure vessel has a pressure difference p between interior and exterior, a radius R, and wall thickness t $(\ll R)$.

(a) By considering sections through the centre of the sphere, and comparing them directly with the analysis in Exercise E5.33 for a cylindrical pressure vessel, find expressions for the biaxial stresses in the wall, in terms of p, R, and t.
(b) In 1960, the Swiss oceanographer and engineer Jacques Piccard reached a depth of 10,910 m in the Pacific Ocean. The pressure vessel of his

submersible craft consisted of an alloy steel sphere of external diameter $d = 2.18$ m and wall thickness $t = 120$ mm. Find the hydrostatic pressure at the maximum depth, and compare this with atmospheric pressure (≈ 0.1 MPa), assuming the density of sea water is 1000 kg/m^3. Estimate the compressive stress in the wall of the vessel using the thin-walled result derived in (a) (noting that it is approximate because the wall is not particularly thin compared to the radius). Compare this stress with the typical value of the yield stress for low alloy steels given in Appendix A.

Exercise E5.35   A large plate carrying a uniform tensile stress σ contains a large circular hole of radius R, and small circular hole of radius r. Which hole do you think acts as the greater stress concentration? Use the formula in Equation (5.30) to check if you are correct.

Exercise E5.36   A plate with a rectangular section 500 mm × 15 mm carries a tensile load of 50 kN. It is made of a ductile metal with a yield strength of 50 MPa. The plate contains an elliptical hole of length 100 mm and a minimum radius of 1 mm, oriented as shown in the diagram. What is:

(a) the nominal stress?
(b) the maximum stress in the plate?

Will the plate start to yield? Will it collapse completely?

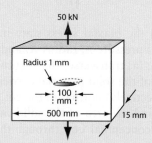

Exercise E5.37   A tie made of a rectangular section strip of metal has a thickness t, and a width b that reduces to width a with a radius of ρ, as shown in the figure. The material has a yield stress σ$_y$.

(a) Find an expression for the value of a that would be needed to avoid yield at the nominal stress level in the thinner section, with a safety factor of S.
(b) The applied load is expected to reach the stress level in (a). Find an expression for the value of the fillet radius ρ that would ensure that the maximum stress in the tie rod is less than the yield stress.

(c) Determine the value of ρ if $a = 30$ mm, $b = 40$ mm, and $S = 3$.

(d) What is the real safety factor of the part with this value of ρ?

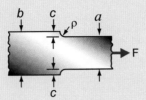

Exercise E5.38 Toilet paper, ring-push drink can openers, and chocolate bars are all examples of the use of stress-concentrating features to give failure in a desired location in products. Take a look around your home for other examples of stress raisers that help the consumer break something by hand in a controlled way.

Chapter 6

Fracture, fatigue, and fracture-limited applications

Ductile and brittle fracture.
(Image of bolt courtesy of Bolt Science; www.boltscience.com)

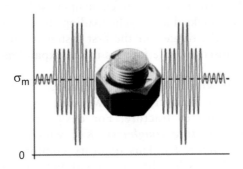

σ_m

0

Fatigue failure of a bolt – the crack initiated at the root of a thread, which acts as a stress concentration.
(Image of bolt courtesy Bolt Science: www.boltscience.com)

https://doi.org/10.1016/B978-0-08-102399-0.00006-6

6.1 Introduction and synopsis

One of the great triumphs of engineering science over the past 100 years or so has been the development of *fracture mechanics*, enabling fail-safe design and the avoidance of many catastrophic failures, notably in transport and power generation systems. Fracture mechanics deals with the interaction of stresses in materials with the cracks that they inevitably contain – often as an undesirable side effect of processing, from casting to rolling, welding, and heat treatment. More challenging still is the problem of *fatigue* – the initiation and growth of cracks due to cyclic loading, at stresses that are often well below the elastic limit; indeed, the trigger for the first systematic investigation of fracture was the serious accidents caused by axle failures in the expanding European railway system of the mid-19th century.

Sections 6.2-6.7 deal with fracture, first defining the mechanics of cracked components and the resulting property data. We start by distinguishing *strength* from *toughness* — the resistance to fracture. For design purposes, this leads to a new material property: the *fracture toughness*, K_{1c}, which is the critical value of the tensile *stress intensity factor*, K_1. This quantity combines the applied stresses from conventional solid mechanics for the given component geometry with the length and location of a crack in the component – the longer the crack, the lower the failure stress. Data for fracture toughness is presented using property charts, in combination with modulus or strength. Drilling down into the underlying materials science then gives insight into the ways we can manage the resistance to fracture via the microstructure.

Cyclic loading and fatigue follow in Sections 6.8 and 6.9. Cracks may be initiated by cyclic stresses, or there may be pre-existing cracks – either way, they grow incrementally on each cycle until the stress intensity factor exceeds the fracture toughness, K_{1c}, leading to fracture. A number of design rules for fatigue, and associated material properties, are presented. This provides strategies for avoiding crack initiation or propagation altogether and for fail-safe design of components, in which the expected cyclic crack growth is explicitly evaluated. The chapter concludes (Section 6.10) with case studies illustrating the application of fracture and fatigue mechanics in design.

6.2 Strength and toughness

Strength and toughness? Why both? What's the difference? *Strength*, when speaking of a material, is usually short-hand for its resistance to plastic flow measured by the initial *yield strength*. Chapter 5 covered this, including the concept that the stress to maintain dislocation motion generally increases with plastic strain, because of work hardening, reaching a maximum at the *tensile strength*.

Toughness is different: it is the resistance of a material to the propagation of a *crack*. Suppose that a sample of material contains a small, sharp crack, as in Figure 6.1(a). The crack reduces the cross-section A and, since stress σ is *F/A*, it increases the stress. A 'tough' material will yield, work harden, and absorb energy by plastic deformation as before – the crack makes no significant difference [Figure 6.1(b)]. If the material is *not* tough (defined in a moment, and referred to as 'brittle' behaviour), then the unexpected happens: the crack suddenly propagates, and the sample fractures at a stress that can be far below the yield strength [Figure 6.1(c)]. Strength-limited elastic design based on yield is common practice, but the possibility of fracture at stresses below the yield strength is really bad news. Historically it has happened on spectacular scales, causing boilers to burst, bridges to collapse, ships to break in half, pipelines to split, and aircraft to crash. Thankfully, due to fracture mechanics, catastrophic failures are now rare – but they do still happen.

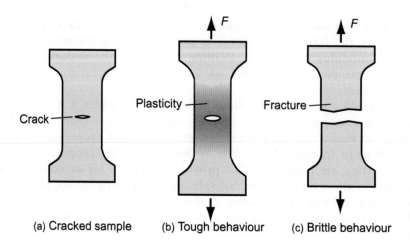

(a) Cracked sample   (b) Tough behaviour   (c) Brittle behaviour

Figure 6.1 Tough and brittle behaviour. The crack in the tough material (b) does not propagate when the sample is loaded; that in the brittle material (c) propagates without general plasticity at a stress below the yield strength.

So what is the material property that measures the resistance to the propagation of a crack? How concerned should you be if you hear that cracks have been detected in the track of the railway on which you commute or in the pressure vessels of a nuclear reactor of a power station?

Fracture behaviour is actually very familiar – everyday products often include design details to enable failure. Think of the packaging around food, tools, and toothbrushes – too strong to simply pull apart, but make a small, sharp nick in the side and it tears easily: the failure is now controlled by the material toughness, not its strength.

Tests for fracture resistance Simple tests to characterise resistance to fracture introduce a notch in a standard sample, then pulling on it, or applying an impact in bending (Figure 6.2). Both approaches measure fracture *energy* (or 'work of fracture'), but neither lead to a true material property, meaning it is independent of the size and shape of the test sample, so these energy measurements do not help with design. So they can only be used for crude empirical ranking of resistance to fracture or impact. However, what they correctly recognise is that fracture has something to do with providing enough energy to overcome the material's resistance to breaking apart. To get at the true, underlying behaviour, and the material properties suitable for design, we need *fracture mechanics*, and the key ideas of *stress intensity* and *fracture toughness*.

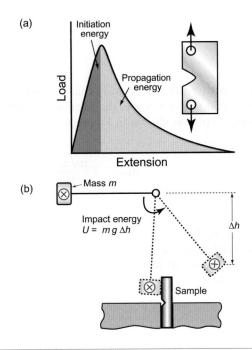

Figure 6.2 (a) The tear test, and (b) the impact test. Both are used as empirical ranking tests, but neither measures a true material property.

6.3 The mechanics of fracture

Stress intensity K_1 and fracture toughness K_{1c} Cracks and notches both concentrate stress, but with subtle differences. For notches we defined (in Chapter 5) a 'stress concentration factor,' which tells us how much greater the peak local stress is compared to the remote stress – noting that this is for changes in cross-section with a well-defined size and radius, such as a circular hole. Now we consider how *cracks* affect the stress field. Figure 6.3 shows a remote stress σ applied to a cracked material. The load has to 'bypass' the crack – we can envisage 'lines of force' that are uniformly spaced in the remote region but bunch together around the crack, giving a stress that rises steeply as the crack tip is approached. Referring to Equation (5.30), the stress concentration factor for a notch does not help here – it is relevant only for features with a finite radius of curvature. Cracks are sharp: the radius at the tip is essentially zero, giving (notionally) an infinite stress by Equation (5.30). So a different approach is needed for cracks. Analysis of the elastic stress field ahead of a sharp crack of length c shows that the local stress at a distance r from its tip, caused by a remote uniform tensile stress σ, is

$$\sigma_{local} = \sigma \left(1 + Y\sqrt{\frac{\pi c}{2\pi r}} \right) \tag{6.1}$$

where Y is a constant with a value near unity that depends on the geometry of the cracked body. Far from the crack, where $r \gg c$, the local stress falls to the value σ; near the tip, where $r \ll c$, it rises steeply (as shown in Figure 6.3) as

$$\sigma_{local} = Y\frac{\sigma\sqrt{\pi c}}{\sqrt{2\pi r}} \tag{6.2}$$

So for any given value of r the local stress scales as $\sigma\sqrt{\pi c}$, which therefore is a measure of the 'intensity' of the local stress field (the inclusion of the π is a convention used universally). This quantity (including the geometric factor Y) is called the *mode 1 stress intensity factor* (the 'mode 1' means tensile loading perpendicular to the crack) and given the symbol K_1 (with units of MPa.m$^{1/2}$):

$$K_1 = Y\sigma\sqrt{\pi c} \tag{6.3}$$

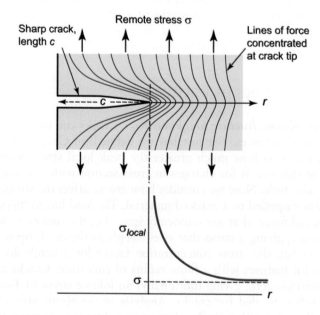

Figure 6.3 Lines of force in a cracked body under load; the local stress increases steeply as the crack tip is approached.

The stress intensity factor is thus a measure of the elastic stress field near the tip of a sharp crack. Equation (6.2) predicts that, as $r \to 0$, the crack tip stress is infinite. In practice the material will yield in a small zone at the crack tip (or some other localised form of damage will occur, such as micro-cracking); this is discussed further later. The important point for now is that this zone is small and contained, while the loading on the surrounding crack tip region is an elastic stress field that scales with K_1, and it is this stress field that drives fracture. Cracks propagate unstably when the stress intensity factor exceeds a critical value, and this is called the *fracture toughness*, K_{1c}.

Figure 6.4 shows a number of simple cracked geometries and their corresponding expression for K_1. Note the geometry correction factor Y in each case, of order unity; also note that the crack length is defined as c for a surface crack, and $2c$ for a contained crack. Formulae for other geometries can be found in the references listed in Further Reading. For a given crack geometry it is straightforward to evaluate K_1 in design, since we will have already found the applied stress σ (using the solid mechanics analysis in Chapters 4 and 5) to ensure the stresses are below the elastic limit, to avoid yielding. Failure by fracture is then expected when

$$K_1 = Y\sigma\sqrt{\pi c} = K_{1c} \qquad (6.4)$$

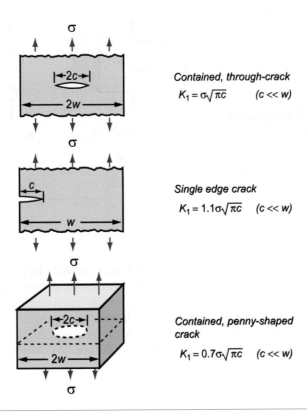

Contained, through-crack
$$K_1 = \sigma\sqrt{\pi c} \qquad (c \ll w)$$

Single edge crack
$$K_1 = 1.1\sigma\sqrt{\pi c} \qquad (c \ll w)$$

Contained, penny-shaped crack
$$K_1 = 0.7\sigma\sqrt{\pi c} \qquad (c \ll w)$$

Figure 6.4 Stress intensity factors K_1 associated with cracks in different loading geometries.

So the stress intensity factor K_1 measures the state of loading on the crack, and the fracture toughness K_{1c} is the critical value of K_1 for crack propagation (i.e. it is a *material property*). This means its value is independent of the way it is measured: tests using different geometries, if properly conducted, give the same value of K_{1c} for any given material.

Figure 6.5 shows two common geometries used to measure K_{1c}, recording the peak tensile stress σ^* or load F^* at which it suddenly propagates. It is essential that the crack be sharp — not an easy thing to achieve — because if it is not, the crack tip stress field is changed. A good way to make sharp cracks for doing this test in the laboratory is to do the thing that often leads to fracture in practice – growing a crack by *cyclic loading*, the process called *fatigue* (Section 6.8). As seen previously, the value of Y for the centre-cracked plate is 1 (provided $c \ll w$); the compact tension test loads the crack in bending, so the calibration for K_1 is written in terms of the applied load, F.

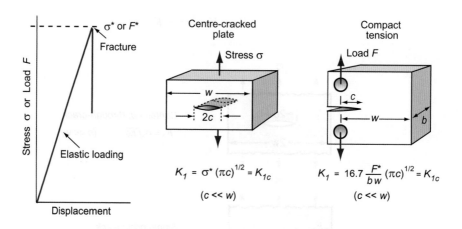

Figure 6.5 Measuring fracture toughness, K_{1c}.

The fracture toughness K_{1c} is used with Equation (6.4) to design against fracture in a number of ways. If we have measured the size of crack that is present (e.g. by *non-destructive testing*), then we can evaluate the failure stress. By reversing the calculation, for a given applied stress we can find the 'critical crack length,' being the maximum crack size that we can tolerate without failure; or if we know both stress and crack length, we can specify the minimum fracture toughness required to avoid failure.

Example 6.1

The stainless steel and polystyrene rulers in Examples 4.7 and 4.8 are loaded as cantilevers of length 250 mm. Both have sharp transverse scratches of depth 0.2 mm, near the base of the cantilever on the tensile side. For this surface crack geometry, assume the crack is subjected to a uniform tensile stress equal to the maximum bending stress, so the factor Y in Equation (6.4) is 1.1. The fracture toughness K_{1c} and yield strength σ_y of the stainless-steel are 80 MPa.m$^{1/2}$ and 540 MPa, respectively; for the polystyrene, the values are 1 MPa.m$^{1/2}$ and 40 MPa, respectively. Calculate the stress needed to cause fast fracture in each case. Will the cantilevers fail by yielding or by fast fracture?

Answer. From Equation (6.4), the stress at fast fracture for the stainless steel is

$$\sigma^* = \frac{K_{1c}}{1.1\sqrt{\pi c}} = \frac{80 \times 10^6}{1.1\sqrt{\pi \times 2 \times 10^{-4}}} = 2.9 \text{ GPa}$$

This is far in excess of the yield strength of 540 MPa – meaning that the ruler will yield before it fractures.

For the polystyrene the fracture stress is

$$\sigma^* = \frac{K_{1c}}{1.1\sqrt{\pi c}} = \frac{1 \times 10^6}{1.1\sqrt{\pi \times 2 \times 10^{-4}}} = 36 \text{ MPa}$$

which is less than its yield strength (40 MPa) – so the ruler will fracture before it yields.

Example 6.2

A compact tension specimen, as in Figure 6.5, is manufactured from an Al alloy with dimensions $w = 50$ mm and $b = 20$ mm, with a sharp pre-crack of length $c = 10$ mm. The maximum load capacity available on the test machine is 10 kN. Is the machine capacity sufficient to fracture the alloy, if its fracture toughness is 25 MPa.m$^{1/2}$?

Answer. From Figure 6.5, the stress intensity factor is

$$K_{1c} = 16.7 \frac{F^*}{bw} \sqrt{\pi c}$$

So for $K_{1c} = 25$ MPa.m$^{1/2}$, the failure load will be

$$F^* = \frac{25 \times 10^6 \times 20 \times 10^{-3} \times 50 \times 10^{-3}}{16.7 \times \sqrt{\pi \times 10 \times 10^{-3}}} = 8.45 \text{ kN}$$

So the test machine is strong enough.

Energy release rate G and toughness G_c Fracture toughness is the property to use in design, while the stress intensity factor identifies the elastic energy in the crack tip region as the driver for fracture. But what determines the resistance of a material to fracture? When a sample fractures, a new surface is created. Surfaces have energy, the *surface energy* γ, with units of Joules[1] per square metre (typically γ = 1 J/m²). If you fracture a sample across a cross-section area *A* you make an area 2*A* of new surface, requiring an energy of at least 2*A*γ J to do so. Consider first the question of the *necessary condition for fracture*. It is that sufficient external work be done, or elastic energy released, to at least supply the surface energy, γ per unit area, of the two new surfaces that are created. We write this as

$$G \geq 2\gamma \qquad (6.5)$$

where *G* is called the *strain energy release rate* (again in J/m²). In practice, it takes much more energy than 2γ to break a material in two, because of plastic deformation also dissipating energy round the crack tip. But the argument still holds: growing a crack costs a characteristic energy per unit area to create the two surfaces – a sort of 'effective' surface energy, replacing 2γ. It is called, somewhat confusingly, the *toughness* G_c (J/m²) and is a material property that measures the *critical strain energy release rate* for fracture from a starter crack. This is why the crude energy absorption tests of Figure 6.1 are not rigorous – notches localise the failure but do not create the stress field associated with a sharp crack.

Toughness G_c describes the same physical event as *fracture toughness* K_{1c} but in a different way, so it should be no surprise to find that the two may be related, as follows. Think of a slab of material of unit thickness carrying a stress σ. The elastic energy per unit volume stored in it (see Chapter 4) is

$$U_v = \frac{\sigma^2}{2E} \qquad (6.6)$$

Now put in a crack of length *c*, as in Figure 6.6. The crack relaxes the stress in a region of approximately a half-cylinder of radius about *c* – the reddish area in the figure – releasing the energy it contained:

$$U(c) = \frac{\sigma^2}{2E} \times \frac{1}{2}\pi c^2 \qquad (6.7)$$

Suppose now that the crack extends by δ*c*, releasing the elastic energy in the yellow segment. This energy must pay for the extra surface created, and the cost of this is $G_c \, \delta c$. Thus differentiating Equation (6.7), the condition for fracture becomes

$$\delta U = \frac{\sigma^2 \pi c}{2E} \, \delta c = G_c \, \delta c \qquad (6.8)$$

---

[1] James Joule (1818–1889), English physicist, who did not work on fracture or on surfaces, but his demonstration of the equivalence of heat and mechanical work linked his name to the unit of energy.

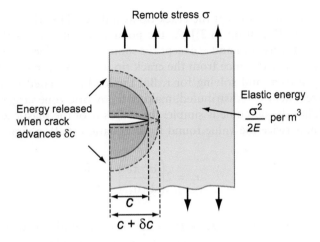

Figure 6.6 The release of elastic energy when a crack extends.

But $\sigma^2 \pi c$ is just K_{1c}^2 , so from Equation (6.4), taking $Y = 1$,

$$\frac{K_{1c}^2}{2E} = G_c \tag{6.9}$$

This derivation is an approximate one. A more rigorous (but much more complicated) one shows that the form of Equation (6.9) is right, but that it is too small by exactly a factor of 2. Thus, correctly, the result we want (taking the square root) is

$$K_{1c} = \sqrt{EG_c} \tag{6.10}$$

Toughness G_c is a material property that is more obviously related to the physics of fracture: a tough material requires more energy to be dissipated when a crack propagates. But quantifying the energy release rate G for a loaded crack is difficult, so for design it is preferable to use $K = K_{1c}$ as the failure criterion. The remote elastic stress state is something we must analyse anyway, and the crack length is a physical quantity that can be directly observed and measured. The physical equivalence of the two criteria, expressed by Equation (6.10), is straightforward: crack growth is driven by the release of elastic stored energy (G), and this scales with the crack tip elastic stress field (K).

The crack tip plastic zone   It was noted previously that the intense stress field at the tip of a crack generates a *process zone*: a plastic zone in ductile solids or a small region of micro-cracking in ceramics; fibre composites, too, will suffer local damage, such as debonding and fibre breaks giving 'pull-out' (discussed later). Within the process zone, work is done against plastic and frictional forces; it is this that accounts for the difference between the measured fracture energy G_c and the true surface energy 2γ. In plastic materials, we can estimate the size of yield zone that forms at the crack tip as follows. The stress rises as $1/\sqrt{r}$ as the crack tip is approached [Equation (6.2)]. At the point where it reaches the yield strength σ_y the material yields (Figure 6.7) and — except for some work hardening — the stress cannot climb higher than this. The distance from the crack tip where $\sigma_{local} = \sigma_y$ is found by setting Equation (6.2) equal to σ_y and solving for r. But the load associated with the truncated part of the elastic stress field is redistributed, making the plastic zone larger. The analysis of this is complicated, but the outcome is simple: the radius r_y of the plastic zone, allowing for stress redistribution, is twice the value found from Equation (6.2), giving

$$r_y = 2\left(\frac{\sigma^2\,\pi c}{2\pi\,\sigma_y^2}\right) = \frac{K_1^2}{Y^2\pi\,\sigma_y^2} \tag{6.11}$$

(with $Y \approx 1$). Note that the size of the zone shrinks rapidly as σ_y increases: cracks in soft metals have large plastic zones; those in ceramics and glasses have small zones or none at all.

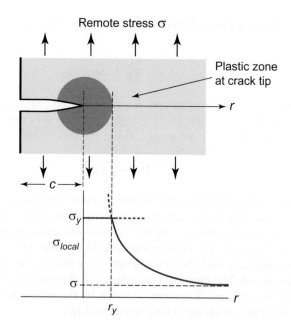

Figure 6.7   A plastic zone forms at the crack tip where the stress would otherwise exceed the yield strength σ_y.

The property K_{1c} has well-defined values for brittle materials and for those in which the plastic zone is small compared to all dimensions of the test sample, so that most of the sample is elastic. When this is not so, a more complex characterisation of fracture is needed. In very ductile materials the plastic zone size exceeds the width of the sample; then the crack does not propagate at all – the sample simply yields [as in Figure 6.1(b)].

So when cracks are small, materials yield before they fracture; when they are large, the opposite is true. But what is 'small'? Figure 6.8 shows how the tensile failure stress varies with crack size, for both yield and fracture. When the crack is small, this stress is equal to the yield stress; when large, it falls off according to Equation (6.4), which we write (taking $Y = 1$ again) as

$$\sigma_f = \frac{K_{1c}}{\sqrt{\pi c}} \tag{6.12}$$

The transition from yield to fracture is smooth, as shown in the figure, but occurs around the intersection of the two curves, when $\sigma_f = \sigma_y$, giving the *transition crack length*

$$c_{crit} = \frac{K_{1c}^2}{\pi \sigma_y^2} \tag{6.13}$$

Note that this is the same as the plastic zone size at fracture [Equation (6.11)], when $K_1 = K_{1c}$.

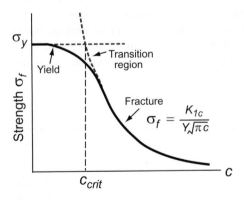

Figure 6.8 The transition from yield to fracture at the critical crack length, c_{crit}.

Example 6.3

Estimate the plastic zone sizes for the cracks in the stainless steel and polystyrene rulers in Example 6.1.

Answer. The plastic zone size is given by

$$r_y = \frac{K_{1c}^2}{Y^2 \pi \sigma_y^2}$$

For stainless steel this is $\dfrac{80^2}{1.1^2 \pi 540^2} = 5.8$ mm – much greater than the crack depth of 0.2 mm and greater than the thickness of the ruler. For polystyrene it is $\dfrac{1^2}{1.1^2 \pi 40^2} = 0.16$ mm, which is less than the crack depth. So the stainless steel ruler yields, and the polystyrene one fractures.

Table 6.1 lists the range of values of c_{crit} for the main material classes. These crack lengths are a measure of the *damage tolerance* of the material. Tough metals are able to contain large cracks but still yield in a predictable, ductile manner. Ceramics (which always contain small cracks) fail in a brittle way at stresses far below their yield strengths. Glass can be used structurally, but requires careful treatment to prevent surface flaws developing. Polymers are perceived as tough, due to their resistance to impact, but this is when they are not cracked. Table 6.1 shows that defects less than 1 mm can be sufficient to cause some polymers to fail in a brittle manner – this is what we exploit in opening packaging by making a small, sharp cut first.

Table 6.1 Approximate crack lengths for transition between yield and fracture

Material class	Transition crack length, c_{crit} (mm)
Metals	1–1000
Polymers	0.1–10
Ceramics and glasses	0.01–0.1
Composites	0.1–10

6.4 Material property charts for toughness

The fracture toughness–modulus chart The fracture toughness K_{1c} is plotted against modulus E in Figure 6.9. The range of K_{1c} is large: from less than 0.01 to over 100 MPa.m$^{1/2}$. At the lower end of this range are brittle materials, which, when loaded, remain elastic until they fracture. For these, linear elastic fracture mechanics works well, and the fracture toughness itself is a well-defined property. At the upper end lie the super-tough materials, all of which show substantial plasticity before they break. For these the values of K_{1c} are approximate but still helpful in providing a ranking of materials. The figure shows one reason for the dominance of metals in engineering; they almost all have values of K_{1c} above 15 MPa.m$^{1/2}$, a value often quoted as a minimum for conventional design.

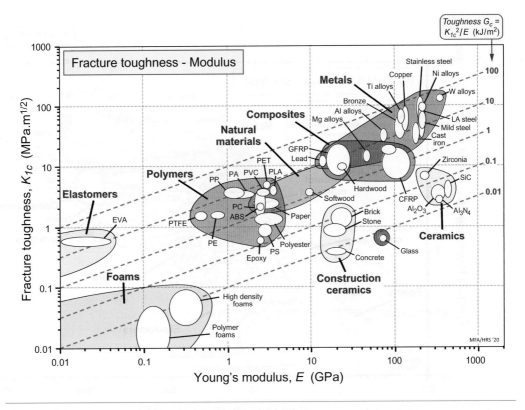

Figure 6.9 A chart of fracture toughness K_{1c} and Young's modulus E. The contours show the toughness, G_c.

The log scales of Figure 6.9 readily allow us to plot contours of *toughness*, G_c (since $G_c \approx K_{1c}^2/E$). The diagonal broken lines on the chart, with a slope of ½, show values of the toughness, starting at 10^{-3} kJ/m² (about equal to the surface energy, γ) and ranging through 5 decades to over 100 kJ/m². On this scale, ceramics (10^{-3} – 0.1 kJ/m²) are much lower than polymers (0.1 – 10 kJ/m²); this is part of the reason polymers are more widely used in engineering than ceramics.

The fracture toughness–strength chart   Strength-limited design relies on the component yielding before it fractures. This involves a comparison between strength and fracture toughness. Figure 6.10 shows them on a property chart. Metals are both strong and tough, which is why they have become the workhorse materials of mechanical and structural engineering.

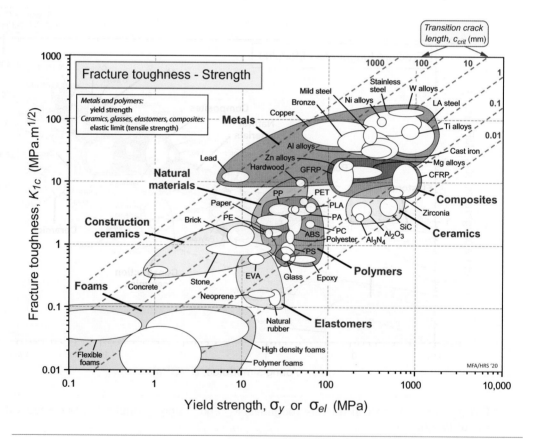

Figure 6.10   A chart of fracture toughness K_{1c} and yield strength, σ_y (or elastic limit σ_{el}). The contours show the transition crack size, c_{crit}.

The stress at which fracture occurs depends on both K_{1c} and the crack length c [Equation (6.12)]. The transition crack length c_{crit} at which ductile behaviour is replaced by brittle is given by Equation (6.13). It is plotted on the chart as dashed lines of slope = 1. The values vary enormously, from near-atomic dimensions for brittle ceramics and glasses to almost a metre for the most ductile of metals like copper or lead. Materials falling on the dashed lines towards the bottom right have high strength and low toughness; they *fracture before they yield*. Those towards the top left do the opposite: they *yield before they fracture*.

The diagram has application in selecting materials for the safe design of load-bearing structures (Section 6.10). The strength–fracture toughness chart is also useful for assessing the influence of composition and processing on properties.

6.5 The physical origins of toughness

Surface energy   As noted previously, the surface energy of a solid is the energy it costs to make it, per unit area (units J/m^2). Think of taking a 1m cube of material and cutting it in half to make two new surfaces, one above and one below, as in Figure 6.11. To do so we have to provide the cohesive energy associated with the bonds that previously connected across the cut. The atoms are bonded on all sides so the surface atoms lose one-sixth of their bonds when the cut is made. This means that we have to provide one-sixth of the cohesive energy H_c (an energy per unit volume) for a double layer of atoms $4r_o$ thick, where r_o is the atom radius, and thus to a volume $4r_o$ m^3. So the surface energy should be:

$$2\gamma = \frac{1}{6}H_c.4r_o \text{ or } \gamma = \frac{1}{3}H_c.r_o \qquad (6.14)$$

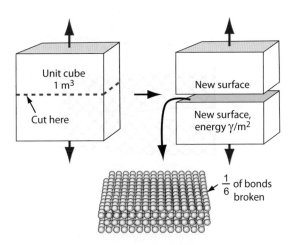

Figure 6.11   The creation of new surface – atomic bonds are broken, requiring some fraction of the cohesive energy, H_c.

The cohesive energy H_c is typically 3×10^{10} J/m³ and r_o typically 10^{-10} m, so surface energies are around 1 J/m², and the toughness G_c cannot be less than 2γ. From the contours of G_c on the chart of Figure 6.9 (in kJ/m²), we see that for most materials its value is hundreds of times larger than 2γ. Where is the extra energy going? The answer is: into plastic work. We will examine that in more detail in a moment. First, let's examine brittle cleavage fracture.

Brittle 'cleavage' fracture   Brittle fracture is characteristic of ceramics and glasses. These have very high yield strengths, giving them no way to relieve the crack tip stresses by plastic flow. This means that, near the tip, the stress reaches the ideal strength (about $E/15$; see Chapter 5). That is enough to tear the atomic bonds apart, allowing the crack to grow as in Figure 6.12. Since $K_1 = \sigma\sqrt{\pi c}$, an increase in c means an increase in K_1, exceeding K_{1c}, causing the crack to accelerate until it reaches the speed of sound, which is why brittle materials fail with a bang, and it is often called 'fast fracture.' Some polymers are brittle, particularly the amorphous ones, like perspex and polystyrene. The crack tip stresses unzip the weak van der Waals bonds between the molecules.

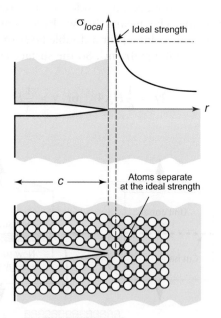

Figure 6.12   Cleavage fracture. The local stress rises as $1/\sqrt{r}$ towards the crack tip. If it exceeds that required to break inter-atomic bonds (the 'ideal strength'), then they separate, giving a cleavage fracture with little energy being absorbed.

Tough 'ductile' fracture To understand how cracks propagate in ductile materials, think first of pulling a sample with no crack (Figure 6.13). Ductile metals deform plastically when loaded above their yield strength, work hardening until the tensile strength is reached. Thereafter, they weaken and fail. What causes the weakening? If ultra-pure, the metal may simply neck down until the cross section goes to zero. However, engineering alloys are not ultra-pure; almost all contain inclusions – small, hard particles of oxides, nitrides, sulphides, and the like. As the material is stretched, it deforms at first in a uniform way, building up stress at the inclusions, which act as stress concentrations. These either separate from the matrix or fracture, nucleating tiny voids [Figure 6.13(c)]. These voids grow as strain increases, linking and weakening the part of the specimen in which they are most numerous until they finally coalesce to give a ductile fracture. Many polymers, too, are ductile. They don't usually contain inclusions, but when stretched they craze [see Figure 5.8(b)] – tiny cracks open up as voids that are initially bridged by molecules, causing whitening if the polymer is transparent. The details differ, but the result is the same: the crazes nucleate, grow, and link to give a ductile fracture.

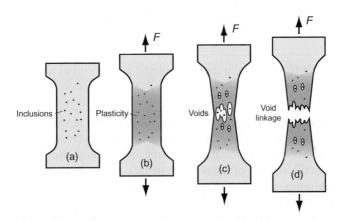

Figure 6.13 Ductile fracture. Plasticity, shown in brown, concentrates stress on inclusions that fracture or separate from the matrix, nucleating voids that grow and link, ultimately causing fracture.

Return now to a cracked sample (Figure 6.14). As we have seen, the stress rises as $1/\sqrt{r}$ as the crack tip is approached, and a plastic zone develops. Within the plastic zone the same sequence as that of Figure 6.13 takes place: voids nucleate, grow, and link to give a *ductile fracture*. As the crack advances the failure process runs just ahead of the crack tip, dissipating energy by plasticity and increasing the toughness G_c.

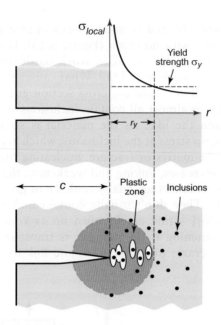

Figure 6.14 If the material is ductile, a plastic zone forms at the crack tip. Within it voids nucleate, grow, and link, advancing the crack in a ductile mode, absorbing energy in the process.

The ductile-to-brittle transition   A cleavage fracture is much more dangerous than one that is ductile: it occurs without warning or any prior plastic deformation. At low temperatures some metals and all polymers become brittle, and the fracture mode switches from one that is ductile to one of cleavage – in fact, only those metals with an FCC structure (copper, aluminium, nickel, and stainless steel, for example) remain ductile to the lowest temperatures. All others have yield strengths that increase as the temperature falls, with the result that the plastic zone at any crack tip becomes smaller and the fracture mode switches, giving a *ductile-to-brittle transition*. For some steels this transition temperature is as high as 0°C (though for most it is considerably lower), with the result that steel ships, bridges, and oil rigs are more susceptible to fracture in winter in high latitudes and in polar regions all year round. Polymers, too, have a ductile-to-brittle transition associated with their glass transition, a consideration in selecting those that are to be used in freezers and fridges.

Embrittlement of other kinds   Change of temperature can lead to brittleness; so, too, can *chemical segregation*. When metals solidify, the grains start as tiny solid crystals suspended in the melt, growing outward until they impinge to form grain boundaries. The boundaries, being the last bit to solidify, end up as the repository for the impurities in the alloy (more details in *Guided Learning Unit 4*). This grain boundary segregation can create a network of low-toughness paths through the material so that, although the bulk of the grains is tough, the material as a whole fails by brittle intergranular fracture (Figure 6.15). The locally different chemistry of grain boundaries causes other problems, such as corrosion (see Chapter 9), which is another way in which cracks can appear in initially defect-free components.

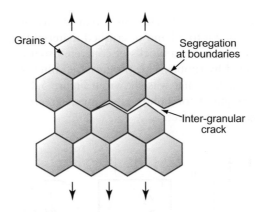

Figure 6.15 Chemical segregation can cause brittle inter-granular cracking.

6.6 Failure of ceramics

Ceramics are brittle materials in tension, with a low fracture toughness (Figures 6.10 and 6.12). Provided we can control the size of the defects, however, ceramics can have useful tensile strength. The difficulty is that failure is not now characterised by a well-defined material property, like a yield stress, but has statistical variability (reflecting the distribution in size and orientation of the underlying cracks in the material). Furthermore, in compression their strength is typically 10 to 15 times greater than their tensile strength – something that we exploit to the full in civil engineering, building with stone, bricks, and concrete. Failure is still controlled by cracking, however, but now as the collective growth of many defects. So the way we design with ceramics doesn't use a fracture toughness criterion, and both compressive and tensile failure processes in ceramics need some further explanation.

Figure 6.16 shows ceramic samples loaded in tension and in compression. Ceramics contain a distribution of micro-cracks, usually of a size related to the solid particles that were compacted to make the component (see Chapter 11). In tension, the 'worst flaw' (in terms of both size and orientation to the applied stress) propagates to failure [Figure 6.16(a)]. Failure in tension must therefore be characterised by a probability of failure at a given stress, reflecting the statistical variability of flaw sizes. Furthermore, this probability displays a dependence on the volume of the sample – larger samples are more likely to contain a large flaw, and therefore on average will fail at lower stresses. In compression, something rather different happens. The cracks can still extend, but in a stable manner, as illustrated in Figure 6.16(b). Any crack that is inclined to the compressive stress axis is subjected to a combination of compression and shear. Shear stresses applied to a sharp crack also generate a crack tip stress field (which we now call 'mode 2'), and part of this field is tensile. The cracks extend at right angles to this induced tension, developing 'wing cracks' that grow, rather surprisingly, *parallel* to the applied compression. In this configuration, crack growth is stable – the crack only continues to extend if the stress is increased. Multiple cracking therefore occurs, with cracks growing different amounts depending on their size and orientation. Eventually the extent of damage to the material is such that an overall sample instability occurs, with the material fragmenting into a band of crushed material that shears away [Figure 6.16(c)]. Compression tests on cylindrical blocks of concrete usually end up as a conical remnant of the sample surrounded by rubble. This progressive failure mechanism also explains the serrated stress–strain response that was shown in Figure 5.1(c). It also explains why the peak compressive stress required for failure is so much greater than that needed in tension to propagate the single worst defect.

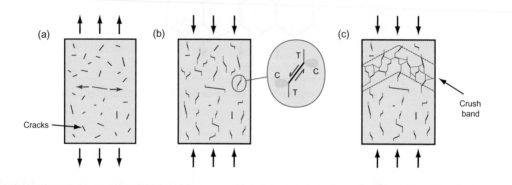

Figure 6.16   Ceramic failure mechanisms: (a) tensile failure from the largest flaw; (b) stable growth of wing cracks in compression, driven by the shear acting on inclined cracks generating crack tip regions of compression (C) and tension (T); (c) final compressive failure by collapse of a crush band of fragmented material.

6.7 Manipulating properties: the strength–toughness trade-off

Metals   It is not easy to make materials that are both strong and tough. The energy absorbed by a crack when it advances, giving toughness, derives from the deformation that occurs in the plastic zone at its tip. Equation (6.11) showed that increasing the yield strength causes the zone to shrink, and the smaller the zone, the smaller the toughness. Figure 6.17 shows fracture toughness and strength for wrought and cast aluminium alloys, indicating the strengthening mechanisms. All the alloys have a higher strength and a comparable or lower fracture toughness than pure aluminium. The wrought alloys show a slight drop in toughness with increasing strength; for the cast alloys the drop is much larger, in part because of intergranular fracture, but also because aluminium casting alloys contain a lot of silicon as a separate 'phase' within the microstructure, and this silicon is brittle (more details in Chapter 11).

Separation within the plastic zone, allowing crack advance, results from the growth of voids that nucleate at inclusions (Figure 6.14). Toughness is increased with no loss of strength if the inclusions are removed, delaying the nucleation of the voids. 'Clean' steels, superalloys and aluminium alloys, made by filtering the molten metal before casting, have significantly higher toughness than those made by conventional casting methods.

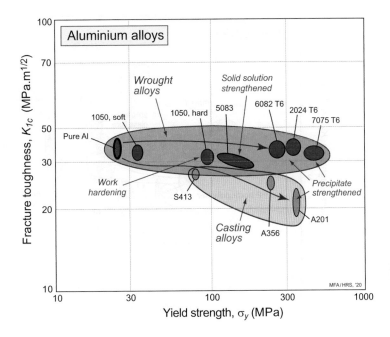

Figure 6.17   The fracture toughness and yield strength of wrought and cast aluminium alloys.

Polymers and composites   The microstructures and properties of polymers are manipulated by cross-linking and by adjusting the molecular weight and degree of crystallinity. Figures 6.9 and 6.10 showed that the strengths of polymers span around a factor of 5, whereas fracture toughness spans a factor of 20. More dramatic changes are possible by *blending*, by adding *fillers*, and by *reinforcement* with chopped or continuous fibres to form composites. Figure 6.18 shows how these influence the modulus E and fracture toughness K_{1c} of polypropylene (PP). Blending or co-polymerisation with elastomers such as EPR or EDPM ('impact modifiers') reduces the modulus but increases the fracture toughness K_{1c} and toughness G_c. Filling with cheap powdered glass, talc, or calcium carbonate more than doubles the modulus, but at the expense of some loss of toughness.

Reinforcement with glass or carbon fibres most effectively increases both modulus and fracture toughness K_{1c}, and in the case of glass, the toughness G_c as well. How is it that a relatively brittle polymer ($K_{1c} \approx 1$ MPa.m$^{1/2}$) mixed with even more brittle fibres (glass $K_{1c} \approx 0.8$ MPa.m$^{1/2}$) can give a composite K_{1c} as high as 10 MPa.m$^{1/2}$? The answer is illustrated in Figure 6.19. The fine fibres contain only tiny flaws and consequently have high strengths. When a crack grows in the matrix, the fibres remain intact and bridge the crack. This promotes multiple cracking – each contributing its own energy and thereby raising the

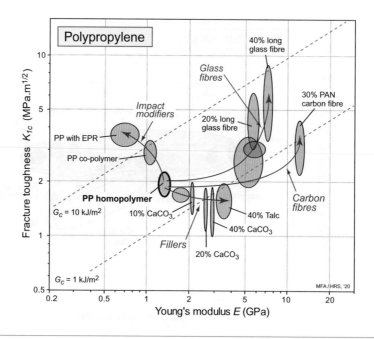

Figure 6.18 The fracture toughness and Young's modulus of polypropylene (PP), showing the effect of fillers, impact modifiers, and fibres.

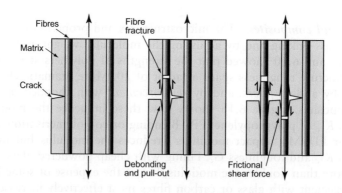

Figure 6.19 Toughening by fibres. The pull-out force opposes the opening of the crack.

overall dissipation. When the fibres do break, the breaks are statistically distributed, leaving ligaments of fibre buried in the matrix. *Fibre pull-out*, as the cracks open up, dissipates more energy by friction. Composites are processed to control the adhesion between the fibres and the matrix to maximise the toughening by these mechanisms.

6.8 Fatigue

The problem of fatigue Low-amplitude vibration causes no permanent damage in materials. Increase the amplitude, however, and the material starts to suffer fatigue. You will, at some time or other, have used fatigue – to flex the lid off a food can, or to break an out-of-date credit card in two. In metals, the cyclic stress hardens the material and causes damage such as dislocation tangles to accumulate; from this a crack nucleates and grows until it reaches the critical size for fracture. There are many potential candidates for fatigue failure – anything that is repeatedly loaded (like an oil rig loaded by the waves, or a pressure vessel that undergoes pressure cycles), or rotates under bending loads (like an axle), or reciprocates under axial load (like an automobile connecting rod), or vibrates (like the rotor of a helicopter).

Figure 6.20 shows in a schematic way how the stress on the underside of an aircraft wing might vary during a typical flight, introducing some important practicalities of fatigue. First, the amplitude of the stress varies as the aircraft takes off, cruises at altitude, bumps its way through turbulence, and finally lands. Second, on the ground the weight of the engines (and the wings themselves) bends them downward, putting the underside in compression; once in flight, aerodynamic lift bends them upward, putting the underside in tension. So during a flight the amplitude of the stress cycles varies, but so does the mean stress. Fatigue failure depends on both of these, and of course on the total number of cycles. Aircraft wings bend to and fro at a frequency of a few hertz. In a trans-Atlantic flight, tens of thousands of loading cycles take place; in the lifetime of the aircraft, it is millions. For this reason, fatigue testing needs to apply tens of millions of fatigue cycles to provide meaningful design data.

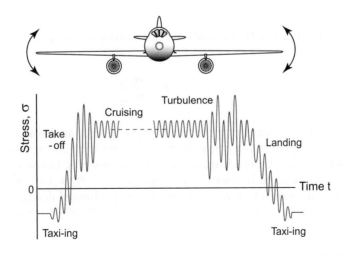

Figure 6.20 Schematic of stress cycling on the underside of an aircraft wing during a flight.

The food can and credit card are examples of *low-cycle fatigue*, meaning that the component survives for only a small number of cycles. It is typical of cycling at stresses above the yield stress, σ_y. More significant in engineering terms is *high-cycle fatigue*: here the stresses remain generally elastic and may be well below σ_y, but cracks nonetheless develop and cause failure, albeit taking many more cycles to do so.

In both cases we are dealing with components that are initially undamaged, containing no cracks. In these cases, most of the fatigue life is spent generating the crack. Its growth to failure occurs only at the end. We call this *initiation-controlled* fatigue. Some structures contain cracks right from the word go, or are so safety-critical (like aircraft) that they are assumed to have small cracks. There is then no initiation stage – the crack is already there – and the fatigue life is *propagation controlled* – that is, it depends on the rate at which the crack grows. A different approach to design is then called for (we return to this later).

High-cycle fatigue and the S–N curve   Figure 6.21 shows how fatigue characteristics are measured and plotted. A sample is cyclically stressed with peak-to-peak amplitude $\Delta\sigma$ about a mean value σ_m, and the number of cycles to cause fracture, N_f, is recorded. The data are presented as $\Delta\sigma - N_f$ ('S – N') curves. Most tests use a sinusoidally varying stress with a mean-to-peak amplitude σ_a such that

$$\sigma_a = \frac{\Delta\sigma}{2} = \frac{\sigma_{max} - \sigma_{min}}{2} \tag{6.15}$$

and a mean stress σ_m of

$$\sigma_m = \frac{\sigma_{max} + \sigma_{min}}{2} \tag{6.16}$$

as illustrated in Figure 6.21. Fatigue data are usually reported for a specified *R-value*:

$$R = \frac{\sigma_{min}}{\sigma_{max}} \tag{6.17}$$

An R-value of -1 means that the mean stress is zero; an R-value of 0 means the stress cycles from 0 to σ_{max}. For many materials there exists a *fatigue* or *endurance limit*, σ_e (units: MPa). It is the stress amplitude σ_a, about zero mean stress, below which fracture does not occur at all, or occurs only after a very large number ($N_f > 10^7$) of cycles. Design against high-cycle fatigue is therefore very similar to strength-limited design but with the maximum stresses limited by the endurance limit σ_e rather than the yield stress σ_y.

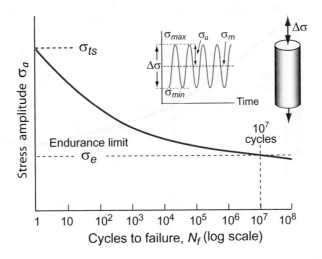

Figure 6.21 An *S–N* curve, with the fatigue strength at 10^7 cycles defining the endurance limit, σ_e.

Experiments show that the high-cycle fatigue life is approximately related to the stress range by what is called Basquin's law:

$$\Delta\sigma\, N_f^b = C_1 \tag{6.18}$$

where b and C_1 are empirical constants; the value of b is small, typically 0.07 and 0.13. Dividing $\Delta\sigma$ by the modulus E gives the strain range $\Delta\varepsilon$ (since the sample is elastic):

$$\Delta\varepsilon = \frac{\Delta\sigma}{E} = \frac{C_1/E}{N_f^b} \tag{6.19}$$

or, taking logs,

$$\log(\Delta\varepsilon) = -b\,\log(N_f) + \log(C_1/E)$$

This is plotted in Figure 6.22, giving the right-hand, high-cycle fatigue part of the curve with a slope of $-b$.

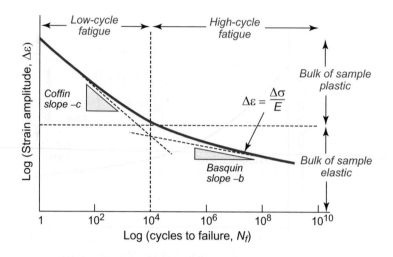

Figure 6.22 The low- and high-cycle regimes of fatigue and their empirical description.

Example 6.4

The fatigue life of a component obeys Basquin's law [Equation (6.18)] with $b = 0.1$. The component is loaded cyclically with a sinusoidal stress of amplitude 100 MPa (stress *range* of 200 MPa) with zero mean, and has a fatigue life of 200,000 cycles. What will be the fatigue life if the stress amplitude is increased to 120 MPa (stress *range* = 240 MPa)?

Answer. From Equation (6.18), $C_1 = \Delta\sigma_1 \, N_1^b = \Delta\sigma_2 \, N_2^b$,

hence $N_2 = N_1 \left(\dfrac{\Delta\sigma_1}{\Delta\sigma_2} \right)^{1/b} = 200{,}000 \left(\dfrac{200}{240} \right)^{10} = 32{,}300$ cycles

This is an 84% reduction in fatigue life for a 20% increase in stress amplitude. Fatigue is very sensitive to stress level.

Low-cycle fatigue. In low-cycle fatigue the peak stress exceeds yield, so at least initially (before work hardening raises the strength), the entire sample is plastic. Basquin is no help to us here; we need another empirical law, this time that of Dr. Coffin:

$$\Delta\varepsilon^{pl} = \frac{C_2}{N_f^c} \tag{6.20}$$

where $\Delta\varepsilon^{pl}$ means the plastic strain range – the total strain minus the (usually small) elastic part. For our purposes we can neglect that distinction and plot it on Figure 6.22 as well – the left-hand part – illustrating that Coffin's exponent, c, is much larger than Basquin's: typically it is 0.5. Figure 6.22 shows that there is a smooth transition between low- and high-cycle behaviour, typically around 10^4 cycles to failure, but this may vary by a factor of 10 either way, depending on the alloy.

High-cycle fatigue: mean stress and variable amplitude   These laws adequately describe the fatigue failure of uncracked components cycled at constant amplitude about a mean stress of zero. But as we saw, real loading histories are often much more complicated (Figure 6.20). How do we make some allowance for variations in mean stress and stress range, particularly for high-cycle fatigue where the effect is strong? Here we need yet more empirical laws, courtesy this time of Goodman and Miner. Goodman's rule relates the stress range $\Delta\sigma_{\sigma_m}$, for failure under a mean stress σ_m, to that for failure at zero mean stress $\Delta\sigma_{\sigma_o}$:

$$\Delta\sigma_{\sigma_m} = \Delta\sigma_{\sigma_o}\left(1 - \frac{\sigma_m}{\sigma_{ts}}\right) \tag{6.21}$$

where σ_{ts} is the tensile strength. So Goodman's rule applies a correction to the stress range, to give the same number of cycles to failure. A tensile mean stress σ_m means that a smaller stress range $\Delta\sigma_{\sigma_m}$ is as damaging as a larger $\Delta\sigma_{\sigma_o}$ applied with zero mean [Figure 6.23(a)]. The corrected stress range may then be plugged into Basquin's law. Note that the Goodman correction can be applied equally well to the stress *amplitude*, as $\sigma_a = \Delta\sigma/2$, so the factor of ½ on both sides of Equation (6.21) will cancel.

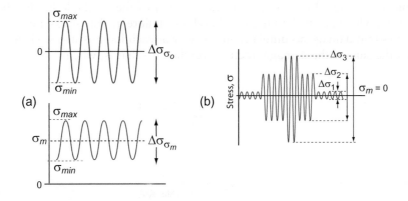

Figure 6.23   (a) *S–N* curves refer to cyclic loading under a zero mean stress. Goodman's rule scales the amplitude to an equivalent value under a mean stress σ_m. (b) When the cyclic stress amplitude changes, the life is calculated using Miner's cumulative damage rule.

Example 6.5

The component in Example 6.4 is made of a material with a tensile strength $\sigma_{ts} = 200$ MPa. If the mean stress is 50 MPa (instead of zero), and the stress amplitude is 100 MPa, what is the new fatigue life?

Answer. Using Goodman's rule [Equation (6.21)] the equivalent stress amplitude for zero mean stress is:

$$\sigma_a = \Delta\sigma_{\sigma_o}/2 = \frac{(\Delta\sigma_{\sigma_m}/2)}{\left(1 - \dfrac{\sigma_m}{\sigma_{ts}}\right)} = \frac{100}{\left(1 - \dfrac{50}{200}\right)} = 133 \text{ MPa}$$

So, following the approach in Example 6.4 with the stress range now 266 MPa, Basquin's law gives the new fatigue life to be

$$N_2 = N_1\left(\frac{\Delta\sigma_1}{\Delta\sigma_2}\right)^{1/b} = 200{,}000\left(\frac{200}{266}\right)^{10} = 11{,}550 \text{ cycles}$$

The variable amplitude problem can be addressed approximately with Miner's *rule of cumulative damage*. Figure 6.23(b) shows an idealised loading history with three stress amplitudes (all about zero mean). Basquin's law gives the number of cycles to failure if each amplitude were to be maintained throughout the life of the component. So if N_1 cycles are spent at stress amplitude $\Delta\sigma_1$, a fraction N_1/N_{f1} of the available life has been used up, so to speak, where N_{f1} is the number of cycles to failure at that stress amplitude. Miner's rule assumes that damage accumulates in this way at each level of stress. Then failure will occur when the sum of the damage fractions reaches 1 – that is, when

$$\sum_{i=1}^{n} \frac{N_i}{N_{f,i}} = 1 \tag{6.22}$$

Example 6.6

The component in Examples 6.4 and 6.5 is loaded for $N_1 = 5000$ cycles with a mean stress of 50 MPa and a stress amplitude of 100 MPa (as in Example 6.5). It is then cycled about zero mean for N_2 cycles with the same stress amplitude (as in Example 6.4) until it breaks. Use Miner's rule [Equation (6.22)] to determine N_2.

> **Answer.** From Equation (6.22)
>
> $$\frac{5000}{11,550} + \frac{N_2}{200,000} = 1$$
>
> So $N_2 = 113,420$: so the 5000 cycles with a mean stress take up less than 5% of the total, but account for half of the fatigue life within Miner's rule.

Goodman's rule and Miner's rule are adequate for preliminary design, but they are approximate; in safety-critical applications, tests replicating service conditions are essential. It is for this reason that new models of cars and trucks are driven over rough 'durability tracks' until they fail – it is a test of fatigue performance.

Fatigue loading of cracked components In fabricating large structures like bridges, ships, oil rigs, pressure vessels, and steam turbines, cracks and other flaws cannot be avoided. Cracks in castings appear because of differential shrinkage during solidification and entrapment of oxide and other inclusions. Welding, a cheap, widely-used joining process, can introduce both cracks and internal stresses caused by the intense local heating. If the cracks are sufficiently large it may be possible to repair them, but finding them is the problem. All non-destructive testing (NDT) methods for detecting cracks have a resolution limit; they cannot tell us that there are no cracks, only that there are none longer than this limit. Thus it is necessary to assume an initial crack exists and design the structure to survive a given number of load cycles. So how is the propagation of a fatigue crack characterised?

Fatigue crack growth is studied by cyclically loading specimens containing a sharp crack of length c like that shown in Figure 6.24. We define the *cyclic stress intensity range*, ΔK, using Equation (6.3), as

$$\Delta K = K_{max} - K_{min} = \Delta \sigma \sqrt{\pi c} \tag{6.23}$$

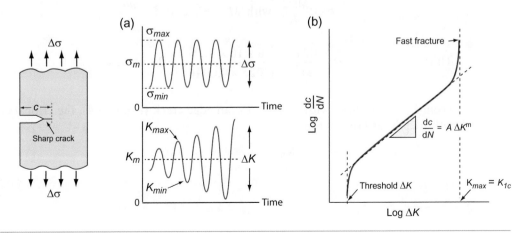

Figure 6.24 Cyclic loading of a cracked component. (a) A constant stress amplitude $\Delta \sigma$ gives an increasing amplitude of stress intensity, $\Delta K = \Delta \sigma \sqrt{\pi c}$ as the crack grows in length. (b) Crack growth rate during cyclic loading.

The range ΔK increases with time under constant cyclic stress because the crack grows in length: the growth per cycle, dc/dN, increases with ΔK in the way shown in Figure 6.24(b). The rate is zero below a threshold cyclic stress intensity ΔK_{th}, useful if you want to make sure it does not grow at all. Above it, there is a steady-state regime described by the Paris law:

$$\frac{dc}{dN} = A\,\Delta K^m \tag{6.24}$$

where A and m are constants. At high ΔK the growth rate accelerates as the maximum applied K approaches the fracture toughness K_{1c}. When it reaches K_{1c} the sample fails at the peak of the final load cycle.

Safe design against fatigue failure in potentially cracked components means calculating the number of loading cycles that can safely be applied without the crack growing to a dangerous length.

Example 6.7

The polystyrene ruler in Example 6.1 is to be loaded as a cantilever with a cyclic force that varies from 0 to 10 N. (a) What depth of transverse crack is needed at the base of the cantilever for fast fracture to occur when the end force is 10 N? (b) If the ruler has an initial transverse scratch of depth $c_i = 0.1$ mm, how many cycles of the force will it take before the ruler breaks? The Paris law constants in Equation (6.24) are $m = 4$ and $A = 5 \times 10^{-6}$ when $\Delta\sigma$ is in MPa. (Assume that the scratch is shallow so that the stress acting on the whole crack is the same as the stress at the surface of the beam).

Answer
(a) The maximum stress in the bending beam is given by Equation (5.25)

$$\sigma = \frac{Mt/2}{I}, \text{ with } M = FL \text{ and } I = \frac{wt^3}{12}$$

Thus
$$\sigma = \frac{6\,FL}{wt^2} = \frac{6 \times 10 \times 0.25}{0.025 \times 0.0046^2} = 28.4 \text{ MPa}$$

From Equation (6.4), with $Y = 1.1$ for an edge crack (Figure 6.4), the critical crack length (depth) at fast fracture is

$$c^* = \frac{K_{1c}^2}{Y^2 \pi \sigma^2} = \frac{1^2}{1.1^2\,\pi\,28.4^2} = 0.33 \text{ mm}$$

(b) From Equation (6.24), the range of stress intensity factor is $\Delta K = Y\Delta\sigma\sqrt{\pi c}$, where $\Delta\sigma = \sigma_{max} - \sigma_{min}$. In this case, $\Delta\sigma = \sigma_{max}$ as $\sigma_{min} = 0$. The crack starts at length c_i and grows steadily according to the Paris law [Equation (6.24)], $dc/dN = A\,\Delta K^m$, assuming that this regime continues until it reaches the critical length c^*, when the ruler breaks.
Combining Equations (6.23) and (6.24) gives:

$$\frac{dc}{dN} = A\,(Y\sigma_{max}\sqrt{\pi c})^m$$

from which, for $m = 4$

$$\int_0^{N_f} dN = \frac{1}{AY^4\,\sigma_{max}^4\,\pi^2}\int_{c_i}^{c^*}\frac{dc}{c^2}$$

where N_f is the number of cycles to failure. Integrating the equation gives:

$$N_f = \frac{-1}{AY^4\,\sigma_{max}^4\,\pi^2}\left[\frac{1}{c^*} - \frac{1}{c_i}\right] = \frac{-1}{5\times10^{-6}\times1.1^4\times28.4^4\,\pi^2}\left[\frac{1}{0.00033} - \frac{1}{0.0001}\right] = 148 \text{ cycles}$$

6.9 The physical origins of fatigue

Fatigue damage and cracking   A perfectly smooth sample with no changes of section, and containing no inclusions, holes, or cracks, would be immune to fatigue provided σ_{max} remains below the yield strength. However, real components contain stress concentrating features (as shown in Figure 5.26) – rivet holes, sharp changes in section, threads, and notches; even surface scratches and roughness can concentrate stress and act as the origins of fatigue cracks. Even though the general stress levels are below yield, the locally magnified stresses can lead to reversing plastic deformation. Dislocation motion is limited to a small volume near the stress concentration, but that is enough to cause damage that develops into a tiny crack.

In high-cycle fatigue, once a crack is present it propagates in the way shown in Figure 6.25(a). During the tensile part of a cycle a tiny plastic zone forms at the crack tip, stretching it open and thereby creating a new surface. On the compressive part of the cycle the crack closes again and the newly formed surface folds forward, advancing the crack. Repeated cycles make it inch forward, leaving tiny ripples on the crack face marking its position on each cycle. These 'striations' are characteristic of a fatigue failure and are useful, in a forensic sense, for revealing where the crack started and how fast it propagated.

In low-cycle fatigue the stresses are higher and the plastic zone larger [Figure 6.25(b)]. It may be so large that the entire sample is plastic, as it is when you flex the lid of a tin to make it break off. The largest strains are at the crack tip, where plasticity now causes voids to nucleate, grow and link, just as in ductile fracture.

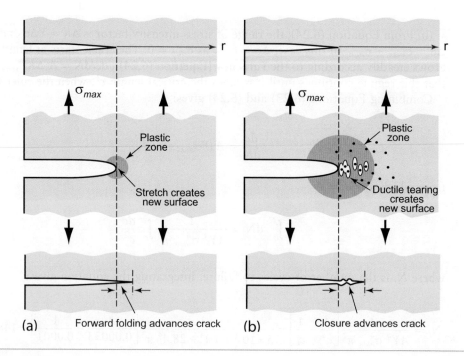

Figure 6.25 (a) In high-cycle fatigue, small-scale crack tip plasticity opens and advances the crack. (b) In low-cycle fatigue, a larger plastic zone nucleates voids, which coalesce to advance the crack.

Manipulating fatigue resistance To improve the fatigue life of a component, the goal is to counter the mechanisms that lead to the generation and growth of cracks. First we should consider higher strength materials – cracks start at the surface by local yielding, so impeding dislocation motion raises the endurance limit. Empirically we find that the endurance limit scales closely with the tensile strength: $\sigma_e \approx \sigma_{ts}/3$. The correlation with yield strength is weaker (due to the differences in work hardening between alloys), but as design is generally based on yield strength, it is more common to define the *fatigue ratio*, F_r, as σ_e/σ_y, and this ratio ranges from 0.3 to 0.9. So in general, the higher the yield strength the greater the fatigue resistance. The second microstructural technique for enhancing fatigue life is to use 'clean' alloys that have been carefully processed to remove unwanted particles, which can act as sites for nucleating fatigue cracks.

Control of conventional shaping, joining, and heat treatment processes also plays a role in managing fatigue. Forming and mechanical joining processes (such as riveting) may produce crack-like defects, bypassing initiation altogether. But in all shaping and machining processes the surface roughness must be considered in fatigue-critical design, due to the role of surface bumps and grooves in starting a crack. Final machining and polishing treatments may specifically be aimed at extending the fatigue life. Heat treatments involving quenching from high temperature, and thermal welding methods, can generate both cracks and thermally induced stresses (see Chapter 7). All processes involving heating and cooling are susceptible to leaving behind 'residual' tensile stresses in a component. These are superimposed on the cyclic service loading, effectively acting as a mean tensile stress and damaging the fatigue life – as we have seen previously in Goodman's rule [Figure 6.23(a)]. Special surface treatments have, however, been developed to deliberately introduce residual *compressive* stresses in the near-surface material, to clamp shut any microcracks developing there under cyclic loading. Shot peening, for example, involves bombarding the surface with small hard spheres to plastically deform a thin layer, which leads to a compressive stress parallel to the surface.

6.10 Fracture and fatigue in design

Selecting materials to resist high-cycle fatigue: con-rods for high-performance engines   The engine of a family car is designed to tolerate speeds up to about 6000 rpm – 100 revolutions per second; in Formula 1 (F1) racing it is three times higher. F1 engines are not designed to last very long before they are replaced – about 30 hours – but 30 hours at around 15,000 rpm is still 3×10^6 cycles, which means design against high-cycle fatigue, with stresses below the endurance limit. The connecting rods of a high-performance engine are therefore fatigue-critical components: if one fails, the engine self-destructs. Yet to minimise inertial forces and bearing loads, each must weigh as little as possible. This implies the use of light, strong materials, stressed near their limits. What, then, are the best materials for such con-rods?

Here we can apply the selection methods introduced in *Guided Learning Unit 2*, for strength-limited design at minimum mass. The only change is that the functional constraint of "must not fail" now sets an upper limit on the applied cyclic stress amplitude being below the endurance limit (rather than the maximum stress being below the yield strength).

Figure 6.26(a) shows an idealised connecting rod, with a uniform shaft of fixed length L, and section area A, which may be varied for each material, as required to meet the design constraint, which is that the con-rod must carry a specified cyclic load $\pm F$ without fatigue failure. The mass of the shaft (the objective) is

$$m = AL\rho \tag{6.25}$$

where ρ is the density. The functional constraint requires that

$$\frac{F}{A} \leq \sigma_e$$

where σ_e is the endurance limit of the material. Using this to eliminate the free variable A in Equation (6.25) gives an equation for the mass:

$$m \geq FL\left(\frac{\rho}{\sigma_e}\right)$$

containing the material index that should be maximised to give the lightest con-rod

$$M = \frac{\sigma_e}{\rho} \qquad\qquad (6.26)$$

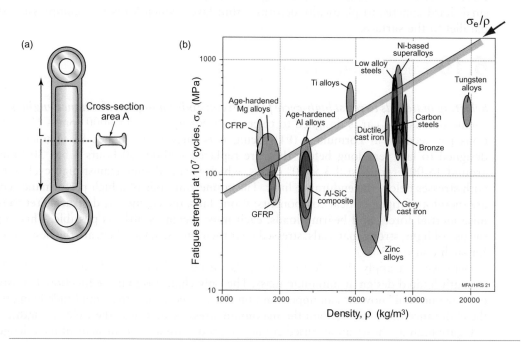

Figure 6.26　(a) A connecting rod; (b) property chart for endurance limit and density, for high-strength metals and composites.

Materials with high values of this index are identified by creating a chart with σ_e and ρ as axes [Figure 6.26(b)]. Here we have applied an additional, standard constraint that the fracture toughness exceeds 15 MPa.m$^{1/2}$ – a common 'rule-of-thumb' in mechanical and structural design (see later). Following the methods of *Guided Learning Unit 2*, a constant value of the index σ_e/ρ has a slope = 1 on this chart, and materials along a given line have equal mass. Materials towards the top-left corner are the lightest candidates: high-strength magnesium, aluminium and titanium alloys, and ultra-high-strength steels. CFRP looks better still – if we can find a way to make the complex shape from a fibre-based material, and to embed inserts of a suitable hard material for the bearing surfaces – manufacturing details such as these often impose limits on the ability to replace metals with fibre composites, outweighing their attractive properties. If instead the con-rod was for a truck, and minimising cost was the objective, then the index is simply modified as in *Guided Learning Unit 2*, replacing density ρ with (ρC_m), where C_m is the material cost/kg. For this index, $\sigma_e/(\rho C_m)$, we find a common material choice for con-rods is cast iron.

Material indices for fracture-safe design As we have seen in previous chapters, many branches of engineering are concerned with *load-limited design*, often under tension or bending: from the structural members of bridges to the wings and fuselage of an aircraft, and from sports equipment and bicycles to mobile device packaging and artificial hip replacements. In all cases, the avoidance of brittle fracture is essential, so a simple rule-of-thumb is applied to the fracture toughness in selecting materials: avoid materials with a fracture toughness $K_{1c} < 15$ MPa.m$^{1/2}$. We can also think of this as a simple performance index M_1 to maximise the load that would cause fracture, and thereby favour yield over fracture (as in Figure 6.8)

$$M_1 = K_{1c} \tag{6.27}$$

Figure 6.27 shows the $K_{1c} - E$ chart with this index plotted at the cut-off value of <15 MPa.m$^{1/2}$. Almost all metals pass: the best have values higher than 100 MPa.m$^{1/2}$, but of the light alloys we find Mg alloys are close to the limit. Fibre composites, too, are on this limit – another reason that replacing metals with composites for lightweight design is not so straightforward. Engineering ceramics have values in the range 1 to 10 MPa.m$^{1/2}$, and glasses are lower still – engineers use these with great caution under tensile or bending loads (while recognising their great value in compression). But engineering polymers have even smaller values of K_{1c} in the range 0.5 to 5 MPa.m$^{1/2}$ and yet engineers use them all the time. To explain this, we need to consider a different design scenario: that of *deflection-limited* design, as in snap-on bottle tops and flexible elastic hinges. These must allow sufficient elastic *deflection* without failure, requiring a large failure *strain* ε_f. The strain is related to the stress by Hooke's law, $\varepsilon = \sigma/E$, and the stress is limited by the fracture Equation (6.4). Thus for a given maximum defect size c_{max}, the failure strain is

$$\varepsilon_f = \frac{1}{Y\sqrt{\pi c_{max}}} \frac{K_{1c}}{E}$$

The best materials for deflection-limited design are therefore those with large values of

$$M_2 = \frac{K_{1c}}{E} \tag{6.28}$$

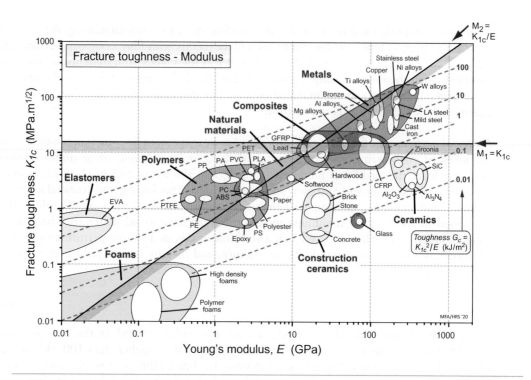

Figure 6.27 The $K_{1c} - E$ chart showing the indices $M_1 = K_{1c}$ and $M_2 = K_{1c}/E$.

Figure 6.27 shows this index, too. This illustrates why polymers find such wide application: when the design is deflection-limited, many of them are better than the best metals. We should also remember that part of the link between each index and the load or strain is the inherent flaw size. Polymers are relatively easy to process without cracks, giving them their robust impact-resistant performance for children's toys and garden furniture (at least when new); ceramics contain cracks whatever we do, which combined with their high modulus and yield strength makes fracture inevitable.

These performance indices give a good indication of the materials to choose for broad fracture resistance, depending on whether load or strain is dominant. But for guaranteed avoidance of fracture under load, we need to apply fracture mechanics at a deeper level, taking account explicitly of the size of cracks that are (or could be) present. We introduce this with an example of a design that went badly wrong, to introduce the need for *fail-safe design*.

Forensic fracture mechanics: pressure vessels   An aerosol or fizzy beverage can is a pressure vessel. So, too, is a propane gas cylinder, the body of an airliner, the boiler of a power station, and the containment of a nuclear reactor. Their function is to contain a gas under pressure – CO_2, propane, air, or steam. Failure can be catastrophic, causing fatalities and significant economic impact, so when working with gases under pressure, design is safety-critical.

Figure 6.28 shows a truck-mounted propane tank. In this failure case study, a full tank was parked in the sun with the engine running, while the driver left the vehicle to have lunch. Without warning the tank exploded, causing considerable property damage and one death. The tank was made of rolled AISI 1030 (medium carbon) steel plate formed into a cylinder, with a longitudinal seam weld, and with two hemispherical domes joined at the ends by circumferential welds. The failure occurred through the longitudinal weld (Figure 6.28), causing the tank to burst. Subsequent examination showed that the weld had contained a surface crack of depth 10 mm that had been there for a long time, as indicated by discolouration by corrosion, and by 'striations' showing that it was growing slowly by fatigue each time the tank was emptied and refilled. At first sight it appeared that the crack was the direct cause of the failure.

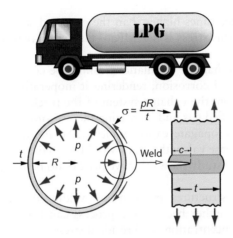

Figure 6.28   A cylindrical pressure vessel with a cracked weld.

Fracture mechanics can help us to explain what happened. Post-mortem measurements on a section of the tank wall gave a fracture toughness of 45 MPa.m$^{1/2}$. The design specification for the tank is listed in Table 6.2. The working pressure was limited for safety by a pressure release valve, set to 1.5 MPa.

Table 6.2   Tank design specification

Working design pressure, p	1.4 MPa
Wall thickness, t	14 mm (thicker at weld)
Outer diameter, $2R$	1680 mm
Length, L	3710 mm

The greatest tensile stress in the wall of a thin-walled cylindrical pressure vessel of radius R and wall thickness t containing a pressure p is in the circumferential (or 'hoop') direction

(Figure 6.28). Stresses in pressure vessels were analysed in Chapter 5 – using Equation (5.29), the working stress in the tank should have been

$$\sigma = \frac{pR}{t} = 84 \text{ MPa} \tag{6.29}$$

A plate with a fracture toughness $K_{1c} = 45$ MPa.m$^{1/2}$ containing a surface crack of depth 10 mm will fail at a stress (assuming $Y = 1.1$) of

$$\sigma = \frac{K_{1c}}{Y\sqrt{\pi c}} = \frac{45}{1.1\sqrt{\pi 0.01}} = 231 \text{ MPa}$$

This is nearly three times higher than the expected stress in the tank wall. The pressure needed to generate this stress is 3.9 MPa – far higher than the limit set by the safety valve of 1.5 MPa.

At first it appears that the calculations cannot be correct. But an inspection of the relief valve showed rust and corrosion, rendering it inoperative. It was concluded that the heat from the sun and from the exhaust system of the truck raised the temperature of the tank, vapourising the liquefied gas and increasing the pressure up to the value of 3.9 MPa needed to make the crack propagate catastrophically. So the direct cause of the failure was the jammed pressure release valve; the crack would not have propagated at the normal operating pressure. But the case study shows that we need to do better, and to strive for *fail-safe design*.

Fail-safe design If structures are made by welding or riveting, it is wise to assume that cracks are present – either because the process can lead to various forms of cracking, or because the stress concentrations and residual stresses associated with welded joints accelerate the formation of fatigue cracks. A number of techniques exist for detecting cracks and measuring their length, without damaging the component or structure; this is called *non-destructive testing* (NDT). X-ray imaging uses contrast in the transmitted beam; ultrasonic testing uses sound waves reflected by a crack; and surface cracks can be revealed with fluorescent dyes. All of these techniques carry a risk of operator error, with cracks being overlooked. When this risk is unacceptable, we turn to *proof testing* (e.g. filling a pressure vessel with water and pressurising it to a level above the planned working pressure). This demonstrates that there are no cracks large enough to propagate at the proof stress, and therefore it is definitely safe during service at a lower pressure. (Note that water is used, as it stores much less energy than gas at a given pressure, and there is a chance that failure will occur during the test, but this avoids an explosion). Proof testing therefore sets an upper limit on the crack size that could be present. So all of these techniques have a resolution limit, or a crack size c_{lim} below which detection is not possible. They do not demonstrate that the component or structure is crack-free, only that there are no cracks larger than the resolution limit. Proof testing is expensive, but it does guarantee a maximum crack size that could be present. How do we now design our pressure vessel, allowing for the possible presence of a crack of known length?

The first condition applies the load-limited index from above, ensuring that the stresses everywhere must be less than that required to make a crack of length c_{lim} propagate. Applying this condition gives the allowable pressure:

$$p \le \frac{t}{R} \frac{K_{1c}}{\sqrt{\pi c_{lim}}} \tag{6.30}$$

So the largest pressure (for a given R, t, and c_{lim}) is carried by the material with the greatest value of

$$M_1 = K_{1c} \tag{6.31}$$

Due to the high cost, proof testing is something we would only wish to conduct infrequently on a large pressure vessel. But cracks can grow slowly during service because of corrosion or cyclic loading – if there is a crack just below c_{lim} at the start of the lifetime of the vessel, then during service this length may be exceeded. Safety can be ensured by arranging that a crack remains stable, and will first penetrate the outer surface of the vessel before it fractures. In other words, the critical crack length c^*, for failure at the working stress, must exceed the wall thickness t. The vessel will leak, but this is not catastrophic and can be detected. This condition on the design is called the *leak-before-break* criterion, and is illustrated in Figure 6.29. The limiting case is where leaking *just* occurs before fracture, and $c^* = t$, which from Equation (6.4) is given by

$$\sigma = \frac{K_{1c}}{Y\sqrt{\pi t}} \tag{6.32}$$

with $Y \approx 1$ for a semi-circular surface crack.

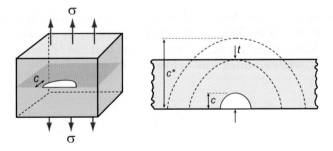

Figure 6.29 A crack of length c in the wall of a pressure vessel, illustrating in plan view *(right)* the leak-before-break criterion: the critical crack length c^* must exceed the wall thickness t.

Example 6.8

A cylindrical steel pressure vessel of 5 m diameter is to hold a pressure p of 4 MPa. The fracture toughness is 110 MPa.m$^{1/2}$, and we assume that a semi-circular crack is oriented normal to the maximum stress σ in the wall. What is the greatest wall thickness that can be used while ensuring that the vessel will fail by leaking rather than fracture, should a crack grow within the wall?

Answer.   For the vessel to leak, the crack length c has to reach the wall thickness t without fracture. Consequently, $c = t$ and $K_{1c} = \sigma\sqrt{\pi t}$, as in Equation (6.32). The hoop stress is the greatest stress in the vessel wall, $\sigma = pR/t$. Combining these gives the maximum wall thickness for which the crack will penetrate the wall when it is just on the point of fast fracture:

$$t = \frac{\pi p^2 R^2}{K_{1c}^2} = \frac{\pi \times 4^2 \times 2.5^2}{110^2} = 0.026 \text{ m} = 26 \text{ mm}$$

Clearly in practice we would require a safety margin, ensuring that the critical crack length is greater than the wall thickness. But for reasons of economy (particularly in a mobile pressure vessel like the propane tank on a truck) we also seek to minimise the mass and hence the wall thickness, increasing the applied stress. So while still ensuring leak-before-break, we must also contain the pressure p without yielding. The maximum pressure that can be carried is when the stress $\sigma = \sigma_y$ giving

$$\frac{pR}{t} = \sigma_y$$

So the pressure that can be carried increases with the wall thickness. By combining this equation with the upper limit on the thickness, given by the leak-before-break criterion [Equation (6.32)], we find the maximum pressure that can be carried when the stress equals the yield stress, while ensuring leak-before-break, is

$$p \leq \frac{1}{Y^2 \pi R} \left(\frac{K_{1c}^2}{\sigma_y} \right)$$

This gives us a material index M_3 indicating the materials that can carry the pressure most safely:

$$M_3 = \frac{K_{1c}^2}{\sigma_y} \tag{6.33}$$

The index M_3 could be made large by using a material with a low yield strength, σ_y: lead, for instance. But we have already noted that we seek to limit the mass, and therefore the thickness, for reasons of economy. The higher the yield strength, the lower the thickness at yield. So we also seek a reasonably high value of

$$M_4 = \sigma_y \tag{6.34}$$

narrowing further the choice of material.

These criteria can be explored using the $K_{1c} - \sigma_y$ chart – Figure 6.30 shows the indices M_1, M_3, and M_4. The fracture toughness is set to the usual lower limit of $M_1 = 15$ MPa.m$^{1/2}$. Applying the 'leak-before-break' criterion, $M_3 = K_{1c}^2 / \sigma_y$ is more restrictive, excluding everything but the toughest metals. A third selection line for yield strength, $M_4 = \sigma_y$, positioned at 200 MPa narrows the selection further. Stainless and other steels and Ni alloys do indeed dominate pressure vessel design, while historically Cu alloys were more common. Tungsten alloys would perform well but are excluded in practice by both mass and cost.

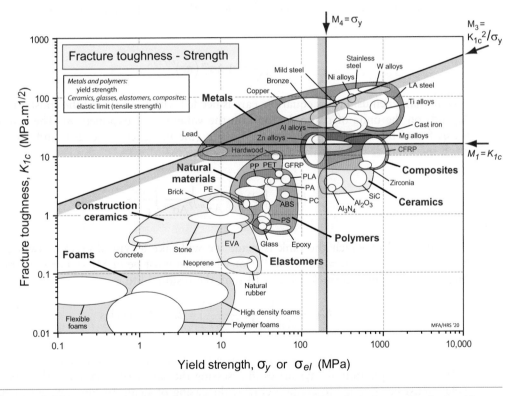

Figure 6.30 Selecting the best materials for leak-before-break design, using index M_3. The additional requirements of acceptable fracture toughness (M_1) and high yield strength (M_4) are also shown. The best choices are the materials in the small area at the top right.

Fatigue crack growth: living with cracks The crack in the LPG tank of the earlier case study had been growing by fatigue, driven by the cyclic pressurisation of the tank. The failure of the safety valve and consequent over-pressure caused the explosion, but if the valve had been maintained, would it have met the leak-before-break criterion, and how long would it have survived?

First we check leak-before-break. To calculate the critical crack length, c^*, we invert Equation (6.4) giving:

$$c^* = \frac{K_{1c}^2}{Y^2 \pi \sigma^2} \tag{6.35}$$

With $K_{1c} = 45$ MPa.m$^{1/2}$, $\sigma = 84$ MPa (the working stress) and $Y = 1$, the answer is $c^* = 91$ mm. This is much greater than the wall thickness $t = 14$ mm, so leak-before-break is comfortably satisfied at the working pressure.

So now we ask, how long would it take the initial crack that was identified in the failure, of length 10 mm, to grow to 14 mm, with failure occurring by leaking? Fatigue-crack growth from a pre-existing crack was described earlier by the Paris law [Equations (6.23) and (6.24)], which may be combined to give

$$\frac{dc}{dN} = A \, \Delta \sigma^m \left(\pi c \right)^{m/2} \tag{6.36}$$

Note that when using this equation, care is needed with the value of the constant A, which depends on the units of c and $\Delta \sigma$, and on the value of the exponent m. The residual life, N_R, is found by integrating this between $c = c_i$ (the initial crack length) and (in this case) $c = t$, the value at which leaking will occur:

$$N_R = \int_0^{N_R} dN = \frac{1}{A \, \pi^{m/2} \Delta \sigma^m} \int_{c_i}^{t} \frac{dc}{c^{m/2}} \tag{6.37}$$

For the steel of which the tank is made we will take a typical crack-growth exponent $m = 4$ and a value of $A = 2.5 \times 10^{-14}$ (for which $\Delta \sigma$ must be in MPa). For the initial crack length $c_i = 10$ mm and thickness $t = 14$ mm, the residual life is

$$N_R = \frac{1}{A \, \pi^2 \Delta \sigma^4} \left(\frac{1}{c_i} - \frac{1}{t} \right) = 2.3 \times 10^6 \text{ cycles}$$

So assuming that the tank is pressurised once per day, it will last forever.

This sort of question always arises in assessing the safety of large plant that (because of welds) must be assumed to contain cracks: the casing of steam turbines, chemical engineering equipment, boilers, and pipework. The cost of replacement can be considerable, with associated downtime. It does not make sense to take it out of service if, despite the crack, it is perfectly safe. This illustrates the significance of *proof testing* to guarantee a safe number of cycles. Note that the guaranteed residual life is increased if the initial crack length in the integral of Equation (6.37) is smaller. This implies increasing the proof stress, but there is a practical limit, as noted earlier: it cannot exceed the yield stress of the material (another reason for favouring high-strength alloys, as in Figure 6.30). However, after the specified number of safe pressurisation cycles, the proof test can be repeated and the same lifetime is guaranteed all over again.

Example 6.9

The cylindrical steel pressure vessel in Example 6.8 is built with a wall thickness of 15 mm, to operate at the same working pressure of 4 MPa. It is to be proof tested to guarantee 2000 safe cycles of pressurisation. The rate of crack growth is given by $dc/dN = A\,(\Delta K)^4$, where $A = 3 \times 10^{-14}$ (with c in metres, and ΔK in units of MPa.m$^{1/2}$).

Use the integral in Equation (6.37) to find the maximum initial crack size that will take 2000 cycles to grow to the wall thickness (giving safe failure by leaking). Hence calculate the pressure required in a proof test to ensure that such a crack does not exist. Use the chart in Figure 6.30 to see if the applied stress looks reasonable relative to the yield stress of steels with fracture toughness of 110 MPa.m$^{1/2}$.

Answer. The stress amplitude in the wall during normal pressurisation cycles is

$$\Delta\sigma = \frac{pR}{t} = \frac{4 \times 2.5}{0.015} = 667 \text{ MPa}$$

From Equation (6.37) with a Paris law exponent $m = 4$

$$N_R = \frac{1}{A\pi^2\Delta\sigma^4}\left(\frac{1}{c_i} - \frac{1}{t}\right) \Rightarrow 2000 = \frac{1}{3 \times 10^{-14} \times \pi^2 \times 667^4}\left(\frac{1}{c_i} - \frac{1}{t}\right) \Rightarrow c_i = 5.4 \text{ mm}$$

The proof stress required to fracture the steel with this crack length is

$$\sigma_{\text{proof}} = \frac{K_{1c}}{Y\sqrt{\pi c_i}} = \frac{110}{1 \times \sqrt{\pi \times 0.0054}} = 845 \text{ MPa}$$

The corresponding proof test pressure is

$$p_{\text{proof}} = \frac{\sigma_{\text{proof}}\,t}{R} = \frac{845 \times 0.015}{2.5} = 5.1 \text{ MPa}$$

From Figure 6.30, the yield strength of steels (stainless, or possibly low alloy [LA]) with fracture toughness of 110 MPa.m$^{1/2}$ is in the range 300 to 1000 MPa, but a high-strength variant must be in use, given the working stress is 667 MPa. So this proof stress may be close to yield and a lower proof pressure may be needed – clearly the limited 'head room' between working and proof stresses governs the value of the starting crack length that must be assumed, and thus the number of cycles that can be guaranteed before a repeat test.

Failures of boilers and other vessels that are repeatedly pressurised used to be common-place, and now they are rare – though when safety margins are pared to a minimum to save weight (e.g. rockets) or maintenance is neglected (as in the propane tank), disasters still occasionally happen. This relative success is one of the major contributions of fracture mechanics to engineering practice and public safety.

6.11 Summary and conclusions

Toughness is the resistance of a material to the propagation of a crack. Tough materials tolerate the presence of cracks, they absorb impact without shattering, and, if overloaded, they yield rather than fracture. The dominance of steel as a structural material derives from its unbeatable combination of low cost, high stiffness and strength, and high toughness.

Resistance to fracture is properly measured by loading a pre-cracked sample with one of a number of standard geometries, measuring the stress or load at which the crack propagates unstably. Fracture is driven by the elastic stress field around the tip of the crack, characterised by the stress intensity factor K_1, with failure occurring when this reaches a critical value K_{1c}: the material property called fracture toughness. Values of K_{1c} above about 15 MPa.m$^{1/2}$ are desirable for damage-tolerant design – design immune to the presence of small cracks. The property charts in this chapter show that most metals and fibre-reinforced composites meet this criterion, but polymers and ceramics fall below it. Polymers compete better in design limited by strain rather than load, but ceramics are always poor in tension; in compression, however, they have considerable strength as tensile fracture is suppressed and material crushing requires much higher stresses.

The physical basis of high toughness is that crack advance absorbs energy – in metals this occurs in the contained plastic zone at the crack tip. In ceramics, with their very high yield strengths, the zone becomes vanishingly small – the crack tip stresses approach the ideal strength and the material fails by brittle cleavage fracture.

Materials are better at supporting static loads than loads that fluctuate. The endurance limit σ_e is the long-term cyclic load amplitude that a material can tolerate and is barely one-third of its tensile strength, σ_{ts}. Empirical rules and data describing fatigue failure give a basis for design to avoid fatigue failure. Different underlying mechanisms and design rules apply for low-cycle fatigue, when cyclic plastic strains dominate, and for high-cycle fatigue in the elastic regime. In uncracked components, the fatigue life is controlled by the time it takes to initiate a fatigue crack, before it runs to failure; with pre-existing cracks, the life is propagation-controlled.

Elastic deformation is recoverable, while plastic deformation is usually gradual and detectable before a catastrophe occurs. Failure by fast fracture, often resulting from fatigue, has been the cause of many great engineering disasters that occurred without warning: bridges collapsing, gas tanks exploding, and aircraft crashing. Large safety-critical structures, particularly those that are welded and then subject to cyclic loading, cannot be assumed to be crack free, and fail-safe design is essential. The chapter closed with case studies on design of pressure vessels to ensure that they leak in a controlled way, rather than by breaking by fracture, and the use of proof testing to guarantee a safe number of pressurisation cycles.

6.12 Further reading

Anderson, T. L. (2021). *Fracture mechanics – fundamentals and applications*. CRC Press. ISBN-13: 978-1498728133. (Anderson introduces the theory and applications of linear and nonlinear fracture mechanics, including computational analysis and modelling).

Broek, D. (2008). *Elementary engineering fracture mechanics* (4th ed.). Martinus Nijhoff. ISBN 978-9024726561. (A standard, well-documented introduction to the intricacies of fracture mechanics).

Hertzberg, R. W. (2012). *Deformation and fracture of engineering materials* (5th ed.). Wiley. ISBN: 978-0-470-52780-1. (A readable and detailed coverage of deformation, fracture, and fatigue).

Kinloch, A. J., & Young, R. J. (1983). *Fracture behaviour of polymers*. Elsevier Applied Science. ISBN 0-85334-186-9. (An introduction both to fracture mechanics as it is applied to polymeric systems and to the fracture behaviour of different classes of polymers and composites).

Suresh, S. (1998). *Fatigue of materials* (2nd ed.). Cambridge University Press. (Suresh presents unified treatment of the mechanics and micromechanisms of fatigue in metals, non-metals, and composites at a level well suited to an introductory course).

Tada, H., Paris, G., & Irwin, G. R. (2000). *The stress analysis of cracks handbook* (3rd ed.) ISBN 978-1860583049. (Here we have another Yellow Pages, like Roark for stress analysis of uncracked bodies. This time it is of stress intensity factors for a great range of geometries and modes of loading).

6.13 Exercises

Exercise E6.1	What is meant by stress intensity factor? How does this differ from the stress concentration factor (defined in Chapter 5)?
Exercise E6.2	What is meant by toughness? How does it differ from strength, and from fracture toughness?
Exercise E6.3	A tensile sample of width 10 mm contains an internal crack of length 0.3 mm. When loaded in tension, the crack suddenly propagates when the stress reaches 450 MPa. What is the fracture toughness K_{1c} of the material of the sample? If the material has a modulus E of 200 GPa, what is its toughness G_c? (Assume the geometric factor $Y = 1$).
Exercise E6.4	Two long wooden beams of square cross section $t \times t = 0.1$ m are butt-jointed using an epoxy adhesive. The adhesive was stirred before application,

trapping air bubbles that, under pressure in forming the joint, deform to flat, penny-shaped cracks of diameter 2 mm (see figure). For these cracks the geometry factor $Y = 0.7$ in the expression for stress intensity factor. The fracture toughness of the epoxy $K_{1c} = 1.3$ MPa.m$^{1/2}$.

(a) If the beams are loaded in axial tension, what is the load F_t that leads to fast fracture?

(b) The beams are now loaded in three-point bending, with a central load F_b and a total span $L = 2$ m. Show that the maximum tensile stress at the surface of the beam is given by $\sigma_b = 3 F_b L / 2t^3$. Assuming that there is a bubble close to the beam surface, find the maximum load that the beam can support without fast fracture.

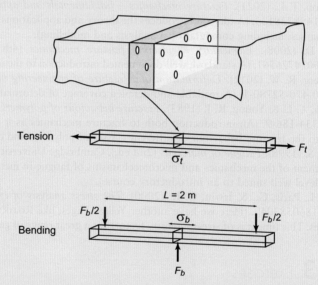

Exercise E6.5   Example 5.9 (Chapter 5) showed the importance of the stress concentrating effect of corners, for the case of the design of an aircraft window. In response to this problem, a revised design changes the corner radius of the window to 20 mm, instead of 5 mm.

(a) Use Equation (5.30) to find the new stress concentration factor at the corner of the window and the resulting maximum stress.

(b) Now suppose there is a small sharp defect near the corner of the window. This means that the effective length of the crack in the fuselage is the size of the window itself. If the fracture toughness of the fuselage material is 25 MPa.m$^{1/2}$ and the window is of length 160 mm, what stress would cause fast fracture? Is the design safe?

Exercise E6.6   A precision instrument is supported on a granite slab to provide a smooth surface and minimise distortion and vibration. The granite has a fracture toughness of 0.9 MPa.m$^{1/2}$ and is known, from NDT procedures, to contain internal cracks up to 3 mm in length. The slab is 2 m long, 1 m wide, and the proposed design has a thickness of 30 mm. It is simply supported at its ends, carrying a uniformly distributed load of 2000 N on its upper surface (as in Figure 5.22).

(a) Find the total distributed load, including the self-weight of the slab. (The density of granite is 2700 kg/m^3).

(b) Find the maximum bending moment at the middle of the slab, and hence find the maximum tensile stress on the lower surface.

(c) Assuming that one of the cracks lies at the surface, perpendicular to this highest tensile stress (so the factor $Y = 1.1$ in calculating K_1), show that the design load is safe, and find the safety factor between the fracture and design loads.

Exercise E6.7   Why does a plastic zone form at the tip of a crack when the cracked body is loaded in tension?

Exercise E6.8   A compact tension specimen (Figure 6.5) with a crack length $c = 10$ mm was used to measure the fracture toughness of a low alloy steel of yield strength 1000 MPa. The width w (from the loading holes) was 50 mm, and the thickness b was 10 mm.

(a) The failure load was 6.5 kN. What was the measured fracture toughness?

(b) For the test result to be valid, the plastic zone size must be small compared to the crack length and the specimen thickness. Check whether this was the case.

Exercise E6.9   Why is there a transition from ductile to brittle behavior at a transition crack length, c_{crit}?

Exercise E6.10   Suppose that the resolution limit of the non-destructive testing (NDT) facility available to you is 1 mm, meaning that it can detect cracks of this length or larger. You are asked to explore which materials will tolerate cracks equal to or smaller than this without brittle fracture. Using data for K_{1c} from Appendix A, calculate the fracture stress $\sigma_f = K_{1c}/\sqrt{\pi c}$ for an internal crack of length $2c = 1$ mm for cast iron, low alloy steel, titanium alloy, silicon carbide, and CFRP. Compare these with the values in Appendix A for the yield strengths σ_y (for the metals) and the tensile strengths σ_{ts} (for SiC and CFRP). Which materials will yield, and which will fracture?

Exercise E6.11   Use the indicative data for fracture toughness and strength in Appendix A to compare the transition crack length c_{crit} of titanium alloy and silicon carbide (using yield strength σ_y for Ti alloy, and tensile strength σ_{ts} for SiC). Comment on the ductility of the two materials.

Exercise E6.12   An engineering rule-of-thumb is that for 'conventional' load-limited design, materials should have a fracture toughness K_{1c} greater than 15 MPa.m$^{1/2}$ – otherwise brittle fracture is a concern. Compare metals, ceramics, and composites using the property chart in Figure 6.9. Which materials need special attention during design because they are brittle? What are some of their uses?

Exercise E6.13   Use the $K_{1c} - E$ chart of Figure 6.9 to establish whether:

- CFRP has a higher fracture toughness K_{1c} than aluminium alloys.
- Polypropylene (PP) has a higher toughness G_c than aluminium alloys.
- Polycarbonate (PC) has a higher fracture toughness K_{1c} than glass.

Exercise E6.14   Find epoxy, glass, and GFRP (epoxy reinforced with glass fibres) on the chart of Figure 6.9 and read off an approximate mean value for the toughness G_c for each. Explain how the toughness of the GFRP is so much larger than that of either of its components.

Exercise E6.15   Use the chart of Figure 6.9 to compare the fracture toughness K_{1c} of the two composites GFRP and CFRP. Do the same for their toughness, G_c. What do the values suggest about applications they might best fill?

Exercise E6.16   Materials with high toughness G_c generally have high modulus. Sometimes, however, the need is for high toughness with low modulus, so that the component has some flexibility. Use the chart of Figure 6.9 to find the material (from among those on the chart) that has a modulus less than 0.5 GPa and the highest toughness G_c. List applications of this material that you think exploit this combination of properties.

Exercise E6.17   Use the $K_{1c} - \sigma_y$ chart of Figure 6.10 to find the range of transition crack sizes for: stainless steel, polycarbonate (PC), and alumina (Al$_2$O$_3$).

Exercise E6.18   Distinguish between low-cycle and high-cycle fatigue. Find examples of engineering components that may fail by high-cycle fatigue. What is meant by the endurance limit, σ_e, of a material?

Exercise E6.19   The figure shows an *S–N* curve for AISI 4340 low alloy steel, hardened to a tensile strength of 1800 MPa.

(a) What is the endurance limit?
(b) If cycled for 100 cycles at an amplitude of 1200 MPa and a zero mean stress, will it fail?
(c) If cycled for 100,000 cycles at an amplitude of 900 MPa and zero mean stress, will it fail?
(d) If cycled for 100,000 cycles at an amplitude of 800 MPa and a mean stress of 300 MPa, will it fail?

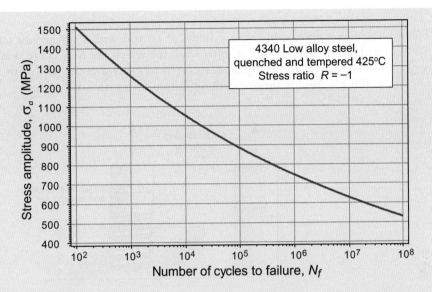

Exercise E6.20   The high-cycle fatigue life, N_f, of an aluminium alloy is described by Basquin's law:

$$\Delta\sigma = 960\left(N_f\right)^{-0.12}$$

(where $\Delta\sigma$ is the stress range in MPa). How many cycles will the material tolerate at a stress amplitude σ_a of $\pm\,70$ MPa and zero mean stress? How will this change if the mean stress is 10 MPa? What if the mean stress is -10 MPa? The tensile strength of the alloy is 200 MPa.

Exercise E6.21   The low-cycle fatigue of an aluminium alloy is described by Coffin's law:

$$\Delta\varepsilon^{pl} = \frac{0.2}{N_f^{0.5}}$$

How many cycles will the material tolerate at a plastic strain amplitude $\Delta\varepsilon^{pl}$ of 2%?

Exercise E6.22   A material with a tensile stress $\sigma_{ts} = 350$ MPa is loaded cyclically about a mean stress of 70 MPa. If the stress amplitude σ_a that will cause fatigue fracture in 10^5 cycles under zero mean stress is $\pm\,60$ MPa, what stress range about the mean of 70 MPa will give the same life?

Exercise E6.23   A component made of the AISI 4340 steel with a tensile strength of 1800 MPa and the S–N curve shown in Exercise E6.19 is loaded cyclically between 0 and 1200 MPa. What is the R-value and the mean stress, σ_m? Use Goodman's rule to find the equivalent stress amplitude for an R-value of -1, and read off the fatigue life from the S–N curve.

Exercise E6.24   The figure shows, in (a), a cylindrical tie rod with diameter 20 mm. It is made from AISI 4340 low alloy steel, for which the *S–N* curve was shown in Exercise E6.19. The plan is to use it to carry a cyclic load with a range ± 200 kN. Will it survive without failure for at least 10^5 cycles?

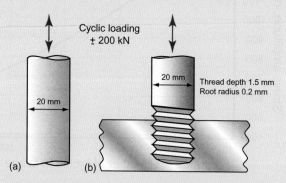

Exercise E6.25   The component of the previous exercise was made and tested. It failed in less than 10^5 cycles. A post-mortem revealed that it had fractured at the root of a threaded end, shown at (b) in the figure, which acted as a stress concentration. The threads have a depth of 1.5 mm and a root radius of 0.2 mm. Given this additional information, how many cycles would you expect it to survive?

Exercise E6.26   The figure shows a component to be made from the high-strength aerospace alloy Ti-6Al-4V. It will be loaded cyclically at ± 210 MPa. Taking account of the stress concentrating corner, predict how long will it last.

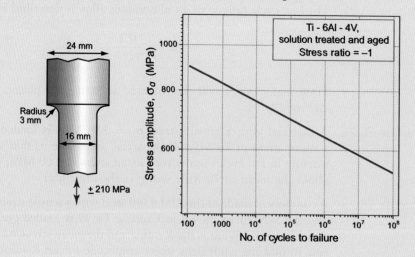

Exercise E6.27   Some uncracked bicycle forks are subject to fatigue loading. Approximate
S–N data for the material used are given in the figure, for zero mean stress.
This curve shows a 'fatigue limit': a stress amplitude below which the life is
infinite.

(a) The loading cycle due to road roughness is assumed to have a constant
stress range $\Delta\sigma$ of 1200 MPa and a mean stress of zero. How many
loading cycles will the forks withstand before failing?

(b) Due to a constant rider load the mean stress is 100 MPa. Use Goodman's
rule to estimate the percentage reduction in lifetime associated with this
mean stress. The tensile strength σ_{ts} of the steel is 1100 MPa.

(c) What practical changes could be made to the forks to bring the stress
range below the fatigue limit and so avoid fatigue failure?

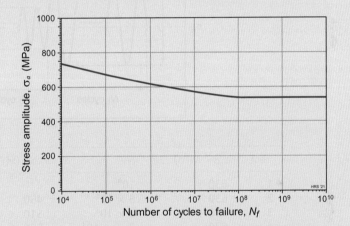

Exercise E6.28   An aluminium alloy for an airframe component was tested in the laboratory
under an applied stress, which varied sinusoidally with time about a mean
stress of zero. The alloy failed under a stress range $\Delta\sigma$ of 280 MPa after 10^5
cycles. Under a stress range of 200 MPa, the alloy failed after 10^7 cycles.
Assume that the fatigue behaviour of the alloy can be represented by:

$$\Delta\sigma \, N_f^b = C_1$$

where b and C_1 are material constants.

(a) Find the number of cycles to failure for a component subject to a stress
range of 150 MPa.

(b) An aircraft using the airframe components has encountered an estimated
4×10^8 cycles at a stress range of 150 MPa. It is desired to extend the life
of the airframe by another 4×10^8 cycles by reducing the performance
of the aircraft. Use Miner's rule to find the decrease in the stress range
needed to achieve this additional life.

Exercise E6.29 A medium carbon steel was tested to obtain high-cycle fatigue data for the number of cycles to failure N_f in terms of the applied stress range $\Delta\sigma$ (peak-to-peak). As the test equipment was only available for a limited time, the tests had to be accelerated. This was achieved by testing the specimen using the loading history shown schematically in the figure. A first set of N_1 cycles was applied with a stress range $\Delta\sigma_1$, followed by a second set of N_2 cycles with a stress range $\Delta\sigma_2$. The table shows the test programme. In every test the mean stress σ_m was held constant at 150 MPa and each test was continued until specimen failure. A separate tensile test gave a tensile strength for the steel of 600 MPa.

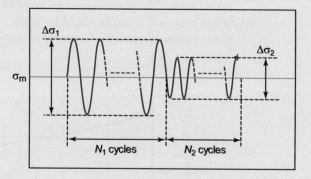

Test	$\Delta\sigma_1$ (MPa)	N_1	$\Delta\sigma_2$ (MPa)	N_2
1	630	10^4	-	-
2	630	5×10^3	460	5×10^5
3	630	5×10^3	510	1.2×10^5
4	630	2.5×10^3	560	4.4×10^4

(a) Using Goodman's rule, show that the stress *range* that would give failure in 10^4 cycles with *zero* mean stress is 840 MPa.
(b) Use Miner's rule to find the expected number of cycles to failure for each of the stress ranges $\Delta\sigma_2$ in the table, with the mean stress of 150 MPa.
(c) Convert the $\Delta\sigma_2 - N_f$ data obtained to the equivalent data for zero mean stress.
(d) Plot a suitable graph to show that all the fatigue life data for zero mean stress are consistent with Basquin's law for high-cycle fatigue, $\Delta\sigma N_f^b = C_1$, and find the constant b.

Exercise E6.30 A material has a threshold cyclic stress intensity ΔK_{th} of 2.5 MPa.m$^{1/2}$. If it contains an internal crack of length 1 mm, will it be safe (meaning, no failure) if subjected to a continuous cyclic range of tensile stress $\Delta\sigma$ of 50 MPa?

Exercise E6.31   A plate of width $a = 50$ mm contains a sharp, transverse edge-crack of length $c = 5$ mm. The plate is subjected to a cyclic axial stress σ that varies from 0 to 50 MPa. The Paris law constants [Equation (6.24)] are $m = 4$ and $A = 5 \times 10^{-9}$, where σ is in MPa. The fracture toughness of the material is $K_{1c} = 10$ MPa.m$^{1/2}$.

(a) At what 'critical' crack length will fast fracture occur? (Assume the geometry factor in the expression for K is $Y = 1.1$).

(b) How many load cycles will it take for the crack to grow to the critical crack length?

Exercise E6.32   Consider the design of a bicycle crank. It is assumed that the design is governed by fatigue failure, and that to avoid this the maximum stress amplitude must be kept below the endurance limit σ_e of the material. The crank is modelled as a beam in bending of square cross section $a \times a$, where a is free to vary. The length of the beam and the amplitude of the applied cyclic moment are fixed.

(a) Derive an expression for the material index, which should be maximised to minimise the mass. Hence choose a shortlist of materials using the property chart [Figure 6.26(b)], commenting on your choice (including price/kg, using the data in Appendix A). Use the most advantageous material properties within the ranges shown on the chart.

(b) An existing crank made of the best performing Al alloy weighs 0.2 kg. A cyclist is prepared to pay extra for a lighter component made of titanium, but only if it saves enough weight. Use the property chart with the material index from (a) to find the ratio of the mass of the best performing Ti alloy to that of the best Al alloy. Hence estimate the mass of the corresponding titanium component.

(c) The as-manufactured costs per kg of aluminium and titanium are 10 and 40 £/kg, respectively. Estimate the costs of the two cranks. How much value (in £) would the cyclist need to associate with each kg of weight saved, in order to go ahead with the switch from aluminium to titanium?

Exercise E6.33   Supersonic wind tunnels store air under high pressure in cylindrical pressure vessels – the pressure, when released, produces hypersonic rates of flow. The pressure vessels are routinely proof tested to ensure that they are safe. If such a cylinder, of diameter 400 mm and wall thickness 20 mm, made of a steel with a fracture toughness $K_{1c} = 42$ MPa.m$^{1/2}$, survives a proof test to 40 MPa (400 atmospheres), what is the length of the largest crack it might contain?

Exercise E6.34   A thin-walled spherical pressure vessel of radius $R = 1$ m is made of a steel with a yield stress of 600 MPa and a fracture toughness K_{1c} of 100 MPa.m$^{1/2}$. The vessel is designed for an internal working pressure $p = 20$ MPa.

(a) Find a suitable wall thickness if the maximum stress is to be kept below 60% of the yield stress. The stress in the wall of a uniform spherical pressure vessel is given by $\sigma = pR/2t$.

(b) The pressure vessel is inspected by a non-destructive method, and an embedded penny-shaped crack of length $2c = 10$ mm is found in the wall (i.e. from Figure 6.4, the geometry factor $Y = 0.7$ in the expression for K). Will the pressure vessel survive at the working pressure, and in a proof test at 90% of the yield stress?

Exercise E6.35 Discuss the factors that determine whether a pressure vessel will fail in a controlled way by yielding or by leaking, or catastrophically by fast fracture, when subjected to repeated applications of internal pressure.

Exercise E6.36 Consider three cylindrical pressure vessels made of the aluminium alloys in the table. Each has a wall thickness of $t = 14$ mm and a radius of $R = 840$ mm. The vessels are each periodically emptied and filled to an internal pressure of $p = 2$ MPa. Cracks can grow through the wall by fatigue, driven by the cycle of stress on pressurisation. Determine the failure mode of each vessel: yield/leak/fracture. Comment on which material would be best for this application.

Material	Yield stress, σ_y (MPa)	Fracture toughness, K_{1c} (MPa.m$^{1/2}$)
Aluminium 2024 T4	300	40
Aluminium 6061 T4	110	35
Aluminium 5454 H111	140	20

Exercise E6.37 (a) Use the Paris law for crack growth rate [Equation (6.24)] to show that the number of cycles to failure N_f of a component cycled through a uniform stress range $\Delta\sigma$ and containing an initial crack of length c_i, is:

$$N_f = \frac{1}{AY^m \pi^{m/2}(1 - m/2)}\left[(c^*)^{1-m/2} - (c_i)^{1-m/2}\right]$$

where c^* is the critical crack length, Y is the geometric factor, A and m are constants in the Paris law (with $m > 2$).

(b) A die fabricated from a low alloy steel is to withstand tensile hoop stresses that cycle between 0 and 400 MPa. Prior to use it has been determined that the length of the largest crack in the steel is 1 mm. Determine the minimum number of times that the die may be used before failing by fatigue, assuming $m = 3$, $A = 1.0 \times 10^{-12}$ (with $\Delta\sigma$ in MPa), and $Y = 1.13$. The fracture toughness of the steel is 170 MPa.m$^{1/2}$.

Exercise E6.38 A cylindrical steel pressure vessel of 7.5 m diameter and 40 mm wall thickness is to operate at a working pressure of 5.1 MPa. The design assumes that small thumb-nail-shaped flaws in the inside wall will gradually extend through the wall by fatigue. Assume for this crack geometry that $K = \sigma\sqrt{\pi c}$, where c is the length of the edge-crack and σ is the hoop stress in the vessel. The fracture toughness of the steel is 200 MPa.m$^{1/2}$ and the yield stress is 1200 MPa.

(a) Would you expect the vessel to fail in service by leaking (when the crack penetrates the thickness of the wall) or by fast fracture?

(b) The vessel will be subjected to 3000 pressurisation cycles during its service life. The growth of flaws by fatigue is given by Equation (6.24) with $m = 4$, $A = 2.44 \times 10^{-14}$ MPa^{-4}m^{-1}, and stress in MPa. Determine the initial crack length c_i that would grow to failure in 3000 cycles. Hence find the pressure to which the vessel must be subjected in a proof test to guarantee against failure in the service life. Check that this test will not yield the vessel.

Chapter 7
Materials and heat: thermal properties

Heat exchangers.
(Image courtesy of the International Copper Association)

Chapter contents

https://doi.org/10.1016/B978-0-08-102399-0.00007-8

7.1 Introduction and synopsis

Heat, until about 1800, was thought to be a physical substance called 'caloric' that somehow seeped into things exposed to a flame. It took the American, Benjamin Thompson[1], backed up by the formidable Carnot[2], to suggest what we now know to be true: that heat is atoms or molecules in motion. In gases, they fly between occasional collisions with each other. In solids, by contrast, they vibrate about their mean positions, interacting with all their neighbours: the higher the temperature, the greater the amplitude of vibrations. This physical interpretation provides our understanding of the thermal properties of solids: *heat capacity, expansion coefficient, thermal conductivity,* even *melting point.*

The chapter cover page shows an example product whose operation depends on thermal properties: a copper heat exchanger, designed to transfer heat efficiently between two circulating fluids. In other situations, thermal responses need to be managed to avoid failure – for example, buckling of railways in a heatwave. This chapter describes thermal properties, their origins, the ways they can be manipulated, and the ways they are used and managed in design and in manufacturing processes. Heat affects mechanical and physical properties, too. As temperature rises, strength in particular falls and materials undergo *creep,* deforming slowly with time at a rate that increases as the melting point is approached – this behaviour we leave for Chapter 8.

7.2 Thermal properties: definition and measurement

Reference temperatures Pure crystalline solids have a sharp melting point, T_m, abruptly changing state from solid to low-viscosity liquid. Alloys usually melt over a temperature range – but the onset and completion of melting are still at precise temperatures. Non-crystalline molecular materials such as thermoplastics and glasses have a more gradual transition, from true solid to viscous liquid, centred on the glass transition temperature T_g. The two temperatures – *melting temperature,* T_m, and *glass transition temperature,* T_g (units: Kelvin[3] [K] or Centigrade [°C]) – are fundamental points of reference in design and manufacturing, but also in interpreting the underlying physics because they relate directly to the strength of the bonds in the solid. It is also helpful in engineering design to define two further temperatures, established empirically from practical experience: the *maximum* and *minimum*

---

[1] Benjamin Thompson (1753–1814), later Count Rumford, while in charge of the boring of cannons at the Watertown Arsenal near Boston, Massachusetts, noted how hot they became and formulated the concept of a 'mechanical equivalent of heat,' which became known later as the heat capacity.

[2] Nicolas Léonard Sadi Carnot (1796–1832), physicist and engineer, formulator of the Carnot cycle and the concept of entropy, the basis of the optimisation of heat engines. He died of cholera, a fate most physicists today, mercifully, are spared.

[3] William Thompson, Lord Kelvin (1824–1907), Scottish mathematician who contributed to many branches of physics; he was known for a self-confidence that led him to claim (in 1900) that there was nothing more to be discovered in physics – he already knew it all.

service temperatures, T_{max} and T_{min} (units again K or °C). The first tells us the highest temperature at which the material can normally be used continuously without problems, due to oxidation, chemical change, excessive distortion, or softening. The second is the temperature below which the material becomes brittle or otherwise unsafe to use.

Specific heat capacity   It takes energy to heat a material. The *specific heat capacity* (or just *specific heat*) is the energy to heat 1 kg of a material by 1 K, measured at constant pressure (atmospheric pressure) and thus given the symbol C_p. Heat is measured in Joules[4] (J), so the units of specific heat are J/kg.K. When dealing with gases, it is more usual to measure the heat capacity at constant volume (symbol C_v), and for gases this differs from C_p. For solids the difference is so slight that it can be ignored, and we shall do so here. C_p is measured by *calorimetry* – Figure 7.1(a) shows how, in principle, this is done. A measured quantity of

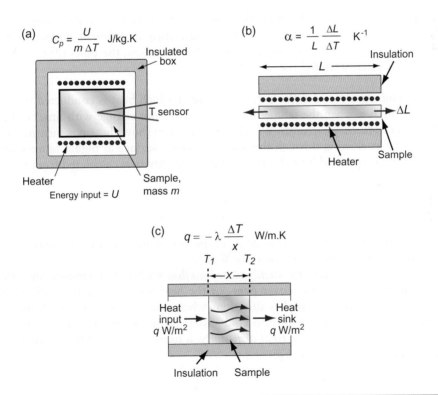

Figure 7.1   Measuring thermal properties: (a) heat capacity, C_p; (b) thermal expansion coefficient, ΔL; (c) thermal conductivity, λ.

---

[4] James Joule (1818–1889), English physicist, who, motivated by theological belief, sought to establish the unity of the forces of nature. His demonstration of the equivalence of heat and mechanical work did much to discredit the caloric theory.

energy U (here, electrical energy) is fed into a sample of material of known mass m, measuring the temperature rise, and hence the heat capacity C_p:

$$C_p = \frac{U}{m \, \Delta T} \tag{7.1}$$

Thermal expansion coefficient   Most materials expand when they are heated – that is, the dimensions of an object change, giving a strain ε that is produced by a stimulus other than stress, as noted in Chapter 4. For an initial specified dimension L of an object this is used directly, as in Figure 7.1(b), to measure the *linear thermal expansion coefficient*, α:

$$\alpha = \frac{1}{L} \frac{\Delta L}{\Delta T} \quad \text{or} \quad \varepsilon = \alpha \, \Delta T \tag{7.2}$$

where ΔL is the change in the linear dimension of the object when the temperature rises by ΔT. If the material is anisotropic it expands differently in different directions, and two or more coefficients are required. Since strain is dimensionless, the units of α are K^{-1} or, more conveniently, 'microstrain/K' (i.e. 10^{-6} K^{-1}).

Thermal conductivity and diffusivity   The rate at which heat is conducted through a solid at steady state (i.e. when the temperature profile does not change with time) is measured by the *thermal conductivity*, λ (units: W/m.K). Figure 7.1(c) shows how it is measured: by recording the heat flux flowing through the material from a surface at higher temperature T_1 to a lower one at T_2 separated by a distance x. The conductivity is calculated from Fourier's[5] law:

$$q = -\lambda \frac{dT}{dx} = \lambda \frac{(T_1 - T_2)}{x} \tag{7.3}$$

where q is the heat flux per unit area, or power density. Power is measured in Watts[6] (J/sec), so the units of power density are W/m^2.

Thermal conductivity governs *steady-state heat flow* with a fixed temperature difference – as in the heat exchanger on the cover page, or the heat lost through the walls of a house in winter. *Transient heat flow* implies situations when temperature varies with time – as in the quenching of a steel component into cold water, an important step in heat treatment. For these problems we need the *thermal diffusivity*, a (units: m^2/s), combining two of the previous thermal properties, λ and C_p:

$$a = \frac{\lambda}{\rho C_p} \tag{7.4}$$

---

[5] Baron Jean Baptiste Joseph Fourier (1768–1830), mathematician and physicist; he nearly came to grief during the French Revolution but survived to become one of the savants who accompanied Napoleon Bonaparte in his conquest of Egypt.

[6] James Watt (1736–1819), instrument maker and inventor of the condenser steam engine (the idea came to him while 'walking on a fine Sabbath afternoon'), which he doggedly developed. Unlike so many of the characters footnoted in this book, Watt, in his final years, was healthy, happy, and famous.

where ρ is the density – the product (ρC_p) is the heat capacity per unit volume, with units of $J/m^3.K$. The thermal diffusivity can be measured directly by measuring the time it takes for a temperature pulse to traverse a specimen of known thickness when a heat source is applied briefly to one side, or it can be calculated from λ (and ρC_p) via Equation (7.4). Solutions to problems of transient heat flow, discussed later, have a characteristic time constant t, for a characteristic heat flow distance x:

$$t \approx \frac{x^2}{a} \quad \text{and} \quad x \approx \sqrt{a\,t} \qquad (7.5)$$

Example 7.1

A room with underfloor heating has 4-mm-thick tiles separating the heating elements from the floor surface. The thermal conductivity λ of a tile is 0.8 W/m.K, its specific heat C_p is 800 J/kg.K, and its density ρ is 2200 kg/m^3. The floor is cold when the heating is switched on. Approximately how long will it take before the floor starts to feel warm?

Answer. The thermal diffusivity of a tile is $a = \lambda/\rho C_p = 4.5 \times 10^{-7}$ m^2/s. The time it takes heat to diffuse through a 4 mm tile is then approximately

$$t \approx \frac{x^2}{a} = \frac{\left(4 \times 10^{-3}\right)^2}{4.5 \times 10^{-7}} = 36 \text{ seconds}$$

7.3 Material property charts: thermal properties

All materials have a thermal expansion coefficient α, a specific heat C_p, and a thermal conductivity λ (and hence a thermal diffusivity, a). Note that all thermal properties themselves vary with temperature. Three charts give an overview of these thermal properties and their relationship to strength (all at room temperature). The charts highlight the underlying physical differences between the materials classes and facilitate choosing materials for applications with thermal constraints or objectives.

Thermal expansion, α, and thermal conductivity λ Metals and technical ceramics have high conductivities and modest expansion coefficients; they lie towards the lower right of the chart in Figure 7.2. Polymers and elastomers have conductivities that are 100 times lower, and their expansion coefficients are 10 times greater than those of metals; they lie at the upper left in Figure 7.2. The chart shows contours of the *material index*[7], λ/α, a quantity important in designing against thermal distortion.

---

[7] Material indices were introduced in *Guided Learning Unit 2*: Material selection in design. They capture the property, or property combination, that should be maximised to give the optimum performance for a given design objective.

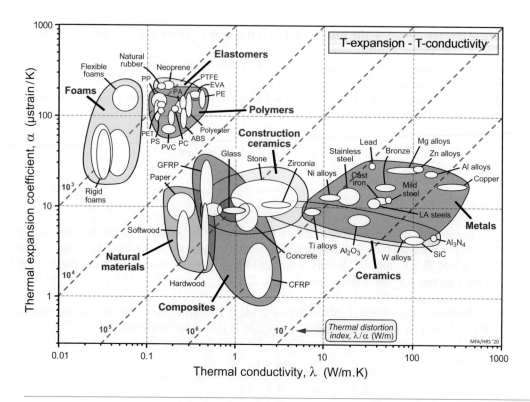

Figure 7.2 The linear thermal expansion coefficient – thermal conductivity, $\alpha - \lambda$, property chart. The contours show the thermal distortion index, λ/α.

Example 7.2

A metal is required to conduct heat away from a high-power density silicon chip. It must conduct heat well, but to avoid thermal stress when the temperature changes, its thermal expansion coefficient must differ as little as possible from that of silicon. Typical values for silicon are: $\lambda = 160$ W/m.K, $\alpha = 3$ µstrain/K. Locate the position of silicon on the $\alpha - \lambda$ chart of Figure 7.2 and suggest a choice of metal for the heat sink.

Answer. On the chart, silicon lies just below the bubble for SiC. The metal with the expansion coefficient closest to silicon is tungsten and its alloys, and these are also good thermal conductors.

Thermal conductivity, λ, and thermal diffusivity, a Good thermal conductors lie at the upper right of the chart in Figure 7.3, good thermal insulators at the lower left. The data span almost 5 decades in λ and a. The diagonal contours show the ratio λ/a, equal to the volumetric heat capacity ρC_p. Solid materials are strung out along the line

$$\rho C_p \approx 3 \times 10^6 \text{ J/m}^3.\text{K} \tag{7.6}$$

meaning that the heat capacity per unit volume, ρC_p, is almost constant for all solids (compared to the huge ranges in λ and a) – a surprising but useful result. To a good approximation (within a factor of 2 each way)

$$\lambda \approx 3 \times 10^6 a \tag{7.7}$$

(λ in W/m.K and a in m^2/s). Some materials deviate from this trend, with lower-than-average volumetric heat capacity – notably foams. Because of their porosity they contain less actual solid per unit volume, so their ρC_p is low. Polymer foams have low *conductivities* (and are widely used for insulation because of this), but their *thermal diffusivities* are not as low as expected – they do not transmit much heat, but they change temperature as quickly as solid polymers.

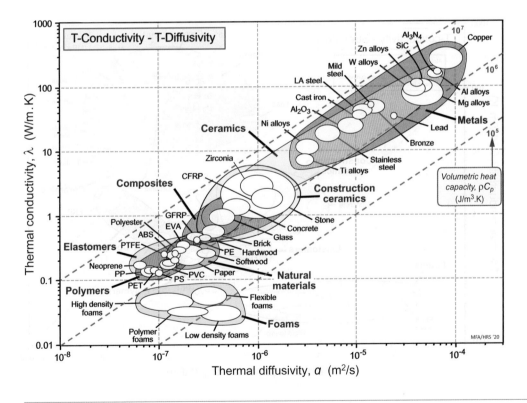

Figure 7.3 The thermal conductivity – thermal diffusivity, λ-a, property chart. The contours show the specific heat per unit volume, $\rho C_p = \lambda/a$.

Example 7.3

A locally available building material has a density $\rho = 2300$ kg/m^3. Without access to further information or equipment, what would you estimate its specific heat capacity, C_p, to be?

Answer. Equation (7.6), based on the data in Figure 7.3, shows that for solids, $\rho C_p \approx 3 \times 10^6$ J/m^3.K. Hence the heat capacity of the material is $C_p \approx 1300$ J/kg.K.

Thermal conductivity, λ, and yield strength, σ_y   Metals as a class are both strong and good thermal conductors, though there is a wide range of both properties (Figure 7.4). A key combination of these properties is captured by the material index ($\lambda\sigma_y$), which we seek to maximise for applications such as heat exchangers (analysed in Section 7.5).

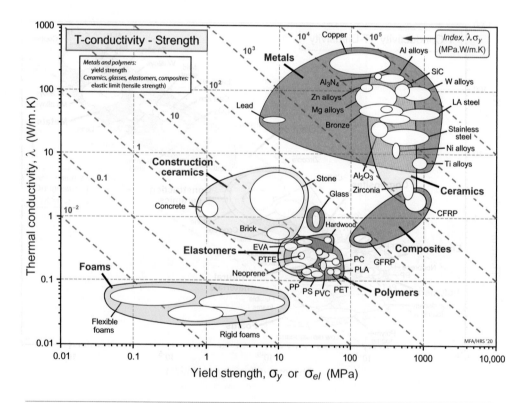

Figure 7.4   The thermal conductivity – yield strength, λ–σ_y, property chart. The contours show the index ($\lambda\sigma_y$).

Example 7.4

A high-performance engine with a pressurised heat exchanger requires a metal with high thermal conductivity and high strength, maximising the material index ($\lambda \sigma_y$). Which three metal systems offer the best combination of properties? Two are routinely used; why do you think the third is not?

Answer. The chart of Figure 7.4 shows that copper, aluminium, and tungsten alloys have the highest conductivities and good strength. Tungsten is not used in practice; compared to copper and aluminium it is expensive, heavy, and much more difficult to form into thin-walled shapes.

7.4 The physical origins of thermal properties

Heat capacity   We owe our understanding of heat capacity to Albert Einstein[8] and Peter Debye[9]. Heat, as already mentioned, is atoms in motion. Atoms in solids vibrate about their mean positions with an amplitude that increases with temperature, but they cannot vibrate independently because they are coupled by their inter-atomic bonds; this leads to vibrations that are like standing elastic waves. Figure 7.5 shows how atoms have three degrees of freedom, so along any row of atoms there can be one longitudinal mode of vibration and two transverse modes, one in the plane of the page and one normal to it. Some have short

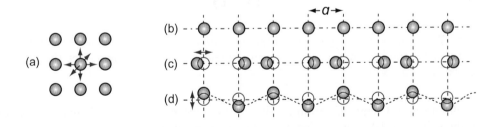

Figure 7.5   (a) An atom vibrating in the 'cage' of surrounding atoms, with three degrees of freedom. (b) A row of atoms at rest, (c) a longitudinal wave, and (d) one of two transverse waves (in the other, the atoms oscillate normal to the page).

---

[8] Albert Einstein (1879–1955), Patent Officer, physicist, and campaigner for peace, one of the greatest scientific minds of the 20th century; he was forced to leave Germany in 1933, moving to Princeton, New Jersey, where his influence on US defence policy was profound.

[9] Peter Debye (1884–1966), Dutch physicist and Nobel Prize winner, did much of his early work in Germany until in 1938, harassed by the Nazis, he moved to Cornell, New York, where he remained until his death.

wavelengths and high energy, others long wavelengths and lower energy (Figure 7.6). The shortest possible wavelength, λ_1, is twice the atomic spacing. Averaged over all wavelengths in a volume of solid, each vibration mode has an average energy $k_B T$ per atom, where k_B is Boltzmann's constant[10], 1.38×10^{-23} J/K. If the volume occupied per atom is Ω, then the number of atoms per unit volume is $1/\Omega$, so the total thermal energy per unit volume of material is $3k_B T/\Omega$. The heat capacity per unit volume, ρC_p, is the *change* in this energy per Kelvin change in temperature, giving:

$$\rho C_p \approx \frac{3k_B}{\Omega} \tag{7.8}$$

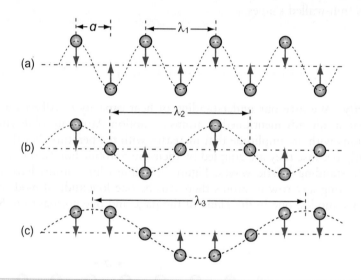

Figure 7.6 Thermal energy involves atom vibrations, illustrated for one of the transverse modes. Each mode has multiple wavelengths, of which three are shown – the shortest possible, $\lambda_1 = 2a$, is in (a).

As discussed in Chapter 3, atomic volumes do not vary much – all lie within a factor of 3 of the value 2×10^{-29}/m³, giving a volumetric heat capacity

$$\rho C_p \approx 2 \times 10^6 \text{ J/m}^3.\text{K} \tag{7.9}$$

This is close to the empirical value identified from Figure 7.3 [Equation (7.6)] and explains the small spread around that value.

[10] Ludwig Boltzmann (1844–1906), born and worked in Vienna at a time when that city was the intellectual centre of Europe; his childhood interest in butterflies evolved into a wider interest in science, culminating in his seminal contributions to statistical mechanics.

Thermal expansion   Almost all solids expand when heated, so the atoms must be moving farther apart, as shown in Figure 7.7. It resembles an earlier one (see Figure 4.9) but with a subtle difference: the force–spacing curve, shown straight in the earlier figure, is not in fact quite straight; the bonds become stiffer when the atoms are pushed together and less stiff when they are pulled apart. The curvature in the force–spacing curve is a consequence of the energy curve being unsymmetrical about its minimum. As temperature rises, atoms vibrate with increasing amplitude about an increasing mean spacing. So thermal expansion is a non-linear effect: If the bonds between atoms were truly linear springs, there would be no expansion.

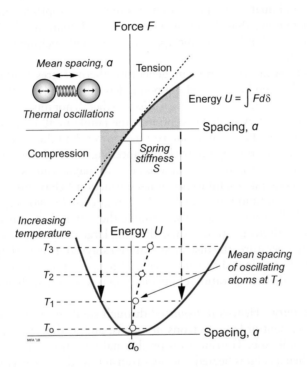

Figure 7.7   Thermal expansion results from the oscillation of atoms in an unsymmetrical energy–spacing curve. Open circles show the mean spacing at different temperatures.

The stiffer the springs, the steeper the force–displacement curve and the narrower the energy–spacing curve, giving less scope for expansion. So we expect materials with high Young's modulus (stiff springs) to have low expansion coefficient, and indeed, to a good approximation

$$\alpha \approx \frac{1.6 \times 10^{-3}}{E} \tag{7.10}$$

(E in GPa, α in K^{-1}). It is also an empirical fact that all crystalline solids expand by about the same amount on heating from absolute zero to their melting point: about 2%. Since the expansion coefficient is the expansion per degree Kelvin:

$$\alpha \approx \frac{0.02}{T_m} \qquad (7.11)$$

Tungsten, with a melting point of around 3330°C (3600 K), has $\alpha = 5 \times 10^{-6}$/K; lead, with a melting point of about 330°C (600 K), has $\alpha = 30 \times 10^{-6}$/K. T_m is six times lower, α is six times greater. Equation (7.11) is a remarkably good approximation for the expansion coefficient (remembering that T_m must be in Kelvin). Equations (7.10) and (7.11) suggest that Young's modulus should also scale approximately with melting point – this investigation is left for the Exercises.

Thermal expansion can be a problem, generating thermal stress and distortion. The expansion coefficient α, like modulus or melting point, depends directly on the stiffness and strength of atomic bonds – it is *microstructure-insensitive*. So there's not much you can do to manipulate the expansion coefficient. A special exception is the family of alloys called *Invars*, with values of $\alpha = 0.5–2 \times 10^{-6}$/K: a factor of about 10 times lower than the other metals on the chart in Figure 7.2. They achieve this by the trick of cancelling some of their normal thermal expansion with a contraction caused by the gradual loss of magnetism as the material is heated.

Another anomalous but useful expansion–contraction behaviour is that found in *shape-memory alloys*: special titanium or nickel alloys that exist in two very different crystal configurations, called *allotropes*. Unusually, straining the material can cause the crystal structure to change from one allotrope to the other, allowing large distortions, particularly in bending. Below a critical temperature, the distorted material stays in its new shape; on warming it up, it springs back to its original shape at the critical temperature. This has obvious application in fire alarms and other temperature-controlled safety systems, such as automatic sprinklers.

Thermal conductivity   Heat is *transmitted* through solids in three ways: by thermal vibrations; by the movement of free electrons in metals; and, if they are transparent to infra-red wavelengths, by radiation. Transmission by thermal vibrations involves the propagation of *elastic waves*. When a solid is heated, the heat transfers as discrete elastic wave packets called *phonons*. Phonons travel through the material, like any elastic wave, at the speed of sound, c_0 $\left(c_0 \approx \sqrt{E/\rho}\right)$. If this is so, why does heat not diffuse at the same (high) speed? It is because phonons travel only a short distance before they are *scattered* by the slightest irregularity in the lattice of atoms through which they move, and even by other phonons. On average they travel a distance called the *mean free path*, ℓ_m, before bouncing off something, and this path is very short: typically less than 0.01 μm (10^{-8} m).

Thermal conductivity arises from a *net flux model*, based on the *difference* between the number of phonons crossing a plane in opposing directions, in a given time. Figure 7.8 shows a rod with unit cross section and a uniform temperature gradient dT/dx between its ends. Phonons within it have three degrees of freedom of motion (they can travel in the $\pm x$, the $\pm y$, and the $\pm z$ directions). Focus on the mid-plane M–M. On average, one-sixth of the phonons are moving in the $+x$ direction, and those within a distance ℓ_m of the plane

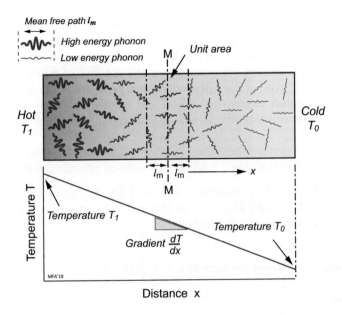

Figure 7.8 Transmission of heat by the motion of phonons.

will cross it from left to right before they are scattered. They carry with them an energy $\rho C_p (T + \Delta T)$, where T is the temperature at the plane M–M, and the temperature difference over the distance ℓ_m is given by the temperature gradient: $\Delta T = (dT/dx)\, \ell_m$. But at the same time, another one-sixth of the phonons move in the $-x$ direction and cross M–M from right to left, carrying an energy $\rho C_p (T - \Delta T)$. So when there is a temperature gradient, there is a difference between the energy transfer in the $\pm x$ directions (while phonons moving in the $\pm y$ or the $\pm z$ directions carry the same energy in opposing directions, as there is no gradient). Thus, the *net energy flux q* $(J/m^2.s)$ across unit area (M–M) per second is

$$q = -\frac{1}{6}\, \rho\, C_p\, c_o \left(T + \frac{dT}{dx}\, \ell_m \right) + \frac{1}{6}\, \rho\, C_p\, c_o \left(T - \frac{dT}{dx}\, \ell_m \right) = -\frac{1}{3}\, \rho\, C_p\, \ell_m\, c_o\, \frac{dT}{dx}$$

Comparing this with the definition of thermal conductivity [Equation (7.3)], we find the conductivity to be

$$\lambda = \frac{1}{3}\, \rho\, C_p\, \ell_m\, c_o \tag{7.12}$$

Sound waves are also elastic waves, like phonons, but these travel through the same bar without much scattering. Why do phonons scatter so readily? It's because waves are scattered most by obstacles with a size comparable with their wavelengths. Audible sound waves have wavelengths of metres, not microns; they are scattered by buildings (which is why you cannot always tell where a vehicle siren is coming from). Phonons are scattered by atom-scale obstacles, as their wavelengths start at two atomic spacings (see Figure 7.6).

Phonons contribute little to the conductivity of pure metals such as copper or aluminium because the heat is carried more rapidly by the net flux of *free electrons*, which have their own characteristic temperature-dependent energy, speed, and mean free path. Since free electrons also conduct electricity, we find that metals with high electrical conductivity also have high thermal conductivity, as will be illustrated in Chapter 10.

Example 7.5 Mean free path of phonons in silicon

A sample of silicon has modulus $E = 1500$ GPa, density $\rho = 2300$ kg/m^3, and specific heat capacity $C_p = 680$ J/kg.K. If its thermal conductivity is $\lambda = 145$ W/m.K, what is the mean free path ℓ_m of its phonons?

Answer. The thermal conductivity is $\lambda = \dfrac{1}{3}\rho\, C_p\, \ell_m\, c_0$ with $c_0 = \sqrt{\dfrac{E}{\rho}} = 8.08 \times 10^3$ m/s.

Thus $\ell_m = \dfrac{3\lambda}{\rho\, C_p \sqrt{E/\rho}} = 0.034$ μm.

Equation (7.12) describes thermal conductivity due to energy transfer by phonons (or electrons), which are scattered by obstacles. Because these obstacles are associated with microstructural features, this means that thermal conductivity is a *microstructure-sensitive property*. So which of the terms in the equation can be manipulated, and which microstructural features are effective? The volumetric heat capacity ρC_p, as we have seen, is almost the same for all solid materials. The sound velocity $c_0 = \sqrt{E/\rho}$ depends on two properties that are not easily changed (in fully dense solids). That leaves the *mean free path*, ℓ_m, of the phonons and, in metals, of the electrons. So to be effective, *scattering centres* for phonons and electrons need to be on the scale of multiple atom spacings – this is the case for solute atoms or finely dispersed particles, which reduce ℓ_m. Increasing alloy content decreases thermal conductivity; for example, pure iron has $\lambda = 80$ W/m.K, whereas stainless steel – iron with up to 30% of nickel and chromium in solution – has $\lambda = 18$ W/m.K. Glassy (non-crystalline) materials are so disordered that every molecule is a scattering centre, reducing the mean free path to a couple of atom spacings; their conductivity is correspondingly low.

Many applications require a combination of good thermal conductivity and strength – pressurised tubing for heat exchangers, for example. Since solute atoms scatter phonons and electrons, solid solution hardening is not the best choice for a high-strength thermal conductor – it is better to use obstacles that are further apart (to preserve thermal conductivity) while being effective against dislocations, giving strength: that is, precipitation hardening. The Exercises explore this property trade-off for different types of Al alloy.

We are not, however, restricted to fully dense solids, making density ρ a possible variable. As noted previously, the best thermal insulators are *porous* materials like polymer foams that take advantage of the low conductivity of still air trapped in the pores (λ for air is 0.02 W/m.K). The same principle explains the insulating benefits of cork, wool, and woven or knitted fabrics.

7.5 Design and manufacture: managing and using thermal properties

Thermal stress Most products and structures, small or large, are made of two or more materials that are clamped, welded, or otherwise bonded together. This causes problems when temperatures change. Railway track will buckle in exceptionally hot weather (see Example 7.6); overhead transmission lines sag – a problem for high-speed electric trains. Bearings seize, doors jam. Thermal distortion is a particular problem in precision measuring equipment and in the precise registration needed to fabricate computer chips. All of these derive from *differential thermal expansion*, where the constraint between different materials generates *thermal stress*. First a simple example, where material is prevented completely from straining while temperature rises: a clamped continuous railway track. For zero total strain in a slender column (the track), the strain due to expansion is equal and opposite to the strain due to the (induced) uniaxial compressive stress.

Example 7.6 Buckling railway track

A steel rail of square cross-section 25 mm × 25 mm is rigidly clamped, preventing movement or rotation, at two points a distance $L = 2$ m apart. If the rail is stress-free at 20°C, at what temperature will thermal expansion cause it to buckle elastically? (The thermal expansion coefficient of steel is $\alpha = 12 \times 10^{-6}$/K).

What steps are taken in railway track design to avoid buckling?

Answer. The elastic buckling load of a column clamped in this way is given in Chapter 4 [Equation (4.29), with $n = 2$, from Figure 4.19]:

$$F_{crit} = \frac{4\pi^2 EI}{L^2}$$ where E is Young's modulus and $I = \frac{t^2}{12} = 3.26 \times 10^{-8}$ m^4 is the second moment of area of the rail.

An unclamped rail heated by ΔT expands to give a strain $\alpha \Delta T$. The clamps keep the total strain equal to zero, so the induced compressive stress that gives the same (negative) strain has a magnitude $\sigma = E \alpha \Delta T$. The axial compressive force $F = \sigma A = E \alpha \Delta T A$, where $A = 6.25 \times 10^{-4}$ m^2 is the area of its cross-section. Equating this to F_{crit} and solving for ΔT gives

$$\Delta T = \frac{4\pi^2 I}{\alpha A L^2} = 43°C$$

This is too close to the temperature changes experienced in heatwaves – some railway tracks have small expansion gaps (responsible for the rhythmic metallic sound on trains as the wheel-sets cross the gaps); continuous welded track is installed under tension so that expansion is accommodated by relaxing this tension, avoiding compressive loading.

As a further example of differential thermal stress, many technologies involve coating materials with a thin surface layer of a different material to impart resistance to wear, corrosion, or oxidation. The coating is applied hot, and thermal stresses appear as it cools because of the difference in expansion coefficients of coating and substrate.

To find the stress in the coating, consider a thin film bonded to a much thicker component, as in Figure 7.9(a). Imagine first that the layer is detached [Figure 7.9(b)]. A temperature drop of ΔT causes the layer to shorten in length by $\delta L_1 = \alpha_1 L_o \Delta T$, while the substrate contracts by $\delta L_2 = \alpha_2 L_o \Delta T$. If $\alpha_1 > \alpha_2$, the surface layer shrinks more than the substrate. So, if we want to stick the film back on the much-more-massive substrate, we must stretch it by applying the strain

$$\varepsilon = \frac{\delta L_1 - \delta L_2}{L_o} = \Delta T (\alpha_1 - \alpha_2)$$

This requires a (biaxial) stress in the film of

$$\sigma_1 = \frac{E_1}{(1 - v)}(\alpha_1 - \alpha_2)\Delta T \tag{7.13}$$

where Poisson's ratio v enters because the stress state in the film is *biaxial* (see Chapter 4). For force equilibrium, there is a balancing stress in the substrate, but as this is much thicker the stress is negligible. The stress in the surface film can be large enough to crack it; the pattern of cracks seen on glazed tiles arises in this way. The way to avoid it is to avoid material combinations with very different expansion coefficients. The α–λ chart in Figure 7.2 has expansion α as one of its axes: benign choices are those that lie close together on this axis, and dangerous ones are those that lie far apart.

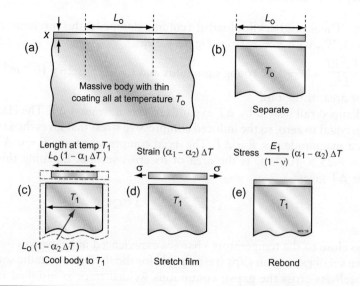

Figure 7.9 Thermal stresses in thin films arise on cooling or heating when the film and substrate have different expansion coefficients, α_1 in the film and α_2 in the substrate.

Avoiding materials with α mismatch is not always possible. Think of joining glass to stainless steel – a common combination in high-vacuum equipment. Heating to as little as 150°C is enough for the mismatch to crack the glass. The answer is to 'grade' the joint with materials whose expansion coefficients lie between the two, giving multiple interfaces with lower mismatch, lowering the stress and avoiding damage. In applications that do not require such a tight bond between metal and glass, an alternative solution is to put a low stiffness layer of rubber in the joint. This is how windows are mounted in cars and in double glazing units. The difference in expansion is taken up by distortion in the rubber, transmitting very little mismatch stress to the glass.

Thermal sensing and actuation Thermal expansion can be used to *sense* (to measure temperature change) and to *actuate* (to open or close valves or electrical circuits, for instance). The direct axial thermal displacement $\delta = \alpha L_o \Delta T$ is small; the *bi-material strip* is a way to magnify it, using bending. Two materials, now deliberately chosen to have different expansion coefficients α_1 and α_2, are bonded together in the form of a bi-material strip of thickness $2t$ as in Figure 7.10(a).

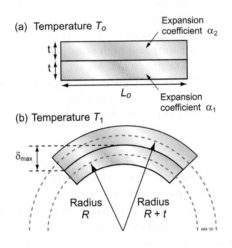

Figure 7.10 A bi-material strip.

When the temperature changes by $\Delta T = (T_1 - T_o)$, one expands more than the other, causing the strip to bend. A full analysis is a little complicated – the resulting curvature of the strip depends on both the thermal expansion coefficients and the two Young's moduli, E_1 and E_2. But we can get the idea if we consider two materials with the *same* value of E (which means the neutral axis of the cross section in bending coincides with the bimaterial interface). In this case, the mid-thickness planes of each strip [dotted in Figure 7.10(b)] differ in length by $L_o(\alpha_2 - \alpha_1)\Delta T$. Simple geometry then shows that

$$\frac{R+t}{R} = 1 + (\alpha_2 - \alpha_1)\Delta T$$

from which

$$R = \frac{t}{(\alpha_2 - \alpha_1)\Delta T} \qquad (7.14)$$

The resulting upward displacement of the centre of the bi-material strip (assumed thin compared to its length) is

$$\delta_{\max} = \frac{L_o^2}{8t}(\alpha_2 - \alpha_1)\Delta T \qquad (7.15)$$

So a large strip aspect ratio, L_o/t, produces a large displacement that is linear in temperature.

Managing thermal distortion   You do not need two materials to get thermal distortion: a *temperature gradient* will cause a single material to distort. A flat plate that becomes warm on one side will bend into a curve for the same reason as the bi-metallic strip does: the warm side expands more than the other. Thermal distortion is a problem in precision equipment in which heat is generated by the electronics, motors, actuators, or sensors that are necessary for its operation. The best materials to use have low expansion α (to minimise the differential expansion) and high conductivity λ (to spread the heat, reducing the steepness of the temperature gradient). So we aim to maximise the material index, λ/α, shown as diagonal contours in Figure 7.2. Materials with the highest values lie towards the bottom right: copper, aluminium alloys, silicon carbide, aluminium nitride, and tungsten distort the least; stainless steel and titanium are significantly less good. Polymers are very poor.

Thermal shock resistance   When thermal expansion is constrained, thermal stress appears. If the temperature of a component is changed abruptly by quenching it when hot into cold water, the surfaces are chilled almost instantly. It takes time for heat to be conducted out of the interior, so straight after quenching the bulk of the component is still hot. If the difference in temperature between the interior and the surface is ΔT, the difference in strain between them is $\varepsilon = \alpha \, \Delta T$. The surface is stuck to the interior, which constrains it, and thermal stress appears just as it did in the thin film of Figure 7.9. Complete constraint results in a surface stress of

$$\sigma = \frac{E}{(1-v)} \alpha \, \Delta T \qquad (7.16)$$

Superficially this looks the same as Equation (7.13), but here only one material is involved. In ductile materials the quenching stresses can cause yielding, permanent distortion, and residual stress; in brittle materials they can cause fracture. The ability of a material to resist fracture, its *thermal shock resistance* ΔT_s (units: K or °C), is the maximum sudden change of temperature to which such a material can be subjected without damage.

Example 7.7 Thermal shock resistance

The temperature of a domestic oven door can reach 200°C. If the hot oven door is opened and cold water accidentally spilt on it, it will encounter a thermal shock. What is the thermal shock resistance of the three glass-based materials listed in the accompanying table?

Material	Tensile strength, σ_t (MPa)	Modulus, E (GPa)	Expansion coefficient, α (1/K)	Poisson's ratio, ν
Soda-lime glass	30	62	9×10^{-6}	0.2
Borosilicate glass	32	63	3.19×10^{-6}	0.19
Glass ceramic	71	86	0.9×10^{-6}	0.25

Answer. From Equation (7.16):

Soda-lime (bottle) glass: $\Delta T = \dfrac{\sigma_t (1 - \nu)}{E \alpha} = \dfrac{30 \times (1 - 0.2)}{62 \times 10^3 \times 9 \times 10^{-6}} = 43°C$

Borosilicate glass (pyrex): $\Delta T = 129°C$

Glass ceramic: $\Delta T = 688°C$. (This material is used for electric hob tops).

Insulation: thermal walls Before 1940 few people had central heating or air conditioning. Now many have both, along with fridges and freezers, and maybe a sauna, too. All consume power, which is both expensive and hard on the environment (see Chapter 12), so it pays to insulate. Consider, as an example, the heated chamber (e.g. of a kiln) of Figure 7.11. The power lost per m^2 by conduction through its wall is

$$q = \lambda \frac{(T_i - T_o)}{t} \ \text{W/m}^2 \qquad (7.17)$$

where t is the thickness of the insulation, T_o is the temperature of the outside, and T_i that of the inside. For a given wall thickness t the power consumption is minimised by choosing materials with the lowest possible λ. The chart of Figure 7.4 shows that these are foams largely because they are about 95% gas and only 5% solid. Polymer foams are usable up to about 170°C; above this temperature even the best of them lose strength and deteriorate. When insulating against heat, then, there is an additional constraint that must be met: that the material has a maximum service temperature T_{max} that lies above the planned temperature of operation. Metal foams are usable to higher temperatures, but they are expensive, and – being metal – they insulate less well than polymers. For high temperatures, ceramic foams are the answer. Low-density firebrick is relatively cheap, can operate up to 1000°C, and is a good thermal insulator. It is used for kiln and furnace linings.

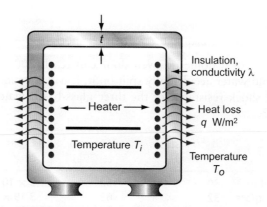

Figure 7.11 A heated kiln.

Conduction with strength Heat exchangers transfer heat from one fluid to another while keeping the fluids physically separated, as in power plants and cooling systems in car and marine engines. Heat is transferred most efficiently if the wall separating the fluids is kept thin and its surface area large – a good solution is a tubular coil. There is a pressure difference between the fluids, as one fluid is pumped along the tube, so the wall must be strong enough to withstand it. Here there is a trade-off in the design: thinner walls conduct heat faster, thicker walls are stronger. Which material gives the best compromise? We will analyse this problem using the techniques shown in *Guided Learning Unit 2*. Figure 7.12 shows an idealised heat exchanger made of thin-walled tubes.

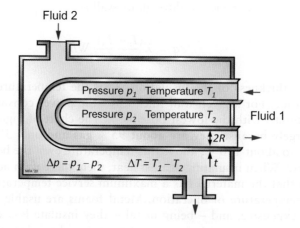

Figure 7.12 A heat exchanger.

The tubes have a given radius R, a wall thickness t (the *free variable*, which may be varied), and must support a pressure difference Δp without failure (a *functional constraint*), with a fixed temperature difference between the fluids of ΔT. The *objective* is to maximise the heat transferred per unit surface area of tube wall. The heat flow rate per unit area, q, is given by

$$q = -\lambda \frac{dT}{dx} = \lambda \frac{\Delta T}{t} \, \text{W/m}^2 \tag{7.18}$$

The circumferential stress due to the internal pressure in a cylindrical thin-walled tube [see Chapter 5, Equation (5.29)] is

$$\sigma = p \, \frac{R}{t}$$

This stress must not exceed the yield strength σ_y of the material of which it is made, so the minimum thickness t needed to support the pressure difference Δp is

$$t = \frac{\Delta p R}{\sigma_y} \tag{7.19}$$

Combining Equations (7.18) and (7.19) to eliminate the free variable t gives

$$q = \lambda \sigma_y \frac{\Delta T}{\Delta p R} \tag{7.20}$$

The best materials are therefore those with the highest values of the index $(\lambda \sigma_y)$. It is shown as a set of diagonal contours on the λ–σ_y chart in Figure 7.4. The best materials for heat exchangers are those at the upper right of the figure: copper and aluminium alloys (excluding tungsten alloys for the reasons noted in Example 7.4). Copper and aluminium both offer good corrosion resistance, but aluminium wins where cost and weight are to be kept low – in automotive radiators, for example. Marine heat exchangers use copper alloys, even though they are heavier and more expensive, because of their better resistance to corrosion in seawater.

Storing heat: storage heaters and phase-change materials (PCMs)   The property name 'heat capacity' suggests applications of materials to store heat for later release. Storage heaters do exactly this, storing heat in a large block of material at night, with heat being extracted from it with a fan or by natural convection during the day. The thinking behind this is that off-peak electricity is cheap – demand is lower at night, but electricity companies prefer to maintain more continuous output, so they smooth demand by charging less at night. The design objective for the heat storage material is therefore to store as much heat per unit cost as possible – we analyse this in the Exercises.

Another type of heat store exploits the physics of *phase changes*: when a material solidifies from the melt, or condenses from gas to liquid, it releases *latent heat* (J/kg), without changing temperature; conversely, heat is absorbed at constant temperature when materials melt or boil. Latent heat is exploited to prevent the temperature of an object falling below a set value, by surrounding it with a liquid that solidifies at that temperature – if it cools to this temperature, the liquid solidifies, and cooling is arrested by the release of latent heat of fusion. The reverse is also practical: using latent heat of melting to keep things chilled, as in the ice blocks used for holiday cool boxes. Phase changes therefore provide a *thermal buffer*, leading to the development of *phase-change materials* (PCMs), mostly based on paraffin and other waxes, spanning target temperatures from 0°C to 100°C. This has been driven in part by the need for greater efficiency in heating and cooling of buildings, in response to concerns about climate change – an example shows how such a 'passive' system might work.

Example 7.8 Phase change materials

Some climates have large changes of temperature between day and night. It is suggested that passive radiators filled with a PCM with a melting point around 20°C could absorb heat when the ambient temperature rises above this temperature, and release it when it fell below, reducing the need for powered cooling and heating. The PCM C17 paraffin wax ($C_{17}H_{36}$) has a melting point close to 20°C, and a latent heat of melting of 240 kJ/kg. How much of it is needed to provide the equivalent of 1 kWh of energy management?

Answer. The required energy output is 1 kWh, or 3600 kJ. As the latent heat is 240 kJ/kg, we need 15 kg of the C17 paraffin wax to provide 1 kWh of buffering.

Transient heat flow problems The temperature distribution is static during steady-state heat flow, governed by Fourier's first law [Equation (7.3)]. When heat flow is transient, the distribution changes with time. Evaluating it (for heat conduction in one dimension) requires a solution for Fourier's second law:

$$\frac{\partial^2 T}{\partial x^2} = \frac{1}{a} \frac{\partial T}{\partial t} \tag{7.21}$$

with appropriate symmetry and boundary conditions (here t is time, x is distance, and a is the thermal diffusivity, as before). The maths has been done for us in many standard situations, including many that are relevant to material processing, such as cooling components from high temperatures. As we have seen, the imposed cooling rates determine distortion and induced stresses but also the evolution of *microstructure*, and hence final properties, the central topic of Chapter 11.

As an example, consider a steel plate initially at uniform high temperature T_1, plunged into a tank of cold water at temperature T_o (Figure 7.13) – the critical *quenching* step in many heat treatments. For a plate that is large compared to its thickness, heat flow is one-dimensional through-thickness. Heat transfer to water is good, so the surface is assumed to immediately reach T_o, leading to the temperature profiles shown in the figure, evolving in time. For short times ($t < t_3$) the two sides cool independently, as if the plate is semi-infinite. For the initial near-linear part of the temperature profile, the solution to Equation (7.21) is approximately

$$\frac{T - T_o}{T_1 - T_o} = \left(\frac{x}{2\sqrt{at}} \right) \tag{7.22}$$

For the point at which the temperature has fallen to halfway between T_1 and T_o, the LH side of Equation (7.22) is equal to 0.5, such that

$$x \approx \sqrt{at} \tag{7.23}$$

Note that this is the same 'rule-of-thumb' from Equation (7.5) for the characteristic distance of transient heat flow in time t. This distance is shown in Figure 7.13 for time $t = t_2$: it indicates a length-scale for the depth over which a significant degree of cooling has occurred.

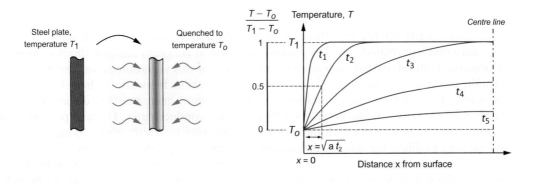

Figure 7.13 Quenching a steel plate – a transient heat flow problem. The temperature profile through the plate evolves with time.

For longer time-scales ($t = t_4$ or t_5), cooling from both faces of the sheet leads to a different solution to Equation (7.21), which gives the temperature at mid-thickness for a plate of thickness ℓ as:

$$\frac{T - T_o}{T_1 - T_o} = \frac{4}{\pi} \exp\left(-\frac{t}{(\ell^2 / \pi^2 a)} \right) \tag{7.24}$$

The temperature falls exponentially with time, with a characteristic 'time constant' $t^* = (\ell^2 / \pi^2 a)$. Comparing this with Equation (7.5), we find that this characteristic cooling time scales with a distance (ℓ/π), about one-third of the plate thickness.

7.6 Summary and conclusions

All materials have thermal properties:

- A specific heat, C_p: the amount of energy it takes to heat unit mass by unit temperature.
- A thermal expansion coefficient, a: the change in its linear dimensions with change of temperature.
- A thermal conductivity, λ, and thermal diffusivity, a: the former characterising how fast heat is transmitted through the solid, the latter how long it takes for the temperature, once perturbed, to settle back to a steady pattern.
- Characteristic temperatures of changes of phase or behaviour – for a crystalline solid, its melting point T_m, and for non-crystalline solids, a glass transition temperature T_g; in addition, there are empirical maximum and minimum service temperatures, T_{max} and T_{min}, above or below which continuous use is impractical, for reasons of creep or degradation (if hot), or brittleness (if cold).
- Latent heats of melting and of vaporisation: the energy absorbed at constant temperature when the solid melts or evaporates (and, conversely, released during solidification and condensation).

Property charts for the first three illustrated the great range of thermal properties, and the ways they are related. The key underlying physical concepts are that thermal energy is atomic-level vibration, that increasing the amplitude of this vibration causes expansion, and that the vibrations propagate as phonons, giving conduction. In metals the free electrons – the ones responsible for electrical conduction – also transport heat, and they do so more effectively than phonons, giving metals their high thermal conductivities. Thermal properties are manipulated by interfering with these processes: alloying to scatter phonons and electrons; or foaming to dilute the solid with low-conductivity, low-specific-heat gases.

The thermal response of a component to temperature change creates both problems and opportunities: thermal contraction can cause cracking or unwanted distortion; on the other hand, bi-material strips use controlled thermal distortion for actuation or sensing. It is the specific heat that makes an oven take 15 minutes to warm up, but it also stores heat in a way that can be recovered on demand. Good thermal conduction is not what you want in a coffee cup, but when it cools the engine of your car or the chip in your computer it is a big help. Thermal stresses and distortion are major issues in manufacturing, too. Furthermore, the transient cooling histories in components, controlled by their geometry and thermal diffusivity, are central to understanding the evolution and control of the microstructure on which so many properties depend (see Chapter 11).

This chapter has been about the thermal properties of materials. Heat, as we have said, also changes other properties, notably strength. We look more deeply into that in Chapter 8.

7.7 Further reading

Cottrell, A. H. (1964). *The Mechanical Properties of Matter.* Wiley. Library of Congress Number 64-14262. (The best introduction to the physical understanding of the properties of materials, now out of print but worth searching for).

Hollman, J. P. (1981). *Heat transfer* (5th ed.). McGraw-Hill. ISBN 0-07-029618-9. (A good starting point: an elementary introduction to the principles of heat transfer).

Tabor, D. (1969). *Gases, Liquids and Solids.* Penguin Books. ISBN 14-080052-2. (A concise introduction to the links between inter-atomic forces and the bulk properties of materials).

7.8 Exercises

Exercise E7.1 Define specific heat. What are its units? How do you calculate the specific heat per unit volume from the specific heat per unit mass? If you wanted to select a material for a compact heat storing device, which of the two would you use as a criterion of choice?

Exercise E7.2 A material has a thermal conductivity $\lambda = 0.3$ W/m.K, a density $\rho = 1200$ kg/m^3, and a specific heat $C_p = 1400$ J/kg.K. What is its thermal diffusivity?

Exercise E7.3 The thermal expansion coefficient was expressed in Equations (7.10) and (7.11) of the text in relation to two other important properties, melting point T_m and Young's modulus E. Use these expressions to find an approximate relationship between point T_m (in K) and E, and test its validity by finding the ratio E/T_m, using the typical data from Appendix A for the following alloys: Mg, Al, Cu, W, and low carbon steel. You could also plot E against T_m on a (linear) graph and compare the data with a line through the origin with the expected slope.

Exercise E7.4 Which two metals would you choose to make a bimetallic strip to maximise the displacement per unit temperature change [using Equation (7.15), neglecting the effect of the difference in Young's modulus]? Use the $\lambda - \alpha$ chart (Figure 7.2) to make the choice. If the bimetallic strip has a thickness of 2 mm and an average thermal diffusivity a of 5×10^{-5} m^2/s, approximately how long will it take to respond when the temperature suddenly changes?

Exercise E7.5 A structural material is sought for a low-temperature device for use at –20°C that requires high strength but low thermal conductivity. Use the $\lambda - \sigma_y$ chart in Figure 7.4 to suggest two promising candidates (rejecting ceramics on the grounds that they are too brittle for structural use in this application).

Exercise E7.6 A new metal alloy has a density of 7380 kg/m^3. Its specific heat has not yet been measured. Use the chart in Figure 7.3 to make an approximate estimate of its value in the normal units of J/kg.K.

Exercise E7.7 The same new alloy of the last exercise has a melting point of 1540 K. Its thermal expansion coefficient has not yet been measured. How would you make an approximate estimate of its value? What value would you then report?

Exercise E7.8 The table shows thermal conductivity data for Cu, Ni, and their alloys. Here the notation Cu-10Ni means 10 weight % Ni (and 90% Cu). Heat conduction theory suggests that thermal conductivity is a microstructure-sensitive property. Show graphically how the data for these alloys deviate from a simple linear rule of mixtures (in weight %).

Alloy	Thermal conductivity, λ (W/m.K)
Pure Cu	395
Cu-2Ni	165
Cu-5Ni	66
Cu-10Ni	42
Cu-30Ni	21
Cu-65Ni	22
Pure Ni	91

Exercise E7.9 Interior wall insulation should insulate well but have as low a heat capacity as possible; that way most of the heat goes into the room, not into the wall. This requires materials with low thermal conductivity, λ, and low volumetric specific heat, ρC_p. Use the $\lambda - a$ chart in Figure 7.3 to find the materials that do this best. (The contours will help).

Exercise E7.10 An external wall of a house has an area of 15 m^2 and is of single brick construction with a thickness 10 cm. The occupant keeps the internal temperature at 22°C. How much heat is lost through the wall per hour, if the outside temperature is a stable 0°C? What percentage reduction in heat loss through the external wall would be achieved by reducing the internal temperature to 18°C? The thermal conductivity of the brick is 0.8 W/m.K.

Exercise E7.11 The house occupant in Exercise E7.10 wishes to reduce heating costs. The wall could be thickened to a double brick construction, with a cavity filled with polyurethane foam. The foam manufacturer quotes an effective thermal conductivity of 0.25 W/m.K for this wall. How much heat would be saved per hour, compared to the single brick, for internal temperatures of 22°C and 18°C?

Exercise E7.12 When a nuclear reactor is shut down quickly, the temperature at the surface of a thick stainless steel component falls from 600°C to 300°C in less than 1 second. Due to the relatively low thermal conductivity of the steel, the bulk of the component remains at the higher temperature for several seconds. The coefficient of thermal expansion α of stainless steel is 1.2×10^{-5} K^{-1}, Poisson's ratio $v = 0.33$, Young's modulus $E = 200$ GPa, and the yield strength $\sigma_y = 585$ MPa.

(a) If the surface were free to contract, what would be the thermal strain?

(b) The surface is in fact constrained by the much more massive interior and as a result suffers no strain; instead, a state of biaxial tensile stress appears in it, giving an equal and opposite strain to the thermal contraction. If this thermally induced strain is assumed to be elastic, the tensile stress in the plane of the surface is

$$\sigma_{thermal} = \frac{1}{(1 - \nu)} E\left(-\varepsilon_{thermal}\right)$$

Will the thermally induced stress, in fact, cause yielding?

Exercise E7.13 Square porcelain tiles are to be manufactured with a uniform layer of glaze, which is thin compared to the thickness of the tile. They are fired at a temperature of 700°C and then cooled slowly to room temperature, 20°C. Two tile types are to be manufactured: the first is decorative, using tension in the surface to form a mosaic of cracks; the second uses compression in the surface to resist surface cracking, improving the strength. Use Equation (7.13) in the text and the following data to select a suitable glaze (A, B or C) for each of these applications.

Young's modulus of glaze, $E = 90$ GPa, Poisson's ratio of glaze, $\nu = 0.2$, tensile failure stress of glaze $= 10$ MPa. The coefficients of thermal expansion for the tile body and the glazes are:

Material	$\alpha \times 10^{-6}$ (K^{-1})
Porcelain	2.2
Glaze A	1.5
Glaze B	2.3
Glaze C	2.5

Exercise E7.14 When hot coffee is poured into a ceramic office coffee mug, it gets too hot to hold in about 10 seconds. The wall thickness of the cup is 2 mm.

(a) What, approximately, is the thermal diffusivity of the ceramic of the mug?

(b) Given that the volume specific heat of solids, ρC_p, is more or less constant at 3×10^6 J/m^3.K, what approximately is the thermal conductivity of the cup material?

(c) If the cup were made of a metal with a thermal diffusivity of 2×10^{-5} m^2/s, roughly how long does it take to become too hot to hold?

Exercise E7.15 A flat heat shield is designed to protect against sudden temperature surges. Its maximum allowable thickness is 10 mm. What value of thermal diffusivity is needed to ensure that, when heat is applied to one of its surfaces, the other does not start to change significantly in temperature for at least 5 minutes?

Use the $\lambda - a$ chart in Figure 7.3 to identify material classes that you might use for the heat shield.

Exercise E7.16 The table lists properties of soda-lime glass and of single crystal diamond. Thermal conduction in both is by phonon transport. Compare the mean free path length of phonons in the two materials. How do they compare with the size of the atoms of which the materials are made? Why do they differ? (Atomic diameter of silicon = 0.24 nm; atomic diameter of carbon = 0.16 nm).

Property	Soda-lime glass	Single crystal diamond
Young's modulus (GPa), E	69	1200
Density (kg/m³), ρ	2450	3500
Specific heat capacity (J/kg.K), C_p	910	510
Thermal conductivity (W/m.K), λ	1.1	2200

Exercise E7.17 Aluminium alloys come in a variety of strengths, depending on composition and processing. Like strength, thermal conductivity is also a microstructure-sensitive property. The table shows a range of data for Al alloys hardened by three of the mechanisms discussed in Chapter 5. Plot the data on a property chart of thermal conductivity against strength, and comment on the 'property trajectory' produced by each of the hardening mechanisms. Use log scales with ranges 50–500 W/m.K for conductivity and 10–1000 MPa for strength. Add a contour of the index $M = \lambda\sigma_y$ (as in Figure 7.4) to see how the Al alloys compete with one another for heat exchanger applications.

	Thermal conductivity, λ (W/m.K)	Yield strength, σ_y (MPa)
Pure Al	240	20
Pure Al, work hardened	240	140
Al alloy, solid solution hardened	125	120
Al alloy, solid solution + work hardened	125	200
Al alloy, precipitation hardened	160	300

Exercise E7.18 A material is sought for a storage heater, with the objective of maximising the heat stored per unit material cost, for a given temperature increase of ΔT. Find an expression for the energy stored in a mass m of the material, and hence the energy stored per unit cost, taking the material cost/kg to be C_m. Identify a *material index* of properties that should be maximised to give the highest value for the objective. Use the typical data in Appendix A to compare the performance of cast iron, Mg alloys, concrete, and PEEK. What other factors would need to be taken into account for this selection?

Exercise E7.19 A large steel plate of thickness $\ell = 5$ mm is quenched from 920°C into a bath of cold water at 20°C. Find the time taken for the temperature at the centre of the plate to fall to 470°C, halfway between the initial and final temperatures. The thermal diffusivity of the steel is a $= 9 \times 10^{-6}$ m²/s. Compare the result with rough estimates of the characteristic cooling time, assuming the heat conduction distance is of order $(\ell/4)$ to $(\ell/2)$.

Exercise E7.20 In the development of a new quenching facility for steel components, scale model tests were conducted on dummy steel samples containing a thermocouple located in the centre of the sample. The geometry of the samples was geometrically similar but one-third the scale of the real components, and the quenching process was identical. The thermocouple readout showed that on quenching a sample from a furnace into cold water, the temperature fell by 50% of the temperature interval in a time of 15 seconds. How long do you estimate it would take to cool by the same amount in the real component? In a second trial, full-sized components were tested using aluminium dummy samples, with a suitably reduced furnace temperature. What time would you expect to find is required to cool the centre to 50% of the new temperature interval? The thermal diffusivities of the steel and aluminium are 9×10^{-6} m²/s and 7×10^{-5} m²/s, respectively.

Chapter 8
Materials at high temperatures: diffusion and creep

Extreme design requirements in a shuttle launch.
(istock.com)

https://doi.org/10.1016/B978-0-08-102399-0.00008-X

8.1 Introduction and synopsis

Material properties change with temperature. Some do so in a simple linear way, easy to allow for in design: the density, the modulus, and the electrical resistivity are examples. But others, particularly the yield strength and the rates of oxidation and corrosion, change in more rapid and complex ways that can lead to disaster if not understood and allowed for in design.

This chapter explores the ways in which properties change with temperature and discusses design methods to deal with materials at high temperature. To do this we must first understand *diffusion*, the movement of atoms through solids, and the processes of *creep* and *creep fracture*, the continuous deformation of materials under load at temperature. Diffusion theory lies directly behind the analysis and procedures for high-temperature design with metals and ceramics. Polymers, being molecular, are more complicated in their creep behaviour, but semi-empirical methods allow safe design with these, too. Diffusion and hot deformation also relate closely to material processing – forging, rolling, and heat treatment exploit the same physical mechanisms – so the chapter also explores this.

8.2 The temperature dependence of material properties

Linear and non-linear temperature dependence   Some properties depend on temperature T in a linear way, meaning that

$$P = P_o \left(1 + \beta \frac{T - T_o}{T_m} \right) \tag{8.1}$$

where P is the value of the property, P_o its value at a reference low temperature T_o (at or close to 0 K), and β is a constant. Figure 8.1 shows three examples: density, modulus, and – for metals – electrical resistivity. The density ρ decreases by about 6% on heating from cold to

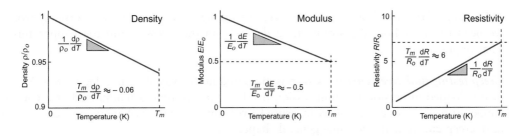

Figure 8.1 Linear dependence of properties on temperature: density, modulus, and electrical resistivity.

the melting point T_m ($\beta \approx -0.06$), the modulus E falls by a factor of 2 ($\beta \approx -0.5$), while the resistivity R increases by a factor of about 7 ($\beta \approx +6$). These changes cannot be neglected but are easily accommodated by using the value of the property at the temperature of the design.

Other properties are less forgiving. Strength falls in a much more sudden way, and the rate of *creep* – the main topic here – increases exponentially (Figure 8.2).

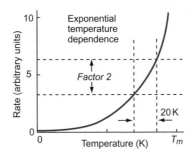

Figure 8.2 The exponential increase of rate with temperature.

Creep At room temperature, most metals and ceramics deform in a way that depends on stress but not on time. As the temperature is raised, loads that are too small to give permanent deformation at room temperature cause materials to *creep*: to undergo slow, continuous deformation with time, ending in fracture. It is usual to refer to the time-independent behaviour as the 'low-temperature' response, and the time-dependent flow as the 'high-temperature' response. But what, in this context, is 'low' and what is 'high'? The melting point of tungsten is over 3000°C; room temperature, for tungsten, is a very low temperature. The melting point of lead is 327°C; room temperature, for lead, is a high temperature.

Creep can be a problem – it is performance-limiting in gas turbines and much power-generating equipment – but it can be useful, too. Hot forging, rolling, and extrusion are carried out at temperatures and stresses at which creep occurs, lowering the force and power required for the shaping operation.

To design for control of creep in service, or to design and operate a hot forming process, we need to know how the strain-rate $\dot{\varepsilon}$ depends on the stress σ and the temperature T. That requires *creep testing*.

Creep testing and creep curves　Creep is measured in the way shown in Figure 8.3. A specimen is loaded in tension or compression inside a furnace that is maintained at a constant temperature, T. The extension is measured over time. Metals, polymers, and ceramics all have creep curves with the same general shape.

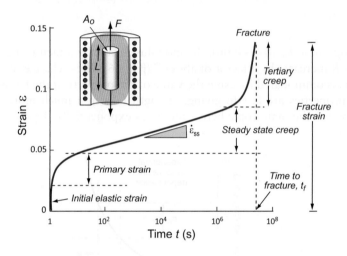

Figure 8.3　Creep testing and the creep curve.

The *initial elastic* and the *primary creep* strains occur quickly and can be treated in much the same way as elastic deflection is allowed for in a structure. After that, the strain increases steadily with time in what is called the *secondary creep* or the *steady-state creep* regime. Plotting the log of the steady-state creep rate, $\dot{\varepsilon}_{ss}$, against the log of the stress, σ, at constant T, as in Figure 8.4(a), shows that

$$\dot{\varepsilon}_{ss} = B\sigma^n \tag{8.2}$$

where n, the *creep exponent*, usually lies between 3 and 8, which is why this is called *power-law creep*. At low σ there is a tail with slope $n \approx 1$ [labeled *diffusional flow* in Figure 8.4(a)] – we'll explain that later on.

A plot of the natural logarithm (ln) of $\dot{\varepsilon}_{ss}$ against the reciprocal of the absolute temperature $(1/T)$ at constant stress [Figure 8.4(b)] gives a straight line, implying that

$$\dot{\varepsilon}_{ss} = C \exp -\left(\frac{Q_c}{RT}\right) \tag{8.3}$$

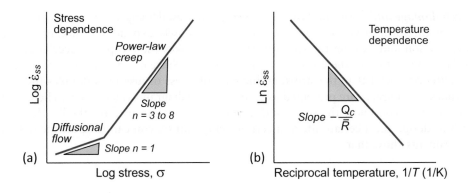

Figure 8.4 Stress and temperature dependence of the creep rate.

Here R is the gas constant (8.314 J/mol.K) and Q_c is called the *activation energy for creep*, with units of J/mol. The creep rate $\dot{\varepsilon}_{ss}$ increases exponentially in the way suggested by Figure 8.2. Combining these two findings gives

$$\dot{\varepsilon}_{ss} = C'\sigma^n \exp-\left(\frac{Q_c}{RT}\right) \tag{8.4}$$

where C' is a constant. Written in this way the constant C' has weird units ($s^{-1}.MPa^{-n}$) so it is more usual and sensible to write instead

$$\dot{\varepsilon}_{ss} = \dot{\varepsilon}_o \left(\frac{\sigma}{\sigma_o}\right)^n \exp-\left(\frac{Q_c}{RT}\right) \tag{8.5}$$

The four constants $\dot{\varepsilon}_o$ (units: s^{-1}), σ_o (units: MPa), n, and Q_c characterize the steady-state creep of a material; if you know these, you can calculate the strain-rate $\dot{\varepsilon}_{ss}$ at any temperature and stress using Equation (8.5). They vary from material to material and have to be measured experimentally.

Example 8.1 Creep extension

A stainless steel tie of length $L = 100$ mm is loaded in tension under a stress $\sigma = 150$ MPa at a temperature $T = 800°C$. The creep constants for stainless steel are reference strain-rate $\dot{\varepsilon}_o = 10^6/s$, reference stress $\sigma_o = 100$ MPa, stress exponent $n = 7.5$, and activation energy for power-law creep $Q_c = 280$ kJ/mol. What is the creep rate of the tie? By how much will it extend in 100 hours?

Answer. The steady-state creep rate is given by $\dot{\varepsilon}_{ss} = \dot{\varepsilon}_o \left(\frac{\sigma}{\sigma_o}\right)^n \exp-\left(\frac{Q_c}{RT}\right)$. Using the data provided, converting the temperature to Kelvin, and using the gas constant $R = 8.314$ J/mol.K gives $\dot{\varepsilon}_{ss} = 4.8\times10^{-7}$/s. After 100 hours ($3.6 \times 10^5$ seconds) the strain is $\varepsilon = 0.173$, and the 100 mm long tie has extended by $\Delta L = 17.3$ mm.

Creep damage and creep fracture As creep continues, damage accumulates in the material. It takes the form of voids or internal cracks that slowly expand and link, eating away the cross-section and causing the stress to rise. This makes the creep rate accelerate as shown in the tertiary stage of the creep curve of Figure 8.3. Since $\dot{\varepsilon} \propto \sigma^n$ with $n \approx 5$, the creep rate goes up even faster than the stress: an increase of stress of just 15% doubles the creep rate.

Times to failure, t_f, are presented as *creep-rupture* diagrams (Figure 8.5). Their application is obvious: if you know the stress and temperature you can read off the life. If instead you wish to design for a certain life at a certain temperature, you can read off the design stress. Experiments show that

$$\dot{\varepsilon}_{ss}\, t_f \approx C \tag{8.6}$$

which is called the Monkman-Grant law. The Monkman-Grant constant, C, is typically 0.05 to 0.3. Knowing it, the creep life (meaning t_f) can be estimated from Equation (8.6).

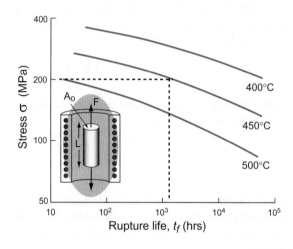

Figure 8.5 Creep rupture life as a function of stress and temperature.

Example 8.2 Creep of a pressure vessel

A component in a chemical engineering plant is made of the stainless steel of Example 8.1. It is loaded by an internal pressure that generates a hoop stress of 150 MPa.

(a) If the hoop creep rate $\dot{\varepsilon}_{ss} = 4.8\times10^{-7}/\text{s}$ at $T = 800°C$ under this stress, by what factor will it increase if the pressure is raised by 20%?
(b) If instead the pressure is held constant but the temperature is raised from $T_1 = 800°C$ to $T_2 = 850°C$, by what factor will the creep life t_f be reduced? (Remember that T in the equation is in Kelvin, not Centigrade).

Answer. The hoop stress is proportional to the pressure [Equation (5.29)]. The factor by which the strain-rate increases when the pressure is raised is

(a) $$\frac{\text{Creep rate at 180 MPa}}{\text{Creep rate at 150 MPa}} = \left(\frac{180}{150}\right)^{7.5} = 3.9;$$

the creep rate increases to $\dot{\varepsilon}_{ss} = 1.9 \times 10^{-6}/\text{s}$

(b)   The factor by which the strain-rate is increased by the increase in temperature is

$$\frac{\text{Creep rate at } T_2 = 1123\,\text{K}}{\text{Creep rate at } T_1 = 1703\,\text{K}} = \exp{-\frac{Q_c}{R}\left(\frac{1}{T_2} - \frac{1}{T_1}\right)} = 4.0$$

The creep life $t_f = C/\dot{\varepsilon}_{ss}$ where C is a constant. The creep life therefore is reduced by the same factor. The message here is that comparatively small changes in stress or temperature cause big changes in creep rate and creep life because of their strongly non-linear dependence on both σ and T.

8.3 Data for key reference temperatures

Melting point   Figure 8.6 shows data for the melting point T_m for metals, ceramics, and polymers. As a general rule, creep starts when $T \approx 0.35\ T_m$ for metals and $0.45\ T_m$ for ceramics (temperatures in Kelvin). Alloying can raise this temperature significantly. Most metals and ceramics have high melting points and, because of this, they creep only at temperatures well above room temperature. Lead, however, has a melting point of 327°C (600 K), so room temperature is almost half its absolute melting point and it creeps – a problem with lead roofs and cladding of old buildings. Crystalline polymers, most with melting points in the range 150° to 200°C, creep slowly if loaded at room temperature; glassy polymers, with T_g of typically 50° to 150°C, do the same.

Service temperatures   Classes of industrial applications tend to be associated with certain characteristic temperature ranges (Figure 8.7). There is the cryogenic range, between –273°C and roughly room temperature, associated with the use of liquid gases like hydrogen, oxygen, or nitrogen and superconducting devices. Here the issue is not creep but the avoidance of brittle fracture. There is the regime at and near room temperature (–20° to +150°C) associated with conventional everyday life. Above this is the range 150° to 400°C, associated with automobile engines and with food and industrial processing. Higher still are the regimes of steam turbines (typically 400°–650°C), and of gas turbines and chemical reactors (650°–1000°C). Special applications (rocket nozzles) require materials that withstand even higher temperatures, extending to 2800°C.

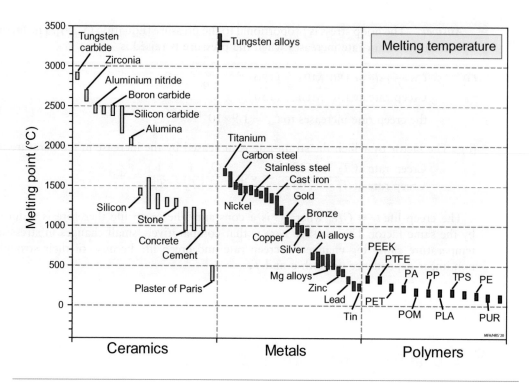

Figure 8.6 Melting points of ceramics, metals, and polymers.

Materials have been developed to fill the needs of each of these temperature ranges (Figure 8.7). Polymers mostly only have useful strength below 150°C, but a select few, and composites based on them, can be used in applications up to 250°C, where they now compete with magnesium and aluminium alloys, and with the much heavier cast irons and steels, traditionally used in those ranges. Temperatures above 400°C require special creep-resistant alloys: ferritic steels, titanium alloys (lighter, but more expensive), and certain stainless steels. Stainless steels and ferrous superalloys really come into their own in the temperature range above this, where they are widely used in steam turbines and heat exchangers. Gas turbines require, in general, nickel-based or cobalt-based superalloys. Above 1000°C, the refractory metals and ceramics become the only candidates. Materials used at high temperatures will, generally, perform perfectly well at lower temperatures, too, but are not used there because of cost.

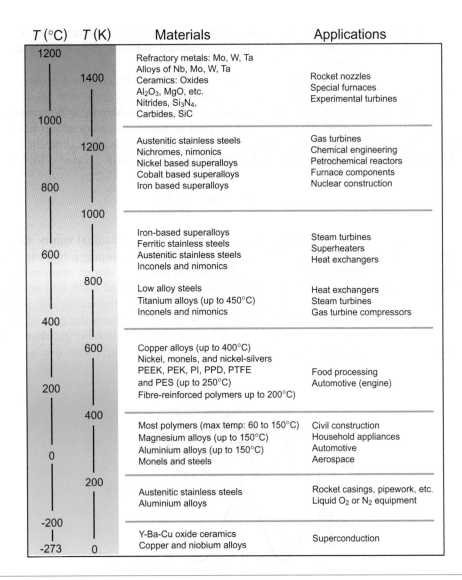

T (°C)	T (K)	Materials	Applications
1200	1400	Refractory metals: Mo, W, Ta Alloys of Nb, Mo, W, Ta Ceramics: Oxides Al_2O_3, MgO, etc. Nitrides, Si_3N_4, Carbides, SiC	Rocket nozzles Special furnaces Experimental turbines
1000	1200	Austenitic stainless steels Nichromes, nimonics Nickel based superalloys Cobalt based superalloys Iron based superalloys	Gas turbines Chemical engineering Petrochemical reactors Furnace components Nuclear construction
800 600	1000 800	Iron-based superalloys Ferritic stainless steels Austenitic stainless steels Inconels and nimonics	Steam turbines Superheaters Heat exchangers
400		Low alloy steels Titanium alloys (up to 450°C) Inconels and nimonics	Heat exchangers Steam turbines Gas turbine compressors
200	600	Copper alloys (up to 400°C) Nickel, monels, and nickel-silvers PEEK, PEK, PI, PPD, PTFE and PES (up to 250°C) Fibre-reinforced polymers up to 200°C)	Food processing Automotive (engine)
0	400	Most polymers (max temp: 60 to 150°C) Magnesium alloys (up to 150°C) Aluminium alloys (up to 150°C) Monels and steels	Civil construction Household appliances Automotive Aerospace
-200	200	Austenitic stainless steels Aluminium alloys	Rocket casings, pipework, etc. Liquid O_2 or N_2 equipment
-273	0	Y-Ba-Cu oxide ceramics Copper and niobium alloys	Superconduction

Figure 8.7   Common materials used and their areas of applications, for various regimes of temperature. Above 0°C the issue is creep (or oxidation); below 0°C, it is brittle fracture.

8.4 The science: diffusion

Creep deformation requires the relative motion of atoms. How does this happen, and what is its rate?

Diffusion   Diffusion is the spontaneous intermixing and migration of atoms over time. In gases and liquids this is familiar – the spreading of smoke in still air, the dispersion of an ink-drop in water – and explained as Brownian motion, the random collision of atoms or molecules. In crystalline solids the atoms are confined to lattice sites, but they can still move and mix if they are warm enough.

Heat, as we have said, is atoms in motion. In a solid at temperature T atoms vibrate about their mean position with a frequency ν (about 10^{13} per second) with an average energy, kinetic plus potential, of $k_B T$ in each mode of vibration, where k_B is Boltzmann's constant (1.38×10^{-23} J/atom.K). This is the average, but at any instant some atoms have less energy, some more. The Maxwell-Boltzmann equation[1], a result of statistical mechanics, describes the probability p that a given atom has an energy greater than a value q Joules:

$$p = \exp -\left(\frac{q}{k_B T}\right) \tag{8.7}$$

Crystals, as we saw in Chapter 3, contain *vacancies* – occasional empty atom sites. It is these that allow diffusive motion. Figure 8.8 shows an atom jumping into a vacancy. To make such a jump, the atom marked in red (though it is the same sort of atom as the rest) must break away from its original comfortable site at A, its *ground state*, and squeeze between neighbours, passing through an *activated state*, to drop into the vacant site at B where it is once again comfortable. There is an energy barrier, q_m, between the ground state and the activated state to overcome if the atom is to move. The probability p_m that a given atom has thermal energy larger than this barrier is just Equation (8.7) with $q = q_m$.

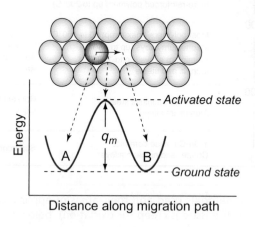

Figure 8.8 The energy change during a diffusive jump from site A to site B.

---

[1] James Clerk Maxwell (1831–1879), pre-eminent Scottish theoretical physicist, considered by many to be the equal of Newton and Einstein (the latter kept a photo of Maxwell in his study, alongside pictures of Newton and Faraday). Maxwell's input to kinetic theory of diffusion was a relative sideshow compared to his electromagnetic theory, which unified the understanding at the time of electricity, magnetism, and optics.

So two things are needed for an atom to switch sites: enough thermal energy and an adjacent vacancy. A vacancy has an energy q_v, so – not surprisingly – the probability p_v that a given site be vacant is also given by Equation (8.7), this time with $q = q_v$. Thus the overall probability of an atom changing sites is

$$p = p_v\, p_m = \exp - \left(\frac{q_v + q_m}{k_B T} \right) = \exp - \left(\frac{q_d}{k_B T} \right) \tag{8.8}$$

where q_d is called the *activation energy for self-diffusion*. Diffusion, and phenomena whose rates are governed by diffusion, are often said to be *thermally activated*. If instead the red atom were chemically different from the green ones (making it a substitutional solid solution), the process is known as *inter-diffusion*. Its activation energy has the same origin, but the barrier will have a different value. Interstitial solutes diffuse, too: they have comfortable locations in their interstitial holes and need thermal activation to hop between them, but without the need for an adjoining vacancy, so their diffusion rates are usually much higher.

Figure 8.9 illustrates how mixing occurs by inter-diffusion. It shows a solid in which there is a *concentration gradient* dc/dx of red atoms: there are more in slice A immediately to the left of the central, shaded plane than in slice B to its right. If atoms jump across this plane at random, there will be a net flux of red atoms to the right because there are more on the left to jump, and a net flux of white atoms in the opposite direction. The number in slice A, per unit area, is $n_A = 2 r_o c_A$, and that in slice B is $n_B = 2 r_o c_B$ where $2 r_o$, the atom size, is the

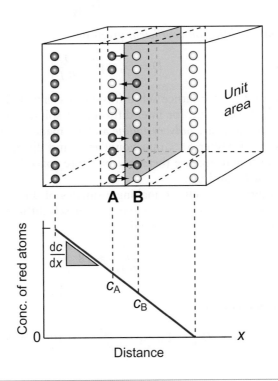

Figure 8.9 Diffusion in a concentration gradient.

thickness of the slices and c_A and c_B are the concentration of red atoms on the two planes expressed as atom fractions. The difference is

$$n_A - n_B = 2r_o(c_A - c_B) = 4r_o^2 \frac{dc}{dx}$$

since $dc/dx = (c_A - c_B)/2r_o$. The number of times per second that an atom on slice A oscillates towards B, or one on B towards A, is $v/6$, since there are six possible directions in which an atom can oscillate in three dimensions, only one of which is in the right direction. Thus the net flux per second of red atoms across unit area from left to right is

$$J = -\frac{v}{6} \exp -\left(\frac{q_d}{k_B T}\right) 4r_o^2 \frac{dc}{dx}$$

It is usual to work with the activation energy per mole, Q_d, rather than that per atom, q_d, so we write $Q_d = N_A q_d$ and $R = N_A k_B$ (where N_A is Avogadro's number, 6.022×10^{23} mol^{-1}), and assemble the terms $4v r_o^2/6$ into a single *diffusion constant* D_o to give

$$J = -D_o \exp -\left(\frac{Q_d}{RT}\right) \frac{dc}{dx} \tag{8.9}$$

This is known as Fick's law[2]:

$$J = -D \frac{dc}{dx} \tag{8.10}$$

where D is called the *diffusion coefficient* given by

$$D = D_o \exp -\left(\frac{Q_d}{RT}\right) \tag{8.11}$$

Values for D_o and Q_d for lattice self-diffusion of elements have been measured experimentally for self-diffusion and for inter-diffusion of solutes in many alloys and ceramics. Most metals and ceramics are *polycrystalline* and are made up of grains, within which there are dislocations (see Chapter 5). Grain boundaries and dislocations provide 'short-circuit' paths for faster diffusion because there is a bit more space between the atoms; these have their own values of D_o and Q_d.

Two rules-of-thumb are useful in making estimates of diffusion rates when data are not available. The first is that the activation energy for self-diffusion in kJ/mol, normalised by RT_m, is approximately constant for metals:

$$\frac{Q_d}{RT_m} \approx 18 \tag{8.12}$$

The second is that the diffusion coefficient of metals, evaluated at their melting point, is also approximately constant:

$$D_{T_m} = D_o \exp -\left(\frac{Q_d}{RT_m}\right) \approx 10^{-12} \text{ m}^2/\text{s} \tag{8.13}$$

---

[2] Adolph Eugen Fick (1829–1901), German physicist, physiologist, and inventor of the contact lens. He formulated the law that describes the passage of gas through a membrane, but which also describes the heat-activated mixing of atoms in solids.

Example 8.3 Fuel-cell membranes

Alumina (Al_2O_3) is suggested for the membrane of a fuel cell. The performance of the cell for oxygen ion transport depends on the diffusion coefficient of oxygen ions in the membrane at the operating temperature, a maximum of $T = 1100°C$. If $D < 10^{-12}$ m²/s, the cell is impractical. The activation energy for oxygen diffusion in Al_2O_3 is $Q_d = 636$ kJ/mol and the pre-exponential diffusion constant $D_o = 0.19$ m²/s. Will the cell work?

Answer. The oxygen diffusion coefficient at $T = 1100°C$ is

$$D = D_o \exp - \left(\frac{Q_d}{RT} \right) = 0.19 \exp - \left(\frac{636,000}{8.314 \times 1373} \right) = 1.17 \times 10^{-25} \text{ m}^2/\text{s}$$

This is very much less than 10^{-12} m²/s – the cell is not practical.

Diffusion driven by other fields Diffusion, as we have seen, can be driven by a concentration gradient. Concentration c is a *field quantity*; it has discrete values at different points in space (the concentration field). The difference in c between two nearby points divided by the distance between them defines the local concentration gradient, dc/dx. Diffusion can be driven by other field gradients. A stress gradient, as we shall see in a moment, drives diffusional flow and power law creep. An electric field gradient can drive diffusion in non-conducting materials. Even a temperature gradient can drive diffusion of matter as well as diffusion of heat.

Transient diffusion problems Fick's law [Equation (8.10)] describes the rate at which atoms flow down a uniform concentration gradient at steady-state (i.e. when the concentration at a given position doesn't change with time). Fick's law is very like Fourier's law for steady-state heat flow [Equation (7.3)] and so, too, are the solutions for transient flow of heat and of atoms. Figure 8.10 shows two solid blocks of pure elements A and B that are brought into contact and heated. We will assume that A and B can form a solid solution in one another in any proportions, as do copper and nickel, for example. The concentration profile evolves with time, heading (after a very long time) to a uniform mixture of 50% A atoms to 50% B atoms, in much the same way as heat flows from a hot to a cold object until both reach the same uniform temperature.

Consider the flux of atoms in and out of a small volume element at a given value of x. The flux of diffusing atoms is proportional to the local concentration gradient [Equation (8.10)], so in a small time step δt, a slightly different number of atoms enter and leave the volume element, reflecting the small difference in the concentration on either side. The concentration of the element will therefore change by δC. The differential equation describing the change in concentration C with position x and time t, when atoms are diffusing in one dimension (the x-direction) is:

$$\frac{\partial^2 C}{\partial x^2} = \frac{1}{D} \frac{\partial C}{\partial t} \tag{8.14}$$

(where the curly '∂' means that this is a partial differential equation).

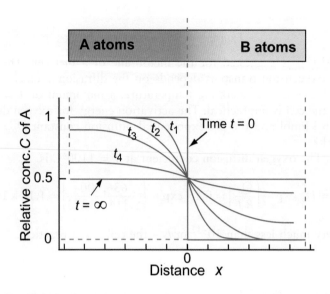

Figure 8.10 Intermixing by diffusion.

This is the same as the heat-flow equivalent [Equation (7.21)] with concentration C replacing temperature T and the diffusion coefficient D replacing the thermal diffusivity, a, so the solution for one is easily adapted for the other. The solution of Equation (8.14) for the configuration of Figure 8.10 gives concentration $C(x,t)$ at the point x after a time t as

$$\left(\frac{C(x,t) - C_o}{C_s - C_o}\right) = 1 - \text{erf}\left(\frac{x}{2\sqrt{Dt}}\right) \tag{8.15}$$

where erf is the error function, C_o is the starting composition, and C_s is the composition after infinite time. Figure 8.10 shows how the concentration changes over time. At time $t = 0$ the relative concentration of A atoms is uniform and equal to 1; that of B is 0. The successive concentration profiles at times t_1, t_2, t_3, t_4 evolve between these limits. The error function is close to linear and equal to its argument, provided $x/2\sqrt{Dt} < 0.7$ (approx.), so this can be simplified to

$$\left(\frac{C(x,t) - C_o}{C_s - C_o}\right) \approx 1 - \left(\frac{x}{2\sqrt{Dt}}\right) \tag{8.16}$$

When the concentration has changed halfway from its initial to its final value, then the left-hand side side of Equation (8.16) = 0.5. So the distance from the interface at which this occurs is given by

$$x \approx \sqrt{Dt} \tag{8.17}$$

This is a useful rule-of-thumb for a characteristic atomic diffusion distance in a time t – directly analogous to that seen in problems of transient temperature change in heat flow

[Equations (7.5) and (7.23)]. These transient solutions apply to many problems in materials processing, examples of which appear in this chapter.

Diffusion in surface hardening of steel: carburising   Some steels are used in applications in which the surfaces are in sliding or rolling contact, requiring that they have a hard, wear-resistant surface, achieved by *surface engineering*. One method is *carburising* – diffusing carbon into the surface of a hot steel component, locally increasing the concentration, and raising the hardness and the resistance to wear (see Chapter 9). Figure 8.11 shows the concentration profile of the carbon from the surface inwards; it is a plot of Equation (8.15) replacing distance x by the normalised (dimensionless) distance $x/\sqrt{Dt}$. The concentration has increased halfway from C_o to C_s at a distance below the surface where $x/\sqrt{Dt} = 1$, or $x = \sqrt{Dt}$.

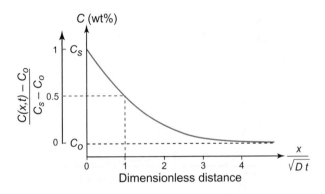

Figure 8.11   Concentration profile of carbon in iron during surface carburising.

Example 8.4 Carburising

It is proposed to harden a component of low carbon steel, of concentration 0.2 wt% C, by diffusing carbon into the surface from a carbon-rich environment at $T = 1000°C$. At this temperature, the maximum carbon concentration in solid solution is $C_s = 1.6$ wt% C. Iron at this temperature has an FCC structure (called austenite) with a diffusion constant for carbon $D_o = 8 \times 10^{-6}$ m^2/s and activation energy $Q_d = 131$ kJ/mol. The objective is to raise the carbon concentration to a minimum of 0.9 wt% C, to a depth of 0.2 mm. Roughly how long will it take to achieve the desired depth of hardening?

Answer.   The diffusion coefficient at 1000°C is

$$D = 8 \times 10^{-6} \exp - \left(\frac{131{,}000}{8.314 \times 1273} \right) = 3.35 \times 10^{-11} \text{ m}^2\text{/s}$$

The target change in concentration is exactly half of the interval between the initial and surface concentration. Hence the time to reach this concentration at a depth $x = 0.2$ mm is approximately

$$t \approx \frac{x^2}{D} = \frac{4 \times 10^{-8}}{3.35 \times 10^{-11}} 1.2 \times 10^{3} \text{ s} \approx 20 \text{ minutes}$$

Diffusion in doping of semiconductors The unusual electrical properties of semiconductors are created by *doping* high-purity silicon with low concentrations of atoms such as boron and phosphorus. One way to do this is to diffuse in a controlled quantity of boron atoms at the surface – 'pre-deposition' – and then to remove the source of boron and to 'drive-in' the dopant to a greater depth by controlled heating. Figure 8.12 shows the evolving concentration profiles. The area under the two curves is the same, being the total number of B atoms diffused in during pre-deposition. In the first stage, the profile is the same as that for carburising. The mathematical solution for the drive-in stage takes a different form because the surface concentration C_s now changes with time, but the extent of the distributions of solute can still be estimated for both stages using $x = \sqrt{Dt}$ with the relevant values for D and t.

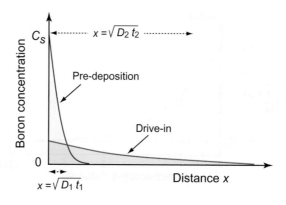

Figure 8.12 Concentration profiles of boron in silicon doping by pre-deposition and drive-in.

8.5 The science: creep

Creep by diffusional flow Liquids and hot glasses are wobbly structures, like a bag loosely filled with beans. If atoms (or beans) move inside the structure, the shape of the whole structure changes in response. Where there is enough free volume, a stress favours the movements that change the shape in the direction of the stress and opposes those that do the opposite, giving viscous flow. The higher the temperature, the more is the free volume and the faster the flow.

But how do jumping atoms change the shape of a crystalline solid? Atoms in a crystal occupy well-defined sites. Atom jumps like that shown in Figure 8.8 allow diffusion, but they don't change the shape of the crystal. But that neglects the presence of surfaces. A *polycrystalline* solid is made up of many crystals meeting at grain boundaries – internal surfaces – that act as sources and sinks for vacancies. If a vacancy joins a boundary, an atom must leave it. Repeat this many times and the face of the crystal is eaten away. If instead a vacancy leaves a boundary, an atom must join it and – when repeated – that face grows. Figure 8.13(a) shows the consequences: the slow extension of the polycrystal in the direction of stress. It is driven by a stress gradient: the difference between the tensile stress σ on the horizontal boundaries from which vacancies flow, and the stress on the others, essentially zero, to which they go. The shape change enables the external stress to do work, which effectively drives the direction of diffusion. If the grain size is d, the stress gradient is σ/d. The flux of atoms, and thus the rate at which each grain extends, $\delta d/\delta t$, is proportional to $D\sigma/d$. The strain-rate, $\dot{\varepsilon}$, is the extension rate divided by the original grain size, d, giving

$$\dot{\varepsilon} = C\,\frac{D\sigma}{d^2} = C\,\frac{\sigma}{d^2}\,D_o\,\exp-\left(\frac{Q_d}{RT}\right) \tag{8.18}$$

where C is a constant. This is a sort of viscous flow, linear in stress. The smaller the grain size the faster it goes.

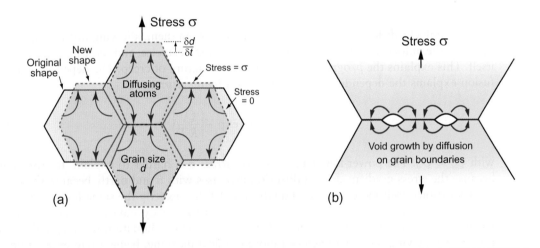

Figure 8.13   (a) Diffusional creep and (b) the diffusive growth of voids on grain boundaries.

Dislocation climb and power-law creep   Plastic flow, as we saw in Chapter 5, is the result of the motion of dislocations. Their movement is resisted by dissolved solute atoms, precipitate particles, grain boundaries, and other dislocations; the yield strength is the stress needed to force dislocations past or between them. Diffusion can unlock dislocations from obstacles in their path, making it easier for them to move. Figure 8.14 shows how it happens. Here a dislocation is obstructed by a particle. The glide force $\sigma_s b$ per unit length is balanced by the reaction f_o from the particle. If the atoms of the extra half-plane diffuse away, making a step in it, the dislocation can continue to glide. The stress gradient this time is less obvious – it is the difference between the local stress where the dislocation is pressed against the particle and that on the dislocation remote from it. The process is called 'climb,' and since it requires diffusion it can occur at a measurable rate only when the temperature is above ~0.35 T_m.

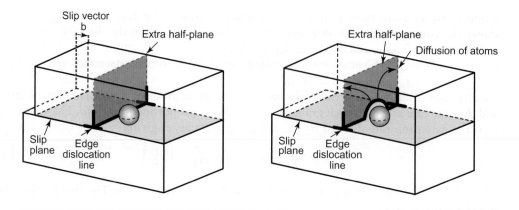

Figure 8.14 Climb of a dislocation leading to *power-law creep*.

Climb unlocks dislocations from the obstacles that pin them, allowing further slip. After a little slip the unlocked dislocations encounter the next obstacles, and the whole cycle repeats itself. This explains the *progressive* and *continuous* nature of creep. The dependence on diffusion explains the dependence of creep rate on *temperature*, with

$$\dot{\varepsilon}_{ss} = A \, \sigma^n \, \exp -\left(\frac{Q_c}{RT}\right) \tag{8.19}$$

with $Q_c \approx Q_d$. The power-law dependence on *stress* σ is harder to explain. It arises partly because the stress-gradient driving diffusion increases with σ and partly because the density of dislocations itself increases, too. Equation (8.19) has exactly the same form as Equation (8.5), the general constitutive model we introduced for steady-state creep. The constant A has just been sub-divided into a ratio of two parameters in Equation (8.5), $\dot{\varepsilon}_o$ and σ_o (to a power of n), giving these parameters a more physical meaning, being a reference strain-rate and stress.

Creep fracture   Diffusion, we have seen, gives creep. It also gives creep fracture. You might think that a creeping material would behave like toffee or chewing gum – that it would stretch a long way before breaking in two – but, for crystalline materials, this is very rare. Creep fracture (in tension) can happen at strains as small as 2% by the mechanism shown in Figure 8.13(b). Voids nucleate on grain boundaries that lie normal to the tensile stress. These are the boundaries to which atoms diffuse to give diffusional creep, migrating from the boundaries that lie more nearly parallel to the stress. But if the tensile boundaries have voids on them, the voids also act as sources of atoms, and in doing so, they grow. The voids cannot support load, so the stress rises on the remaining intact bits of boundary, making the voids grow more and more quickly until finally they link, giving creep fracture.

Many engineering components (e.g. tie bars in furnaces, superheater tubes, high-temperature pressure vessels in chemical reaction plants) must withstand loads for long times (say, 20 years) at high temperatures without failure. The loads or pressure they can safely carry are calculated by methods such as those we have just described. One would like to be able to test new materials for these applications without having to wait for 20 years to get the results. It is thus desirable to speed up the tests by increasing the load or the temperature. This is done in practice, but there are risks associated with a change in creep mechanism. On Figure 8.4(a), we see that if test data are extrapolated from the high stress end (where $n = 3$–8) down to lower stresses (where $n \approx 1$), the predicted rate will be a significant underestimate of the actual creep rate.

Deformation mechanism diagrams   Let us now pull all of this together. Materials can deform by dislocation plasticity (see Chapter 5), or, if the temperature is high enough, by diffusional flow or power-law creep. If the stress and temperature are too low for any of these, the deformation is elastic. This competition between mechanisms may be summarised in *deformation mechanism diagrams*, of which Figure 8.15 is an example. It shows the ranges of stress and temperature in which we expect to find each sort of deformation is dominant, and the strain-rate that they produce (the blue contours). Diagrams like these are a useful summary of deformation behaviour, helpful in selecting a material for high-temperature applications, and in estimating the stresses and temperatures for hot working, discussed next.

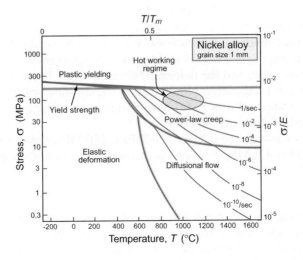

Figure 8.15   Deformation mechanism map for a nickel alloy.

Creep in hot forming processes   The ductility of most metals and alloys allows a diverse range of shaping methods – rolling, extrusion, and forging, for example (see Figure 5.27). The power expended in shaping a component scales with the material strength. Raising the temperature is one way to reduce the strength – something blacksmiths have known for centuries. Hot forming temperatures are typically 0.5 to 0.7 T_m, where T_m is the melting point of the alloy in Kelvin, while economic production requires strain-rates of order 10^{-2} to 1/s, giving the hot working region shown shaded in Figure 8.15.

Example 8.5 Hot-forging an alloy billet

A Ni alloy billet is to be hot forged between flat parallel dies at a temperature of 1200°C. The deformation mechanism map for the alloy is that shown in Figure 8.15. The initial billet is a cylinder of diameter 20 cm, and height 50 cm, with its axis vertical. The top platen descends at a speed of 5 cm/s and comes to rest when the billet has been shortened by 2.5 cm. Assume that the deceleration is constant with time – that is, the speed varies linearly.

(a) Estimate the initial and the average strain-rate during the process, and the time that the deformation takes.

(b) Locate the hot working regime for this process on the deformation mechanism map, and estimate the stress needed to perform the operation. By what factor is this below the room temperature yield stress?

(c) Estimate the load required for this process, assuming that the axial stress is approximately 1.5 times the uniaxial yield stress (a consequence of the friction acting between the workpiece and the platen).

Answer.

(a) On initial contact the strain-rate is given by the speed of the platen divided by the height of the cylinder = 0.05/0.5 = 0.1 s^{-1}. For a linear deceleration in speed with time, the average speed is 2.5 cm/s, so the average strain-rate is approximately half of the initial value, and the deformation takes about 1 second.

(b) From Figure 8.15, the stress required for this strain-rate at 1200°C is approximately 100 MPa, compared to a room temperature yield stress around 600 MPa, so the hot strength is about six times lower.

(c) The axial stress required is 1.5 × 100 MPa = 150 MPa, giving a force of 150 × 10^6 × (π × 0.2^2)/4 = 4.7 MN. Forging on this scale needs a big, expensive machine.

The high deformation rates mean that power-law creep is the expected mechanism operating in hot forming. One special forming process exploits diffusional flow. The map suggests that this mechanism would normally be much too slow for a manufacturing process, but Equation (8.18) shows that a factor of 10 in grain size gives a factor of 100 in strain-rate. If the grain size is made very small, the strain-rate in diffusional flow rises to a level acceptable for a shaping process. *Superplastic forming*, which relies on this, is used to shape lightweight titanium and aluminium alloys for aerospace applications.

Creep in polymers   Creep in crystalline materials, as we have seen, is closely related to diffusion. The same is true of polymers, but because most of them are partly or wholly amorphous, diffusion is controlled by 'free volume' – the space between the molecules in a polymer (the equivalent to vacancies in a crystal). This, too, increases with temperature (the fractional change per degree is just the volumetric thermal expansion, α_v), and it does so most rapidly at the glass transition temperature T_g. Thus polymers start to creep as the temperature approaches T_g, and that, for many simple amorphous polymers, is a low temperature: 50° to 150°C. Increased molecular weight, cross-linking, and crystallinity all raise T_g, and with it the creep resistance. It is increased further by filling with powdered glass, silica (sand), or talc, roughly in proportion to the volume fraction of filler. Composites – polymers reinforced with chopped or continuous fibres of glass or carbon – resist creep even more effectively because much of the load is now carried by the fibres, which do not creep at all.

In the temperature range around T_g, polymers are neither simple elastic solids nor viscous liquids; they are *visco-elastic* solids. The physics of molecular sliding are more difficult to describe analytically than the atomic diffusion and dislocation behaviour in crystals. *Polymer rheology* is used to develop constitutive models, by fitting non-linear equations to experimental data for the stress-strain responses as a function of time, at strain-rates and temperatures relevant to the design.

A simpler approach is to use graphical data for what is called the *creep modulus, E_c,* to provide an estimate of the deformation during the life of the component. Figure 8.16 shows data for E_c as a function of temperature T and time t for the polymer PMMA. Creep modulus data allow conventional elastic solutions to be used for design against creep. The service temperature and design life are chosen, the resulting creep modulus is read from the graph, and this is used instead of Young's modulus in any of the solutions for elastic design listed in Chapter 4.

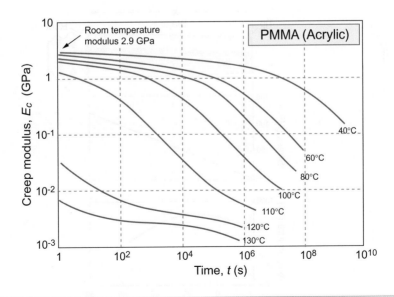

Figure 8.16 The creep modulus of PMMA.

8.6 Materials and design to cope with creep

Turbine blades   Throughout the history of its development, the gas turbine has been limited in thrust and efficiency by the availability of materials that can withstand high stress at high temperatures. The stress comes from the centripetal load carried by the rapidly spinning turbine disk and rotor blades. Where conditions are at their most extreme, in the first stage of a gas turbine, high melting point nickel- and cobalt-based superalloys are used because of their unique combination of high-temperature strength, toughness, and oxidation resistance. Typical of these is MAR-M200, an alloy based on nickel, strengthened by a solid solution of tungsten and cobalt and by precipitates of $Ni_3(Ti, Al)$, and containing chromium to improve its corrosion resistance.

When a turbine is running at a steady angular velocity ω, centripetal forces subject each rotor blade to an axial tension. Consider a short blade of constant cross-section, attached to a rotor disk of radius much larger than the length of the blade. If the radius to the tip of the blade is R, the tensile stress rises linearly with distance x from the tip as

$$\sigma = R\omega^2 \rho x \qquad (8.20)$$

where ρ is the density of the alloy, about 8000 kg/m^3 for typical blade alloys. As an example, a medium-duty engine has blades of length 50 mm, with a radius to the blade tip of $R = 0.3$ m, rotating at an angular velocity ω of 10,000 rpm ($\approx$ 1000 radians/s). This induces an axial stress $\approx$ 120 MPa at its root – Figure 8.17 shows typical stress and temperature profiles for the blade.

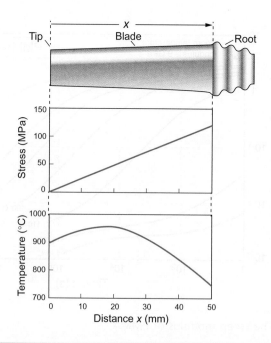

Figure 8.17   The stress and temperature profiles for a turbine blade.

The service conditions are plotted as a shaded box on two deformation mechanism maps in Figure 8.18. If made of pure nickel [Figure 8.18(a)] the blade would deform by *power-law creep*, at a totally unacceptable rate. The strengthening methods used in MAR-M200 with a typical as-cast grain size of 0.1 mm [Figure 8.18(b)] reduce the rate of power-law creep by a factor of 10^6 and the dominant mechanism changes from power-law creep to diffusional flow. Further solution strengthening or precipitation hardening is now ineffective. Equation (8.18) shows that the way to avoid diffusional flow is to use a large grain size, d, so that diffusion distances are long. Single crystals are best of all; they have no grain boundaries to act as sinks and sources for vacancies, suppressing diffusional creep completely. This is the strategy used today: most gas-turbine powered aircraft have single-crystal turbine blades.

You might think that this fixes everything, but nature isn't like that. Diffusional creep, now suppressed, was the dominant mechanism, but it wasn't the only one. Power-law creep now dominates, more slowly, but still there. Generations of alloy development culminating in the series of which MAR-M200 is an example have delivered materials with extraordinary high-temperature properties. If you want to go higher, you need to think about a change of material class. Many ceramics have melting points above those of metals. Among them are the technical ceramics alumina (Al_2O_3), silicon carbide (SiC), and silicon nitride (Si_3N_4). Turbine blades made of these can survive higher burn temperatures and allow higher engine efficiency. Development programs are exploring these – small turbines have run with ceramic blades – but using them is not easy: ceramics are brittle, even at high temperatures. Today's commercial aircraft don't have ceramic blades – but ceramics have greatly increased their efficiency. The next case study explains how.

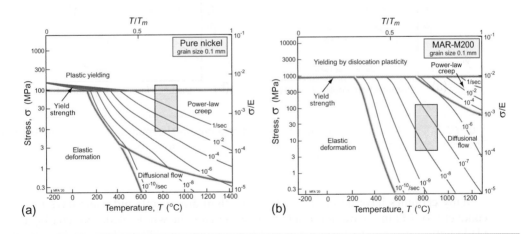

Figure 8.18 (a) The stress-temperature profile of the blade plotted onto a deformation map for pure nickel. (b) The same profile plotted on a map for the alloy MAR-M200.

Thermal barrier coatings The efficiency and power-to-weight ratio of advanced gas turbines, as already said, is limited by the burn temperature, and this in turn is limited by the materials of which the rotor and stator blades are made. Heat enters the blade from the burning gas (Figure 8.19). Blades are cooled by pumping air through internal channels. The surface temperature of the blade is set by a balance between the heat transfer coefficient between gas and blade (determining the heat in) and conduction within the blade to the cooling channel (determining the heat out), leading to the temperature profile shown at Figure 8.19(a). The temperature step at the surface is increased by bleeding a trickle of the cooling air through holes in the blade surface. This technology is already quite remarkable: the air-cooled blades operate under significant stresses in a gas stream at a temperature above the melting point of the alloy of which they are made. Could this burn temperature be increased yet further?

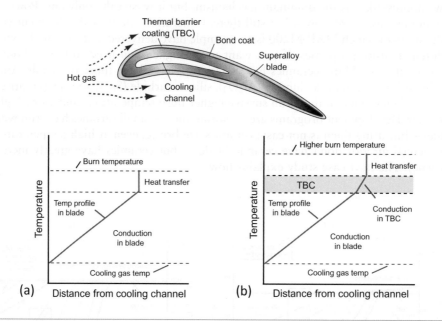

Figure 8.19   A cross section of a turbine blade with a thermal barrier coating (TBC). (a) The temperature profile for the uncoated blade left; (b) that for the coated blade.

Many ceramics have higher melting points than any metal, and some have low thermal conductivity. Ceramics are not tough enough to make the whole blade, but coating the metal blade with a ceramic to form a *thermal barrier coating* (TBC) allows an increase in burn temperature with no increase in that of the blade [Figure 8.19(b)]. In choosing the ceramic, the first considerations are those of a low thermal conductivity and a maximum service temperature above that of the gas. The technical ceramic zirconia (ZrO_2) has a melting point of 2600°C (Figure 8.6) and exceptionally low thermal conductivity (see Figure 7.4), and this is used to make TBCs.

8.7 Summary and conclusions

All material properties depend on temperature. Some, like the density and the modulus, change relatively little and in a predictable way, making compensation easy. *Transport properties*, meaning thermal conductivity and the equivalent for matter flow (diffusion), change in more complex ways. Diffusion has a profound effect on mechanical properties when temperatures are high. To understand it we need the idea of *thermal activation* – the ability of atoms to jump from one site to another, using thermal energy as the springboard. In crystals, atoms jump from vacancy to vacancy, and this atom motion allows the intermixing of atoms in a concentration gradient.

Diffusion plays a fundamental role in creep and creep fracture, and is also central to processing of materials, the subject of Chapter 11. Creep rates increase exponentially with temperature, presenting a major challenge for design: that of predicting the rates with useful accuracy. Exponential rates require precise data – small changes in temperature and stress give large changes in strain-rate – and these data are hard to measure and are sensitive to composition and microstructure. A second challenge is that there is more than one mechanism of creep. The mechanisms compete, the one giving the fastest rate being dominant. If you test and design using data and formulae for one, but another is operative under service conditions, you are in trouble. Deformation mechanism maps help here, identifying both the mechanism and the approximate rate of creep. The chapter ends with case studies in material innovation to manage the extremes of high temperature performance in jet engines.

8.8 Further reading

Frost, H.J., & Ashby, M.F. (1982). *Deformation mechanism maps*. Pergamon Press. ISBN 0-08-029337-9. (A monograph compiling deformation mechanism maps for a wide range of materials, with explanation of their construction).

Penny, R. K., & Marriott, D. L. (1971). *Design for creep*. McGraw-Hill. ISBN 07-094157-2. (A monograph bringing together the mechanics and the materials aspects of creep and creep fracture).

Waterman, N. A., & Ashby, M. F. (1991). *The Elsevier materials selector*. Elsevier Applied Science. ISBN 1-85166-605-2. (A three-volume compilation of data and charts for material properties, including creep and creep fracture).

Yang, Z. G., Stevenson, J. W., Paxton, D. M., Singh, P., & Weil, K. S. (2002). *Materials properties database for selection of high-temperature alloys and concepts of alloy design for SOFC applications*. Pacific Northwest National Laboratory, Batelle Memorial Institute. https://pdfs.semanticscholar.org/3713/e545d4ae46b898509601cf3577a7cae881f3.pdf (A compilation of properties of high-temperature alloys).

8.9 Exercises

Exercise E8.1   The temperature-dependence of density and Young's modulus is approximately linear, and can be estimated using Equation (8.1) in the text:

$P = P_o\left(1 + \beta\dfrac{T - T_o}{T_m}\right)$, where P is the value of the property, P_o its value at temperature T_o (= 0 K), β is a constant, and T and T_m are the temperature and melting point, respectively (in Kelvin). The values of β for density ρ and Young's modulus E are −0.06 and −0.5, respectively. Use the data at room temperature (20°C) for these properties in Appendix A, Tables A2 and A3, and data for melting point from Table A5 to estimate values of ρ and E for aluminium alloys at $T = 400$°C.

Exercise E8.2   The temperature dependence of electrical resistivity is approximately linear, and can be estimated using Equation (8.1) in the text: $P = P_o\left(1 + \beta\dfrac{T - T_o}{T_m}\right)$, where P is the value of the property, P_o its value at temperature T_o (= 0 K), β is a constant, and T and T_m are the temperature and melting point, respectively (in Kelvin). The value of β for electrical resistivity is +6.0. Use data for the electrical resistivity of tungsten alloys at room temperature (20°C) from Appendix A, Table A7, and for melting point from Table A5 to estimate the resistivity of tungsten alloys at 1200°C.

Exercise E8.3   The diffusion constants for lattice self-diffusion of aluminium are $D_o = 1.7 \times 10^{-4}$ m^2/s and $Q_d = 142$ kJ/mol. Calculate the diffusion coefficient D in aluminium at 200°C and 500°C.

Exercise E8.4   The diffusion constants for lattice self-diffusion of copper are $D_o = 2.0 \times 10^{-5}$ m^2/s and $Q_d = 197$ kJ/mol. Calculate the diffusion coefficient D in copper at room temperature (20°C) and 600°C.

Exercise E8.5   The element lead (Pb) has a number of isotopes. The most stable and commonest is Pb (207). A thin layer of Pb (204) is plated onto a thick plate of Pb (207), and the couple is held at 200°C for 4 hours. What is the approximate depth of penetration of the Pb (204) into the plate? The diffusion constants for lead are $D_o = 1.4 \times 10^{-4}$ m^2/s and $Q_d = 109$ kJ/mol.

Exercise E8.6   A steel component is nickel plated to give corrosion protection. To increase the strength of the bond between the steel and the nickel, the component is heated for 4 hours at 1000°C. If the diffusion parameters for nickel in iron are $D_o = 1.9 \times 10^{-4}$ m^2/s and $Q_d = 284$ kJ/mol, how far would you expect the nickel to diffuse into the steel in this time?

Exercise E8.7   The diffusion coefficient at the melting point for materials is approximately constant, with the value $D = 10^{-12}$ m^2/s. What is the diffusion distance if a material is held for 12 hours just below its melting temperature? This distance gives an idea of the maximum distance over which concentration gradients can be smoothed by diffusion.

Exercise E8.8   In the fabrication of a silicon semiconductor device, boron is diffused into silicon in two stages: a pre-deposition stage lasting 10 minutes at a temperature of 1000°C, followed by a drive-in stage for 2 hours at 1100°C. Compare the characteristic diffusion distances of these two stages. For the diffusion of boron in silicon, $D_o = 3.7 \times 10^{-6}$ m^{2}s^{-1}, and the activation energy $Q = 333$ kJ/mol.

Exercise E8.9   (a) The diffusion constant for carbon diffusion in FCC iron is $D_o = 2.3 \times 10^{-5}$ m^{2}s^{-1}, and the activation energy $Q = 148$ kJ/mol. Find the diffusion coefficient D for carbon in iron at 1000°C, and its approximate distance of diffusion in time t.
(b) Compare this with the thermal diffusivity of steel at 1000°C, $a = 9 \times 10^{-6}$ m^{2}s^{-1}, and the corresponding distance of heat diffusion in time t.

Exercise E8.10   A design for a sodium-sulphur battery involves separating the anode (molten sodium) from the cathode (molten sulphur) with a solid alumina (Al$_2$O$_3$) electrolyte through which the sodium must diffuse. Measurements of the diffusion rate of sodium through alumina indicate diffusion constants of $D_o = 1.2 \times 10^{-16}$ m^{2}s^{-1}, and activation energy $Q_d = 50$ kJ/mol. What is the rate of sodium diffusion through the electrolyte at 350°C, the operating temperature of such a battery?

Exercise E8.11   Molten aluminium is cast into a copper mould having a wall thickness of 5 mm. The diffusion coefficient D of aluminium in copper is 2.6×10^{-17} m^2/s at 500°C and 1.6×10^{-12} m^2/s at 1000°C. Use this information to calculate the activation energy Q_d and the pre-exponential D_o for the diffusion of aluminium in copper. Then use the results to calculate the coefficient of diffusion of aluminium in copper at 750°C.

Exercise E8.12   A useful rule-of-thumb in diffusion problems states that the characteristic diffusion distance in time t is given by $\sqrt{Dt}$, where D is the diffusion coefficient for the temperature concerned. Make an order-of-magnitude estimate of the time taken for an aluminium atom to diffuse through the 5 mm thick wall of the copper mould of Exercise E8.11 if the mould temperature is 750°C.

Exercise E8.13   Use your knowledge of diffusion and solid solutions to account for the following observations:
(a) The rate of diffusion of oxygen from the atmosphere into an oxide film is strongly dependent on the temperature and the concentration of oxygen in the atmosphere.
(b) Diffusion is more rapid in polycrystalline silver with a small grain size than in coarse-grained silver.
(c) Carbon diffuses far more rapidly than chromium through iron.

Exercise E8.14   A high-temperature alloy, loaded at 1000°C, extends by power-law creep when the applied tensile stress σ is 70 MPa. The creep exponent n is 5. By what factor will the strain-rate $\dot{\varepsilon}$ increase if the stress is raised to 80 MPa at the same temperature?

Exercise E8.15 The high-temperature alloy of Exercise E8.14 undergoes power-law creep when the applied stress is 70 MPa and the temperature is 1000°C. The activation energy for creep Q_c is 300 kJ/mol. By what factor will the strain-rate increase if the temperature is increased to 1100°C at the same stress?

Exercise E8.16 Pipework with a radius of 20 mm and a wall thickness of 4 mm made of 2¼ Cr Mo steel contains a hot fluid under pressure. The pressure is 10 MPa at a temperature of 600°C. The table lists the creep constants for this steel. Calculate the creep rate of the pipe wall, assuming steady-state power-law creep.

Material	Reference strain-rate $\dot{\varepsilon}_o$ (1/s)	Reference stress σ_o (MPa)	Creep exponent n	Activation energy Q_c (kJ/mol)
2¼ Cr Mo steel	4.23×10^{10}	100	7.5	280

Exercise E8.17 There is concern that the pipework described in Exercise E8.16 might rupture in less than the design life of 1 year. If the Monkman-Grant constant for 2¼ Cr Mo steel is 0.06, how long will it last before it ruptures?

Exercise E8.18 If the creep rate of a component made of 2¼ Cr Mo steel must not exceed 10^{-8}/sec at 500°C, what is the greatest stress that it can safely carry? Use the data in Exercise E8.16 to find out.

Exercise E8.19 Polycrystalline copper has a steady-state creep rate of $10^{-4}\,s^{-1}$ at 560°C when subjected to a given tensile stress. The activation energy Q_c for creep in copper is 197 kJ/mol. Calculate the steady-state creep rate of copper at 500°C at the same stress.

Exercise E8.20 The designers of a chemical plant are concerned about the creep of a critical alloy tie bar. Creep tests on specimens of the alloy under the nominal service conditions of a stress $\sigma = 25$ MPa at 620°C, gave a steady-state creep rate rate $\dot{\varepsilon}$ of 3.1×10^{-12}/s. In service it is expected that for 30% of the running time the stress may increase to 30 MPa, while for the remaining time the stress remains at the nominal service values. Calculate the expected *average* creep rate in service. Assume that the alloy creeps according to the equation:

$$\dot{\varepsilon} = \dot{\varepsilon}_o \left(\frac{\sigma}{\sigma_o} \right)^n \exp\left(-\frac{Q_c}{RT} \right)$$

where $\dot{\varepsilon}_o$, σ_o, and Q_c are constants, R is the universal gas constant, and T is the absolute temperature. For this alloy, $n = 5$.

Exercise E8.21 The designers of the chemical plant that was the subject of Exercise E8.20 remain concerned about the creep of a critical alloy tie bar. Creep tests on specimens of the alloy under the nominal service conditions of a stress $\sigma = 25$ MPa at 620°C gave a steady-state creep rate $\dot{\varepsilon}$ of 3.1×10^{-12}/s. Managers propose to increase the operating temperature from 620°C to 650°C of the plant to increase efficiency, but they guarantee that the stress in the tie bar will not fluctuate. How much faster will the tie bar creep at the higher operating temperature? Assume that the alloy creeps according to the equation:

$$\dot{\varepsilon} = \dot{\varepsilon}_o \left(\frac{\sigma}{\sigma_o} \right)^n \exp\left(-\frac{Q_c}{RT} \right)$$

where $\dot{\varepsilon}_o$, σ_o, and Q_c are constants, R is the universal gas constant, and T is the absolute temperature. For this alloy, $Q_c = 160$ kJ/mol and $n = 5$.

Exercise E8.22 The steady-state creep strain-rate at constant temperature $\dot{\varepsilon}_{ss}$ depends on the applied stress according to

$$\dot{\varepsilon}_{ss} = \dot{\varepsilon}_o \left(\frac{\sigma}{\sigma_o} \right)^n \exp\left(-\frac{Q_c}{RT} \right)$$

where $\dot{\varepsilon}_o$, σ_o, n, and Q_c are constants. The figure shows the steady-state creep response of a 12% Cr steel at 500°C. Determine the creep exponent n for this material.

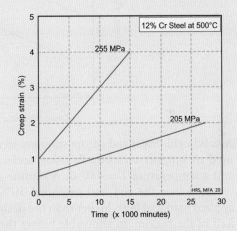

Exercise E8.23 A turbine blade, made of the 12% Cr steel of Exercise E8.22, is 10 cm long and runs with an initial clearance of 100 μm between its tip and the housing in a generator operating at 515°C. Determine the lifetime of the blade (meaning the time before it fouls the housing) if it runs continuously, assuming a uniform tensile stress of 60 MPa is applied along its axis. (The creep activation energy Q for 12% Cr steel is 270 kJ/mol, $n = 6$, $\dot{\varepsilon}_o = 6 \times 10^8$/s, and $\sigma_o = 100$ MPa).

Exercise E8.24   Creep data for a stainless steel at two temperatures are shown in the figure. The steel obeys the steady-state creep equation

$$\dot{\varepsilon} = C' \, \sigma^n \, \exp\left(-Q/RT\right)$$

where $C' = \dot{\varepsilon}_o / \sigma_o^n$ and $R = 8.314$ J/mol.K.

(a) A specimen of the steel of length 750 mm was subjected to a tensile stress $\sigma = 40$ MPa at a temperature $T = 538°C$. Use the data plotted in the figure to estimate the extension after 5000 hours.

(b) Use the data plotted in the figure to evaluate the constants n and Q for the stainless steel.

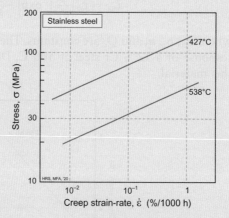

Exercise E8.25   A solid cylindrical rod of the stainless steel described in Exercise E8.24 was used to support a turbine housing under a uniform tensile load $P = 6$ kN and a temperature $T = 450°C$. A routine check after 10,000 hours in service revealed that the rod had extended 50% more than expected because the applied load had been higher than the design specification. This extension was half of the maximum allowable before the component required replacement.

(a) Evaluate the load that had actually been applied, given the operating temperature had been correctly maintained.

(b) The loading on the component could not be corrected, but a drop in operating temperature of 20°C was possible. Would this be sufficient to allow a further 40,000 hours of operation before replacement?

Exercise E8.26   The figure shows a deformation-mechanism map for nickel. The contours are lines of equal strain-rate for steady-state creep in units of s^{-1}. They are spaced by factors of 10.

(a) Estimate by eye the maximum allowable stress at 600°C if the maximum allowable strain-rate is $10^{-10}\,s^{-1}$.

(b) Estimate the creep rate if the stress were increased to 3 MPa and the temperature raised by 200°C.

(c) What is the dominant creep mechanism at 1000°C under a stress of 10 MPa?

(d) What, approximately, is the creep rate at 1000°C and a stress of 10 MPa?

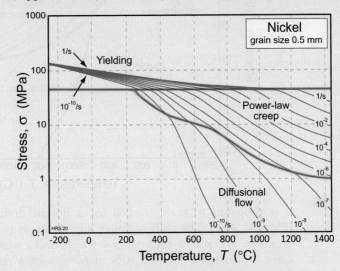

Exercise E8.27 The strain-rate contours on the map in Exercise E8.26 for both diffusional flow and power-law creep are based on curve fits to the creep equation:

$$\dot{\varepsilon} = C'\sigma^n \exp(-Q_c/RT)$$

where $C' = \dot{\varepsilon}_o/\sigma_o^n$. Use the strain-rate contours on the map to estimate the value of the stress exponent n, for diffusional flow and for power-law creep.

Exercise E8.28 (a) A stainless steel suspension cable in a furnace is subjected to a stress of 100 MPa at 700°C. Its creep rate is found to be unacceptably high. By what mechanism is it creeping? What material variation would you suggest to tackle the problem? The figure shows the deformation mechanism map for the stainless steel.

(b) The wall of a pipe of the same stainless steel carries a stress of 3 MPa at the very high temperature of 1000°C. In this application it, too, creeps at a rate that is unacceptably high. By what mechanism is it creeping? What material variation would you now suggest to tackle the problem?

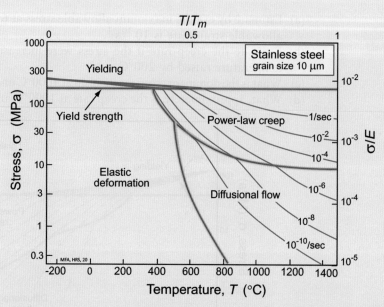

Exercise E8.29 It is proposed to make a shelf for a hot-air drying system from Acrylic (PMMA) sheet. The shelf is simply supported, as in the diagram, and has a span $w = 300$ mm, a thickness $t = 8$ mm, and a width $b = 200$ mm. It must carry a distributed load of 50 N at 60°C with a design life of 8000 hours (about a year) of continuous use. Use the creep modulus plotted in Figure 8.16 and the solution to the appropriate elastic problem [see Chapter 4, Equation (4.28), and Figure 4.18] to find out how much it will sag in that time.

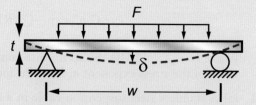

Exercise E8.30 Chemical engineering and power generation plant have pipework that carries hot gases and liquids under high pressure. A little creep can be accepted but rupture cannot. Pipes in a typical 600 MW steam turbine unit carry steam at temperature $T = 650$°C and pressure $p = 15$ MPa. Temporary pipework is to be installed while modifications are made to the plant, with a design life of 6 months. Type 304 stainless steel pipe with a diameter of 300 mm and a wall thickness of 10 mm is available. Will it function safely for the design life? Use the Stress–Time to rupture plot for 304 stainless steel to find out.

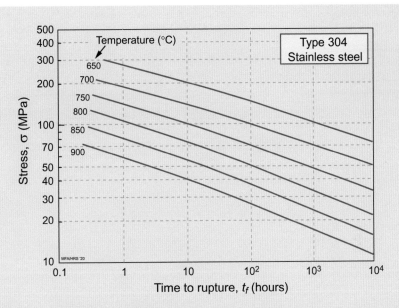

Exercise E8.31   (a) Explain briefly, with sketches as appropriate, the *differences* between the following pairs of mechanisms in metal deformation:
(i) room temperature yielding and power-law creep;
(ii) power-law creep and diffusional flow.

(b) Explain the following manufacturing characteristics:
(i) superplastic forming requires a fine-grain size material;
(ii) creep-resistant alloys are often cast as single crystals.

Chapter 9
Surfaces: friction, wear, oxidation, corrosion

A disc brake. (Image courtesy Paul Turnbull, SEMTA, Watford, UK: www.gcseinenginering.com)

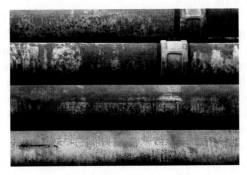

Corrosion. (istock.com)

https://doi.org/10.1016/B978-0-08-102399-0.00009-1

9.1 Introduction and synopsis

God, it is said, created materials; but their surfaces are (in some respects) the work of the devil. Surfaces are certainly the source of many problems. When surfaces touch and slide, there is *friction*; and where there is friction, there is *wear*. Tribologists – the collective noun for those who study friction and wear – are fond of citing the enormous cost, through lost energy and worn equipment, for which these two phenomena are responsible. It is certainly true that, if friction could be eliminated, the efficiency of engines, gearboxes, drive trains, and the like would increase enormously; and if wear could be eradicated, they would also last much longer. But before accepting this negative image, we should remember that, without friction and wear, pencils would not write on paper, and we would struggle to walk or drive anywhere – particularly on slopes.

And that is not all. The Gospel according to St. Matthew (6:19) reminds us that we live in a world in which 'moth and rust doth corrupt' – a world of degradation and decay – not a happy thought with which to start a chapter. Atoms inside a material are only in contact with atoms like themselves. Atoms at the surface come into contact with the rest of the atomic world – including aggressive gases and corrosive liquids and chemicals – leading to *oxidation* and *corrosion*. These, too, can be useful – it's the way that a decorative patina is applied to bronze and patterns are etched into glass – but much of the time it is a problem, reducing product lifetime and being as costly as wear.

So surfaces have a mixed reputation. But by understanding degradation at surfaces we can, to a degree, control it. This chapter is about wear, oxidation and corrosion, the underlying mechanisms, and what can be done to manage them to provide *durability* – a key material attribute, central to the safety and economy of products. It is one of the more difficult attributes to characterise, quantify, and use for selection for the following reasons:

- It is a function of both the material and the environment with which it is in contact.
- There are many mechanisms, some general, some peculiar to particular materials and environments.

9.2 Tribological properties

Three properties characterise sliding couples: the coefficient of friction μ, the Archard wear-rate constant k_a, and the limiting pressure-velocity product Pv.

Coefficient of friction   When two surfaces are placed in contact under a normal load F_n and one is made to slide over the other, a minimum force F_s is needed to overcome the friction that opposes the motion. This limiting force F_s is proportional to F_n but does not depend on the nominal contact area of the surfaces, A_n. These facts were discovered by none other than Leonardo da Vinci,[1] then forgotten, then rediscovered 200 years later by Amontons,[2] whose name is carried by the law defining the *coefficient of friction* μ, illustrated in Figure 9.1, and given by

$$\mu = \frac{F_s}{F_n} \tag{9.1}$$

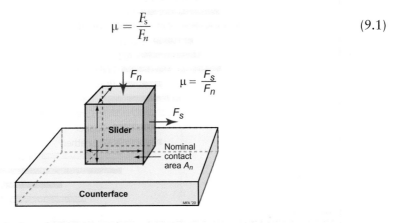

Figure 9.1   The definition of the coefficient of friction, μ.

Figure 9.2 shows approximate values for the coefficient of friction for dry – that is, unlubricated – sliding of materials on a steel surface. The values of μ against other materials differ from those for steel, but the range is much the same: typically, μ ≈ 0.4-0.5. Certain combinations show much higher values, either because they seize when rubbed together (a soft metal rubbed on itself in a vacuum with no lubrication, for instance) or because one surface has a sufficiently low modulus that it conforms to the rough surface of the other, giving full contact (rubber on rough concrete). At the other extreme are sliding combinations

---

[1] Leonardo da Vinci (1452–1519), Renaissance painter, architect, engineer, mathematician, philosopher, and prolific inventor.
[2] Guillaume Amontons (1663–1705), French scientific instrument maker and physicist, perfector of the *clepsydra* or water clock – a device for measuring time by letting water flow from a container through a tiny hole.

with exceptionally low coefficients of friction, such as PTFE, or bronze bearings loaded with graphite, sliding on polished steel. Here the coefficient of friction falls as low as 0.04, though this is still high compared with friction for lubricated surfaces, as noted in the upper right of Figure 9.2.

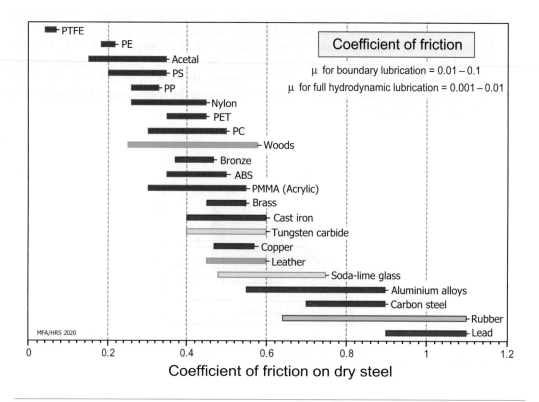

Figure 9.2 The coefficient of friction μ of materials sliding on an unlubricated steel surface.

Example 9.1 Sliding friction

Wood, sliding on concrete, has a coefficient of friction of 0.4. How much force is required to slide a wooden palette carrying equipment weighing 300 kg across a horizontal concrete floor?

Answer. The normal force F_n is the weight times the acceleration due to gravity (9.81 m/s^2), giving F_n = 2943 N. The sliding force $F_s = \mu F_n$ = 1177 N.

Archard wear-rate constant When surfaces slide, they wear. Material is lost from both surfaces, even when one is much harder than the other. The *wear rate*, W, is conventionally defined as

$$W = \frac{\text{Volume of material removed from contact surface (m}^3)}{\text{Distance slid (m)}} \qquad (9.2)$$

and thus has units of m^2. A more useful quantity, for our purposes, is the wear rate per unit nominal area (A_n) of surface, called the *specific wear rate*, Ω:

$$\Omega = \frac{W}{A_n} \qquad (9.3)$$

which is dimensionless. It increases with bearing pressure P (the normal force F_n divided by A_n), such that

$$\Omega = k_a \frac{F_n}{A_n} = k_a P \qquad (9.4)$$

where k_a is the *Archard wear constant*, with units of $(MPa)^{-1}$. It is a measure of the propensity of a sliding couple for wear: high k_a means rapid wear at a given bearing pressure. If H is the hardness of the softer surface (expressed in MPa), the product $K = k_a H$ becomes the *dimensionless wear constant*.

Example 9.2 Wear of sliding bearings

A dry sliding bearing oscillates with an amplitude $d = 2$ mm at a frequency $f = 1$ Hz under a normal load of $F_n = 10$ N. The nominal area of the contact is $A_n = 100$ mm^2. The Archard wear constant for the materials of the bearing is $k_a = 10^{-8}$ MPa^{-1}. If the bearing is used continuously for 1 year, how much will the surfaces have worn down?
  Answer. Define Δx as the distance the surfaces have worn down. Then

$$\Delta x = \frac{\text{Volume removed}}{\text{Nominal contact area}} = \Omega. \text{ Distance slid} = k_a \frac{F_n}{A_n}. \text{ Distance slid}$$

The distance slid per cycle is $4d$, and 1 year is $t = 3.15 \times 10^7$ s. So the distance slid in 1 year is $4d \times f \times t = 2.5 \times 10^5$ m, giving a wear depth $\Delta x = 2.5 \times 10^{-4}$ m $= 0.25$ mm.

Limiting pressure-velocity product If the slider of Figure 9.1 slides with a velocity v (m/s), the power dissipated per unit area against the friction force F_s is $F_s v / A_n = \mu P v$ (Watt/m^2), and this appears as heat. Excessive heating softens the lower-melting component of the sliding couple, causing the surfaces to seize and setting an upper limit for the product $\mu P v$.

9.3 The physics of friction and wear[3]

Friction Surfaces, no matter how meticulously honed and polished, are never perfectly flat. The *roughness* depends on how the surface was made: a sand-casting has a very rough surface, while one that is precision-machined has a much smoother one. But even a mirror finish is not perfectly smooth. Magnified vertically, surfaces look like Figure 9.3(a) – an endless range of 'asperity' peaks and troughs. So when two surfaces are placed in contact, they touch only at points where asperities meet [Figure 9.3(b)], and these contacts solely support the load F_n.

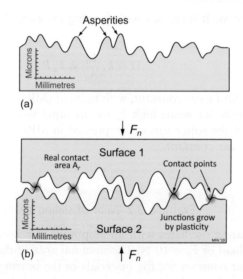

Figure 9.3 (a) The profile of a surface, much magnified vertically. (b) Two surfaces touch only at asperities.

The real contact area A_r in Figure 9.3(b) is only a tiny fraction of the apparent, nominal area A_n. How much is this fraction? The asperities deform elastically when they first touch. Even small loads generate large contact stresses, enough to cause plastic deformation. The contact points flatten, forming junctions with a total area A_r. Localised contacts on a nearly flat surface resemble hardness indentations (see Chapter 6) such that the total load transmitted across the surface is approximately

$$F_n = A_r H \qquad (9.5)$$

[3] We owe much of our understanding of friction and wear to the work of many engineers and scientists, above all to the remarkably durable and productive collaboration between the Australian Philip Bowden (1903–1968) and David Tabor (1913–2005), English of Lithuanian descent, between 1936 and 1968 in Melbourne, Australia, and Cambridge, England.

where the hardness $H \approx 3\sigma_y$ and σ_y is the yield strength. Thus a lower limit on the real contact area is

$$A_r = \frac{F_n}{3\sigma_y} \qquad (9.6)$$

To see how small this is, return to the cube of Figure 9.1. Its mass is $m = \rho a^3$ so the force needed to support it is $F_n = \rho g a^3$, where g is the acceleration due to gravity (9.81 m/s²). Inserting this into Equation (9.6) and setting $A_n = a^2$ gives the ratio of real to nominal contact areas:

$$\frac{A_r}{A_n} = \frac{\rho g a}{3\sigma_y}$$

Thus a 100 mm cube of structural steel (density 7900 kg/m³, yield strength 250 MPa) weighing a hefty 7.9 kg, contacts any hard surface only over a fraction of 1% of its nominal area – in this example, around 0.1 mm².

Now think of sliding one surface over the other. If the junctions weld together (as they do when surfaces are clean) it will need a shear stress equal to the shear yield strength k of the material ($k \approx \sigma_y/2$) or, if the materials differ, the shear strength of the softer one. Thus for unlubricated sliding

$$\frac{F_s}{A_r} = k \qquad (9.7)$$

Combining this with Equation (9.6) gives an estimate for the coefficient of friction

$$\mu = \frac{F_s}{F_n} = \frac{1}{6} = 0.17 \qquad (9.8)$$

This estimate gives a lower limit on μ because sliding makes the asperity contact area to grow. As Figure 9.2 showed, the more typical value is about 0.4 to 0.5.

So far we have spoken of *sliding* friction: F_s is the force to maintain a steady rate of sliding. If surfaces are left in static contact, the junctions tend to grow by creep, and the bonding between them becomes stronger. The force to start sliding is larger than that to maintain it, giving the stick-slip behavior that leads to vibration in brakes.

Wear When surfaces of equal hardness slide, asperity tips can stick together and shear off to give wear particles (Figure 9.4). When one surface is much harder than the other (steel

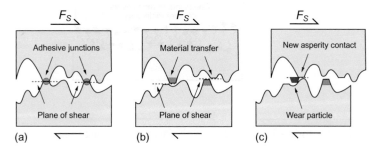

Figure 9.4 Wear: adhesion at the work-hardened junctions causes shear-off of material when the surfaces slide.

on plastic, say), the asperity tips of the harder material plough through the softer one, abrading it like sandpaper. To minimise the rate of wear we need to minimise the size of the fragments; that means minimising the real area of contact. Since $A_r \propto F_n / \sigma_y$, reducing the load or increasing σ_y (i.e. making the surfaces harder) reduces wear.

9.4 Friction in design

Bearings allow two components to move relative to each other in one or two dimensions while constraining them in the others. The key criteria are low friction coupled with the ability to support the bearing load F_n without damage. *Brakes* and *clutches*, by contrast, rely on high friction, but there is more to their design than that, as we shall see in a moment.

Dry sliding, as we have seen, typically results in a friction coefficient μ of 0.4 to 0.5. If the bearings of the drive train of your car had a μ of this magnitude, very little power would reach the wheels; almost all would be lost as heat in the bearings. In reality, about 15% of the power of the engine is lost in friction because of two innovations: *lubrication* and the replacement of sliding by *rolling bearings*.

Lubrication Most lubricants are viscous oils containing additives with polar groups (fatty acids with $-OH^-$ groups, not unlike ordinary soap) that bind to metal surfaces to create a chemisorbed boundary layer (Figure 9.5). A layer of the lubricant between the surfaces allows easy shear, impeded only by the viscosity of the oil, giving a coefficient of friction in the range 0.001 to 0.01. When the sliding surfaces are static, the bearing load F_n tends to squeeze the oil out of the bearing, but the chemisorbed boundary layer is enough to keep the surfaces separated. This protects the surfaces during start-up, but it is not something to rely on for continuous use. For this you need *hydrodynamic lubrication*.

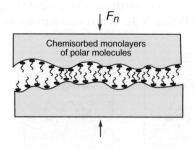

Figure 9.5 The formation of a boundary layer of chemisorbed, polar molecules.

The viscosity of the oil is both a curse and a blessing – a curse because it exerts drag (and thus friction) on the sliding components; a blessing because, without it, hydrodynamic lubrication would not be possible. Sliding bearings are designed so that the surfaces are slightly inclined. Oil is dragged along the wedge-shaped region between them, becoming compressed and forcing the surfaces apart. The same principle governs the self-alignment of a rotating journal bearing. If the journal goes off-centre, lubricant is dragged by viscosity into the side of the gap and swept toward the narrow side, where the rise of pressure pushes the journal back onto the axis of the bearing. The result is that the surfaces never touch but remain separated by a thin film of oil or other lubricating fluid.

Rolling bearings   In rolling element bearings the load is carried by a set of balls or rollers trapped between an inner and an outer *race* by a *cage* (Figure 9.6). When the loading is radial the load is carried at any instant by half the balls or rollers, but not equally: the most heavily loaded one carries about $5/Z$ times the applied load, where Z is the total number of balls or rollers in the bearing. Their contact area with the race is small, so the contact stresses are high.

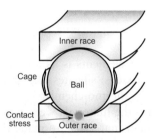

Figure 9.6   A rolling-contact bearing.

The main requirement is that the deformation of the components of the bearing remains elastic at all times, so the criterion for material choice is that of high hardness. That usually means a high-carbon steel or a tool steel. If the environment is corrosive or the temperature high, stainless steel or ceramic (silicon carbide or silicon nitride) is used. When the loads are very light, the races can be made from thermoplastic – polypropylene or acetal – with the advantage that no lubrication is needed (cheap bike pedals sometimes have such bearings).

High friction: materials for brakes and clutches   Applying friction sounds easy – just apply a normal force to the sliding surfaces. Brakes for horse-drawn carriages did just that – spoon-shaped blocks of wood, sometimes faced with leather, were wound down with a screw to rub on the rim of the carriage wheel. Brake materials today do much more.

The friction coefficient of materials for brake linings and pads is – surprisingly – not especially large: μ is about 0.4. The key thing is not μ but the ability to apply it without vibration, wear, or 'fade.' This is not easy when you consider that to stop a 1-tonne car from 100 kph (60 mph) requires a heat-sink for about 0.4 MJ. That's enough heat energy (ignoring losses to the air) to take 1 kg of steel to 800°C. Brake materials must apply friction at temperatures that can approach such values, and they must do so consistently and without the stick-slip that causes 'brake squeal' – not acceptable in even the cheapest car. This means materials must tolerate heat, conduct it away to limit temperature rise, and – curiously – lubricate, since it is this that suppresses squeal.

Today's brake materials are all *hybrids*: a matrix containing an abrasive additive to increase friction, a conducting reinforcement to help remove heat, and a lubricant to suppress vibration. Details are trade secrets, but some of the simpler mixes are well known. The brake pads on a bike are made up of a synthetic rubber often filled with cork dust and particles of a cheap silicate to reduce pad wear and give wet friction. Those on a car have a phenolic matrix with particles of silicates, silicon carbide (sandpaper grit), and graphite or MoS_2 as a lubricant to stop squeal. Those on military or passenger jets, or an F1 car, are carbon or ceramic, with frequent replacement being acceptable – here it is high temperature that is the problem (try calculating how much aluminium you could melt with the energy dissipated in bringing a 250-tonne jet to rest from 300 kph).

9.5 Chemical degradation: oxidation

The most stable state of most elements is as an oxide. The earth's crust is almost entirely made of simple or complex oxides: silicates like granite, aluminates like basalt, carbonates like limestone. Techniques of thermo-chemistry, electro-chemistry, cracking, and synthesis allow these to be refined into the materials we use in engineering, but these are not, in general, oxides. From the moment materials are made they start to re-oxidise, some extremely slowly, others more quickly; and the hotter they are, the faster it happens. For safe high-temperature design we need to understand rates of oxidation.

Oxidation mechanisms To measure oxidation rates, a thin sheet of the material is held at temperature T and the gain or loss in weight is measured over time t. The rate of oxidation of most metals in dry air at room temperature is too slow to be an engineering problem; indeed, slight oxidation can be beneficial in protecting metals from corrosion of other sorts. But heat them up, and the rate of oxidation increases, bringing problems. The driving force for a metal to oxidise is its *free energy of oxidation* – the energy released when it reacts with oxygen – but a big driving force does not necessarily mean rapid oxidation. The rate of oxidation is determined by the *kinetics* (rate) of the oxidation reaction, and that has to do with the nature of the oxide. When any metal (except for gold, platinum, and a few others that are even more expensive) is exposed to air, an ultra-thin surface film of oxide forms on it immediately, following the oxidation reaction

$$x\text{M (metal)} + \frac{1}{2}y\,O_2 \text{ (oxygen)} = M_xO_y \text{ (oxide)} + \text{energy} \qquad (9.9)$$

The film now coats the surface, separating the metal beneath from the oxygen. If the reaction is to go further, oxygen must get in or the metal must get out.

For some metals the mass gain per unit area, Δm, is linear, meaning that the oxidation is progressing at a constant rate:

$$\frac{d(\Delta m)}{dt} = k_l \;\; \text{giving} \;\; \Delta m = k_l t \tag{9.10}$$

where k_l (kg/m^2.s) is the *linear kinetic constant*. This is because the oxide film cracks (and, when thick, spalls off) and does not screen the underlying metal [Figure 9.7(a)]. Some metals behave better than this. The film that develops on their surfaces is compact, coherent, and strongly bonded to the metal. For these the weight gain per unit area of surface is parabolic, slowing up with time [Figure 9.7(b)]. This implies an oxidation rate with the form

$$\frac{d(\Delta m)}{dt} = \frac{k_p}{\Delta m} \;\; \text{giving} \;\; \Delta m^2 = k_p t \tag{9.11}$$

where k_p (kg^2/m^4.s) is the parabolic kinetic constant.

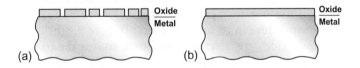

Figure 9.7   (a) Non-protective and (b) protective oxide films.

Example 9.3 Rates of oxidation

A metal oxidises at 500°C with a parabolic rate constant $k_p = 3 \times 10^{-10}$ kg^2/m^4.s. What is the mass gain per unit area after 1000 hours?

Answer. The mass gain is $\Delta m = (k_p t)^{1/2} = 0.033$ kg/m^2.

Kinetics of oxidation   What determines the rate of oxidation? Figure 9.8 shows a coherent, strongly bonded oxide film, thickness x. The reaction creating the oxide MO

$$M + O = MO$$

goes in two steps. The metal first forms an ion, say M^{2+}, releasing electrons:

$$M = M^{2+} + 2e^-$$

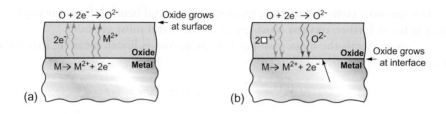

Figure 9.8   Oxidation mechanisms.

The electrons are then captured by an oxygen atom to give an oxygen ion:

$$O + 2e^- = O^{2-}$$

The problem is that the first of these reactions occurs at the metal side of the oxide film, whereas the oxygen reaction is on the other side. Either the metal ions and the electrons must diffuse out to meet the oxygen [Figure 9.8(a)] or the oxygen ions and electron holes must diffuse in to find the metal [Figure 9.8(b)]. If the film is an electrical insulator, as many oxides are, electrons are stuck so it is the oxygen that must diffuse in. The concentration gradient of oxygen is that in the gas C_o divided by the film thickness, x. The rate of growth of the film is proportional to the flux of atoms diffusing through the film, giving

$$\frac{dx}{dt} \propto D\,\frac{C_o}{x} \approx D_o \exp - \left(\frac{Q_d}{RT}\right)\frac{C_o}{x} \tag{9.12}$$

where D is the diffusion coefficient, and D_o and Q_d are the pre-exponential and the activation energy for oxygen diffusion in the oxide. Integrating gives

$$x^2 \propto D_o \exp - \left(\frac{Q_d}{RT}\right) C_o\, t$$

This has the same form as Equation (9.11) with

$$k_p \propto D_o\, C_o \exp - \left(\frac{Q_d}{RT}\right) \tag{9.13}$$

This explains why the growth is parabolic and why oxidation rates rise so steeply with temperature – it's the exponential temperature dependence.

The most protective films are those that are both electrical insulators and have low diffusion coefficients, and thus have high melting points. This is why the Al_2O_3 oxide film on aluminium, the Cr_2O_3 film on stainless steel and chrome plate, and the SiO_2 film on high-silicon cast iron are so protective.

Example 9.4 Rates of oxidation

If heated in air, magnesium (Mg) oxidises to MgO. The activation energy for oxygen diffusion in MgO is 370 kJ/mol. A magnesium sample, heated in air at 350°C for 1000 hours, develops an oxide film that is 1.5 microns thick. If the test in Example 9.2 had been carried out at the temperature $T_1 = 400$°C rather than at $T_o = 350$°C, how thick would the oxide be after 1000 hours? The gas constant $R = 8.314 \times 10^{-3}$ kJ/mol.K.

Answer. The thickness scales with temperature as $x \propto \left[\exp - \left(\dfrac{Q_d}{RT} \right) \right]^{1/2}$, so the thickness x_1 after 1000 hours at 450°C is greater than that, x_o, at 350°C by the factor

$$\frac{x_1}{x_o} = \left[\exp - \frac{Q_d}{R} \left(\frac{1}{T_1} - \frac{1}{T_o} \right) \right]^{1/2} = \left[\exp - 4.45 \times 10^4 \left(\frac{1}{673} - \frac{1}{623} \right) \right]^{1/2} = 13.0$$

The thickness after the test at 400°C is 13 times larger than before, at 19.4 microns.

Creating resistance to oxidation Elements for heaters, furnace components, power generation, and chemical engineering plants all require materials that can tolerate high temperatures. If you want the ultimate in high temperature and long life, it has to be platinum. If you want something more affordable, it must be an alloy.

Oxides, of course, are stable in air at high temperature – they are already oxidised. One way to provide high-temperature protection is to coat metals like cast irons, steels, or nickel alloys with an oxide coating. Stoves are protected by enameling (a glass coating, largely SiO_2); turbine blades are coated with plasma-sprayed thermal barrier coatings (TBCs) based on the oxide zirconia, ZrO_2 (see Chapter 8, Section 8.6). But coatings of this sort are expensive, and, if damaged, they cease to protect.

There is another way to give oxidation resistance to metals, and it is one that repairs itself if damaged. The oxides of chromium (Cr_2O_3), of aluminium (Al_2O_3), of titanium (TiO_2), and of silicon (SiO_2) have high melting points and are electrical insulators, so the diffusion of metal, oxygen, and electrons through them is slow, giving low rate constants k_p. The result: the oxide stops growing when it is a few molecules thick. These oxides adhere well and are very protective. The films can be artificially thickened by *anodising* – an electro-chemical process for increasing their thickness and thus their protective power. Anodised films accept a wide range of coloured dyes, giving them a decorative as well as a protective function.

If enough chromium, aluminium, or silicon can be dissolved in a metal like iron or nickel – 'enough' means 18% to 20% – a similar protective oxide grows on the alloy. And if the oxide is damaged, more chromium, aluminium, or silicon immediately oxidises, repairing the damage. Stainless steels (typical composition Fe–18% Cr–8% Ni), widely used for high-temperature equipment, and nichromes (nickel with 10–30% chromium), used for heating elements, get their oxidation resistance in this way.

9.6 Corrosion: acids, alkalis, water, and organic solvents

Acids and alkalis   Dunk a metal into an acid and you can expect trouble. The sulphates, nitrates, chlorides, and acetates of most metals are more stable than the metal itself. The question is not whether the metal will be attacked, but how fast. As with oxidation, this depends on the nature of the corrosion products and on any surface coating the metal may have. The oxide films on aluminium, titanium, and chromium protect them from some acids provided the oxide itself is not damaged. Alkalis, too, attack some metals. Zinc, tin, lead, and aluminium react with NaOH, for instance, to give zincates, stannates, plumbates, and aluminates, as anyone who has heated washing soda in an aluminium pan will have discovered.

Oils and organic solvents   Oily liquids are everywhere. We depend on them for fuels and lubricants, for cooking, for removing stains, as face cream, as nail varnish, and much more – any material in service will encounter them. Metals, ceramics, and glasses are largely immune to them, but not all polymers can tolerate organic liquids without problems.

Fresh and impure water   Corrosion is the degradation of a metal by an electro-chemical reaction with its environment [Figure 9.9(a)]. If a metal is placed in a conducting solution like salt water, its surface atoms dissociate into ions, releasing electrons (as the iron is shown doing in the figure) via the *anodic reaction*

$$Fe \rightarrow Fe^{2+} + 2e^- \tag{9.14}$$

The electrons accumulate on the iron giving it a negative charge that grows until the electrostatic attraction starts to pull the Fe^{2+} ions back onto the metal surface, stifling further dissociation. At this point the iron has a potential (relative to a standard, the *hydrogen standard*) of –0.44 volt. Each metal has its own characteristic potential (called the *standard reduction potential*), shown on the left of Figure 9.10.

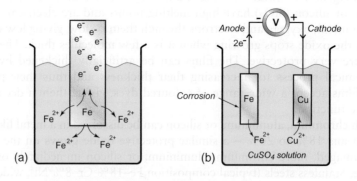

(a)

(b)

Figure 9.9   (a) Ionisation; (b) a bi-metal corrosion cell.

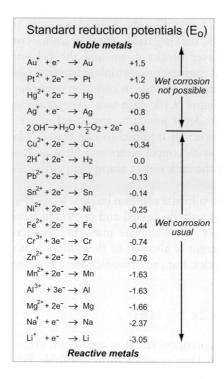

Standard reduction potentials (E_0)	
Noble metals	
$Au^+ + e^- \rightarrow Au$	+1.5
$Pt^{2+} + 2e^- \rightarrow Pt$	+1.2 *Wet corrosion*
$Hg^{2+} + 2e^- \rightarrow Hg$	+0.95 *not possible*
$Ag^+ + e^- \rightarrow Ag$	+0.8
$2\,OH^- \rightarrow H_2O + \frac{1}{2}O_2 + 2e^-$	+0.4
$Cu^{2+} + 2e^- \rightarrow Cu$	+0.34
$2H^+ + 2e^- \rightarrow H_2$	0.0
$Pb^{2+} + 2e^- \rightarrow Pb$	-0.13
$Sn^{2+} + 2e^- \rightarrow Sn$	-0.14
$Ni^{2+} + 2e^- \rightarrow Ni$	-0.25
$Fe^{2+} + 2e^- \rightarrow Fe$	-0.44 *Wet corrosion*
$Cr^{3+} + 3e^- \rightarrow Cr$	-0.74 *usual*
$Zn^{2+} + 2e^- \rightarrow Zn$	-0.76
$Mn^{2+} + 2e^- \rightarrow Mn$	-1.63
$Al^{3+} + 3e^- \rightarrow Al$	-1.63
$Mg^{2+} + 2e^- \rightarrow Mg$	-1.66
$Na^+ + e^- \rightarrow Na$	-2.37
$Li^+ + e^- \rightarrow Li$	-3.05
Reactive metals	

Galvanic series in sea water
Noble metals
Platinum
Gold
Graphite
Titanium
Silver
Stainless steel (passive)
Nickel based alloys
Cupro-nickel
Bronze
Copper
60-40 brass
Tin
Lead
Stainless steel (active)
Lead-tin solder
Steel and iron
Aluminium alloys
Cadmium
Zinc
Magnesium alloys
Reactive metals

MFA, 20

Figure 9.10 Standard reduction potentials of metals *(left)* and galvanic series in sea water *(right)*, for engineering alloys.

If two metals are connected together in a cell, like the iron and copper samples in Figure 9.9(b), a potential difference equal to their difference in the reduction potential appears between them. The reduction potential of iron, –0.44, differs from that of copper, +0.34, by 0.78 volt, so if no current flows in the connection the voltmeter will register this difference. If a current is now allowed, electrons flow from the iron (the *anode*) to the copper (the *cathode*); the iron recommences to ionise (that is, it corrodes), following the anodic reaction of Equation (9.14) and, if the solution contains copper ions Cu^{2+} (copper sulphate, for example), these plate out onto the copper following the *cathodic reaction*

$$Cu^{2+} + 2\,e^- \rightarrow Cu \qquad (9.15)$$

Reduction potentials tell you which metal will corrode if you connect them together in a conducting solution. Electroplating, you might say, is 'un-corrosion,' and that is a helpful analogy. If you want to reverse the corrosion path, returning ions to the surface they came from, applying a potential of opposite sign and larger than the corrosion potential will do it.

Engineering metals are almost all alloys. Some, like those of aluminium, titanium, and stainless-steel, form protective surface layers, making them 'passive' (more corrosion-resistant). But in some solutions the passive layer breaks down and the behaviour becomes 'active,' corroding more quickly. This has led engineers to formulate empirical *galvanic series*, ranking engineering alloys by their propensity to corrode in common environments such as sea water when joined to another metal (Figure 9.10, right side). The ranking is such that any metal will become the anode (and corrode) if joined to any metal above it in the list, and it will become the cathode (and be protected) if joined to one below, when immersed in sea water. If you make a mild-steel tank with copper rivets and fill it with sea water, the steel will corrode. But if, instead, you made the tank out of aluminium with mild steel rivets, the aluminium corrodes.

Suppose now that the liquid is not a copper sulphate solution but just water [Figure 9.11(a)]. Water dissolves oxygen, so unless it is previously de-gassed and protected from air, there is oxygen in solution. The iron and the copper still dissociate until their reduction potential difference is established but now, if the current is allowed to flow, there is no reservoir of copper ions to plate out. The iron still corrodes, but the cathodic reaction changes; it is now the *hydrolysis reaction*

$$H_2O + O + 2e^- \leftrightarrow 2OH^- \tag{9.16}$$

While oxygen can reach the copper, the corrosion reaction continues, creating Fe^{2+} ions at the anode and OH^- ions at the cathode. They react to form insoluble $Fe(OH)_2$, which ultimately oxidises further to $Fe_2O_3.H_2O$, which is what we call *rust*.

Connecting dissimilar metals in oxygen-containing water, or water with dissolved salts, is a bad thing to do. Corrosion cells appear that eat up the metal with the lower (more

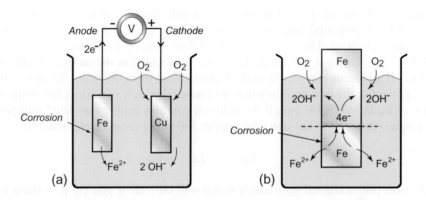

Figure 9.11   (a) A bi-metal cell containing pure water with dissolved oxygen; (b) a corrosion cell created by differential access to oxygen.

negative) reduction potential. Worse news is to come: it is not necessary to have two metals – both anodic and cathodic reactions can take place on the *same* surface. Figure 9.11(b) shows how this happens. Here an iron sample is immersed in water with access to air. The part of the sample nearest the water surface has an easy supply of oxygen; that further away does not. On the remoter part, the ionisation reaction of Equation (9.14) takes place, corroding the iron and releasing electrons that flow through the metal to the near-surface part where oxygen is plentiful, where they enable the hydrolysis reaction of Equation (9.16). The hydrolysis reaction has a reduction potential of +0.4 volt – it is shown on Figure 9.10 – and the difference between this and that of iron, –0.44 volt, drives the corrosion. If the sample could be cut in two along the broken line in Figure 9.11(b) and a tiny voltmeter inserted, it would register the difference: 0.84 volt.

Differential oxidation corrosion is one of the most difficult forms of corrosion to prevent: where there is water and a region with access to oxygen and one that is starved of it, a cell is set up. Only metals above the hydrolysis reaction potential of +0.4 volt in Figure 9.10 are immune.

Selective corrosion   Wet corrosion can be uniform, but it can also be selective – when it is, it can lead to failure much more rapidly than the uniform rate would suggest. Stress and corrosion acting together ('stress-corrosion' and 'corrosion-fatigue') frequently lead to localised attack, as do local changes in microstructure.

Intergranular corrosion occurs because grain boundaries have chemical properties that differ from those of the grain; the lower packing density at the boundary gives these atoms higher energy. Grain boundaries can also be electrochemically different because precipitates can form on the boundaries, sucking protective elements out of the surrounding matrix. As an example, stainless steels derive their corrosion resistance from the chromium they contain, giving a protective surface layer of Cr_2O_3. It takes about 18% Cr to give full protection; less than 10% and protection fails. Incorrect heat treatment, or the heat of a weld, can cause particles of the carbide, $Cr_{23}C_6$, to form on grain boundaries, draining the surrounding metal of chromium and leaving it vulnerable to attack [Figure 9.12(a)].

Stress-corrosion cracking (SCC) and corrosion-fatigue are accelerated forms of corrosion that generate local cracks. As we saw in Chapter 6, fracture becomes possible when the energy released by the growth of a crack exceeds the energy required to make the two new surfaces, plus that absorbed in the plastic zone at the crack tip. A corrosive environment provides a new source of energy – the chemical energy released when the material corrodes. Crack advance exposes new, unprotected surfaces; it is also at the crack tip that the elastic strain energy is highest, so the corrosive reaction is most intense there. The result is that cracks grow under a stress intensity K_{scc} that can be far below K_{1c}. Figure 9.12(b) shows a beam loaded in bending while submersed in a corrosive environment. The bending moment M creates a tensile stress in the top face, which is further concentrated by any scratch or defect. Material at the tip of the crack is now doubly vulnerable, a victim of both its chemical and its mechanical energy. The crack tunnels downward, branching as it goes, until the beam fails. In this example stress derives from external loads, but internal (residual) stresses left by working or welding have the same effect.

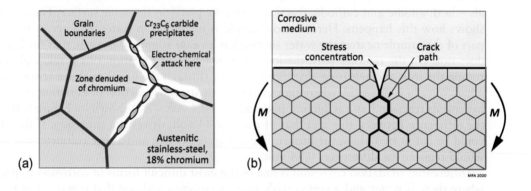

Figure 9.12 (a) Grain boundary precipitates causing intergranular corrosion; (b) stress corrosion cracking.

9.7 Fighting corrosion

In fighting corrosion, there are four broad strategies:

- *Judicious design*, meaning informed choice of material, shape, and configuration
- *Protective coatings*, either passive (meaning that the coating simply excludes the corrosive medium) or active (meaning that the coating protects even when incomplete)
- *Corrosion inhibitors*, chemicals added to the corrosive medium that slow the rate of corrosion
- *Monitoring*, with protective maintenance or regular replacement

We start with the first: design.

Design: material choice Materials differ greatly in their vulnerability to attack by corrosive media, some surviving unscathed while others, in the same environment, are severely attacked. There are no simple rules or indices for predicting susceptibility. The best that can be done is a ranking, based on experience. The best choice of material depends not just on the environment in which it must operate, but also – for economic reasons – on the application: it may be cheaper to live with some corrosion (providing regular replacement) than to use expensive materials with better intrinsic resistance. And there is the potential for protection of vulnerable materials by coatings, by inhibitors, or by electro-chemical means. Experience in balancing these influences has led to a set of 'preferred' materials for use in any given environment. Table A10 of Appendix A lists these for 23 fluids.

Example 9.5

Use Table A10 to identify materials for pipework to handle 10% sodium hydroxide. Consider material combinations as well as single materials.

Answer. Table A10 identifies nickel alloys, stainless steels, and magnesium alloys as able to withstand NaOH. Polyethylene, PVC, and PTFE have good resistance, so they could be used to line a simple steel pipe.

Design: Material choice for implants in the human body The human body is at its best for only about 50 years. From there on it's downhill, with an increasing chance that bits (arteries, bones, teeth, lenses) will need replacing. That's where materials come in.

What is required of a material that will be implanted in the human body? There are two overriding considerations, and corrosion is a factor in both:

- Bio-compatibility – that is, not exhibiting any toxicity to the surrounding biological system
- Durability – the body does not take kindly to repeated replacement of its parts.

Not many materials meet these requirements well. The most successful are listed under *Body fluids* in Table A10 of Appendix A. Let's take a closer look at them, starting with metals.

The human body is not a friendly place for metals. Many body fluids are oxygenated saline electrolytes at a temperature of 98.6°F (37°C) – perfect conditions for corrosion. Variations in alloy compositions, heat treatment, cold work, and surface finish create local differences in electrochemical potential, providing the conditions for galvanic cells to form. Electrolytic activity implies dissolution of one component of the cell and the release of metal ions from it into the body – unacceptable in an implant that may be in place for years. To overcome these electrochemical problems, implants are pre-passivated by anodising to thicken their protective oxide film. Implants involved in joints with sliding surfaces must meet restrictions for wear resistance to minimise the dispersion of damaging wear-debris into the surrounding tissue. Those that work well appear in Table A10.

Polymers do better on corrosion issues, less well on stiffness and strength. Almost all polymers are good insulators, so electro-chemical problems disappear. Some polymers – particularly the polyolefins (PE, PP) and the fluoropolymers (e.g. PTFE) – are chemically very stable, with zero solubility in body fluids. Silicones and polyurethanes are the polymers of choice for replacement of soft tissue (facial and breast implants). Ultra-high molecular weight polyethylene (UHMW-PE) is the best choice for sliding surfaces with minimal friction and wear (hip and knee replacements). Acrylic (PMMA), with optical clarity, is used for lens replacement, silicone hydrogel for contact lenses.

Biocompatible 'bio-ceramics' and 'bio-glasses' are standard engineering ceramics (alumina and zirconia, for instance) or glasses made with special attention to purity and sterility. Others are bio-active, stimulating the growth of bone-like material through chemical reaction with surrounding body fluid. A few, based on hydroxyapatite and calcium phosphate, are also bio-resorbable. Their bone-like chemistry allows them to break down slowly in body fluids; the chemicals released in the breakdown are processed via the normal metabolic pathways of the body. Like all ceramics, bio-ceramics have high compressive strength but are less strong in tension or bending because of their low fracture toughness.

Design: Shape and configuration   Corrosion can't be stopped, but it can be managed. Here are the dos and don'ts for best results:

Allow for uniform attack. Nothing lasts forever. Some corrosion can be tolerated, provided it is uniform, simply by allowing in the initial design for the loss of section over life.

Avoid fluid trapping [Figure 9.13(a)]. Simple design changes permit drainage and minimise the trapping of water and potentially corrosive dirt.

Suppress galvanic attack [Figure 9.13(b)]. Two different metals, connected electrically and immersed in water with almost anything dissolved in it, create a corrosion cell like that of Figure 9.11(b). The metal that lies lower on the scale of reduction potential is the one that is attacked. The rate of attack can be large and localised at or near the contact. Attaching an aluminium body shell to a steel auto chassis risks the same fate, as incautious car makers have found. Even the weld metal of a weld can differ from the parent plate in reduction potential enough to set up a cell. The answer is to avoid bi-metal couples if water is around or, when this is impossible, to insulate them electrically from each other and reduce the exposed area of the nobler metal to minimise the rate of the cathodic reaction.

Avoid crevices [Figure 9.13(c)]. Crevices trap moisture and create the conditions for differential-aeration attack. The figure shows riveted lap joints (the rivet, of course, made from the same metal as the plate). If water can get under the plate edge, it will. The water-air surface has access to oxygen, but the metal between the plates does not. The result is corrosion of the joint surface, just where it is least wanted. The answer is to join by welding or soldering, or to put sealant in the joint before riveting.

Consider cathodic protection [Figure 9.13(d)]. If attaching a metal to one lower in reduction potential causes the lower one to corrode, it follows that the upper one is protected. So connecting buried steel pipework to zinc plates sets up a cell that eats the zinc but spares the pipes. The expensive bronze propellers of large ships carry no surface protection (the conditions are too violent for it to stay in place). Copper and bronze lie above steel in the reduction-potential table and electrical isolation is impossible. Bronze connected to steel in salt water is a recipe for disaster. The solution is to shift the disaster elsewhere by attaching zinc plates to the hull around the propeller shaft. The zinc, in both examples, acts as a *sacrificial anode* protecting both the bronze and the steel. The term 'sacrificial' is accurate – the zinc is consumed – so the protection only lasts as long as the zinc, requiring that it be replaced regularly.

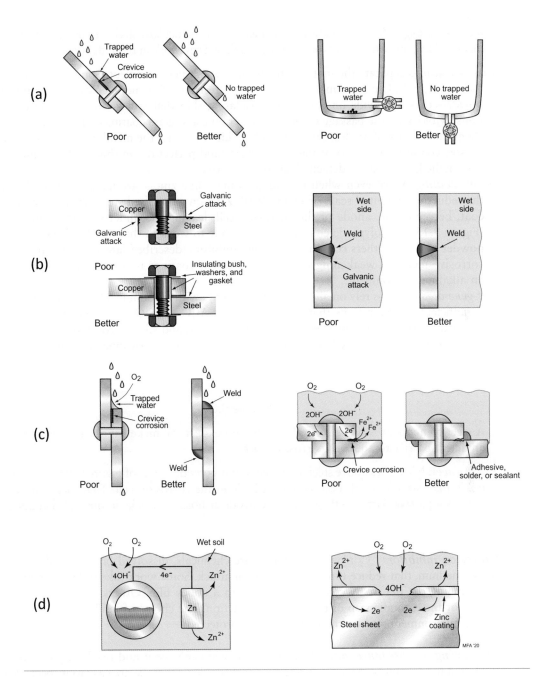

Figure 9.13 Design to avoid corrosion. (a) Avoid trapped fluids. (b) Avoid galvanic couples. (c) Avoid crevice corrosion. (d) Cathodic protection with sacrificial zinc.

Protective coatings   Coatings allow reactive metals – and particularly steels – to be used in environ-ments in which, uncoated, they would corrode rapidly. They work in more than one way:

Passive coatings separate the material from the corrosive environment. Some – like chrome, nickel, or gold plate – are inherently corrosion-resistant. Polymeric coatings – paints and powder-coats – provide an electrically insulating skin that stifles electro-chemical reac-tions by blocking the passage of electrons. Ceramic and glass coatings rely on the extreme chemical stability of oxides, carbides, and nitrides to encase the metal in a refractory shell. Passive coatings only work if they are perfect (and perfection, in this world, is unusual). If scratched, cracked, or detached, corrosion starts.

Active coatings work even when damaged. Some – like zinc – are metals that lie low on the reduction-potential scale of Figure 9.10. Zinc on galvanised steel acts as a sacrificial coat, becoming the anode of the corrosion cell, leaving the underlying steel unscathed [Figure 9.13(d)]. Polymeric coatings can be made to work in this way by dispersing zinc powder in them. Others contain corrosion inhibitors (described later) that leach into the corrosive medium, weakening its potency. Cement coatings inhibit attack by providing an alkaline environment.

Self-generated coatings rely on alloying, almost always with chromium, aluminium, or silicon, in sufficient concentrations that a film of Cr_2O_3, Al_2O_3 or SiO_2 forms spontaneously on the metal surface. Stainless steels, aluminium bronzes, and silicon iron rely on protection of this sort. When scratched, the film immediately re-grows, providing ongoing protection.

Example 9.6

A steel component is nickel-plated. Would you expect the nickel to act as a passive or an active barrier to aqueous corrosion? Why?

Answer. Nickel lies above steel in the reduction-potential chart of Figure 9.10. It is more noble than steel and will become the cathode in the corrosion cell. That means it acts only as a passive barrier – the steel will corrode at holes or cracks in the nickel coating.

Corrosion inhibitors   Corrosion inhibitors are chemicals, dissolved or dispersed in a cor-rosive medium, that reduce ('inhibit') the rate of attack. Some work (like indigestion tablets) by changing the pH or by coating the part with a gooey film. Others suppress either the anodic or the cathodic step of the corrosion reaction. The choice of inhibitor depends both on the environment and on the material to be protected.

Monitoring, maintenance, and replacement   If you can't see it and reach it, you can't fix it. Design for durability means ensuring that parts liable to corrode are easy to inspect, clean, and replace. Picking up signs of corrosion early minimises the risk of failure and cost of down-time. For safety-critical components the best strategy is replacement at intervals too short for corrosion to have caused any serious damage.

9.8 Summary and conclusions

Friction and wear   When surfaces slide, there is friction and there is wear. Engineers have learnt to live with both, to use them when they are useful and to minimise the problems they cause when they are not.

The remarkable thing about friction is that, if you slide one surface over another, the force to do so depends only on the normal force pushing the surfaces together, not on the area of contact. That is because surfaces are not perfectly smooth; they touch only at points that are flattened a little by the contact pressure, until there are just enough of them to support it. Changing the area does not change the number of contact points. The sliding force is the force to shear these contact points, so it, too, is independent of area.

The points of contact are where wear happens. When surfaces slide the points shear off, becoming wear particles. This process can be useful – it helps surfaces bed down and conform by what is called 'run-in' provided the wear particles are swept away and filtered out. But if the wear particles or particles from elsewhere (dust, grit) become trapped between the surfaces, the grinding action continues and the surfaces wear.

Soft materials – lead, polyethylene, PTFE, graphite – shear easily. Impregnating a sliding material with one of these reduces friction. Skis and snowboards have polyethylene surfaces for just this reason. Better still is hydrodynamic lubrication: keep the sliding surfaces apart by dragging fluid between them and raising the pressure. This mechanism requires that the surfaces slide; if they stop, the fluid is squeezed out. Clever lubricants overcome this by bonding to the surfaces, leaving a thin boundary layer that is enough to separate the surfaces until hydrodynamic lubrication gets going.

Oxidation and corrosion   Oxygen, acids, alkalis, aqueous solutions, aerated water, and organic solvents can all attack materials, particularly metals, at an enormous cost to the economy. Managing the damage is a design challenge, as there are few broad physical rules to apply, but many ways in which localised attack can occur, requiring expertise.

In oxidation of metals, the metal ionises, releasing electrons; the electrons combine with oxygen molecules to give oxygen ions, which in turn combine with metal ions to form the oxide. As the oxide film grows, ions and electrons must diffuse through it to enable further growth; its resistance to this diffusion, and its ability to adhere to the metal surface, determine how protective it is.

In corrosion, the electro-chemical reactions occur in the corrosive fluid rather than in the solid. Most metals ionise when placed in water, to an extent that determine their reduction potentials. Those with the lowest (most negative) reduction potential are the most vulnerable to attack and will corrode spontaneously even in pure water if oxygen is present. A metal with a low reduction potential in electrical contact with one of higher potential creates a galvanic cell, and the first corrodes while the second is protected.

In designing materials to resist corrosion we rely on the ability of a few of them – chromium, aluminium, silicon, and titanium – to form a protective oxide film that adheres strongly to the surface. By alloying with these elements, other metals can be given the same protection. Alternatively, metals can be protected by coating their surfaces with a corrosion-resistant film of polymer or ceramic.

Wear and corrosion cause more damage in engineering than anything else. They are the ultimate determinants of the life of products – or they would be if we kept things until they wore out. But in today's society, many products are binned before they are worn out, often when they still work perfectly well.

9.9 Further reading

Friction and wear

ASM International. (2017). *Friction, lubrication, and wear technology* (vol. 18.). Author. ISBN: 978-1-62708-141-2. (The fundamental physical principles of friction, lubrication, and wear and their application to friction- or wear-critical components).

Bowden, F. P., & Tabor, D. (1950, 1964). *The friction and lubrication of solids, parts 1 and 2*. Oxford University Press. ISBN 0-19-851204-X. (Two volumes that establish the foundations of tribology).

Engineer's Handbook. *Coefficient of friction.* http://www.engineershandbook.com/Tables/frictioncoefficients.htm

Engineering Toolbox. *Friction and friction coefficients.* https://www.engineeringtoolbox.com/friction-coefficients-d_778.html

Hutchings, I. M., & Shipway, P. (2017). *Friction and wear of engineering materials* (2nd ed.). Butterworth Heinemann. ISBN: 978-0-08-100910-9. (An introduction to the mechanics and materials science of friction, wear, and lubrication).

Neale, M. J. (1995). *The tribology handbook* (2nd ed.). Butterworth-Heinemann. ISBN: 0-7506-1198-7. (A tribological bible: materials, bearings, drives, brakes, seals, lubrication failure and repair).

Ruff, A. W. (2001). *Friction and wear data bank.* http://home.ufam.edu.br/berti/nanomateriais/8403_PDF_CH15.pdf (A compilation of some 350 data sets for tribological properties of materials).

Oxidation and corrosion

Bradford, S. A. (1993). *Corrosion control.* Van Nostrand Reinhold. ISBN 0-442-01088-5. (A readable text emphasising practical ways of preventing or dealing with corrosion).

Fontana, M. G. (2005). *Corrosion engineering* (3rd ed.). McGraw Hill. ISBN 0-07-021463-8. (A text focusing on the practicalities of corrosion engineering rather than the science, with numerous examples and data).

Ford, F. P., & Andresen, P. L. (1997). Design for corrosion resistance *Materials selection and design* (vol. 20, pp. 545–572). ASM International. ISBN 0-97170-386-6. (A guide to design for corrosion resistance, emphasizing fundamentals, by two experts in the field).

McCafferty, E. (2010). *Introduction to corrosion science.* Springer Verlag. ISBN 978-1-4419-0454-6. (A good starting point for readers who do not have backgrounds in electrochemistry).

National Physical Laboratory. (2000). *Stress corrosion cracking.* http://www.npl.co.uk/upload/pdf/stress.pdf (A concise summary of stress corrosion cracking and how to deal with it).

Schweitzer, P. A. (1995). *Corrosion resistance tables* 4th ed., vols. 1–3. Marcel Dekker. (The ultimate compilation of corrosion data in numerous environments. Not bedtime reading).

Schweitzer, P. A. (1998). *Encyclopedia of corrosion technology.* Marcel Dekker. ISBN 0-8247-0137-2. (A curious compilation, organised alphabetically, that mixes definitions and terminology with tables of data).

Tretheway, K. R., & Chamberlain, J. (1995). *Corrosion for science and engineering* (2nd ed.). Longman Scientific and Technical. ISBN 0-582-238692. (An unusually readable introduction to corrosion science, filled with little bench-top experiments to illustrate principles).

9.10 Exercises

Friction and wear

Exercise E9.1 Give examples, based on your experience in sport (tennis, golf, cycling, skiing, rock climbing, etc.) of instances in which friction is wanted and when it is not.

Exercise E9.2 Now a more challenging one. Give examples, again based on your experience in sport, of instances in which wear is desirable and in which it is not.

Exercise E9.3 Define the coefficient of friction. Explain why it is independent of the nominal area of contact of the sliding surfaces.

Exercise E9.4 A crate is placed on a slipway with gradient 1 in 5 (20%). The slipway has a coating of dry seaweed, coefficient of friction $\mu = 0.6$. Show that the crate is stable. It then starts to rain, the seaweed absorbs water, and the coefficient of friction falls to 0.1. Will the crate slide into the sea?

Exercise E9.5 You are involved in a project to build wind-powered pumps to provide water for remote isolated villages. Maintenance may be sporadic and dependent on local skills. It is suggested that the bearings for the pump be made of wood. Do woods make good bearings? Which wood would be the best choice? Use the internet to explore the use of wood for bearings.

Exercise E9.6 A bronze statue weighing 4 tonnes with a base of area 0.8 m^2 is placed on a granite museum floor. The yield strength of the bronze is 240 MPa. What is the true area of contact A_r between the base and the floor?

Bronze statue

Exercise E9.7 The statue of the previous example was moved to a roof courtyard with a lead floor. The yield strength of the lead is 6 MPa. What now is the true area of contact?

Exercise E9.8   How might you measure the true area of contact A_r between two surfaces pressed into contact? Use your imagination to devise ways.

Exercise E9.9   A pickle jar is vacuum sealed with a HDPE seal with a static coefficient of friction of 0.55. The cap is $R = 20$ mm in radius, and the thickness of the lip of the jar (onto which the seal presses) is much smaller than the radius. Assume the difference in pressure between the inside and outside of the jar is 1 atmosphere (0.1 MPa). What torque T is needed to uncap the pickle jar?

Pickle jar

Exercise E9.10   What materials are used for brake pads in different types of vehicles? Use the internet to find out.

Exercise E9.11   A prototype disc brake for a bicycle is to be mounted on the hub of the rear wheel and consists of a pair of pads of area 5 cm × 1 cm, which are pressed against opposing faces of the steel disk. The pads are aligned with their longer dimension in the tangential rotation direction, with the pad centres at a distance of 10 cm from the centre of the disk. The brake is to be tested on a rotating machine to simulate the loading and wear in service. The disk rotates at a suitable constant speed, and the brakes are applied for 10 seconds every minute, with each pad applying a normal load of 1 kN.

Brake pad

(a) Find a suitable rotation speed (in rpm) if the test is to simulate braking a bicycle travelling at a speed of 20 km/hour, with a wheel diameter (including the tyre) of 70 cm.

(b) When the brakes are applied, the machine registers an increase in torque of 30 Nm. Find the friction force that is being applied to each pad, and hence estimate the friction coefficient.

(c) Estimate the depth of material removed per hour of testing, if the Archard wear constant for the materials is $k_a = 2 \times 10^{-8}$ MPa^{-1}.

(d) What other factors would influence the wear rate in service?

Oxidation and corrosion

Exercise E9.12   Many metallic structures are partly or wholly buried in soil. Which materials tolerate this best? Use Table A10 of Appendix A to find out.

Exercise E9.13   Where, in daily life, might you encounter carbon tetrachloride, CCl_4? Where might you encounter formaldehyde? Use the internet to find out. Then use Table A10 of Appendix A to find the materials that best resist corrosion by them.

Exercise E9.14   Where, in daily life, might you find ethanol? Which materials are found to resist attack by ethanol well? Use Table A10 of Appendix A to find out.

Exercise E9.15   You want to etch your partner's initials into an elegant glass bowl. That needs HF – hydrofluoric acid. What plastic would you choose to contain the HF during the etching? Use Table A10 of Appendix A to find out.

Exercise E9.16   You need a flexible tube to connect a tank of diesel fuel to a burner. Use Table A10 of Appendix A to find a suitable plastic.

Exercise E9.17   A wholesale supplier offers nail-varnish remover at a cut-down price if you bring your own container to take it away. You decide on a thermoplastic rather than glass container to avoid breakage. What is nail-varnish remover? Which thermoplastics would you choose to contain it? Use Table A10 of Appendix A to find out.

Exercise E9.18   A brain surgeon consults you. She has the brain of a recently deceased person and wishes to preserve it in formaldehyde for later analysis. Would a PET container tolerate formaldehyde? Use Table A10 of Appendix A to find out.

Exercise E9.19   Body fluids are corrosive. Why? Use the internet to find out. Materials for implants and body-part replacements must meet restrictive constraints since any corrosion releases potentially toxic chemicals into the body. Most choice is based on past experience. Use Table A10 of Appendix A to identify three polymers that are routinely used for implants.

Exercise E9.20   Buildings near coastlines are exposed to salt as wind-carried spray. What are the preferred metals to minimise corrosive attack on the cladding of buildings exposed to marine environments? Use Table A10 of Appendix A to find out.

Exercise E9.21   The steel wire frames in your fridge or dishwasher are hot-dip coated with a polymer to protect them from water and food-related fluids. Hot-dipping involves plunging the frame when hot into a fluidised bed of polymer powder, which melts and sticks to the frame. What's the polymer likely to be? Use Table A10 of Appendix A to find out.

Exercise E9.22   A food processing plant uses dilute acetic acid for pickling onions. The acid is piped to and from holding tanks. Use Table A10 of Appendix A to select a suitable material for the pipes and tanks, given that to have sufficient strength and toughness to tolerate external abuse they must be made of a metal.

Exercise E9.23 A polymer coating is required to protect components of a microchip processing unit from attack by hydrogen fluoride (HF). Use Table A10 of Appendix A to identify possible choices.

Exercise E9.24 What is meant by the standard reduction potential? A copper and a platinum electrode are immersed in a bath of dilute copper sulphate. What potential difference would you expect to measure between them? If they are connected so that a current can flow, which one will corrode?

Exercise E9.25 Mild steel sheet guttering is copper-plated to protect it from corroding. The guttering acts as a drain for sea water. If the coating is damaged, exposing the steel, will the guttering corrode in a damaging way? If instead the guttering is zinc-plated, will it be better or less well protected? Use the Galvanic Series in sea water (Figure 9.10 of the text) to find out.

Exercise E9.26 The figure shows a proposed design for an outdoor water tank. What aspects of the design might cause concern for corrosion?

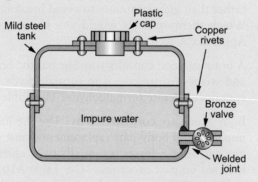

Exercise E9.27 A sample of a metal, heated in air, gains mass per unit area Δm by oxidising, at a rate that declines with time t in a parabolic way, meaning that $\Delta m^2 = k_p t$, where k_p is a constant. Why does the rate fall off in this way? What does it imply about the nature and utility of the oxide?

Exercise E9.28 A 10 mm square sheet of a metal, exposed on both front and back surfaces, gains mass by 4.2 mg when heated for 20 hours in air at 300°C. If the kinetics of oxidation are linear, what is the linear rate constant k_ℓ for this metal at 300°C?

Exercise E9.29 A 10 mm square sheet of a metal, exposed on both front and back surfaces, gains mass by 3.6 mg when heated for 10 hours in air at 700°C. If the kinetics of oxidation are parabolic, what is the parabolic rate constant k_p for this metal at 700°C?

Exercise E9.30 Copper oxidises in air at 1000°C with parabolic kinetics, forming a surface film of Cu_2O. The parabolic rate constant at this temperature, k_p, is 2.1×10^{-6} $kg^2/m^4.s$. What is the mass gain per unit area of copper surface after 1 hour? How thick will the oxide be after this time? (The atomic weight of copper is 63.5 kg/kmol, that of oxygen is 16 kg/kmol, and the density of Cu_2O is 6000 kg/m^3).

Exercise E9.31 The oxidation kinetics of titanium to TiO_2 is limited by oxygen diffusion, with an activation energy Q_d of 275 kJ/mol. If the oxide film grows to a thickness of 0.08 microns after 1 hour at 800°C, how thick a film would you expect if it had been grown at 1000°C for 30 minutes?

Exercise E9.32 A study of slowly propagating cracks in high strength steel plates under constant stress in moist air revealed that the crack growth-rate increased with temperature as follows:

Growth rate (μm s^{-1})	0.70	2.20	8.70	29.1
Temperature (°C)	5	25	55	87

Show, using an appropriate plot, that for these conditions crack propagation is a thermally activated process, and determine the activation energy. It is believed that diffusion of one of the elements listed in the table is the rate-controlling mechanism. Decide which of these elements is likely to be involved.

Diffusing element in α iron	Activation energy for diffusion (kJ/mol)
Hydrogen	38
Nitrogen	72
Carbon	84
Iron	285

Chapter 10

Functional properties: electrical, magnetic, optical

Electrical power transmission: conductors, insulators, and magnetic materials (istock.com).

https://doi.org/10.1016/B978-0-08-102399-0.00010-8

10.1 Introduction and synopsis

Electrons provide the glue that binds atoms together in solids (see Chapter 3). They are also the agent responsible for the electrical, optical, and magnetic properties of materials. Electrons don't just swirl around in solids. They are stored in neat bands like a stack of shelves. It is the structure of these bands, their overlap, and the gaps between them that are key to understanding electrical and optical properties. Electrons move, and they have a quantum mechanical character called *spin*: when they move and have spin they create magnetic fields, so we need to understand this, too.

10.2 Electrical materials and properties

Electrical conduction   (as in circuit wiring, and lightning conductors) and *insulation* (as in electric plug casings) are familiar properties. Superconductivity, semiconduction, and dielectric behaviour may be less so. *Superconductivity* is the characteristic of some materials to lose all resistance at very low temperatures so that a current flows without loss. *Semiconductors*, the materials of transistors and silicon chips, fall somewhere between conductors and insulators, as the name suggests. A *dielectric* is an insulator. It is usual to use the word 'insulator' when referring to its inability to conduct electricity, and to use 'dielectric' when referring to its behaviour in an electric field. Three properties are of importance here. The first, the *dielectric constant* (or *relative permittivity*), descibes the way the material acquires an induced dipole moment (it *polarises*) in an electric field. The second, the *dielectric loss factor*, measures the energy dissipated when radio-frequency waves pass through a material, the energy appearing as heat (the principle of microwave heating). The third is the *dielectric breakdown potential*, the potential (difference) at which the insulator stops insulating.

These different kinds of electrical behaviour are all useful: Figure 10.1 gives an overview, with examples of materials and applications.

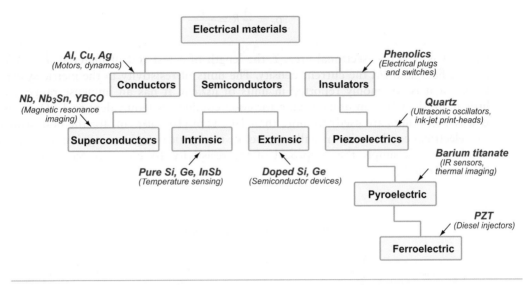

Figure 10.1 The hierarchy of electrical behaviour, with examples of materials and applications.

Resistivity and conductivity The electrical resistance R (unit: ohms,[1] Ω) of a rod of material is the potential drop V (unit: volts,[2] V) across it, divided by the current I (unit: amps,[3] A) passing through it, as in Figure 10.2. This relationship is Ohm's law:

$$R = \frac{V}{I} \tag{10.1}$$

[1] George Simon Ohm (1789–1854), dropped out of his studies at the University of Erlangen, preferring billiards to mathematics, but later became absorbed in the theory of electricity and magnetism.

[2] Alessandro Guiseppe Antonio Anastasio Volta (1745–1827), Italian physicist, inventor of the battery, or, as it was then known, the Voltaic Pile. His face used to appear on the 10,000 lire note.

[3] André Marie Ampère (1775–1836), French mathematician and physicist, largely self-taught and said to have mastered all known mathematics by the age of 14 and went on to contribute to the theories of electricity, light, magnetism, and chemistry (he discovered fluorine).

Resistance also determines the power P (unit: watts,[4] W) dissipated ('Joule heating') when a current passes through a conductor

$$P = I^2 R \tag{10.2}$$

The material property that determines resistance is the *electrical resistivity*, ρ_e. It is related to the resistance by

$$\rho_e = \frac{A}{L} R = \frac{E}{I/A} \tag{10.3}$$

where A is the cross-sectional area, L the length of a test rod of material, $E = V/L$ the electric field, and I/A the current density. The units of resistivity in the metric system are $\Omega.m$, but it is commonly reported in units of $\mu\Omega.cm$. It has an immense range, from a little more than 10^{-8} in units of $\Omega.m$ for good conductors (equivalent to $1\ \mu\Omega.cm$, which is why these units are used) to more than $10^{16}\ \Omega.m$ ($10^{24}\ \mu\Omega.cm$) for the best insulators. The electrical conductivity (usually symbol κ_e in electrical engineering, but σ is also commonly used) is simply the reciprocal of the resistivity. Its units are Siemens per metre [S/m or $(\Omega.m)^{-1}$].

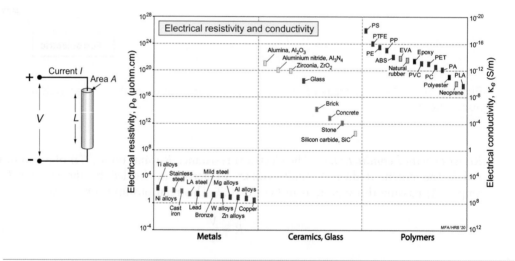

Figure 10.2 Electrical resistivity and conductivity of common engineering materials.

---

[4] James Watt (1763–1819), instrument maker and inventor of the steam engine, which he doggedly developed. Unlike so many of the footnoted characters of this book, Watt, in his final years, was healthy, happy, and famous.

Example 10.1 Electrical current and power

A potential difference of 0.05 volt is applied across a nickel wire 100 mm long and 0.2 mm in diameter. What current flows in the wire? How much power is dissipated in the wire? The resistivity ρ_e of nickel is 9.5 $\mu\Omega$.cm.

Answer. The current is given by Ohm's law: $V = IR$, where V is the potential difference, I the current, and R the resistance. Converting the resistivity from $\mu\Omega$.cm to Ω.m by multiplying by 10^{-8} gives $\rho_e = 9.5 \times 10^{-8}$ Ω.m, so the resistance R of the wire is

$$R = \frac{\rho_e L}{A} = 9.5 \times 10^{-8} \times 0.1 / \pi \left(10^{-4}\right)^2 = 0.3\,\Omega\,(\text{ohms})$$

The current in the wire is

$$I = \frac{V}{R} = 0.05/0.3 = 0.165\,\text{A}\,(\text{amps})$$

The power dissipated in the wire is $P = I^2 R = 8.33 \times 10^{-3}$ W (watts).

Temperature dependence of resistivity The resistivity of metals increases with temperature because thermal vibrations (phonons) scatter electrons. Resistance decreases as temperature falls, which is why very high-powered electro-magnets are pre-cooled in liquid nitrogen. As absolute zero is approached most metals retain some resistivity, but a few suddenly lose all resistance and become *superconducting* between 0 and 10 K. The modern 'high temperature' superconductors operate at the much higher temperatures of liquid nitrogen (77 K), but these are based on complex metal oxides and other compounds.

Effect of alloying and working on resistivity Figure 10.3 shows how the strength and resistivity of copper and aluminium are changed by alloying and deformation. Solute atoms act as scattering centres (discussed in Section 10.5), increasing the electrical resistivity. Dislocations, too, scatter electrons, though much less than solute atoms. Precipitates offer the greatest gain in strength with least loss of conductivity because they reduce the amount of residual solute left in the lattice.

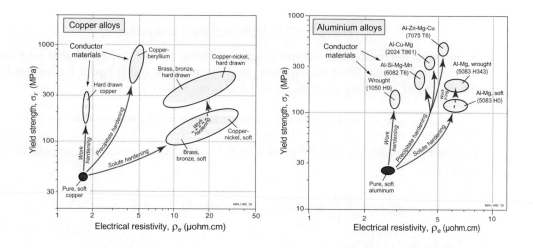

Figure 10.3   Strengthening mechanisms increase resistivity – illustrated here for Cu and Al alloys. Work hardening and precipitation have less effect on resistivity than the addition of solute atoms.

Dielectrics   An *electric field*, E, in a region of space causes charged objects to experience a force. The electric field strength between two oppositely charged parallel plates, separated by a distance t that is small compared to their area, and with a potential difference V between them is

$$E = \frac{V}{t} \tag{10.4}$$

and is independent of position, except near the edge of the plates. Two conducting parallel plates separated by a dielectric make a *capacitor* (Figure 10.4). Capacitors (or *condensers*) store energy as separated charges. The charge Q (unit: coulombs,[5] C) is proportional to the voltage difference between the plates, V:

$$Q = CV \tag{10.5}$$

where C (unit: farads,[6] F) is the capacitance.

The capacitance of a parallel plate capacitor of area A and spacing t, separated by empty space (or by air) is

$$C = \varepsilon_o \frac{A}{t} \tag{10.6}$$

---

[5] Charles Augustin Coulomb (1736–1806), military physicist, laid the foundations of both the mathematical description of electrostatic forces and the laws of friction. Despite his appointment as Intendant des Eaux et Fontaines de Paris, he made no known contributions to hydrodynamics.

[6] Michael Faraday (1791–1867), brilliant experimentalist both in physics and chemistry, discoverer of electro-magnetic induction and inspiring lecturer at the Royal Institution, London.

where ε_o is the *permittivity of free space* (8.85×10^{-12} F/m, where F is farads). If the empty space is replaced by a dielectric, the dielectric *polarises* and the capacitance increases. The field created by the polarisation opposes the field E, reducing the voltage difference V needed to support the charge. Thus the capacitance is increased to the new value

$$C = \varepsilon \frac{A}{t} \qquad (10.7)$$

where ε is the *permittivity of the dielectric* with the same units as ε_o. It is usual to cite the *relative permittivity* or *dielectric constant*, ε_r:

$$\varepsilon_r = \frac{C_{with\ dielectric}}{C_{no\ dielectric}} = \frac{\varepsilon}{\varepsilon_o} \qquad (10.8)$$

making the capacitance

$$C = \varepsilon_r\, \varepsilon_o\, \frac{A}{t} \qquad (10.9)$$

Being a ratio, ε_r is dimensionless. Its value for empty space and, for practical purposes, for most gases, is 1. Most dielectrics have values between 2 and 20, though low-density foams approach the value 1 because they are largely air. Ferroelectrics are special: they have values of ε_r as high as 20,000. We return to ferroelectrics later in this chapter.

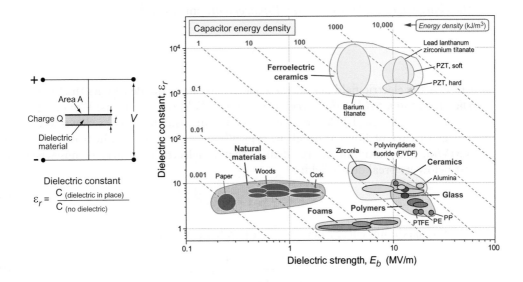

Figure 10.4   Dielectric constant and dielectric strength.

Example 10.2 Capacitance

(a) A parallel plate capacitor has a plate area of $A = 0.1$ m^2 separated by an air gap of $t = 100$ µm. What is its capacitance?

(b) The air gap is reduced and replaced by a dielectric of thickness $t = 1$ µm with a dielectric constant $\varepsilon_r = 20$. What is its capacitance now?

Answer.

(a) The capacitance is

$$C_1 = \varepsilon_o \frac{A}{t} = 8.85 \times 10^{-12} \times \frac{0.1}{100 \times 10^{-6}} = 8.85 \times 10^{-9} \, \text{F} = 8.85 \, \text{pF}$$

(b) The capacitance is

$$C_2 = \varepsilon_r \varepsilon_o \frac{A}{t} = 20 \times 8.85 \times 10^{-12} \times \frac{0.1}{10^{-6}} = 1.77 \times 10^{-5} \, \text{F} = 17.7 \, \mu\text{F}$$

When charged, the energy stored in a capacitor is

$$E_c = \frac{1}{2}QV = \frac{1}{2}CV^2 = \frac{1}{2}Q^2/C \tag{10.10}$$

Supercapacitors, with capacitances measured in farads, store enough energy to power a hybrid car.

The *charge density*, D (or *electric displacement*, units: C/m^2), for a dielectric in a parallel plate capacitor is

$$D = \varepsilon_r \varepsilon_o E \tag{10.11}$$

where E is the electric field across the dielectric (voltage divided by its thickness).

The *breakdown field* or *dielectric strength of a dielectric*, E_b (units: V/m or, more usually MV/m), is the electrical field gradient at which an insulator breaks down and a damaging surge of current flows through it, typically at a potential gradient of between 1 and 100 MV/m. The maximum charge density that a dielectric can carry is thus when the field is just below its breakdown field:

$$D_{\text{max}} = \varepsilon_r \varepsilon_o E_b \tag{10.12}$$

and so the maximum energy density for a parallel plate capacitor is

$$\frac{1}{2} \frac{CV^2}{At} = \frac{1}{2} \varepsilon_r \varepsilon_o E_b^2 \tag{10.13}$$

The chart of Figure 10.4 gives an overview of dielectric behaviour. Molecules containing polar groups, like certain polymers and ionic-bonded ceramics, react strongly to an electric field and polarise, giving them high dielectric constants. Those with purely covalent or van der Waals bonding do not polarise easily and have low values. The dielectric strengths differ, too – polymers can have higher values than ceramics or glasses. The contours show values of the upper limiting energy density [Equation (10.13)].

Example 10.3 Energy in a capacitor

A supercapacitor is required that is able to store an energy of $E_c = 1$ kJ when a potential difference $V = 100$ V is applied to its plates. The dielectric constant of the dielectric is $\varepsilon_r = 10{,}000$ and its breakdown potential is $E_b = 20$ MV/m. What is the minimum area of plate required?

Answer. Breakdown occurs if $\dfrac{V}{t} \geq E_b$, requiring that the dielectric have a thickness $t \geq 5$ µm. The energy stored in a capacitor is $E_c = \dfrac{1}{2}CV^2$, so the capacitance needed is $C = \dfrac{2 \times 1000}{(100)^2} = 0.2$ F. To achieve this with a dielectric of thickness $t \geq 5$ µm and dielectric constant 10,000 requires an area

$$A = \frac{Ct}{\varepsilon_r \, \varepsilon_o} = 11.3\,\text{m}^2$$

The *loss tangent* and the *loss factor* take a little more explanation. Polarisation involves the small displacement of charge (either of electrons or of ions) or of molecules that carry a dipole moment when an electric field is applied. An oscillating field causes the charge to move between two extreme configurations. This charge-motion is like an electric current that − if there were no losses − would be 90° ($\pi/2$ radians) out of phase with the voltage. In real dielectrics this current dissipates energy, just as a current in a resistor does, giving a small shift in the phase between voltage and current. This phase shift is called the *loss angle*, δ (Figure 10.5). The *loss tangent*, tan δ, or *dissipation factor*, *D*, is the tangent of the

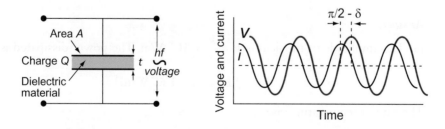

Figure 10.5 A parallel capacitor with an alternating applied field, giving a phase shift between current and voltage. The loss angle δ characterises the shift from the ideal value of $\pi/2$ (for a lossless dielectric).

loss angle. The *power factor*, P_f, is the sine of the loss angle. When δ is small, as it is for the materials of interest here, all the measures are essentially equivalent:

$$\tan\delta \approx \sin\delta \approx \delta \approx P_f \approx D \qquad (10.14)$$

More useful, for our purposes, is the *loss factor L*, which is the loss tangent times the dielectric constant:

$$L = \varepsilon_r \tan\delta \qquad (10.15)$$

It measures the energy dissipated by a dielectric in an oscillating field, produced by an alternating current. If you want to select materials to minimise or maximise dielectric loss, then the parameter of interest is L.

When a dielectric material is placed in a cyclic electric field of amplitude E and frequency f, power P is dissipated, and the field is correspondingly attenuated. The average power dissipated per unit volume, in W/m^3, is

$$P \approx \pi f E^2 \varepsilon_o L \approx \pi f E^2 \varepsilon_o \varepsilon_r \tan\delta \qquad (10.16)$$

This power appears as heat and is generated uniformly (if the field is uniform) through the volume of the material. Dielectric loss is the basis of microwave cooking and is exploited in materials processing – the radio-frequency welding of polymers is an example.

Example 10.4 Dielectric loss

(a) An oscillating voltage with a peak amplitude of 100 V at a frequency $f = 60$ Hz is applied across a nylon insulator of thickness $x = 1$ mm. Nylon has a dielectric loss factor $L \approx 0.1$. How much power is dissipated per unit volume in the nylon? Assuming no heat loss, how fast will the temperature of the nylon rise? (Nylon has a volumetric specific heat $C_p = 3 \times 10^6$ J/m^3.K).

(b) If the frequency was, instead, $f = 1$ GHz, what is the heating rate?

Answer.

(a) The amplitude of the field is $E = \dfrac{V}{x} = 10^5$ V/m. The power dissipated is

$$P \approx \pi f E^2 \varepsilon_o L = 1.67 \text{ W/m}^3$$

The rate of temperature rise is

$$\frac{dT}{dt} = \frac{P}{C_p} = 5.6 \times 10^{-7} \, {}^\circ\text{C/s}$$

(b) If the frequency is raised to $f = 1$ GHz the heating rate becomes $\dfrac{dT}{dt} = 9{}^\circ$C/s

All dielectrics change shape in an electric field, a consequence of the small shift in charge that allows them to polarise; the effect is called *electrostriction*. Electrostriction is a one-sided relationship in that an electric field causes deformation, but deformation does not produce an electric field. *Piezoelectric* materials, by contrast, display a two-sided relationship between polarisation and deformation: a field induces deformation, and deformation induces charge differences between its surfaces, thus creating a field. The piezoelectric coefficient is the strain per unit of electric field, and although it is very small, it is a true linear effect, and this makes it useful, enabling movement with nano-scale precision. *Pyroelectric* materials contain molecules with permanent dipole moments that, in a single crystal, are aligned, giving the crystal a permanent polarisation. When the temperature is changed, the polarisation changes, creating surface charges or, if the surfaces are connected electrically, a pyroelectric current – the principle of intruder-detection systems and of thermal imaging. *Ferroelectric* materials, too, have a natural dipole moment: they are polarised to start with, and the individual polarised molecules line up so that their dipole moments are parallel, like magnetic moments in a magnet. Their special feature is that the direction of polarisation can be changed by applying an electric field, and the change causes a change of shape. Applications of all of these are described in Section 10.4.

10.3 Drilling down: the origins of electrical properties

Electrical conductivity   Solids are made up of atoms containing electrons that carry a charge $-e$ and a nucleus containing protons, each with a positive charge $+e$. An electric field, E (volts/m), exerts a force Ee on a particle carrying a charge e. If the charge-carriers can move, the force Ee causes them to flow through the material – that is, it conducts. Metals are *electron conductors*, meaning that the charge-carriers are the electrons. In ionic solids (which are composed of negatively and positively charged ions like Na^+ and Cl^-) the diffusive motion of ions allows *ionic conduction*, but this is only possible at temperatures at which diffusion is rapid. Many materials have no mobile electrons, and at room temperature they are too cold to be ionic conductors. The charged particles they contain still feel a force in an electric field that displaces the charges slightly, but they are unable to move more than a tiny fraction of the atom spacing. These are insulators; the small displacement of charge gives them dielectric properties.

Chapter 3 (Section 3.3) explained the way electrons of an atom occupy discrete energy states or orbitals, arranged in shells (designated 1, 2, 3, etc., from the innermost to the outermost); each shell is made up of sub-shells (designated s, p, d, and f), each of which contains 1, 3, 5, or 7 orbitals, respectively. When n atoms (a large number) are brought together to form a solid, the inner electrons remain associated with the atom on which they started, but the outer ones interact. Each atom now sits in the field created by the charges of its neighbours. This has the effect of decreasing slightly the energy levels of electrons with spin in a direction favoured by the field of its neighbours and raising that of those with spins in the opposite direction, splitting each energy level. The discrete levels of an isolated atom broaden into *bands* of very closely spaced levels. The number of electrons per atom depends only on the atomic number; they fill the bands from the bottom, starting from the lowest, until all are on board, so to speak. The top-most filled energy level at 0 K is called the Fermi[7] energy.

Above 0 K, thermal energy allows electrons to occupy higher levels with a certain probability; the Fermi level is then defined as the energy level having the probability that it is exactly half filled with electrons. An electron at the Fermi level still has an energy that is lower than it would have if it were isolated in vacuum far from the atoms. This energy difference is called, for historical reasons, the *work function* because it is the work that is required to remove an electron from the Fermi level to infinity.

Whether the material is a conductor or an insulator depends on how full the bands are, and whether they overlap. In Figure 10.6 the central column describes an isolated atom, and the outer ones illustrate the possibilities created by bringing atoms together into an array, with the energies spread into energy bands. Conductors like copper, shown on the left, have an unfilled outer band; there are many very closely spaced levels just above the last full one, and − when accelerated by a field − electrons can use these levels to move freely through the material. In insulators, shown on the right, the outermost band with electrons in it is full, and the nearest empty band is separated from it in energy by a *band gap*. Semiconductors, too, have a band gap, but it is narrower − narrow enough that thermal energy can pop a few electrons into the empty band, where they conduct. Deliberate doping (adding trace levels of other elements) creates new levels in the band gap, reducing the energy barrier to entering the empty states and thus allowing more carriers to become mobile.

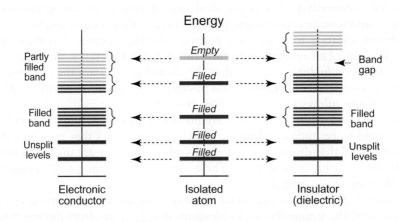

Figure 10.6 Conductors *(left)* have a partly filled outer band. Insulators *(right)* have a filled outer band, separated from the nearest unfilled band by a band gap.

---

[7] Enrico Fermi (1901–1954), devisor of the statistical laws known as Fermi statistics governing the behaviour of electrons in solids. He was one of the leaders of the team of physicists on the Manhattan Project for the development of the atomic bomb.

Electrical resistance If a field E exerts a force Ee on an electron, why does it not accelerate forever, giving a current that continuously increases with time? Instead, the current almost immediately reaches a steady value. Referring back to Equation (10.3), the current density I/A is proportional to the field E

$$\frac{I}{A} = \frac{E}{\rho_e} = \kappa_e E \tag{10.17}$$

where ρ_e is the resistivity and κ_e, its reciprocal, is the electrical conductivity.

Broadly speaking, the picture is this. The outermost or conduction electrons are able to move through the solid. Their thermal energy $k_B T$ (k_B = Boltzmann's constant, with T in Kelvin) causes them to move like gas atoms in all directions. In doing this they collide with *scattering centres*, bouncing off in a new direction. Impurity or solute atoms are particularly effective scattering centres (which is why alloys always have a higher resistivity than pure metals), but electrons are scattered also by imperfections such as dislocations and by the thermal vibration of the atoms themselves. When there is no field, there is no *net* transfer of charge in any direction even though all the conduction electrons are moving freely. A field causes a drift velocity $v_d = \mu_e E$ on the electrons, where μ_e is the electron mobility, and it is this that gives the current (Figure 10.7). The drift velocity is small compared with the thermal velocity; it is like a breeze in air − the thermal motion of the air molecules is far greater than the 'drift' that we feel as the breeze. The greater the number of scattering centres, the shorter is the mean-free path, λ_{mfp}, of the electrons between collisions, and the slower, on average, they move. Just as with thermal conductivity, the electrical conductivity depends on a mean-free path, on the density of carriers (the number n_v of mobile electrons per unit volume), and the charge they carry. Thus the current density, I/A, is given by

$$\frac{I}{A} = n_v e\, v_d = n_v e\, \mu_e E$$

Comparing this with Equation (10.17) gives the conductivity:

$$\kappa_e = n_v e\, \mu_e \tag{10.18}$$

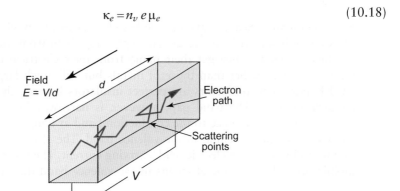

Field
$E = V/d$
d
Electron path
Scattering points
V

Figure 10.7 The movement of an electron through a material with scattering centres.

Thus the conductivity is proportional to the density of free electrons and to the drift velocity, and this is directly proportional to the mean-free path. The purer and more perfect the conductor, the higher the conductivity and the lower its reciprocal, the resistivity. The conduction of heat and electricity in metals and alloys is so similar physically that the electrical and thermal conductivities scale approximately as $\kappa_e \approx 10^5 \lambda$ (a relationship known as the Wiedemann-Franz law).

Example 10.5 Electron mobility

The resistivity of pure copper is $\rho_e = 2.32 \ \mu\Omega.\text{cm}$. Each copper atom provides one free electron. The atomic volume of copper is $1.18 \times 10^{-29} \ \text{m}^3$. What is the electron mobility in copper?

Answer.   If each atom contributes one electron the electron concentration in copper is

$$n_v = \frac{1}{\text{Atomic volume}} = 8.47 \times 10^{28} \ \text{per m}^3$$

The resistivity ρ_e is related to this by

$$\rho_e = \frac{1}{\kappa_e} = \frac{1}{n_v \, e \, \mu_e} \text{ , where } e = 1.6 \times 10^{-19} \ \text{C is the charge on an electron.}$$

Converting the resistivity from $\mu\Omega.\text{cm}$ to $\Omega.\text{m}$ by multiplying by 10^{-8} gives the electron mobility in copper as

$$\mu_e = \frac{1}{n_v \, e \, \rho_e} = 0.0032 \ \text{m}^2/\text{V.s}$$

Semiconductors   Semiconductors are based on elements in which four valence electrons fill the valence band, with an energy gap separating them from the next available level (in the conduction band). They get their name from their electrical conductivities, 10^{-4} to 10^{+4} $(\Omega.\text{m})^{-1}$, much smaller than those of metals but still much larger than those of insulators. At 0 K pure semiconductors are perfect insulators because there are no electrons in their conduction band. The band gap, however, is narrow, allowing thermal energy to excite electrons from the valence to the conduction band at temperatures above 0 K, leaving positively charged holes behind in the valence band. The electrons in the conduction band are mobile and move in an electric field, carrying current. The holes in the valence band are also mobile (in reality it is the electrons in the valence band that move cooperatively, allowing the hole to change position), and this charge movement also carries current. Thus the conductivity of a semiconductor has an extra term

$$\kappa_e = n_v\, e\, \mu_e + n_h\, e\, \mu_h \tag{10.19}$$

where n_h is the number of holes per unit volume and μ_h is the hole mobility. In pure semiconductors each electron promoted to the conduction band leaves a hole in the valence bend, so $n_e = n_h$, and the conductivity becomes

$$\kappa_e = n_v\, e\,(\mu_e + \mu_h) \tag{10.20}$$

This semiconducting behaviour is said to be *intrinsic* (a fundamental property of the pure material).

The number of charge carriers can be controlled in another way: by doping with controlled, very small additions of elements that have a different valence than that of the semiconductor itself. This creates free electrons and holes that simply depend on the concentration of dopant, and the semiconducting behaviour is said to be *extrinsic* (a characteristic of the doping, not of the base material).

Example 10.6 Intrinsic conductivity

The electron mobility μ_e in silicon at 300 K is 0.14 m^2/V.s and that of holes μ_h is 0.04 m^2/V.s. The carrier density is 1.5×10^{16}/m^3 for both. What is the intrinsic conductivity of silicon at 300 K?

Answer. The conductivity is

$$\kappa_e = n_v\, e\,(\mu_e + \mu_h) = 1.5 \times 10^{16} \times 1.6 \times 10^{-19} \times (0.14 + 0.04) = 4.3 \times 10^{-4}\ (\Omega.\mathrm{m})^{-1}$$

Dielectric behaviour   In the absence of an electric field, the electrons and protons in most dielectrics are symmetrically distributed, and the material carries no net charge or dipole moment. In the presence of a field the positively charged particles are pushed in the direction of the field and negatively charged particles are pushed in the opposite direction. The effect is easiest to see in ionic crystals, since here neighbouring ions carry opposite charges, as on the left of Figure 10.8. Switch on the field and the positive ions (charge $+\, q$) are pulled in the field direction, the negative ones (charge $-\, q$) in the reverse, until the restoring force of the inter-atomic bonds just balances the force due to the field at a displacement of Δx, as on the right of the figure. Two charges $\pm\, q$ separated by a distance Δx create a dipole with dipole moment, d, given by

$$d = q\,\Delta x \tag{10.21}$$

The polarisation of the material, P, is the volume-average of all the dipole moments it contains:

$$P = \frac{\sum d}{\text{Volume}} \qquad (10.22)$$

Even in materials that are not ionic, like silicon, a field produces a dipole moment because the nucleus of each atom is displaced a tiny distance in the direction of the field and its surrounding electrons are displaced in the opposite direction. The resulting dipole moment depends on the magnitude of the displacement and the number of charges per unit volume, and it is this that determines the dielectric constant. The bigger the shift, the bigger the dielectric constant. Thus compounds with ionic bonds and polymers that contain polar groups like $-OH^-$ and $-NH^-$ (e.g. nylon) have larger dielectric constants than those that do not.

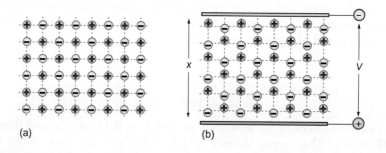

Figure 10.8   An ionic crystal (a) in zero applied field and (b) when a field V/x is applied. The electric field displaces charge, causing the material to acquire a dipole moment.

Dielectric loss   In an alternating electric field, the ions oscillate. If their oscillations were exactly in phase with the field, no energy would be lost, but this is never exactly true. Materials with high dielectric loss usually contain awkwardly shaped molecules that themselves have a dipole moment – the water molecule is an example. These respond to the oscillating field by rotating, but because of their shape they interfere with each other (you could think of it as molecular friction), and this dissipates energy that appears as heat; it is how microwave heating works. As Equation (10.16) showed, the energy that is dissipated depends on the frequency of the electric field. Generally speaking, the higher the frequency, the greater the power dissipated, but there are peaks at certain frequencies that are characteristic of the material structure.

Example 10.7 Polarisation in an electric field

An ionic crystal with the rock-salt structure (see figure) is made up of equal numbers of positive and negative ions each carrying or lacking the charge carried by a single electron. The lattice parameter of the crystal is 0.45 nm. An electric field displaces the ions relative to each other by 1% of the lattice parameter. What is the resulting polarisation?

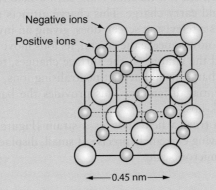

Negative ions
Positive ions
←—0.45 nm—→

Answer. The dipole moment due to one pair of ions is

$$d = q\,\Delta x = 1.6 \times 10^{-19} \times 0.0045 \times 10^{-9}$$
$$= 0.72 \times 10^{-30} \text{ C.m}$$

The volume of the unit cell is $V = (0.45 \times 10^{-9})^3 = 9.1 \times 10^{-29} \text{ m}^3$, and each cell contains four pairs of ions. Thus the polarisation is

$$P = \frac{4d}{V} = 0.032 \text{ C/m}^2$$

Dielectric breakdown In metals, as we have seen, even the smallest field causes electrons to flow. In insulators they cannot because of the band gap. But if, at *some* weak spot, one electron is torn free from its parent atom, the force $E.e$ exerted by the field E accelerates it, giving it kinetic energy; it continues to accelerate, imparting energy to other electrons and triggering a breakdown.

The **critical** field strength to make this happen, called the *breakdown potential*, is large, typically 1 to 15 MV/m. That sounds like a lot, but such fields are found in two very different circumstances: when the voltage V is very high or the distance x is very small. In power transmission, the voltages are sufficiently high — 20,000 V or so — that breakdown can occur at the insulators that support the line; while in microcircuits and thin-film devices the distances between components are very small: a 1 V difference across a distance of 1 micron gives a field of 1 MV/m.

Piezoelectric materials Piezoelectric behaviour is found in crystals in which the ions are not symmetrically distributed, so that each molecule carries a permanent dipole moment [Figure 10.9(a)]. If you cut a little cube from such a material, with faces parallel to crystal planes, the faces would carry charge. This charge attracts ions and charged particles from the atmosphere just as a television screen does, giving an invisible surface layer that neutralises the charge. If you now squeeze the crystal, its ions move relative to each other, the dipole moment changes, and the charge on the surface changes, too [Figure 10.9(b)]. Given time, the newly appeared charge would attract neutralising ions, but this does not happen immediately, giving a potential difference. This provides the basis for the operation of electric microphones and pick-ups.

The inverse is also true: a field induces a strain [Figure 10.9(c)]. The strain is a linear function of field, allowing extremely precise, if small, displacements used for positioning and actuation at the sub-micron scale.

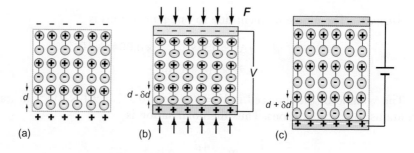

Figure 10.9 (a) A piezoelectric material has a natural dipole moment. (b) When deformed, the dipole moment changes, and the surfaces become charged. (c) The inverse: a field induces a change of shape, the basis of piezoelectric actuation.

Pyroelectric materials Thermal expansion changes the dipole moment of a piezoelectric material, and thermal vibration can misalign its dipoles. Both effects change the surface charge. If conductive electrodes are attached to the surfaces and connected through an ammeter, a current flows until the charge-balance is re-established. If the temperature of the sample remains constant, no current flows. A change in temperature causes expansion or contraction and a change in disorder among the dipoles, generating a pyroelectric current; it's the way that intruder alarms, automatic doors, and safety lights are activated.

Ferroelectric materials   Ferroelectrics are a special case of piezoelectric behaviour. They, too, do not have a symmetric structure but have the special ability to switch asymmetry. Barium titanate ($BaTiO_3$), shown schematically in Figure 10.10, is one of these. Below a critical temperature, the Curie[8] temperature (~120°C for barium titanate), the titanium atom, instead of sitting at the centre of the unit cell, is displaced up, down, to the left, or to the right, as in Figure 10.10(a) and (b). Above the Curie temperature the asymmetry disappears and with it the dipole moment [Figure 10.10(c)]. In ferroelectrics, these dipoles spontaneously align so that large volumes of the material are polarised even when there is no applied field.

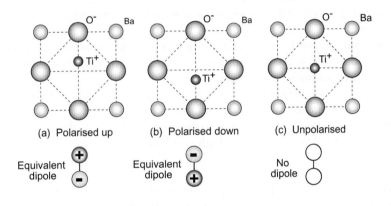

Figure 10.10   Ferroelectric materials have a permanent dipole moment that can switch: here the Ti^+ ion can flip from the upper to the lower position.

Ferroelectric materials have enormous dielectric constants; they appear in the upper part of the chart of Figure 10.4. Those of normal materials with symmetric charge distributions lie in the range 2 to 20. Those for ferroelectrics can be as high as 20,000. It is this that allows their use to make supercapacitors that can store 1000 times more energy than conventional capacitors. Such is the energy density that supercapacitors now compete with batteries in kinetic-energy recovery systems for vehicles.

---

[8] Pierre Curie (1859–1906), French physicist, discoverer of the piezoelectric effect and of magnetic transformations; and husband of the yet more famous Marie Curie. He was killed in a street accident in Paris, a city with dangerous traffic even in 1906.

10.4 Design: using electrical properties

Transmitting electrical power The objective in power transmission is to minimise resistive losses, P_L [Equations (10.2) and (10.3), combined – here for a direct current, but the same holds for alternating currents]:

$$P = I^2 R = I^2 \rho_e \frac{L}{A} = \frac{I^2}{\kappa_e} \frac{L}{A} \qquad (10.23)$$

The best choice, for a given cross-section A and length L, is that of a material with the largest electrical conductivity κ_e; if the line is buried or supported along its entire length, the only other constraint is that of material cost. The chart of Figure 10.2 shows that, of the materials plotted on it, copper (price $\approx$ \$6.5/kg) has the highest conductivity. Aluminium comes next (price $\approx$ \$2.5/kg), and it is these two materials that are most used for power transmisssion. If instead the line is above ground and supported by widely spaced pylons, strength and density become important: high strength, to support self-weight and wind loads; low density, to minimise self-weight.

Here we have a conflict: the materials with the highest conductivities have low strength; those with high strength do not conduct very well. The answer is a hybrid cable in which two materials are combined, the first to give the conductivity, the second to give the strength [Figure 10.11(a)]. The core is pure, high-conductivity copper or aluminium; it is wrapped in a cage of high-strength carbon-steel wires.

Electric energy storage The energy density of a capacitor with a single dielectric layer [Figure 10.11(b)] increases with the square of the field E:

$$\text{Energy density} = \frac{1}{2} \varepsilon_r \, \varepsilon_o \, E^2$$

where ε_r is the relative permittivity (dielectric constant) and ε_o the permittivity of vacuum (8.854×10^{-12} F/m). An upper limit for E is the dielectric strength, E_b, setting an upper limit for energy density:

$$\text{Maximum energy density} = \frac{1}{2} \varepsilon_r \, \varepsilon_o \, E_b^2$$

This upper limit appears as diagonal contours in Figure 10.4, which illustrates the enormous gains that ferroelectric ceramics offer. PZT (lead zirconium titanate) has the highest value, and it is used for high-performance capacitors, but barium titanate is the more usual selection because it is cheaper, easier to process, and lead-free.

Electrical insulation with thermal conduction: keeping microchips cool As microchips shrink in scale and clock-speeds rise, thermal management becomes a problem. To conduct heat away, the chip is attached to a heat sink [Figure 10.11(c)]. The heat sink must have a high thermal conductivity, λ. To minimise electrical coupling, it must also be an electrical insulator. Polymers and many ceramics are good insulators (Figure 10.2), but polymers soften at the temperatues of interest here. That leaves insulator ceramics that are also good thermal

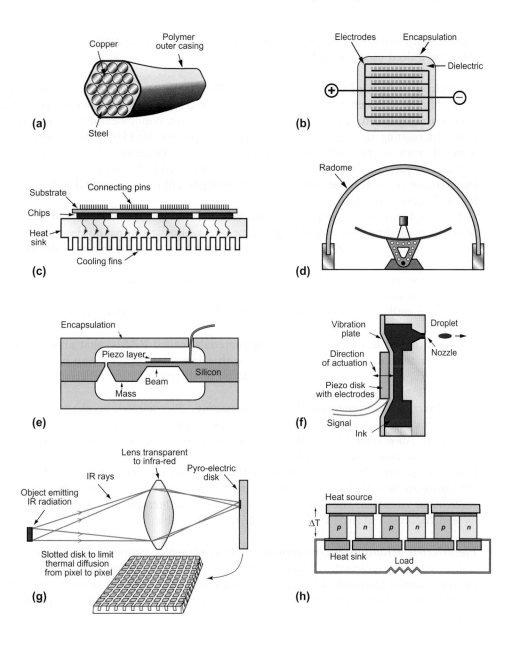

Figure 10.11 Applications of electrical materials.

conductors: Figure 7.2 suggests alumina (Al_2O_3) or, better, aluminium nitride (Al_3N_4), and these are indeed used for such heat sinks.

Minimising dielectric loss: radomes and stealth technologies   Radar works by transmitting bursts of radio-frequency radiation that are reflected back by objects in their path. The function of a *radome* is to shield a microwave antenna from wind, weather, and structural loads while attenuating the signal as little as possible [Figure 10.11(d)]. The answer is to seek materials with exceptionally low values of $L = \varepsilon_r \tan \delta$ [Equation (10.15)]. Since the radome carries structural loads, they should also be strong. Most polymers have low loss factor, but they are not very strong. Reinforcing them with glass fibres, creating GFRPs, combines low loss with good strength.

The metal body of an aircraft is an excellent radar reflector, and its curved shape causes it to reflect in all directions, which is good for safe air-traffic control but less good if avoiding detection is the aim. Stealth technology combines three tricks to minimise detection:

• A shape made up of flat planes that reflect the radar signal away from the detector unless it lies normal to the surface
• Non-metallic structural materials – usually composites – that are semi-transparent to microwaves
• Surface coatings of radar-absorbing materials (called RAMs) that absorb microwaves rather than reflecting them. A RAM with a large loss factor $\varepsilon_r \tan \delta$ – a ferroelectric, for example – attenuates the incoming wave, diminishing the reflected signal.

Piezoelectric accelerometers and energy harvesting   Piezoelectric accelerometers rely on the inertia of a mass supported on a thin beam to which a thin layer of piezoelectric material is bonded. An acceleration or deceleration flexes the beam, stretching or compressing the piezoelectric layer [Figure 10.11(e)]. The resulting charge that appears on the piezoelectric is used to quantify the acceleration or to trigger a response above a threshold level (such as deploying an airbag). The same configuration allows electrical energy to be harvested from mechanical vibration. For this, a material with a high *piezoelectric charge coefficient*, such as lead zirconium titanate or barium titanate, is desirable.

Piezoelectric actuation: ink-jet printers   Ink-jet printers have piezoelectric print heads, one unit of which is sketched in Figure 10.11(f). An electric field applied to the disk of piezoelectric material transmits a kick to the vibration plate, creating a pressure pulse that spits a droplet of ink from the nozzle. On relaxation, more ink is drawn in from the head reservoir to replace it.

Pyroelectric thermal imaging   Night vision systems make use of the pyroelectric effect [Figure 10.11(g)]. A lens made of a material such as germanium that is transparent to infrared (IR) focusses the IR radiation from an object to form an image on a pyroelectric disk. The focused IR radiation warms points on the disk, and the change of temperature changes the polarisation. The image creates a pattern of charge on the surface of the disk that can be 'read' recreating the image.

Energy harvesting from waste heat The *Seebeck*[9] *effect* describes the appearance of an electrical potential when two materials are joined and placed in a thermal gradient. In a thermoelectric material free electrons or holes transport both charge and heat. To a first approximation, the electrons and holes behave like a gas of charged particles. If a normal (uncharged) gas is placed in a box within a temperature gradient, the gas molecules at the hot end move faster than those at the cold one. The faster hot molecules diffuse further than the cold ones, leading to a net buildup of molecules at the cold end. The density-gradient drives the molecules to diffuse back toward the hot end so that, at steady state, the two effects exactly compensate and there is no net flow. Electrons and holes behave in a similar way. Because they carry charge, the temperature gradient creates a charge gradient between the hot and cold ends of the material; the charge gradient tends to push the charges back to the hot end. Since p-type and n-type materials create opposite charge gradients under the same thermal gradient, these can be connected up to give small-scale thermoelectric generators [Figure 10.11(h)].

Thermoelectric generation is used in space probes, drawing heat from a radioactive source. The conversion efficiency is low, which has limited the use of thermoelectrics for energy harvesting on earth. Despite this, the automobile industry expresses interest in using them. A third of the energy used by a car is lost as waste heat. Some of this can be converted back to useful electrical energy using thermoelectric generators attached to the exhaust manifold.

10.5 Electrical properties: summary and conclusions

All materials contain charged particles. An electric field E exerts a force qE on a charge q. If the particles can move freely, as electrons can in metals, an electric current flows through the material, which we refer to as conduction. Special types of conduction are also found — superconductivity at very low temperatures and semiconduction in the materials of electronics such as silicon and germanium. If, instead, the particles cannot move freely, as is the case in insulators, they are still displaced slightly by the field, leading to dielectric polarisation.

Materials that polarise easily have a high dielectric constant. If used as the insulating film of a capacitor, the polarisation screens the electrodes from each other, increasing the capacitance. The slight displacement of charge has other consequences. One of these is electro-striction: the charge movement causes the material to change shape slightly. If the material is made up of ions and these are arranged in a way that lacks a centre of symmetry, a further effect appears: a piezoelectric response. The lack of symmetry means that each molecule carries a dipole moment, and within a single crystal these are aligned. An electric field interacts with the dipole moments, causing a change of shape, and a change of shape changes the molecular dipole moment, thus inducing a change in polarisation, generating a field. A change of temperature, too, changes the molecular dipole moment, so materials that

---

[9] Thomas Johann Seebeck (1770–1831) was a Baltic (now Estonian) physicist who studied medicine but succumbed to a fascination with physics. He discovered and quantified the Seebeck effect, worked with Johann von Goethe, the writer and statesman, on the Theory of Colours and dabbled with experiments on early photography.

are piezoelectric are also pyroelectric. Ferroelectric materials have a further property: that of molecular asymmetry that can switch direction because ions can occupy one of several equivalent sites in the crystal. In the relaxed state the domains take up orientations such that the field of one is cancelled by those of its neighbours, but when 'poled' by placing it in an electric field, the domains align, giving a large net polarisation that is retained when the field is removed. All these effects have practical utility, examples of which were described in Section 10.4.

10.6 Magnetic materials and properties

Migrating birds, some think, navigate using the earth's magnetic field. This may be debatable, but what is beyond question is that sailors, for centuries, have navigated in this way, using a natural magnet, *lodestone*, to track it. Lodestone is a mineral, magnetite (Fe_3O_4), that sometimes occurs naturally in a magnetised state. Today we know lodestone is a *ferrimagnetic* ceramic, or *ferrite*, examples of which can be found in every radio, television, tablet, and microwave oven. *Ferromagnetic* materials, by contrast, are metals typified by iron, but include also nickel, cobalt, and alloys of all three. Placed in a magnetic field, these materials become magnetised, a phenomenon called *magnetic induction*; on removal, some, called soft magnets, lose their magnetisation, while others, the hard magnets, retain it.

There are many different kinds of magnetic behaviour. Figure 10.12 gives an overview, with examples of materials and applications.

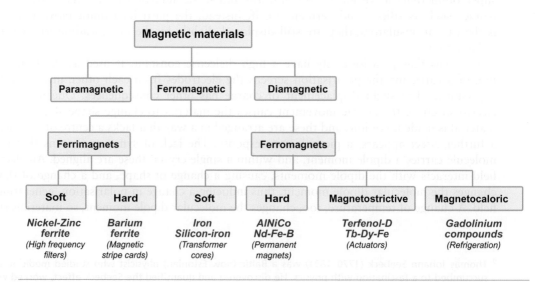

Figure 10.12 Magnetic materials with examples of materials and applications.

Magnetic fields in vacuum   When a current i passes through a long, empty coil of n turns and length L as in Figure 10.13(a), a magnetic field is generated. The magnitude of the field, H, is given by Ampère's law as

$$H = \frac{n\,i}{L} \tag{10.24}$$

and thus has units of amps/metre (A/m). The *magnetic induction* or *flux density*, B, for vacuum or non-magnetic materials, is defined as

$$B = \mu_o\,H \tag{10.25}$$

where μ_o is called the *permeability of vacuum*, $\mu_o = 4\pi \times 10^{-7}$ henry/metre (H/m). The units of B are *tesla*[10], so 1 Tesla = 1 H A/m^2.

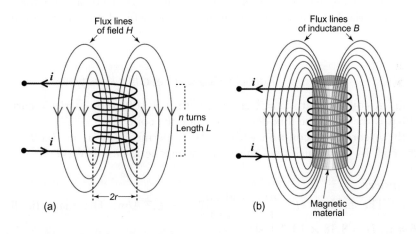

Figure 10.13   (a) A solenoid creates a magnetic field, H. (b) A magnetic material exposed to a field H becomes magnetised.

Magnetic fields in materials   If the space inside the coil is filled with a material, as in Figure 10.13(b), the induction within it changes because its atoms respond to the field by forming little magnetic dipoles (in ways that are explained later). The material acquires a macroscopic dipole moment or *magnetisation*, M (its units are A/m, like H). The induction becomes

---

[10] Nikola Tesla (1856–1943), Serbian-American inventor, discoverer of rotating magnetic fields, the basis of most alternating current machinery, inventor of the Tesla coil and of a rudimentary flying machine (never built).

$$B = \mu_o(H + M) \tag{10.26}$$

The simplicity of this equation is misleading because it suggests that M and H are independent; in reality, M is the response of the material to H, so the two are coupled. If the material of the core is ferromagnetic, the response is a very strong one, and it is non-linear, as we shall see in a moment. It is usual to rewrite Equation (10.26) in the form

$$B = \mu_R \mu_o H$$

where μ_R is called the *relative permeability*, and like the relative permittivity (the dielectric constant) of Section 10.2, it is dimensionless. The magnetisation, M, is thus

$$M = (\mu_R - 1)H = \chi H \tag{10.27}$$

where χ is the *magnetic susceptibility*. Neither μ_R or χ are constants — they depend not only on the material but also on the magnitude of the field, H, for the reason just given.

Example 10.8 Magnetic flux density

(a) A coil of length $L = 3$ cm and with $n = 100$ turns carries a current $i = 0.2$ A. What is the magnetic flux density B inside the coil?

(b) A core of silicon-iron with a relative permeability $\mu_R = 10,000$ is placed inside the coil. By how much does this raise the magnetic flux density?

Answer.

(a) The field is $H = \dfrac{ni}{L} = \dfrac{100 \times 0.2}{3 \times 10^{-2}} = 666.7$ A/m. The magnetic flux density is $B = \mu_o H = 8.38 \times 10^{-4}$ Tesla.

(b) The silicon-iron core raises the magnetic flux density to $B = \mu_R \mu_o H = 8.38$ Tesla.

Example 10.9 Magnetic flux density

A cylindrical coil with a length $L = 10$ mm with $n = 50$ turns carries a current $i = 0.01$ A. What is the field and flux density in the magnet? A ferrite core with a susceptibility $\chi = 950$ is placed in the coil. What is the flux density and what is the magnetisation of the ferrite?

Answer.
The electro-magnet produces a field

$$H = \frac{ni}{L} = \frac{50 \times 0.01}{0.01} = 50 \, \text{A/m}$$

The flux density is

$$B_o = \mu_o H = 4\pi \times 10^{-7} \times 50 = 6.3 \times 10^{-5} \, \text{Tesla}$$

The relative permeability of the ferrite is $\mu_R = \chi + 1 = 951$. When it is inserted into the core the induction increases to

$$B = \mu_R \mu_o H = 951 \times 4\pi \times 10^{-7} \times 50 = 0.06 \, \text{Tesla}$$

The magnetisation is

$$M = \frac{B}{\mu_o} - H = \frac{0.06}{4\pi \times 10^{-7}} - 50 = 4.8 \times 10^4 \, \text{A/m}$$

(This is well below the saturation magnetisation of ferrites, see later).

The magnetic pressure P_{mag} (force per unit area) exerted by an electro-magnet on a section of core material is:

$$P_{mag} = \frac{1}{2} \mu_o H^2 = \frac{1}{2} \frac{B^2}{\mu_o} \tag{10.28}$$

provided that the core does not saturate. Iron has a high saturation magnetisation H_s of about 1.8×10^6 A/m, so the maximum pressure exerted by an electro-magnet with an iron core is

$$P_{mag} = \frac{1}{2} \times 4\pi \times 10^{-7} \times \left(1.8 \times 10^6\right)^2 = 2.9 \times 10^6 \, \text{N/m}^2 = 2 \, \text{MPa}$$

The energy density stored in a magnetic field per unit volume, U/V, has the same form as the magnetic pressure:

$$\frac{U}{V} = \frac{1}{2} \mu_o H^2 = \frac{1}{2} \frac{B^2}{\mu_o} \, \text{J/m}^3 \tag{10.29}$$

Example 10.10 Magnetic energy density

An electro-magnet generates a field of $H = 10,000$ A/m. What is the magnetic energy density?

Answer.   The magnetic energy density is:

$$\frac{U}{V} = \frac{1}{2}\mu_o H^2 = \frac{1}{2}(4\pi \times 10^{-7} \times 10^8) = 62.8 \, \text{J/m}^3$$

Magnetic properties   All materials respond to a magnetic field by becoming magnetised, but most are paramagnetic with a response so faint that it is of no practical use. A few, however, contain atoms that have large dipole moments and the ability to spontaneously magnetise — that is, to align their dipoles in parallel. The magnetic behaviour of a material is characterised by its $M - H$ curve (Figure 10.14). If an increasing field H is applied to a previously demagnetised sample, starting at A on the figure, its magnetisation increases, slowly at first and then faster, following the broken line, until it finally tails off to a maximum, the *saturation magnetisation* M_s at point B. If the field H is now backed off, M does not retrace its original path; some magnetisation is retained when H reaches zero at point C. The residual magnetisation is called the *remanent magnetisation* or *remanence*, M_r, and is usually only a little less than M_s. To decrease M further, we must increase the field in the opposite direction until M finally passes through zero at point D when the field is $-H_c$, the *coercive field*, a measure of the resistance to demagnetisation. Beyond point D the magnetisation M starts to increase in the opposite direction, eventually reaching saturation again at point E. If the field is now decreased again M follows the curve through F and G back to full forward magnetic saturation again at B to form a closed $M - H$ circuit called a *hysteresis loop*.

Magnetic materials differ greatly in the shape and area of their hysteresis loop, the greatest difference being that between *soft magnets*, which have thin loops, and *hard magnets*, which have fat ones. Each full cycle of the hysteresis loop dissipates an energy per unit volume equal to the area of the loop multiplied by μ_o, the permeability of vacuum. This energy appears as heat; it is like magnetic friction. Many texts plot the curve of inductance B against H, rather than the $M - H$ curve. Equation (10.26) says that B is proportional to $(M + H)$. Since the value of M for any magnetic materials worthy of the name is very much larger than the H applied to it, $B \approx \mu_o M$ and the $B - H$ curve of a ferromagnetic material looks very like its $M - H$ curve (it's just that the M axis has been scaled by μ_o).

The remanence–coercive field chart   The differences between families of soft and hard magnetic materials are illustrated via the chart in Figure 10.14. The axes are remanent magnetisation M_r and coercive field H_c. The saturation magnetisation M_s, more relevant for soft magnets, is only slightly larger than M_r, so an $M_s - H_c$ chart looks almost the same as this one. There are 12 distinct families of magnetic materials, each enclosed in a coloured envelope. Soft magnets require high M_s and low H_c; they are the ones on the left, with the best near the top. Hard magnets must hold their magnetism, requiring a high H_c, and to be powerful they need a large M_r; they are the ones on the right, with the best again at the top. The electrical resistivity is also important because of eddy-current losses. Changing magnetic fields induce eddy currents in conductors but not in insulators. The orange and blue envelopes enclose metallic ferromagnetic materials, which are good electrical conductors. The green ones describe ferrites, which are electrical insulators.

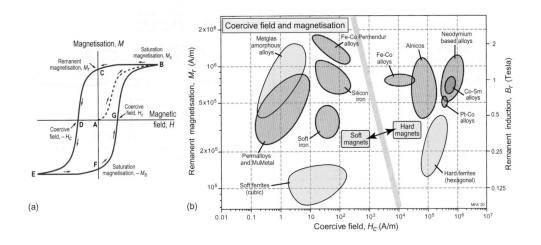

Figure 10.14 (a) A magnetic hysteresis curve. (b) Remanent magnetisation and coercive field. Soft magnetic materials lie on the left, hard magnetic materials on the right. The orange and blue envelopes enclose electrically conducting materials, the green ones enclose insulators.

Example 10.11 Selecting materials for permanent magnets

A material is required for a powerful permanent magnet that must be as small as possible and be resistant to demagnetisation by stray fields. Use the chart of Figure 10.14 to identify your choice.

Answer. The requirement that the magnet be small and powerful implies a high remanent magnetisation M_r. The need to resist demagnetisation implies a high coercive field H_c. The choice, read from the chart, is the neodymium-based family of hard magnetic materials. (These are the magnets of choice for hybrid and electric car motors and for wind turbine generators).

Temperature dependence of magnetic properties Magnetisation decreases with increasing temperature, disappearing completely at the *Curie temperature* T_c [Figure 10.15(a)]. Its value for the materials we shall meet here is well above room temperature, typically 300° to 1000°C [Figure 10.15(b)], but making magnets for use at really high temperatures is a problem.

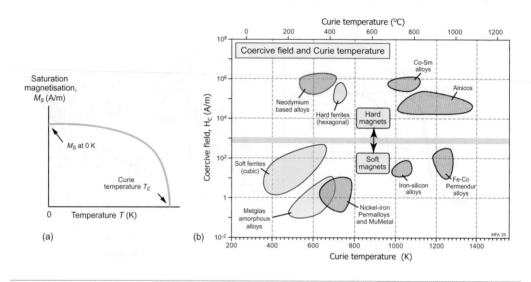

Figure 10.15   (a) Saturation magnetisation decreases with increasing temperature. (b) Coercive field and Curie temperature. The orange and blue envelopes enclose electrically conducting materials, the green ones enclose insulators.

Example 10.12 Selecting materials for high temperatures and high fields

A permanent magnet is needed for an aerospace application. In use the magnet may be exposed to demagnetising fields as high as 3×10^5 A/m (0.38 Tesla) and temperatures of 600°C. What material would you recommend for the magnet?

Answer. The coercive field must exceed the limit of 3×10^5 A/m. Figure 10.15 shows that this fact rules out Alnicos and most hard ferrites. The Curie temperature must comfortably exceeds the operating temperature of 600°C. The figure shows that neodymium-boron and hard ferrites cannot meet this requirement. This leaves the cobalt-samarium group of hard magnets, which comfortably meet both requirements.

10.7 Drilling down: the origins of magnetic properties

The classical picture of an atom is that of a nucleus around which revolve electrons (Figure 10.16). An electric current flowing in a loop creates a magnetic dipole. That is not all. Each electron has spin, creating an additional dipole moment (its *spin moment*) – and this turns out to be large. The total moment of the atom is the sum of the two.

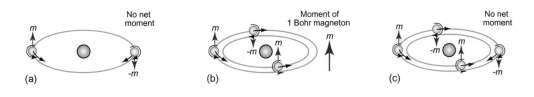

Figure 10.16 Orbital and electron spins create a magnetic dipole. Even numbers of electrons filling energy levels in pairs have moments that cancel, as in (a) and (c). An unpaired electron gives the atom a permanent magnetic moment, as in (b).

A simple atom like that of helium has two electrons per orbital, and they configure themselves such that the moment of one exactly cancels the moment of the other [Figure 10.16(a)], leaving no net moment. The same goes for Figure 10.16(c), but now think of an atom with three electrons as in Figure 10.16(b). The moments of two may cancel, but there remains the third, leaving the atom with a net moment represented by the red arrow at the right. One unpaired electron gives a magnetic moment of 9.27×10^{-24} A.m^2, called a Bohr[11] magneton; two unpaired electrons give two Bohr magnetons, three give three, and so on.

Think now of the magnetic atoms assembled into a crystal. In most materials the atomic moments interact so weakly that thermal energy is enough to randomise their directions [Figure 10.17(a)], and the structure as a whole has no magnetic moment; these materials are *paramagnetic*. In a few materials, though, something quite different happens. The fields of neighbouring atoms interact such that their energy is reduced if their magnetic moments line up. This drop in energy is called the *exchange energy*, and it is strong enough that it beats the randomising effect of thermal energy so long as the temperature is not too high. If they line up anti-parallel, head to tail, so to speak [Figure 10.17(b)], there is still no net magnetic moment; such materials are called *anti-ferromagnets*. But in a few elements, notably iron, cobalt, and nickel, the opposite happens: the moments spontaneously align so that – if all are parallel – the structure has a net moment that is the sum of those of all the atoms it contains [Figure 10.17(c)]. These materials are *ferromagnetic*. Iron has three unpaired electrons per

---

[11] Niels Henrik David Bohr (1885–1962), Danish theoretical physicist, elucidator of the structure of the atom, contributor to the Manhattan Project, and campaigner for peace.

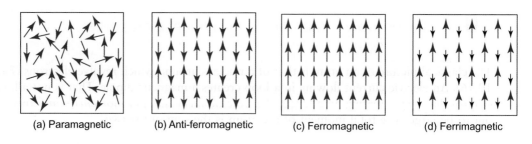

| (a) Paramagnetic | (b) Anti-ferromagnetic | (c) Ferromagnetic | (d) Ferrimagnetic |

Figure 10.17 Types of magnetic behaviour.

atom, cobalt has two, and nickel just one, so the net moment if all the spins are aligned is greatest for iron, less for cobalt, and still less for nickel.

Compounds give a fourth possibility. The materials we have referred to as ferrites are oxides; one class of them has the formula MFe_2O_4, where M is also a magnetic atom, such as cobalt, Co. Both the Fe and the Co atoms have dipoles, but they differ in strength. They line up in the anti-parallel configuration, but because the moments differ, the cancellation is incomplete, leaving a net moment M; these are *ferrimagnets*, ferrites for short [Figure 10.17(d)]. The partial cancellation and the smaller number of magnetic atoms per unit volume means they have lower saturation magnetisation than, say, iron. But as they are oxides they have other advantages, notably that they are electrical insulators.

Domains If atomic moments line up, shouldn't every piece of iron, nickel, or cobalt be a permanent magnet? Although they are magnetic materials, they do not necessarily have a magnetic moment. Why not?

A uniformly magnetised rod creates a magnetic field, H, like that of a solenoid. The field has an energy

$$U = \frac{1}{2}\mu_o \int_V H^2 \, dV \tag{10.30}$$

where the integral is carried out over the volume V within which the field exists. Working out this integral is not simple, but we don't need to do it. All we need to note is that the smaller H is and the smaller the volume V that it invades, the smaller the energy. If the structure can arrange its moments to minimise its overall H or get rid of it entirely (remembering that the exchange energy wants neighbouring atom-moments to stay parallel), then it will try to do so. Figure 10.18 illustrates how this can be done. The material splits up into *domains* within which the atomic moments are parallel, but with a switch of direction between mating domains to minimise the external field. The domains meet at *domain walls*, regions a few atoms thick in which the moments swing from the orientation of one domain to that of the other. Splitting into parallel domains of opposite magnetisation, as in Figure 10.14(b) and (c), reduces the field substantially; adding end regions magnetised perpendicular to both, as in (d), kills it almost completely. The result is that most magnetic materials, unless manipulated in some way, adopt a domain structure with minimal external field, which is the same as saying that, while magnetic, they are not magnets.

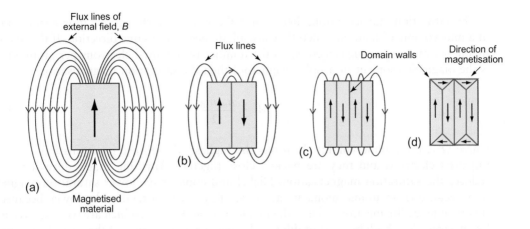

Figure 10.18 Domains cancel the external field while retaining magnetisation of the material itself.

How can they be magnetised? Figure 10.19, left, shows the starting domain-wall structure. Placed in a magnetic field, created, say, with a solenoid, the domains already aligned with the field have lower energy than those aligned against it. The domain wall separating them feels a force, the *Lorentz*[12] *force*, pushing it in a direction to make the favourably oriented domains grow at the expense of the others. As they grow, the magnetisation of the material increases, moving up the $M - H$ curve, finally saturating at M_s when the whole sample is one domain, oriented parallel to the field, as in Figure 10.19, right.

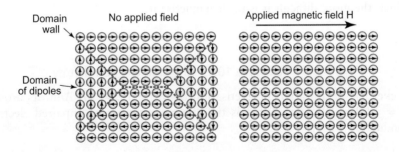

Figure 10.19 An applied field causes domain boundaries to move. At saturation the sample has become a single domain, as on the right.

---

[12] Hendrik Antoon Lorentz (1853–1928), Dutch mathematical physicist, friend and collaborator with Rayleigh and Einstein, originator of key concepts in optics, electricity, relativity, and hydraulics (he modelled and predicted the movement of water caused by the building of the Zuyderzee).

The saturation magnetisation, then, is just the sum of all the atomic moments contained in a unit volume of material when they are all aligned in the same direction. If the magnetic dipole per atom is $n_m m_B$ (where n_m is the number of unpaired electrons per atom, and m_B is the magnetic moment of a Bohr magneton) then the saturation magnetisation is

$$M_s = \frac{n_m\, m_B}{\Omega} \tag{10.31}$$

where Ω is the atomic volume. Iron has the highest M_s because its atoms have the most unpaired electrons and they are packed close together. Anything that reduces either one reduces the saturation magnetisation. Nickel and cobalt have lower saturation magnetisation because their atomic moments are less; alloys of iron tend to be lower because the non-magnetic alloying atoms dilute the magnetic iron. Ferrites, which are oxides, have much lower saturation, both because of dilution by oxygen and because of the partial cancellation of atomic moments sketched in Figure 10.17(d).

There are good reasons for alloying even if it dilutes the dipoles. A good permanent magnet retains its magnetisation even when placed in a reverse field. That's achieved by pinning the domain walls to stop them from moving. Impurities, precipitates, and porosity all interact with domain walls, tending to pin them in place. They act as obstacles to dislocation motion, too (see Chapter 5), so magnetically hard materials are mechanically hard as well. There are also subtler barriers to domain-wall motion. One is *magnetic anisotropy*, arising because certain directions of magnetisation in the crystal are more favourable than others. Another relates to the shape and size of the sample itself: small, single-domain magnetic particles are particularly resistant to having their magnetism reversed, which is why they can be used for information storage.

The performance of a hard magnet is measured by its *energy product*, roughly proportional to the product $B_r H_c$, essentially the axes of Figure 10.14. The higher the energy product, the more difficult it is to demagnetise it.

Example 10.13 Unpaired electron density

Nickel has a saturation magnetisation of $M_s = 5.2 \times 10^5$ A/m. Its atomic volume is $\Omega = 1.09 \times 10^{-29}$ m³. What is the effective number of unpaired electrons n_m in its conduction band?

Answer. The saturation magnetisation is given by $M_s = \dfrac{n_m\, m_B}{\Omega}$. The value of the Bohr magneton is $m_B = 9.27 \times 10^{-24}$ A.m². The resulting value is $n_m = 0.61$.

Example 10.14 Calculating saturation magnetisation

Iron has a density ρ of 7870 kg/m^3, and an atomic weight A_{wt} of 55.85 kg/kmol. The net magnetic moment of an iron atom, in Bohr magnetons per atom, is $n_m = 2.2$. What is the saturation magnetisation M_s of iron?

Answer. The number of atoms per unit volume in iron is

$$n = \frac{1}{\Omega} = \frac{N_A \rho}{A_{wt}} = 8.49 \times 10^{28} / m^3$$

(here N_A is Avogadro's number, 6.022×10^{26} atoms/kmol). The saturation magnetisation M_s is the net magnetic moment per atom $n_m\, m_B$ times the number of atoms per unit volume, n:

$$M_s = n\, n_m\, m_B$$

where $m_B = 9.27 \times 10^{-24}$ A.m^2 is the magnetic moment of one Bohr magneton. Thus the saturation magnetisation of iron is

$$M_s = 8.49 \times 10^{28} \times 2.2 \times 9.27 \times 10^{-24} = 1.73 \times 10^6 \text{ A/m}$$

10.8 Design: using magnetic properties

The key selection issues in choosing magnetic materials are as follows:

- Whether a soft or hard magnetic response is required
- What frequency of alternating field the material must carry
- Whether the material must operate above room temperature or in a large external field
- The size of the magnetically functional part of a machine or device

Soft magnetic devices Electro-magnets, transformers, and electron lenses have magnetic cores that must magnetise easily when the field switches on, be capable of creating a high flux density, yet lose their magnetism when the field is switched off. They do this by using soft, low H_c, magnetic materials, shown on the left of Figure 10.14. Soft magnetic materials are used to 'conduct' and focus magnetic flux. The magnetic circuit conducts the flux from one coil to another, or from a coil to an air-gap, where it is further enhanced by reducing the section of the core [Figure 10.20(a)]. Permendur (Fe-Co-V alloys) and Metglas amorphous alloys are particularly good, but being expensive they are used only in small devices. Silicon-iron (Fe 1–4% Si) is much cheaper and easier to make in large sections; it is the staple material for large transformers.

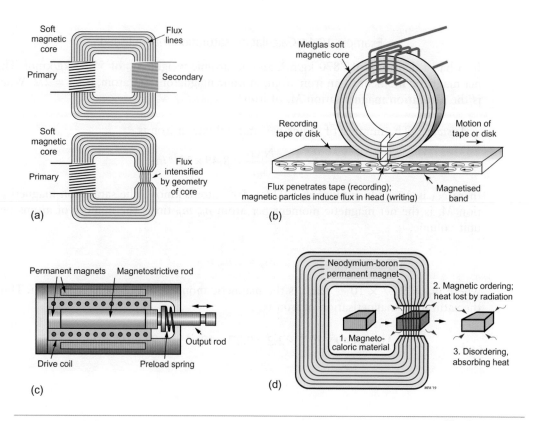

Figure 10.20 Applications of magnetic materials.

Most soft magnets are used in AC devices, where energy loss becomes a consideration. Energy loss is of two types: hysteresis loss, proportional to switching frequency f, and eddy-current loss, caused by the currents induced in the core itself by the AC field, proportional to f^2. In cores of transformers operating at low frequencies (meaning f up to 100 Hz or so) the hysteresis loss dominates, so for these we seek high M_s and a thin loop with a small area. Eddy-current losses are present but are suppressed when necessary by *laminating*: making the core from a stack of thin sheets interleaved with thin insulating layers that interrupt the eddy currents. At audio frequencies ($f < 20$ kHz) losses of both types become greater, requiring an even narrower loop and thinner laminates; here the more expensive Permalloys and Permendurs are used. At higher frequencies still (f in the MHz range) eddy-current loss dominates, and the only way to stop it is to use magnetic materials that are electrical insulators. The best choice is ferrites, shown in green envelopes on the charts, even though they have a lower M_s. Above this ($f > 100$ MHz) only the most exotic ceramic magnets will work. Table 10.1 lists the choices.

Table 10.1   Materials for soft magnetic applications

Application	Frequency f	Material requirements	Material choice
Electro-magnets	< 1 Hz	High M_s, high χ	Silicon iron Fe–Co alloys (Permendur)
Motors, low-frequency transformers	< 100 Hz	High M_s, high χ	Silicon-iron AMA (amorphous Ni–Fe–P alloys)
Audio amplifiers, speakers, microphones	< 100 kHz	High M_s, very low H_c, and high χ	Ni–Fe (Permalloy, Mumetal)
Microwave and UHF applications	< 1 MHz	High M_s, very low H_c, electrical insulator	Cubic ferrites: MFe_2O_4 with M = Cu/Mn/Ni
Gigahertz devices	> 100 MHz	Ultra-low hysteresis, excellent insulator	Garnets

Example 10.15 Selecting soft magnetic materials

A magnetic material is needed to act as the driver for an acoustic frequency fatigue-testing machine. Which classes of magnetic materials would be good choices?

Answer.   The application requires high saturation magnetisation, M_s, to provide the loads required for fatigue testing. It also requires low hysteresis loss at acoustic frequencies to minimise power consumption and avoid heating. The table indicates that the Ni-Fe Permalloys or Mumetals would be good choices.

Hard magnetic devices   Many devices use permanent magnets: magnetic clutches, loud-speakers and earphones, DC motors (as in hybrid and electric cars), and power generation (as in wind turbines). For these, the key property is the remanence, M_r. High M_r, however, is not all. A permanent magnet must resist demagnetisation by its own or other fields. A measure of its resistance is its coercive field, H_c, and this too must be large.

Hard magnetic materials lie on the right of the $M_r - H_c$ chart of Figure 10.14 and the upper half of Figure 10.15. The workhorses of the permanent magnet business are the Alnicos and the hexagonal ferrites. An Alnico (there are many) is an alloy of aluminium, nickel, cobalt, and iron with values of M_r as high as 10^6 A/m and a large coercive field: 5×10^4 A/m. The hexagonal ferrites, too, have high M_r and H_c and, like the cubic soft-magnet ferrites, they are insulators.

The charts of Figures 10.14 and 10.15 show three families of hard magnets with coercive fields that surpass those of Alnicos and hexagonal ferrites: the cobalt-samarium family based on the inter-metallic compound $SmCo_5$, and the neodymium-iron-boron family. Table 10.2 lists the choices.

Table 10.2   Materials for hard magnetic applications

Application	Material requirements	Material choice
Large permanent magnets: electric propulsion, wind turbine generators	High M_r, high H_c, and high-energy product $(MH)_{max}$; low cost	Alnicos, hexagonal ferrites
Small high-performance magnets: acoustic, electronic, and miniature mechanical applications	High M_r, high H_c, and high-energy product $(MH)_{max}$	Cobalt-samarium, Neodymium-iron-boron
Information storage	Thin, 'square' hysteresis loop	Elongated particles of Fe_2O_3, Cr_2O_3, or hexagonal ferrite

Flexible magnets   Flexible magnets are made by mixing ferrite powders into an elastomeric resin. They are relatively cheap, and because of this they are used in signs, promotional magnets, and door seals. Typical of these is the range of magnets with strontium ferrous oxide particles *embedded* in styrene butadiene rubber (SBR) or vinyl (flexible PVC). Table 10.3 lists typical magnetic properties.

Table 10.3   Properties of flexible magnets at room temperature

Property	Value range
Remanent induction B_r	0.15–0.25 T
Coercive force H_c	1×10^5–2×10^5 A/m
Maximum energy product	4.5–15 kJ/m^3

Example 10.16 Using flexible magnets

A flexible magnet with a remanent induction of 0.2 T is used as a seal on a refrigerator door. If the area of contact A of the magnet with the steel frame of the fridge is 0.005 m^2, what clamping force does it exert?

Answer.   From Equation (10.28) the force exerted by the flexible magnet is

$$F = P_{mag} A = \frac{1}{2} A \frac{B^2}{\mu_0} = \frac{1}{2} \times 0.005 \times \frac{0.2^2}{4\pi \times 10^{-7}} = 80 \text{ N}$$

Magnetic information storage   Magnetic information storage requires hard magnets, but those with an unusual loop shape: they are rectangular (called 'square'). The squareness gives binary storage: a unit of the material can flip in magnetisation from $+M_r$ to $-M_r$ and back when exposed to fields above $+H_c$ or $-H_c$. The information is read by moving the magnetised unit past a read-head where it induces an electric pulse. The choice of the word 'unit' was deliberately vague because it can take several forms: discrete particles embedded in a polymer tape or disk, or as a thin magnetic film that has the ability to carry small stable domains called 'bubbles' that are magnetised in one direction, embedded in a background that is magnetised in another direction.

Figure 10.20(b) shows one sort of information storage system. A signal creates a magnetic flux in a soft magnetic core, which is often an amorphous metal alloy (AMA, the blue envelopes on the charts) because of its large susceptibility. The direction of the flux depends on the direction of the signal-current; switching this reverses the direction of the flux. An air gap in the core allows field to escape, penetrating a tape or disk and magnetising the sub-micron particles of a hard magnetic material embedded in it. When the signal is in one direction, all particles in the band of tape or disk passing under the write-head are magnetised in one direction; when the signal is reversed, the direction of magnetisation is reversed. The same head is used for reading.

Magnetostriction   The electrical hum of transformers is due to magnetostriction of the core. Magnetostriction is shape change caused by a magnetic field. The saturation magnetostriction is the maximum strain that can be induced in a material by applying a magnetic field. For most materials this is small; the exceptions are the 'giant magnetostrictive' materials: Terfenol-D (an alloy of terbium, dysprosium, and iron) and Galfenol (an alloy of gallium and iron). This ability to convert magnetic energy into kinetic energy and vice versa allows magnetic actuation and sensing [Figure 10.20(c)] with nanometre resolution and millisecond response times. The effect is used in fuel injectors, atomic-force microscopes, microcircuit fabrication, vibration damping, and sonar detection.

Magnetocaloric materials   When ice melts, entropy increases and heat is absorbed (the latent heat of melting). When a magnetic material is placed into a magnetic field, its magnetic moments tend to align with the applied field, decreasing entropy and releasing heat. Removing the magnetic field allows disordering, increasing the magnetic entropy and absorbing heat [Figure 10.20(d)]. Magnetic refrigeration has the potential for conventional refrigeration and air conditioning, but expense is at present an obstacle.

10.9 Magnetic properties: summary and conclusions

The classical picture of an atom is that of a nucleus around which revolve electrons in discrete orbitals, each electron having a quantum mechanical character called spin. Moving electrons with spin create magnetic moments that, if parallel, add up, but if opposed, cancel to a greater or lesser degree. Most materials achieve near-perfect cancellation either within the atomic orbitals or, if not, by stacking the atomic moments head to tail or randomising them so that, when added, they cancel. A very few — most of them based on iron, cobalt, or nickel — have atoms with residual moments and an inter-atomic interaction that causes them to line up to give a net magnetic moment or magnetisation. Even these materials can find a way to screen their magnetisation by segmenting themselves into domains, each with a magnetisation that is oriented such that it tends to cancel that of its neighbours. A strong magnetic field can override the segregation, creating a single unified domain in which all the atomic moments are parallel, and if the coercive field is large enough, they remain parallel even when the driving field is removed, giving a 'permanent' magnetisation.

Aligning and randomising atomic magnetic moments gives other functionality. The microscopic shape-change induced by magnetisation (magnetostriction) allows ultra-precise actuation. The entropy change of randomising aligned moments allows solid-state cooling devices.

10.10 Optical materials and properties

It was at one time thought that the fact that light could travel through space – from the Sun to Earth, for instance – must mean that space was not really empty but filled with 'luminiferous ether.' It was not until the experiments of Michelson[13] and Morley in 1887 that it was realised light did not need a 'material' for its propagation but could propagate through totally empty space at what is now seen as the ultimate velocity: 3×10^8 m/s.

Light is electromagnetic radiation. Electromagnetic (e-m) radiation permeates the entire universe. Observe the sky with your eye and you see the visible spectrum, the range of wavelengths we call 'light' (0.40–0.77 μm). Observe it with a detector of X-rays or γ-rays and you see radiation with far shorter wavelengths (as short as 10^{-4} nm, one thousandth the size of an atom). Observe it instead with a radio telescope and you pick up radiation with wavelengths measured in millimetres, metres, or even kilometres, known as radiowaves and microwaves. The range of wavelengths of radiation is vast, spanning 18 orders of magnitude (Figure 10.21).

The intensity I of an e-m wave, proportional to the square of its amplitude, is a measure of the energy it carries. When radiation with intensity I_o strikes a material, a part I_R of it is reflected, a part I_A absorbed, and a part I_T may be transmitted. Conservation of energy requires that

$$\frac{I_R}{I_o} + \frac{I_A}{I_o} + \frac{I_T}{I_o} = 1 \tag{10.32}$$

The first term is called the *reflectivity* of the material, the second the *absorptivity*, and the last the *transmittability* (all dimensionless). Each depends on the wavelength of the radiation, the nature of the material, the state of its surfaces, and the angle of incidence.

Materials that reflect or absorb all visible light, transmitting none, are called *opaque* even though they may transmit in the near visible (IR or ultra-violet [UV]). Those that transmit a little diffuse light are called *translucent*. Those that transmit light sufficiently well that you can see through them are called *transparent*; and a subset of these that transmit almost

---

[13] Albert A. Michelson (1852–1937), Prussian-American experimental physicist, who, with E. W. Morley, first demonstrated that the speed of light is independent of Earth's motion, a finding central to the establishment of the theory of relativity.

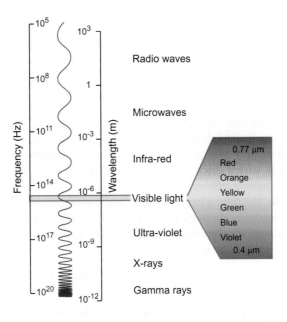

Figure 10.21 The spectrum of electromagnetic (e-m) waves.

perfectly, making them suitable for lenses, light-guides, and optical fibres, are given the additional title of *optical quality*.

Specular and diffuse reflection   *Specular* surfaces are microscopically smooth and flat, meaning that any irregularities are much smaller than the wavelength of light. A beam of light striking a specular surface at an incident angle θ_1 undergoes specular reflection, which means that the angle of reflection is equal to the angle of incidence [Figure 10.22(a)]. *Diffuse* surfaces are irregular; the law of reflection still holds locally, but the incident beam is reflected in many different directions because of the irregularities.

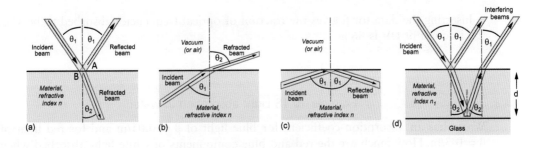

Figure 10.22   (a) Reflection and refraction. (b) Refraction. (c) Total internal reflection. (d) Waves reflected from the front and back surface of thin films interfere.

Absorption   If radiation penetrates a material, some is absorbed. The greater the thickness x through which the radiation passes, the greater the absorption. The intensity I, starting with the initial value I_o, decreases such that

$$I = I_o \exp - \beta x \qquad (10.33)$$

where β is the absorption coefficient, with dimensions of m^{-1} (or, more conveniently, mm^{-1}). The absorption coefficient depends on wavelength with the result that white light passing through a material may emerge with a colour corresponding to the wavelength that is least absorbed – that is why a thick slab of ice looks blue.

Transmission   By the time a beam of light has passed completely through a slab of material, it has lost some intensity through reflection at the surface at which it entered, some in reflection at the surface at which it leaves and some by absorption in between. Its intensity when it emerges at the other side is

$$I_T = I_o \left(1 - \frac{I_R}{I_o} \right)^2 \exp - \beta x \qquad (10.34)$$

The term $(1 - I_R/I_o)$ occurs to the second power because intensity is lost through reflection at both surfaces.

Example 10.17 Light absorption in materials

The polymer EVA has an absorption coefficient for light in the optical range of $\beta = 5/m$. Its absorption coefficient for UV light is $\beta = 1000/m$. If a window is covered with a 2 mm thick sheet of EVA, what fraction of optical and UV light will be absorbed in it?

Answer.   The fraction of light intensity that is absorbed in passing through a thickness x of a material is

$$\frac{I_{absorbed}}{I_o} = 1 - \exp - \beta x$$

Inserting the data for β gives the fraction of optical frequencies absorbed to be 1%; the fraction of UV is 86%.

Example 10.18 Light absorption in water

Water has an absorption coefficient for blue light of $\beta = 0.01/m$ and for red light of $\beta = 0.3/m$. How much are the red and blue components of white light absorbed when a beam of white light passes through 10 m of water? What is the consequent colour of the light perceived by an observer 10 m from the source?

Answer.   The transmitted fraction of the blue component is $\dfrac{I}{I_o} = \exp - \beta x = 0.9$; thus 10% is absorbed. The transmitted fraction of the red component is $\dfrac{I}{I_o} = \exp - \beta x = 0.05$; thus 95% is absorbed. The water, in consequence, appears blue.

Refraction   The velocity of light in vacuum, $c_o = 3 \times 10^8$ m/s, is as fast as it ever goes. When it (or any other electromagnetic radiation) enters a material, it slows down. The *index of refraction*, n, is the ratio of its velocity in vacuum, c_o, to that in the material, c:

$$n = \frac{c_o}{c} \tag{10.35}$$

This retardation makes a beam of light bend or *refract* when it enters a material of different refractive index. When a beam with an angle of incidence θ_1 passes from a material 1 of refractive index n_1 into a material 2 of index n_2, it deflects to an angle θ_2, such that

$$\frac{\sin \theta_1}{\sin \theta_2} = \frac{n_2}{n_1} \tag{10.36}$$

as in Figure 10.22(a); the equation is known as Snell's[14] law. The refractive index depends on wavelength, so each of the colours that make up white light is diffracted through a slightly different angle, producing a spectrum when light passes through a prism. When material 1 is vacuum or air, for which $n_1 = 1$, the equation reduces to

$$\frac{\sin \theta_1}{\sin \theta_2} = n_2 \tag{10.37}$$

Equation (10.37) says that when light passes from a material with index $n_1 = n$ into air with $n_2 = 1$, it is bent away from the normal to the surface, like that in Figure 10.22(b). If the incident angle, here θ_1, is slowly increased, the emerging beam tips down until it becomes parallel with the surface when $\theta_2 = 90°$ and $\sin \theta_2 = 1$. This occurs at an incident angle, from Equation (10.36), of

$$\sin \theta_1 = \frac{1}{n} \tag{10.38}$$

For values of θ_1 greater than this, the equation predicts values of $\sin \theta_2$ that are greater than 1 — and that can't be. The ray is totally reflected back into the material [Figure 10.22(c)].

---

[14] Willebrord Snell (1591–1626), Dutch astronomer, also known as Snell van Royen, or Snellius, who first derived the relationship between the different angles of light as it passes from one transparent medium to another. The lunar crater Snellius is named after him.

This *total internal reflection* has many uses, one being to trap light within an optical fibre, an application that has changed the way we communicate.

Refactive index values mostly fall in the range 1.3 to 1.6, as seen in Figure 10.23, showing materials that are transparent to visible light.

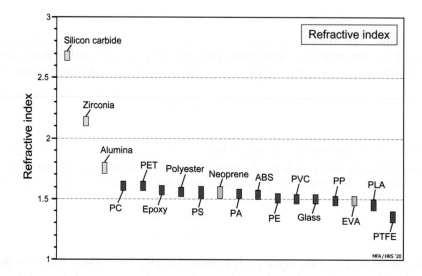

Figure 10.23   Refractive index of selected ceramics, glass, and polymers, all transparent to visible light.

Example 10.19 Refraction

A material has a refractive index of $n = 1.3$. At what internal incident angle is a beam of light first totally reflected?

Answer.   The angle of internal reflection is $\theta_{crit} = \sin^{-1}\left(\dfrac{1}{n}\right) = 50°$.

Interference   The rainbow sheen of a compact disc when viewed obliquely, the iridescence of soap bubbles and oil films, even the colours of peacock feathers and butterfly wings are caused by interference. When a transparent layer coats a surface, the impinging light is both reflected from it and transmitted through. If the transmitted portion is reflected at the back surface of the film, it returns to the front surface and interferes with the incoming light [Figure 10.22(d)]. The nature of the interference depends on the difference in the *optical path length* between the incoming and the reflected light. Optical path length L_{opt} is defined by

$$L_{opt} = \sum_i n_i d_i \qquad (10.39)$$

where n_i is the refractive index of the ith material along the path, and d_i is the physical length of the segment of path. The difference in optical path length between two rays determines whether the interference is constructive or destructive. If this difference is ΔL_{opt}, the phase difference between the two rays is

$$\Delta\phi = \frac{2\pi\,\Delta L_{opt}}{\lambda}$$

The two waves are in phase with

$$\frac{\Delta L_{opt}}{\lambda} = 0, \pm 1, \pm 2,\ldots \qquad (10.40\mathrm{a})$$

and they are half a cycle out of phase when

$$\frac{\Delta L_{opt}}{\lambda} = \pm\,\frac{1}{2}, \pm\,\frac{3}{2}, \pm\,\frac{5}{2},\ldots \qquad (10.40\mathrm{b})$$

Interference is constructive when the waves are in phase, destructive interference when they are half a cycle out of phase. If the two interfering rays have the same amplitude, constructive interference doubles it, destructive interference kills it completely. All this only works if the differences in optical path length are small. If they are more than a few wavelengths the phase relationship between the rays is lost.

Example 10.20 Interference

The track separation on a DVD is $d = 0.74$ micron. Suppose white light falls on a DVD at right angles to its surface. The DVD is viewed from an angle $\theta = 45°$. Which wavelength of the white light will be reinforced most strongly in the direction of viewing?

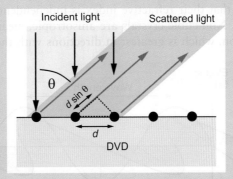

Answer. The illumination is a plane wave normal to the surface so that the incoming rays are in phase. Each line of dots on the DVD scatters light – think of it as a light source. Then the difference in optical path for rays scattered from adjacent lines is

$$\Delta L_{\text{opt}} = n_o\, d \sin\theta$$

where n_o is the refractive index of air, which we take to be 1. Reinforcement (constructive interference) occurs when this is an integral number of wavelengths, that is when

$$d \sin\theta = i\,\lambda$$

where i is an integer. Inserting $d = 0.74$ micron, $\theta = 45°$ and $i = 1$ gives

$$\lambda = 0.52 \text{ micron}$$

This corresponds to a vivid green colour. Any larger value of i gives wavelengths that are outside the visible spectrum.

10.11 Drilling down: the origins of optical properties

Light, like all radiation, is an electromagnetic (e-m) wave. The coupled electric and magnetic fields are sketched in Figure 10.24. The electric part fluctuates with a frequency ν that determines where it lies in the spectrum of Figure 10.21. A fluctuating electric field induces a fluctuating magnetic field that is exactly $\pi/2$ out of phase with the electric one because the induction is at its maximum when the electric field is changing most rapidly. A plane-polarised beam looks like this one: the electric and magnetic fields lie in fixed planes. Natural light is not polarised; the wave also rotates so that the plane containing each wave continuously changes.

Light slows down when it enters a dielectric because the electric field polarises the ions or molecules in the way that was illustrated in Figure 10.8. It is this that gives the dielectric a refractive index greater than 1. Dielectrics with an amorphous or cubic structure are optically isotropic, meaning that their refractive index is independent of the direction that the light enters the material. Non-cubic crystals are anisotropic, meaning that their refractive index depends on direction, which is greatest in directions with the greatest ion-density.

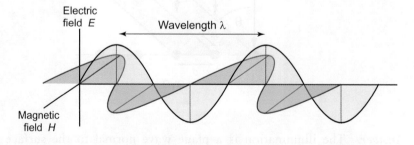

Figure 10.24 An electromagnetic wave. The electric component *(yellow)* is $\pi/2$ out of phase with the magnetic component *(blue)*.

Many aspects of radiation are most easily understood by thinking of it as a wave. Other phenomena need a different picture – that of radiation as discrete packets of energy, or *photons*. The energy E_{ph} of a photon of radiation of frequency ν or wavelength λ is

$$E_{ph} = h\nu = \frac{hc}{\lambda} \tag{10.41}$$

where h is Planck's[15] constant (6.626×10^{-34} J.s) and c is the speed of the radiation. Thus radiation of a given frequency has photons of fixed energy, regardless of its intensity – an intense beam simply has more of them.

Example 10.21 Photon energy

What is the photon energy of light of wavelength $\lambda = 0.5$ μm in vacuum?

Answer. Planck's constant is $h = 6.626 \times 10^{-34}$ J.s and the speed of light in vacuum as $c_o = 3 \times 10^8$ m/s. The photon energy is $E_{ph} = h\nu = hc/\lambda = 4.0 \times 10^{-19}$ J. It is more usual to give photon energies in electron volts (eV). 1 eV is the energy of an electron (charge 1.6×10^{-19} C) after it has fallen through a potential difference of 1 V: thus 1 eV = 1.6×10^{-19} J. The energy of a photon of light of wavelength 0.5 μm is $E_{ph} = 2.5$ eV.

Why aren't metals transparent? Metals have a large number of very closely spaced levels in their conduction band (Section 10.3). The electrons in the metal fill the bands up to the Fermi level. Radiation excites electrons, and in metals there are plenty of empty levels above the Fermi level into which they can be excited. A photon with energy $h\nu$ can excite an electron only if there is an energy level that is exactly $h\nu$ above the Fermi level – and in metals there is. So all the photons of a light beam are captured by electrons regardless of their wavelength. When an excited electron falls back to the Fermi level it releases a photon with exactly the same energy that excited it in the first place – that is, it is reflected [Figure 10.25(a)]. Many metals (e.g. silver, aluminium, and stainless steel) reflect all wavelengths almost equally well, so if exposed to white light they appear silver. Others (copper, brass, bronze) reflect some colours better and appear coloured because, in penetrating this tiny distance into the surface, some wavelengths are slightly absorbed.

How does light get through dielectrics? Non-metals interact with radiation in a different way. Part may be reflected but much enters the material, inducing both dielectric and magnetic responses. Not surprisingly, the velocity of an e-m wave depends on the

---

[15] Max Karl Ernst Ludwig Planck (1858–1947), a central figure in the development of quantum theory; it was he who formulated Equation (10.41).

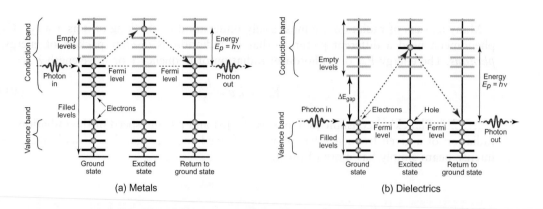

Figure 10.25 (a) The band structure of a metal. (b) The band structure of a dielectric.

dielectric and magnetic properties of the material through which it travels. In vacuum the velocity is

$$c_o = \frac{1}{\sqrt{\varepsilon_o \mu_o}} \qquad (10.42)$$

where ε_o is the electric permittivity of vacuum and μ_o its magnetic permeability, as defined in Equations (10.6) and (10.25). Within a material its velocity is

$$c = \frac{1}{\sqrt{\varepsilon \mu}} \qquad (10.43)$$

where $\varepsilon = \varepsilon_R \varepsilon_o$ and $\mu = \mu_R \mu_o$ are the permittivity and permeability of the material. The refractive index is

$$n = \frac{c_o}{c} = \sqrt{\varepsilon_R \mu_R} \qquad (10.44)$$

The relative permeability μ_R of most dielectrics is very close to unity — only magnetic materials have larger values. Thus

$$n = \sqrt{\varepsilon_R} \qquad (10.45)$$

Example 10.22 Refractive index and dielectric constant

A dielectric material has a high-frequency dielectric constant of $\varepsilon_R = 3.5$. What would you estimate its refractive index to be?

Answer.   The refractive index is approximately $n \approx \sqrt{\varepsilon_R} = 1.87$.

The reason that radiation can enter a dielectric is that its Fermi level lies at the top of the valence band, just below a band gap [Figure 10.25(b)]. The conduction band, with its vast number of empty levels, lies above it. To excite an electron across the gap requires a photon with an energy at least as great as the width of the gap, ΔE_{gap}. Radiation with photon energy less than ΔE_{gap} cannot excite electrons – there are no energy states within the gap for the electron to be excited into. The radiation sees the material as transparent, offering no interaction.

Photons with energies greater than ΔE_{gap} have enough energy to pop electrons into the conduction band, leaving a 'hole' in the valence band from which they came. When they jump back, filling the hole, they emit radiation of the same wavelength that first excited them, and for these wavelengths the material is not transparent [Figure 10.25(b) again]. The critical frequency v_{crit}, above which interaction starts, is given by

$$h v_{crit} = \Delta E_{gap} \tag{10.46}$$

Thus, for example, bakelite is transparent to IR light because its frequency is too low and its photons too feeble to kick electrons across the band gap. The visible spectrum has higher frequencies, with photon-energies exceeding the band-gap energy; they are captured and reflected.

Although dielectrics cannot absorb radiation with photons of energy less than that of the band gap, they are not all transparent. Most are polycrystalline and have a refractive index that depends on direction; then light is *scattered* as it passes from one crystallite to another. Imperfections, particularly porosity, do the same (Figure 10.26). Scattering explains why some polymers are translucent or opaque: their microstructure is a mix of crystalline and amorphous regions with different refractive indices. It explains, too, why some go white when you bend them: it is because light is scattered from internal micro-cracks or crazes.

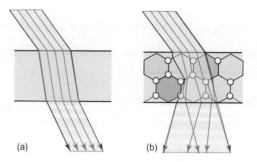

(a)   (b)

Figure 10.26 (a) A pure glass with no internal structure and a wide band-gap is completely transparent. (b) Light entering a polycrystalline ceramic or a partly crystalline polymer suffers multiple refraction and is scattered by porosity, making it translucent.

Example 10.23 Transparency

An insulator has a band gap of $\Delta E_{gap} = 1$ eV. For which wavelengths of radiation will it be transparent?

Answer. The material will be transparent to frequencies below $v_{crit} = \dfrac{\Delta E_{gap}}{h}$ where Planck's constant is $h = 6.6 \times 10^{-34}$ J.s. Thus $v_{crit} \leq 2.4 \times 10^{14}$ Hz. This corresponds to wavelengths $\lambda \geq \dfrac{c}{v_{crit}} = 1.24$ μm.

Colour If a material has a band-gap with an energy ΔE_{gap} that lies within the visible spectrum, the wavelengths with energy greater than this are absorbed and those with energy that is less are not. The absorbed radiation is re-emitted when the excited electron drops back into a lower energy state, but this may not be the one it started from, so the photon it emits has a different wavelength than the one that was originally absorbed. The light emitted from the material is a mix of the transmitted and the re-emitted wavelengths, giving it a characteristic colour.

More specific control of colour is possible by doping – the deliberate introduction of impurities that create a new energy level in the band gap (Figure 10.27). Radiation is absorbed as before, but it is now re-emitted in two discrete steps as the electrons drop first into the dopant level, emitting a photon of frequency $v_1 = \Delta E_1/h$ and from there back into the valence band, emitting a second photon of energy $v_2 = \Delta E_2/h$. Particularly pure colours are created when glasses are doped with metal ions: copper gives blue; cobalt, violet; chromium, green; manganese, yellow.

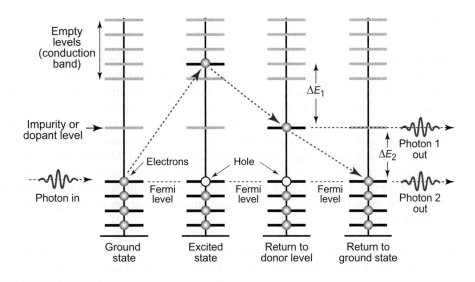

Figure 10.27 Impurities or dopants create energy levels in the band-gap. Electron transitions to and from the dopant level emits photons of specific frequency and colour.

Example 10.24 Colour

An insulator has a dopant-created energy level that lies $\Delta E = 2.1$ eV below the conduction band. If electrons are excited into the conduction band and fall back into this level, what is the wavelength of the radiation that is emitted? Is it in the optical spectrum? If so, what is its colour?

Answer. The frequency of the emitted radiation is

$$\nu = \frac{\Delta E}{h} = \frac{2.1 \times 1.6 \times 10^{-19}}{6.6 \times 10^{-34}} = 5.1 \times 10^{14}/s$$

Its wavelength is $\lambda = c_o/\nu = 0.59$ μm. Referring to Figure 10.21, this corresponds to a colour in the green part of the spectrum.

10.12 Design: using optical properties

Using reflection and refraction   Telescopes, microscopes, cameras, and car headlights all rely on the focussing of light. Reflecting telescopes and car headlights use metallised glass or plastic surfaces, ground or moulded to a concave shape; the metal, commonly, is silver because of its high reflectivity across the entire optical spectrum.

Refracting telescopes, microscopes, and cameras use lenses. Here the important property is the refractive index and its dependence on wavelength (since if this is large, different wavelengths of light are brought to focus in different planes). Most elements and compounds have a fixed refractive index. Glasses are different: their refractive index can be tuned to any value between 1.5 and 2.2 by adding light elements like sodium (as in soda-lime glass) to give a low refractive index, or heavy elements like lead (as in lead glass or 'crystal') to give a high one.

Using total internal reflection   Optical fibres have revolutionised the digital transmission of information. A single fibre consists of a core of pure glass contained in a cladding of another glass with a larger refractive index [Figure 10.28(a)]. A digitised signal is converted into optical pulses by a light-emitting diode (LED). The stream of pulses is fed into the fibre where it is contained within the core, even when the fibre is bent, because any ray striking the core-cladding interface at a low angle suffers total internal reflection. The purity of the core – a silica glass – is so high that absorption is very small; occasional repeater stations are needed to receive, amplify, and retransmit the signal to cover long distances.

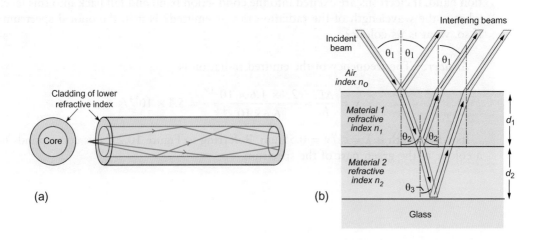

Figure 10.28   (a) An optical fibre. (b) Multiple layers allow fine tuning of the reflectance and transmittance of glass.

Using interference   The total amount of light impinging on a surface must in sum be reflected, transmitted, or absorbed [Equation (10.32)]. If the reflection is reduced, more must be transmitted. Anti-reflective coatings consist of single or multiple layers of transparent thin films with different refractive indices, chosen to produce destructive interference for reflected light and constructive interference for light that is transmitted. The effect of layering was shown in Figure 10.22(d). The light reflects twice, once from the surface to the air and once from that between the layer and the glass. If the layer thickness, d, is chosen to give destructive interference between the two reflected waves when the incident wave is normal to the surfaces, this maximises the transmitted intensity for one wavelength. Cancellation for a wider range of wavelengths and viewing angles is possible with multiple layers like that shown in Figure 10.28(b), often achieved with alternating layers of silica (SiO_2, which has a low refractive index) and titanium dioxide (TiO_2, which has an exceptionally large one).

Fluorescence, phosphorescence, and chemo-luminescence   Electrons can be excited into higher energy levels by incident photons of sufficient energy. In most materials the time delay before they drop is extremely short, but in some there is a longer delay. If, on dropping back, the photon they emit is in the visible spectrum, the effect is generally called *luminescence* – the material continues to glow even when the incident beam is removed. When the time delay is a fraction of a second, it is called *fluorescence*, used in fluorescent lighting where it is excited by UV light from a gas discharge, and in LED lighting where it is excited by the UV of light-emitting diodes. When the time delay is longer, it is called *phosphorescence*, used for static displays like watch faces, where it is excited by electrons released by a mildly radioactive ingredient in the paint. In *chemo-luminescence* the excitation derives instead from a chemical reaction, as in fireflies and the luminous fish of the deep ocean.

Photoconductivity   Dielectrics are true insulators only if there are no electrons in the conduction band. Dielectrics with a band gap that is sufficiently narrow that the photons of visible light excite electrons across it, become conducting (though with high resistance) when exposed to light. The greater the intensity of light, the greater the conductivity. Photo light meters use this effect; the meter is simply a bridge circuit measuring the resistance of a photo-conducting element such as cadmium sulphide.

10.13 Optical properties: summary and conclusions

Materials interact with electromagnetic radiation in several ways: reflecting it, refracting it, and absorbing it. This is because radiation behaves both as a wave of frequency ν and as a stream of discrete photons with energy $E_{ph} = h\nu$. Electrons in materials capture photons, grabbing their energy, provided there are energy levels exactly E_{ph} above their ground state for them to occupy. After a time delay that is usually very short, the electrons drop back down to their ground state, re-emitting their energy as new photons. In metals there are always usable energy levels so every photon hitting a metal is captured and thrown back out, giving reflection. In a dielectric, the atoms ignore low-frequency, low-energy photons and capture only those with energy above a critical level set by the band gap.

Those that are not caught pass straight through; for these frequencies, the material is transparent. Doping introduces extra energy states within the band gap for electrons to fall into as they release their energy, and in doing so they emit photons of specific frequency and colour. When radiation enters a dielectric, it slows down. A consequence of this is that a beam entering at an angle is bent – the phenomenon of refraction. It is this that allows light to be focussed by lenses, reflected by prisms, and trapped in fine, transparent fibres to transmit information. Of all the transparent materials at our disposal, glasses offer the greatest range of refractive index and colour. Glasses are based on amorphous silica (SiO_2). Silica is an extremely good solvent, allowing a wide range of other oxides to be dissolved in it over a wide range of concentrations. It is this that allows the optical properties of glasses to be adjusted and fine-tuned to match design needs.

10.14 Further reading

Electrical materials

Braithwaite, N., & Weaver, G. (1990). *Electronic materials*. The Open University and Butterworth Heinemann. ISBN 0-408-02840-8. (One of the excellent Open University texts that form part of their Materials program).

Callister, W. D., & Rethwisch, D. G. (2011). *Materials science and engineering: An introduction* (8th ed.). John Wiley. ISBN 978-0-470-50586-1. (A well-established and comprehensive introduction to the science of materials).

Jiles, D. (2001). *Introduction to the electronic properties of materials* (2nd ed.). Nelson Thornes. ISBN 0-7487-6042-3. (The author develops the electronic properties of materials, relating them to thermal and optical as well as electrical behaviour).

Solymar, L., & Walsh, D. (2004). *Electrical properties of materials* (7th ed.). Oxford University Press. ISBN 0-19-926793-6. (A mature introduction to electrical materials).

Magnetic materials

Braithwaite, N., & Weaver, G. (1990). *Electronic materials*. The Open University and Butterworth Heinemann. ISBN 0-408-02840-8. (One of the excellent Open University texts that form part of their Materials program).

Callister, W. D., & Rethwisch, D. G. (2014). *Materials science and engineering: An introduction* (9th ed.). John Wiley & Sons. ISBN 978-1-118-54689-5. (A well-established text taking a science-led approach to the presentation of materials teaching).

Campbell, P. (1994). *Permanent magnetic materials and their applications*. Cambridge University Press.

Douglas, W. D. (1995). Magnetically soft materials (9th ed.) *ASM metals handbook* (vol. 2, pp. 761–781). ASM.

Fiepke, J. W. (1995). Permanent magnet materials (9th ed.) *ASM metals handbook* (vol. 2, pp. 782–803). ASM.

Jakubovics, J. P. (1994). *Magnetism and magnetic materials* (2nd ed.). The Institute of Materials. ISBN 0-901716-54-5. (A simple introduction to magnetic materials, short and readable).

Optical materials

Callister, W. D., & Rethwisch, D. G. (2014). *Materials science and engineering: An introduction* (9th ed.). John Wiley & Sons. ISBN 978-1-118-54689-5. (A well-established text taking a science-led approach to the presentation of materials teaching).

Jiles, D. (2001). *Introduction to the electronic properties of materials*. Nelson Thompson. ISBN 0-7487-6042-3. (A refreshingly direct approach to the thermal, electrical, magnetic, and optical properties of materials and their electronic origins).

Locharoenra, K. (2016). *Optical properties of solids, an introductory textbook* (4th ed.). Taylor and Frances: CRC Press. ISBN-13 978-9814669061.

Shackelford, J. F. (2014). *Introduction to materials science for engineers* (8th ed., chap. 8). Prentice Hall. ISBN 978-0133789713. (A well-established materials text with a design slant).

10.15 Exercises

Electrical properties

Exercise E10.1 A potential difference of 4 V is applied across a tungsten wire 200 mm long and 0.1 mm in diameter. What current flows in the wire? How much power is dissipated in the wire? The electrical conductivity κ_e of tungsten is 1.8×10^7 S/m.

Exercise E10.2 A potential difference of 1 Volt is applied across a coil of silver wire 2 m long and 50 μm in diameter. What current flows in the wire? How much power is dissipated in the wire? Find the electrical resistivity of silver from Table A7 of Appendix A.

Exercise E10.3 A gold interconnect 1 mm long and with a rectangular cross section of 10 μm × 1 μm has a potential difference of 1 mV between its ends. What is the current in the interconnect? What is the current density (amps/m^2)? How much power is dissipated in it? The resistivity ρ_e of gold is 2.5×10^{-8} Ω.m.

Exercise E10.4 The gold interconnect of the previous question (length 1 mm and rectangular cross section of 10 μm × 1 μm with a potential difference of 1 mV across its length) carries power for 10 seconds. If no heat is lost by conduction, radiation, or convection, how much will the interconnect temperature rise? The resistivity ρ_e of gold is 2.5×10^{-8} Ω.m, its specific heat C_p is 130 J/kg.K, and its density ρ is 19,300 kg/m^3.

Exercise E10.5 A 0.5 mm diameter wire must carry a current $i = 10$ amps. The system containing the wire will overheat if the power dissipation, $P = i^2 R$, exceeds 5 W/m. The table lists the resistivities ρ_e of four possible candidates for the wire. Which ones meet the design requirement?

Material	ρ_e (Ω.m)
Aluminium	2.7×10^{-8}
Copper	1.8×10^{-8}
Nickel	9.6×10^{-8}
Tungsten	5.7×10^{-8}

Exercise E10.6 What is a dielectric? What is meant by polarisation? Define the dielectric constant.

Exercise E10.7 Why do some dielectrics absorb microwave radiation more than others? What property measures this ability to absorb microwave radiation?

Exercise E10.8 It is much easier to measure the electrical conductivity κ_e of a material than to measure its thermal conductivity λ. Thermal conduction in metals, like electrical conduction, occurs by the motion of electrons. The two conductivities are related by the Weidemann-Franz law

$$\lambda \approx 10^{-5} \, \kappa_e$$

with λ in W/m.K and κ_e in S/m. Use data from Table A7 of Appendix A to estimate the thermal conductivity of stainless steel. How does this compare with the thermal conductivity of stainless steel listed in Table A6 of Appendix A?

Exercise E10.9 In metals, thermal conductivity λ and electrical conductivity κ_e both depend on electron mobility. The two conductivities are related by the Weidemann-Franz law

$$\lambda \approx 10^{-5} \, \kappa_e$$

with λ in W/m.K and κ_e in S/m. Use data from Table A7 of Appendix A to estimate the thermal conductivity of titanium alloys. How does the estimate compare with the thermal conductivity of titanium alloys listed in Table A6 of Appendix A?

Exercise E10.10 The table shows electrical resistivty data for Cu, Ni, and their alloys. The notation Cu-10Ni means 10 wt% Ni (and 90% Cu). Electrical resistivity is a microstructure-sensitive property. Convert the data to electrical conductivity, then plot the values against weight % Ni. Does the data deviate from a simple linear rule of mixtures, as expected? Compare your graph with the solution to Exercise E7.8, for thermal conductivity. What do you observe?

Alloy	Electrical resistivity, ρ_e ($\times 10^{-9}$ Ω/m)
Pure Cu	19
Cu-2Ni	54
Cu-5Ni	123
Cu-10Ni	172
Cu-30Ni	348
Cu-65Ni	610
Pure Ni	85

Exercise E10.11 A power line is to carry 5 kA at 11 kV using pylons 400 m apart. The dip d, in a wire of weight w_ℓ per unit length strung between pylons L apart at a tension T is given by $d = L^2 w_\ell / 8T$. The maximum tension allowed is 0.8 of the yield stress, σ_y. If the maximum allowable dip is 6 m, which of the materials in the table could be used?

Material	Yield stress, σ_y (MPa)	Density, ρ (kg/m^3)
Aluminium	125	2700
Copper	120	8900
Steel	380	7800

Exercise E10.12 A material is required for a transmission line that gives the lowest full-life cost over a 20-year period. The total cost is the sum of the material cost and the cost of the power dissipated in Joule heating. The cost of electricity C_E is 6×10^{-3} \$/MJ. Material prices are listed in the table. Derive an expression for the total cost per metre of cable in terms of the cross-sectional area A (which is a free parameter), the material and electrical costs, and the material parameters. Show that the minimum cost occurs when the two contributions to the cost are equal. Hence derive a performance index for the material and decide on the best of the materials in the table.

Material	Electrical resistivity, ρ_e ($\Omega.m$)	Density, ρ (kg/m^3)	Price, C_m (\$/kg)
Aluminium	1.7×10^{-8}	2700	1.6
Copper	1.5×10^{-8}	8900	5.2
Steel	55×10^{-8}	7800	0.5

Exercise E10.13 In the discussion of conductors, a 50-50 mix of copper and steel strands was suggested for transmission cables. Using the values for resistivity and strength of copper and steel in the table, calculate the effective values of both for the cable, assuming for two materials in parallel a rule of mixtures for both strength and conductivity (*not* resistivity). Plot this on a copy of the $\sigma_y - \rho_e$ chart for copper alloys of Figure 10.3 to explore its performance.

Material	Strength, σ_y (MPa)	Resistivity, ρ_e ($\mu\Omega.cm$)
High-strength steel, cold drawn	1700	22
High-conductivity copper, cold drawn	300	1.7

Exercise E10.14 A 50-50 mix of high-purity aluminium and steel strands is suggested for power transmission cables. Using the values for resistivity and strength for aluminium and steel, calculate the effective values of both for the cable, assuming for two materials in parallel a rule of mixtures for both strength and conductivity (*not* resistivity). Plot this on a copy of the $\sigma_y - \rho_e$ chart for aluminium alloys of Figure 10.3 to explore its performance.

Material	Strength, σ_y (MPa)	Resistivity, ρ_e ($\mu\Omega$.cm)
High-strength steel, cold drawn	1700	22
1050 grade aluminium, H19 (cold drawn)	160	2.8

Exercise E10.15 Use Table A7 of Appendix A to find the dielectric constants of nylon (polyamide, PA) and of polypropylene (PP). Why do they differ so much?

Exercise E10.16 (a) A parallel plate capacitor has a plate area of $A = 0.3$ m^2 separated by an air gap of $t = 50$ μm. What is its capacitance? (b) The air gap is replaced by a dielectric of thickness $t = 0.5$ μm with a dielectric constant $\varepsilon_r = 8$. What is its capacitance now? The permittivity of free space ε_o is 8.85×10^{-12} F/m.

Exercise E10.17 A supercapacitor is required that is able to store an energy of $E_c = 0.5$ kJ when a potential difference $V = 110$ V is applied to its plates. The dielectric constant of the dielectric separating the plates is $\varepsilon_r = 7000$ and its dielectric strength (breakdown potential gradient) is $E_b = 14$ MV/m. What is the minimum area of plate required?

Exercise E10.18 (a) An oscillating voltage with an amplitude of 110 V at a frequency $f = 50$ Hz appears across a PET insulator of thickness $x = 0.5$ mm. PET has a dielectric loss factor $L = \varepsilon_r \tan\delta \approx 0.025$. How much power is dissipated per unit volume in the PET? Assuming no heat loss, how fast will the temperature of the PET rise? (PET has a volumetric specific heat $C_p = 1.56 \times 10^6$ J/m^3.K). (b) If instead the frequency is $f = 1$ MHz, what is the heating rate?

Exercise E10.19 Roughly 50% of all cork that is harvested in Portugal ends up as cork dust, a worthless by-product. As a materials expert, you are approached by an entrepreneur who has the idea of making useful products out of cork dust by compacting it and heating it, using microwave heating. The dielectric loss factor $L = \varepsilon_r \tan\delta$ of cork is 0.21. The entrepreneur reckons he needs a power density P of at least 2 kW/m^3 for the process to work. If the maximum field E is limited to 10^2 V/m, what frequency f of microwaves will be needed?

Exercise E10.20 Derive the expression [Equation (10.13)]

$$\text{Max energy density} = \frac{1}{2}\varepsilon_r\,\varepsilon_o\,E_b^2$$

for the maximum limiting electrical energy density that can be stored in a parallel plate capacitor with plate of area A separated by a dielectric of thickness t. Here ε_r is the dielectric constant of the dielectric, ε_o is the permittivity of free space, and E_b is the field at which the dielectric breaks down.

Exercise E10.21 You are asked to suggest a dielectric material for a capacitor with the highest possible energy density. What materials would you suggest? Use Figure 10.4 of the text to find out.

Exercise E10.22 Why do metals conduct electricity? Why do insulators not conduct electricity, at least when cold?

Exercise E10.23 The metal zinc has free electron concentration per unit volume, n_v of $1.3 \times 10^{29}/m^3$ and an electron mobility μ_e of 8×10^{-4} $m^2/V.s$. The charge carried by an electron, e, is 1.6×10^{-19} C. Based on this information, what is the electrical conductivity of zinc? How does this compare with the value given for zinc alloys in Table A7 of Appendix A? (Watch the units).

Exercise E10.24 The resistivity of pure silver is $\rho_e = 1.59$ $\mu\Omega.cm$. Each silver atom provides one free electron. The atomic volume of silver is 1.71×10^{-29} m^3. What is the electron mobility in silver?

Exercise E10.25 Estimate the drift velocity of free electrons when a potential difference of 3 V is applied between the ends of a copper wire 100 mm long. The electron mobility in copper is $\mu_e = 0.0032$ $m^2/V.s$.

Exercise E10.26 The resistivity of brass with 60 atom% copper and 40 atom% of zinc is $\rho_e = 6.8$ $\mu\Omega.cm$. Each copper atom contributes one free electron. Each zinc atom contributes two. The atomic volumes of copper and zinc are almost the same (assume them to be equal) and the electron concentration in copper is $n_v = 8.45 \times 10^{28}/m^3$. What is the electron mobility μ_e in brass? How does it compare with that of pure copper ($\mu_e = 0.0032$ $m^2/V.s$)?

Exercise E10.27 Graphene is a semiconductor with remarkable electrical properties. It is reported that the free-electron mobility μ_e in graphene is 1.5 $m^2/V.s$ and that the effective free electron concentration per unit volume, n_v, is $5 \times 10^{25}/m^3$. If these values are accepted, what would you expect the resistivity of graphene to be?

Exercise E10.28 The electron mobility μ_e in undoped germanium at 300 K is 0.36 $m^2/V.s$ and that of holes μ_h is 0.19 $m^2/V.s$. The carrier density n_v is $2.3 \times 10^{19}/m^3$. What is the conductivity of germanium at 300 K? What is its resistivity?

Exercise E10.29 The atomic volume of silicon is 2.0×10^{-29} m^3 and the free electron density n_v at 300 K is $1.5 \times 10^{16}/m^3$. What fraction of silicon atoms have provided a conduction electron?

Magnetic properties

Exercise E10.30 A cylindrical coil with a length $L = 30$ mm with $n = 75$ turns carries a current $i = 0.1$ A. What is the field in the magnet? A silicon-iron core with a relative permeability $\mu_R = 3000$ is placed in the core of the coil. What is the magnetisation of the silicon-iron? The permeability of free space $\mu_o = 4\pi \times 10^{-7}$ H/m.

Exercise E10.31 Sketch a $M - H$ curve for a ferromagnetic material. Identify on it the saturation magnetisation, the remanent magnetisation, the coercive field, and the susceptibility. Which combination is desirable in a soft magnet for a transformer core? Which combination for the permanent magnets of the DC motor of a hybrid car?

Exercise E10.32 A neodymium-boron magnet has a coercive field $H_c = 1.03 \times 10^6$ A/m, a saturation magnetisation $M_s = 1.25 \times 10^6$ A/m, a remanent magnetisation $M_r = 1.1 \times 10^6$ A/m, and a large energy product. Sketch approximately what its $M - H$ hysteresis curve looks like.

Exercise E10.33 What are the necessary conditions for a material to be ferromagnetic at a temperature T?

Exercise E10.34 What is a ferrite? What are its characteristics?

Exercise E10.35 What is a Bohr magneton? A magnetic element has two unpaired electrons and an exchange interaction that causes them to align such that their magnetic fields are parallel. Its atomic volume, Ω, is 3.7×10^{-29} m^3. What would you expect its saturation magnetisation, M_s, to be?

Exercise E10.36 Cobalt has a density ρ of 8900 kg/m^3 and atomic weight A_{wt} of 58.93 kg/kmol. The net magnetic moment of a cobalt atom, in Bohr magnetons per atom, is $n_m = 1.8$. What is the saturation magnetisation M_s of cobalt?

Exercise E10.37 The element with the largest saturation magnetisation is holmium, with $M_s = 3.0 \times 10^6$ A/m (although it has a miserably low Curie temperature of 20 K). The density ρ of holmium is 8800 kg/m^3 and its atomic weight A_{wt} is 165 kg/kmol. What is the atomic magnetic moment n_m, in Bohr magnetons per atom, of holmium?

Exercise E10.38 The soft iron laminations of transformers are made of 3 wt% (6 at%) silicon-iron. Silicon-iron has a density ρ of 7650 kg/m^3 and a mean atomic weight A_{wt} of 54.2 kg/kmol. The net magnetic moment of an iron atom, in Bohr magnetons per atom, is $n_m = 2.2$; that of silicon is zero. What is the saturation magnetisation M_s of 3 wt% silicon-iron?

Exercise E10.39 Nickel has a density ρ of 8900 kg/m^3, atomic weight A_{wt} of 58.7 kg/kmol, and a saturation magnetisation M_s of 5.2×10^5 A/m. What is the atomic magnetic moment n_m, in Bohr magnetons per atom, of nickel?

Exercise E10.40 A coil of 50 turns and length 10 mm carries a current of 0.01 A. The core of the coil is made of a material with a susceptibility $\chi = 10^4$. What is the magnetisation M and the induction B?

Exercise E10.41 An inductor core is required for a low frequency harmonic filter. The requirement is for low loss and high saturation magnetisation. Use Figure 10.14 as data sources to choose the classes of magnetic material that best suit this application.

Exercise E10.42 A magnetic material is required for the core of a transformer that forms part of a radar speed camera. It operates at a microwave frequency of 500 kHz. Which class of material would you choose?

Exercise E10.43 The product $B_r H_c$ is a crude measure of the 'power' of a permanent magnet (if the $B - H$ curve were a perfect rectangle, it becomes equal to what is called the energy product, a more realistic measure of the power). Draw contours of $B_r H_c$ onto a copy of the remanent induction–coercive field chart (Figure 10.14) and use these to identify the most powerful permanent magnet materials.

Exercise E10.44 A soft magnetic material is needed for the core of a small high frequency power transformer. The transformer gets hot in use; forced air cooling limits the rise in temperature to 200°C. Eddy current losses are a problem if the core is electrically conducting. What material would you recommend for the core? The coercive field–Curie temperature chart (Figure 10.15) can help.

Exercise E10.45 A material is required for a flexible magnetic seal. It must be in the form of an elastomeric sheet and must be electrically insulating. How would you propose to make such a material?

Exercise E10.46 What are the characteristics required of materials for magnetic information storage?

Exercise E10.47 An electromagnet is designed to pick up and move car parts – the car part becomes a temporary part of the core, completing the magnetic circuit. The field H is created by a coil 100 mm long with a cross section of 0.1 m^2 with 500 turns of conductor carrying a current of 25 A. What force can the electromagnet exert?

Exercise E10.48 The figure shows the characteristics of magnetocaloric materials. The vertical axis shows the temperature drop on demagnetisation from a 2 T field; the horizontal axis shows the limited temperature range over which this magnetic order/disorder transformation is possible. Recommend one or more materials for use in a solid-state refrigerator to operate (as most fridges do) in the temperature range 2° to 6°C.

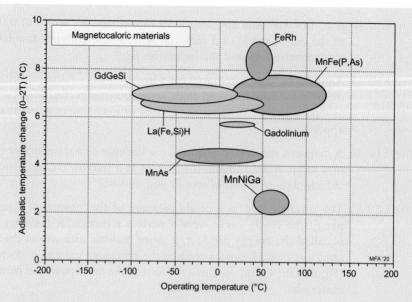

Exercise E10.49 The figure shows the characteristics of magnetostrictive materials. The vertical axis is the maximum strain that results from magnetisation – it is reversible, so recovers when the field is removed. The horizontal axis is the field needed to cause the magnetisation. You are asked to select a material for a magnetostrictive actuator, maximising the strain, but with the limitation that the field for switching is provided by a cylindrical coil with a length $L = 30$ mm with $n = 75$ turns and a maximum current $i = 1$ A.

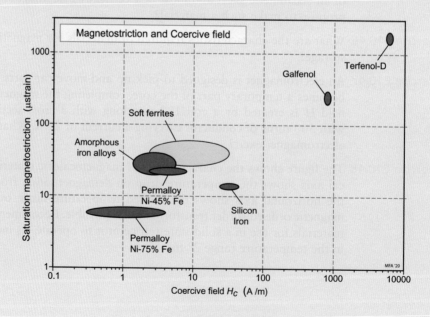

Optical properties

Exercise E10.50 The absorption coefficient β of polyethylene for optical frequencies is $86.6\,\mathrm{cm}^{-1}$. How thick a slab of polyethylene is required to reduce the transmitted light intensity by absorption to one-half of its initial value?

Exercise E10.51 The fraction of light that is transmitted through a 100 mm panel of PMMA is 0.97. Neglecting reflection at both surfaces, what is the absorption coefficient β of PMMA?

Exercise E10.52 Derive the expression for the fraction of light that passes through a transparent panel, allowing for both reflection and absorption:

$$I = I_o \left(1 - \frac{I_R}{I_o} \right)^2 \exp - \beta\, x$$

where β is the absorption coefficient and x is the length of the optical path in the material.

Exercise E10.53 Define refractive index. How is it measured? Give examples of devices that make use of refraction.

Exercise E10.54 The prism for a reflex camera is made from a silica glass with a coefficient of refraction of 1.46. To present a 'through the lens' image to the user of the camera, light rays entering the prism must be totally internally reflected. At what range of angles can they strike the internal surfaces of the prism?

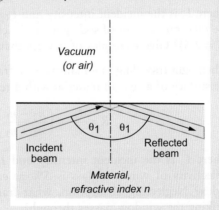

Exercise E10.55 A design for an underwater camera calls for a prism viewfinder that will be immersed in water. If the prism is made from a glass with a coefficient of refraction of 1.46, what range of incident angles will lead to total internal reflection? The refractive index of water is 1.33.

Exercise E10.56 A 30 mm thick block of glass of refractive index 1.66 is placed over a line on a sheet of paper so that only part of the line is covered. The line is viewed from above at an angle of 45° to the glass and normal to the line. Refraction in the glass displaces the image of the part of the line it covers. By how much is the part of the line viewed through the glass displaced relative to the part that is not covered?

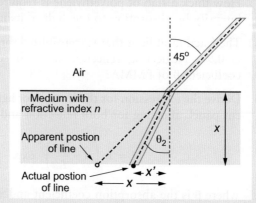

Exercise E10.57 Define reflectivity. Reflectivity is related to refractive index. When light travelling in a material of refractive index n_1 is incident normal to the surface of a second material with a refractive index n_2, the reflectivity is

$$R = \frac{I_R}{I_o} = \left(\frac{n_2 - n_1}{n_2 + n_1} \right)^2$$

where I_o is the incident intensity and I_R that of the reflected beam. What is the reflectance in air of soda glass (refractive index $n = 1.5$) and of diamond ($n = 2.4$)? Give examples of devices that make use of reflection.

Exercise E10.58 When light travelling in a material of refractive index n_1 is incident normal to the surface of a second material with a refractive index n_2, the reflectivity is

$$R = \frac{I_R}{I_o} = \left(\frac{n_2 - n_1}{n_2 + n_1} \right)^2$$

where I_o is the incident intensity and I_R that of the reflected beam. The reflectivity of window glass is 5% at each surface. What is its refractive index? Neglecting absorption, what fraction of the intensity of the light striking the window passes through?

Exercise E10.59 The refractive index of PMMA is 1.52. What would you estimate its dielectric constant ε_R to be?

Exercise E10.60 A crown glass sheet is partly coated by an acrylic film with thickness $t_1 = 4$ microns and refractive index $n_1 = 1.49$. A beam of red light strikes the plate at right angles to its surface. What is the difference in optical path length between a ray passing through the coated and the uncoated part of the glass? If the wavelength of red light is $\lambda = 0.65$ microns, is there potential for constructive or destructive interference between the two rays when they emerge from the glass?

Exercise E10.61 A filter is required to screen out UV light (wavelengths λ around 0.35 microns) from crown glass windows. It is suggested that magnesium fluoride, MgF_2 (refractive index $n_1 = 1.38$), be used as an interference coating because it is hard wearing and easily deposited by physical vapour deposition. What thickness of MgF_2 coating is required to give maximum cancellation?

Exercise E10.62 A textured metal plate has a grid of fine parallel scratches with a regular spacing of $d = 1.6$ microns. Suppose white light falls normal to the surface of the plate. At what angle will the red component of the light, frequency $\nu = 5 \times 10^{14}$ per second, be diffracted most strongly when viewed normal to the scratches? Take the velocity of light in air to be $c = 3 \times 10^8$ m/s.

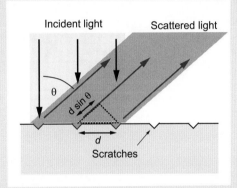

Exercise E10.63 Why are metals good reflectors of radiation?

Exercise E10.64 An X-ray system has a beryllium window to transmit the beam to the patient. The absorption coefficient of beryllium for the wavelength of X-rays of interest is 3.02×10^2 m^{-1}. If the window is 2 mm thick, what fraction of the incident beam intensity will pass through the window? The casing of the equipment is shielded with 4 mm of lead, with absorption coefficient for X-rays of 3.35×10^6 m^{-1}. What fraction of the intensity of the incident beam will escape through the casing?

Exercise E10.65 You are asked to design a light-sensing switch for a greenhouse. What principle would you base it on? What materials would you choose for the sensor?

Exercise E10.66 You are asked to design a heat-sensing switch to turn the lights off in the garage when no one is moving around in it. What principle would you base it on? What materials would you choose for the sensor?

Exercise E10.67 An optical fibre has a glass core with a refractive index of $n_1 = 1.48$, clad with a glass with refractive index $n_2 = 1.45$. What is the maximum angle that the incoming optical signal can deviate from the axis of the core while still remaining trapped in it?

Chapter 11
Manufacturing processes and microstructure evolution

Two parallel sets of run-out rollers supporting steel slabs during continuous casting. The slabs are being cut to length as they move over the rollers.
(istock.com)

https://doi.org/10.1016/B978-0-08-102399-0.00011-X

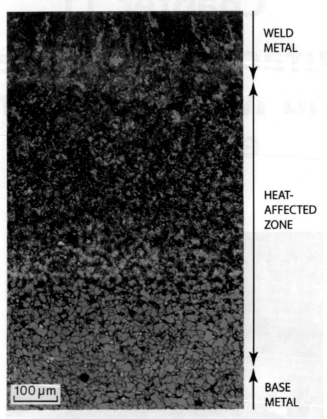

Microstructural change in a steel weld: different colours reveal different phases (or fine-scale phase mixtures) - both the weld metal and heat-affected zone are completely transformed compared to the original base metal.
(Image courtesy: Dr John Ion)

Chapter contents

11.1 Introduction and synopsis

Materials and processing are inseparable. Ores, minerals, and oil are processed to create raw materials in clean, usable form. These are shaped into components, then finished and joined to make products – the process hierarchy introduced in Chapter 2. New processes continue to emerge (e.g. the rapid growth in 'additive manufacturing'). And increasingly, products must be recycled at the end of product life – extracting usable material from waste streams is in itself a technical processing challenge, but brings with it the need for existing processes to tolerate higher levels of compositional impurity.

The strategy for choosing a process in some respects parallels that for materials: *screen out* those that cannot do the job, and *rank* those that can, usually seeking those with the lowest cost. Going further than this initial assessment can be complex – it mainly relates to the interactions between processing and design parameters with the properties of the materials themselves. Some characteristics of a process – its *process attributes* – can readily be described by numeric and non-numeric data (e.g. compatibility with material, the size of part a process can produce, or the surface finish it delivers). This narrows the classes of process that are suitable, and (for shaping processes) may be followed by estimation of the cost per part. This strategy for process selection is introduced in Section 11.2, and then developed separately in *Guided Learning Unit 3*, to give direct practice in applying the methods (as for materials in *Guided Learning Unit 2*).

To refine the choice further, we need to know *how to design well* using each process – to avoid defects, reduce scrap rates, and improve productivity – a strategy often referred to as *design for manufacture*. This stage of *technical evaluation* requires knowledge of the process physics, supported by data – now we are not just identifying which process to use, but specifying suitable process operating conditions, refining the choice of material composition, and optimising design details. This goes well beyond the scope of this text – but one important aspect of process-material interaction in design relates closely to earlier chapters: the role of processing in controlling the material properties, particularly those that are microstructure-sensitive (such as strength and toughness), introduced by example in Section 11.3.

Chapters 3-10 have shown how microstructure across the length-scales is key to engineering properties – from the nanometre scale of atoms and dislocations, to precipitate particles, grains, and cracks, ranging in size from microns to millimetres. Microstructure is not often accessible to direct observation during processing, so controlling it requires the ability to understand and predict how a given process history will cause it to evolve. That is the goal of this chapter, captured concisely by the statement:

$$\text{Composition} + \text{Processing} \rightarrow \text{Microstructure} + \text{Properties}$$

To understand microstructure evolution during processing, particularly for metals, two central topics are *phase diagrams* and *phase transformations*. Section 11.4 provides a concise overview of these topics. Reading phase diagrams, and interpreting the microstructural evolution in common processes, are skills that are again best developed by immediate application and practice – this is provided in *Guided Learning Unit 4: Phase diagrams and phase transformations*. A wide range of manufacturing processes for metals and non-metals are then considered in Sections 11.5-11.10, in many cases highlighting the key role of thermal history and cooling rate in determining the microstructure evolution. As in earlier chapters, property charts are used to illustrate the changes imparted by composition and processing.

The key questions in this chapter (and the associated *Guided Learning Units 3* and *4*) are therefore: how do we choose which process to use to meet the requirements of a design, and how do the chosen materials and processes interact in determining the microstructure and properties.

11.2 Process selection in design

The selection strategy The taxonomy of manufacturing processes was presented in Chapter 2, categorising processes into three broad families: *shaping*, *joining*, and *finishing* (see Figures 2.2 and 2.3). Each is characterised by a set of *attributes* listed, in part, in Figures 2.4 and 2.5 – these are the equivalent to material properties, characterising the capabilities of each process. Chapter 2 also introduced a strategy for materials selection (see Figure 2.10): translating the design requirements into a set of constraints and objectives, used respectively to screen and rank the options. The resulting shortlist of materials then requires technical evaluation – particularly in relation to their processability and performance in service, and the implications for the end of the product life.

The strategy for selecting processes follows a broadly similar path, but with some differences – Figure 11.1 summarises the strategy in this case. Design requirements again

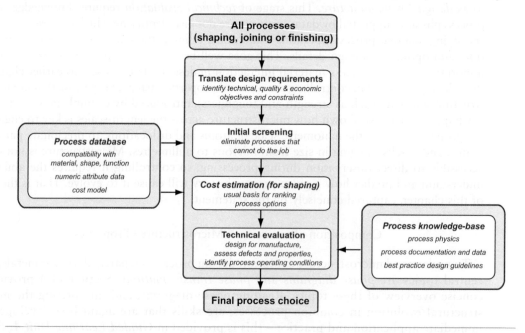

Figure 11.1 The selection strategy applied to processes. It partly mirrors that for materials, with screening and ranking (by cost) guided by process databases, while subsequent technical evaluation of the options draws on a process 'knowledge-base'.

determine objectives and constraints, and initial screening of process attributes identifies the viable process classes. For batch shaping processes, we may then be able to estimate cost per part, to rank possible solutions. Much of this can be implemented in software, using databases of process characteristics. But as with materials, technical evaluation is then needed to refine the choice of material, process, and design detail simultaneously – for this we need expertise, drawing on the cumulative 'knowledge-base' of manufacturing industry and materials science. So what are the key design requirements that impinge on processing, and where are they addressed in the selection strategy?

Translation   For processing, it is helpful to think of three types of design requirement: (i) *technical* – can the process do the job at all? (ii) *quality* – can it do so sufficiently well? and (iii) *economic* – can it be done cheaply enough? Translation of some design requirements into *constraints* under each heading is straightforward. For example, one technical constraint applies across all process families: the compatibility of material and process. Size and shape, too, are technical constraints, although we need to understand how the processes work, to know which aspects of the component geometry are important in each case. Quality includes dimensional precision and surface finish, which can also be quantified in a database. Avoiding defects and achieving target properties are also quality requirements, but these are complex and require technical evaluation. The usual *objective* in processing is to minimise cost – though economic characteristics are also commonly viewed as constraints (e.g. a target production rate of parts per hour, or per day).

In material selection, the identification of *free variables* was an important part of translation – due to the subtle role they take in determining the correct measure of performance. In process selection, there are very many free variables, including the process variant and its operating parameters (such as temperatures, flow rates, etc.), the composition of the material within the selected material class, and details of the component geometry (all of which govern the processability and final properties). Only a few of these can be listed up-front – mostly these considerations are an integral part of the technical evaluation step.

Initial screening   Simple technical, quality, and economic constraints may be compared directly with data for the *process attributes*, eliminating processes that cannot meet one or more of them. Many attributes are non-numeric – lists of materials to which the process can be applied, for example, or (for finishing processes) a list of the purposes of surface treatment. Other attributes have numeric ranges – the size or mass of component that a process can handle, and the precision or the surface smoothness it can deliver. Requirements such as 'made of magnesium alloy and weighing about 3 kg' are easily compared with the process attributes to eliminate those that cannot shape magnesium alloy and/or cannot handle a component as large (or as small) as 3 kg. Even simple screening steps may bring in processing knowledge – for example, knowing that failure to meet a dimensional or surface finish requirement in shaping may be rectified by specifying a secondary machining process. More discriminating requirements are beyond the scope of screening – for example, 'must not contain large pores or cracks' or 'minimum yield stress of 250 MPa' require technical knowledge and expertise, backed by data.

Cost estimation   Ranking a shortlist of processes requires an *objective*, the most obvious in selecting a process being that of minimising cost. Cost modelling is highly proprietary within companies and can be very difficult to itemise for multi-stage production lines. For batch shaping processes, however, it is possible to make simple estimates using generic data for each process. In certain demanding applications, cost may be replaced by the objective of maximising part performance almost regardless of cost – though more usually it is a trade-off between the two that is sought. Similarly, the increasing drive towards a circular economy requires a balance between cost and the need to reuse material recovered from products at their end of life. Whatever the main drivers are, cost will feature somewhere, but a full analysis will need a detailed picture of the material source and composition, process operating conditions, and the final design details – once again calling for technical expertise.

Technical evaluation   The most important technical expertise relates to: (a) *processability* (the ability for the process to work without producing defects, at maximum productivity), and (b) *properties*, determined by the coupling of process-material-design. Machines all have optimum ranges of operating conditions under which they work best and produce products with uncompromised quality. Failure to operate within this window can lead to manufacturing defects, such as excessive porosity, cracking, or residual stress. This in turn leads to scrap and lost productivity, or if passed on to the user, it may cause premature failure (and potential liability and litigation). At the same time material properties depend, to a greater or lesser extent, on process history through its influence on microstructure. Processes are chosen not just to make, join, and finish shapes, but simultaneously to deliver their properties.

In summary, choosing a manufacturing process requires both data and the application of knowledge of how processes work (or don't work). The key skill to cultivate is *knowing what questions to ask*, when consulting experts, handbooks, or software. This does not lend itself to computer implementation in databases, but it does require structured thinking in just the same way as the selection of material. Here we show a visualisation tool to guide our thinking about manufacturing, design, and process selection: the 'manufacturing triangle.'

The manufacturing triangle[1]   Design requirements fall under the headings of technical, quality, and economic, so we can picture these as three corners of a triangle (Figure 11.2). Each corner is surrounded by the dominant factors to consider in product design, providing a checklist for shaping, joining, and surface treatment. In the first instance, we can pick out from the triangle the individual attributes of processes that can be used for screening – material and shape compatibility, part size (or mass), tolerance and roughness, and economic production rate. Cost

---

[1] We owe the concept of the 'manufacturing triangle' to Erik Tempelman, formerly of Delft University of Technology in the Netherlands. For a more detailed exposition of its application, see *Manufacturing and Design* by Tempelman et al., listed under Further Reading in Section 11.11.

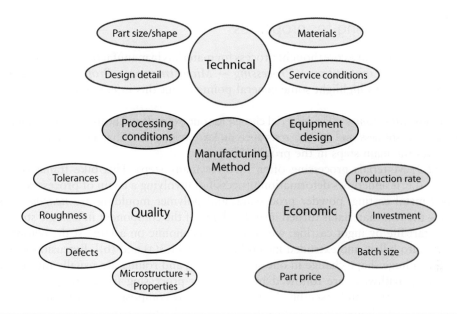

Figure 11.2 The manufacturing triangle – a checklist of the key factors that influence the selection of manufacturing process, under the three categories of technical, quality, and economic requirements. The design and operating parameters of the manufacturing method itself, at the centre, also play a role in achieving a successful outcome. (Adapted from *Manufacturing and Design* by E. Tempelman, H.R. Shercliff, and B. Ninaber van Eyben).

modelling also draws on parameters shown on the triangle, bringing together several factors in a single function (e.g. material and equipment costs, and production rates). Finally, the triangle provides prompts for the more challenging design considerations, such as defects or properties, pulling together multiple facets of the triangle and often combining material, design, and process parameters.

Introduction to Guided Learning Unit 3: Process selection in design

As for material selection, understanding the strategy for choosing a process is best appreciated by doing it, and this is the purpose of *Guided Learning Unit 3*. Once again, we exploit a graphical format as far as possible, through *process attribute charts*. Worked examples and exercises focus on selecting shaping processes, but the same principles are applied in brief to surface treatment and joining. Cost modelling is introduced for shaping. Technical evaluation is beyond the scope of this text – brief notes only are provided – but the interaction of process history with microstructure and properties is the topic of the remainder of Chapter 11. *Guided Learning Unit 3* is free-standing, and complements the rest of the chapter, but is not pre-requisite for understanding what follows.

11.3 Processing for properties

We noted earlier that a core thought process for materials in manufacturing may be summarised by the mantra: *Composition + Processing → Microstructure + Properties*. Here is an illustrative example to highlight some general points about this interaction.

Aluminium bike frames The material chosen for our example is a heat-treatable aluminium alloy, for its good stiffness and fatigue resistance at low weight (see Chapters 4 and 6). Figure 11.3 illustrates the main steps in the process history.

Point 1: Materials processing often involves many steps. This bike frame is a *wrought product* (i.e. it undergoes deformation processing) involving a chain of processes. In contrast, shaped metal casting, powder processing, and polymer moulding are *near-net-shape processes* – the raw material is turned into the shape of the component in a single step. Wrought alloys start life by ingot casting, which is most economic on a large scale, before the ingot is sliced into billets and extruded into tube [Figure 11.3(a) and (b)]. Extrusion is unusual in enabling a huge shape change in one step – forging and rolling are usually multi-stage – but all shaping pathways are followed by secondary processing (machining, heat treatment), surface treatments, and assembly. Note that the alloy composition is adjusted at the outset, while it is in the liquid state, and cannot be changed after casting.

Point 2: Each process step has a characteristic *thermal history*, determined by parameters of the process, material, and design. Casting by definition involves melting, solidification, and cooling. Extrusion is done hot to reduce the strength and to increase the ductility, enabling the material to undergo large plastic strain with a fast throughput. Hot deformation and the subsequent 'age hardening' heat treatment [Figure 11.3(b) and (c)] lead to microstructure evolution – changes in the dislocation and grain structure, and solute atoms dissolving and precipitating – key to determining the final strength. The interplay between deformation and thermal histories is critical for controlling the shape and final properties simultaneously.

Point 3: Designers should beware of unintended side effects in *joining* [Figure 11.3(d)]. Bike frame assembly here is by arc welding, imposing a further temperature cycle. Apart from producing a re-melted zone (the 'weld metal'), the surrounding 'heat-affected zone' doesn't melt but is hot enough for the microstructure to evolve; it ends up with different, commonly inferior properties to the rest of the frame. Differential thermal histories in welding can also do further damage by causing residual stresses, distortion, and cracking.

Point 4: Design focuses on the properties of finished products, but some properties (strength, ductility, thermal conductivity, etc.) play key roles *during processing* – indeed, some properties are only really relevant for processing, such as the melt viscosity in metal casting or polymer moulding. Design and processing can often be in conflict – strong alloys are more difficult and slower to process, and are thus more expensive. So in choosing materials for a component it is important to examine their suitability for processing as well as their performance in service. Clever processing tricks are exploited to resolve the conflicts – the bike frame is hot-formed (when it is soft) and then heat-treated (after shaping) to produce the high strength needed for service.

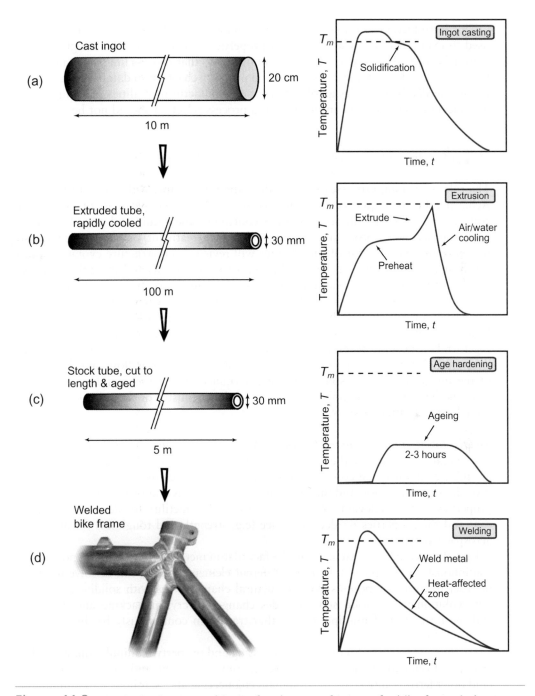

Figure 11.3 A schematic process history for the manufacture of a bike frame in heat-treatable aluminium alloy: (a) casting of large ingot, cut into billets for extrusion; (b) hot extrusion into tube, incorporating the solution heat treatment and quench; (c) sections cut to length and age hardened; (d) assembly into the frame by arc welding.

This introductory example shows how combinations of composition and processing are used to manipulate microstructure and properties. Further examples are given later for all material classes. The bicycle frame case study shows that process limits may make it difficult to achieve the ideal 'database' properties on property charts or in databases, and compromise is often needed between design objectives and manufacturing realities. The importance of the 'technical evaluation' stage in material and process selection should not be underestimated!

11.4 Microstructure evolution in processing

The role of shaping processes is to produce the right shape, with the right final properties. Achieving the first requires control of viscous flow of a liquid, plasticity of a solid, or powder compaction. Achieving the second means controlling the nature and rate of microstructural evolution. Secondary processing (joining and surface treatment) may cause the microstructure to evolve further. The rest of this chapter will review microstructure evolution and control for the main manufacturing classes. The emphasis is on metals, reflecting their great diversity in alloy chemistry, process variants, and solid-state microstructures. Ceramics are limited to solid-state powder compaction, with little or no change in the microstructure within the powder particles themselves. Polymers offer a fairly wide range of discrete compositions, with freedom to mix in fillers and other additives, but only rarely are different polymers mixed together. For most purposes, understanding the sensitivity of properties to processing is relatively straightforward in polymers, with the physics of viscous flow and the configuration of the molecules as they solidify being determined by a small number of characteristics (such as the length of the polymer chains and the chemistry of the side groups). For metals though, it is a different story.

Metals: processing and microstructure As *Guided Learning Unit 3* demonstrates, metals offer great versatility in the manufacturing processes and alloys available to make and assemble components of complex shape. In parallel, as seen in earlier chapters, there is great diversity in composition and the resulting microstructures and properties. This includes both properties that are relevant to the process itself (e.g. ductility for deep drawing a beverage can) and the properties needed in service (e.g. strength and toughness for automotive and aerospace alloys).

Part of the complexity stems from the fact that in metallic alloys, the composition can vary continuously, with up to 10 or more different elements, which can dissolve in one another or react to form compounds. Microstructural change starts with solidification of a liquid, but evolution in the solid state includes changes in crystal packing and grain structure, and the formation of multiple *phases* that transform continuously by diffusion (discussed further below).

To illustrate the breadth of metals processing and property manipulation, we will draw on examples from casting, deformation, heat treatment, joining, and surface engineering. Steels (and other ferrous alloys) are the dominant engineering alloys, followed by aluminium alloys, so these feature in most of the examples. Similar principles apply to all alloy systems – those based on Cu, Ti, Ni, Zn, Mg, and so on – and most offer both cast and wrought variants, and for special products can be processed as powders (like ceramics). Casting and wrought alloys in a given system have quite different compositions: good castability requires higher levels of alloying additions (to lower the melting point), compared to the relatively dilute wrought alloys. Usually either cast or wrought variants dominate an alloy system. This reflects subtle differences in the ease of deformation of the underlying crystal structures in different elements. For example, Zn, Mg, and Ti have the 'hexagonal close-packed (HCP)' crystal structure (see Chapter 3 and GLU1). This is inherently less ductile than the FCC or BCC crystal structures found in Al, Fe, and Cu. So Zn, Mg, and Ti are mostly cast, while ductile Al, Fe, and Cu are mostly wrought. This distinction also impacts on the properties in a given alloy system – casting leads to coarser microstructures and poorer strength and toughness than in wrought alloys (as illustrated for aluminium alloys in Chapter 6).

Most metals processing therefore involves thermal cycles (plus deformation in some shaping processes), with the temperature history determining both the material state (liquid and solid phases) and the evolution of that state (by diffusion). So first we need to spend some time on the underpinning theory of *phase transformations* – noting that some of the underlying science and concepts will also apply to polymer and ceramic processing.

Phase diagrams, phase transformations, and other microstructural change A *phase* is defined as a region of a material with a specified atomic arrangement, giving it certain characteristic physical properties, such as density. Steam, water, and ice are the vapour, liquid, and solid phases of H_2O. Engineering alloys such as steel also melt and then vaporise if heated enough. More interestingly, they can contain different phases simultaneously in the solid state, depending on the composition and temperature. It is this that gives us precipitation-hardened alloys, made up of one or more precipitate phases in a matrix of another phase. Precipitates may contain only a handful of atoms, but they are nonetheless a separate single phase with a defined crystallographic structure.

Phase diagrams are essential tools in processing of metals and ceramics: they are maps showing the phases expected as a function of composition and temperature if the material is in a state of *thermodynamic equilibrium*. By this, we mean the state of lowest *free energy* (defined later), such that there is no tendency to change. A *phase transformation* occurs when the phases present change (e.g. melting of a solid, solidification of a liquid, or a solid solution forming a dispersion of precipitates within a crystal). Phase diagrams show us when these changes are expected as the temperature rises and falls. Phase transformations are thus central to microstructural control in metal processing.

A change in the phases present requires a *driving force* and a *mechanism*. The first is determined by thermodynamics, the second by kinetic theory (i.e. atomic diffusion). The term *driving force*, somewhat confusingly, means the change in free energy between the starting point (the initial phase or phases) and the endpoint (the final state). Solids melt on heating because being a liquid is the state of lowest free energy above the melting point. Chemical compounds can have lower free energy than solid solutions, so an alloy may decompose into a mixture of phases (both solutions and compounds) if, in so doing, the total free energy is reduced.

Microstructures can therefore only evolve from one state to another if it is energetically favourable to do so. But having a free energy change available is not sufficient in itself to make the change happen – microstructure evolution needs a *mechanism* for the atoms to rearrange.

For example, liquids solidify by atomic jumps across an interface between the liquid and the growing solid. Virtually all solid-state phase changes also involve diffusion to move the atoms around to enable new phases to form, and we know from Chapter 8 that this is strongly dependent on temperature. The driving force, too, depends on temperature, so the overall rate of microstructural evolution is strongly linked to the thermal history, and this is of great significance in processing. First, the rate of transformation governs the *length-scale* of the important microstructural features (grains, precipitates, etc.) and thus the properties. But it also affects process economics via the time it takes for the desired structural change to occur.

Solidification and precipitation involve changes in the phases present. Some important types of structure evolution do not involve a phase change, but the principles are the same: there must be a reduction in free energy, and a kinetic mechanism, for the change to occur. For example, if there is a gradient in concentration in a solid solution, atoms tend to diffuse from regions of high concentration to low – a uniform solid solution has lower free energy than a non-uniform one. Recrystallisation changes the grain structure of a crystal, even though the phases making up the grains do not change; here the driving force comes from eliminating dislocations with their associated energy per unit length. We revisit these examples later in the chapter, in relation to manufacturing processes in which they occur.

Thermodynamics of phases Materials processing principally deals with liquids and solids; only a few special coating processes deal with gases (e.g. vapour deposition directly to solid on a substrate). In earlier chapters, we have seen that solid phases can form as pure crystals of a single element, as solid solutions of two or more elements, or as compounds with well-defined crystal structure and stoichiometry (by which we mean the relative proportions of the different elements are simple integers: alumina Al_2O_3, iron carbide Fe_3C, and so on).

So if we take an alloy of a given composition and hold it at a fixed temperature, what determines whether it will be solid, liquid, or a mixture of the two? Or a solid solution, a compound, or a mixture of solid phases? The answer lies in thermodynamic equilibrium and the Gibbs[2] free energy, G, defined as

$$G = U + pV - TS = H - TS \tag{11.1}$$

The thermodynamic variables in Equation (11.1) require some explanation. The *internal energy U* is the intrinsic energy of the material associated with the chemical bonding between the atoms and their thermal vibration. We have encountered it before in the cohesive energy responsible for recoverable elastic strain and stored energy under load (see Chapter 4), or thermal expansion (see Chapter 7). The pressure p, volume V, and temperature T are the thermodynamic state variables familiar from the universal gas law. The combination $U + pV$ is so common that it is frequently combined as the *enthalpy, H*. Pressure is usually at or close

---

[2] Josiah Willard Gibbs (1839–1903), a seventh-generation American academic, spent almost his entire career at Yale, receiving its first PhD in Engineering (on the form of gear teeth). A formative visit to Europe, working with scientists such as Kirchhoff and Helmholtz, set him on the path towards his great work *On the Equilibrium of Heterogeneous Substances*. This established him as a founding father of physical chemistry and chemical thermodynamics.

to atmospheric, and volume changes in metals are modest in the liquid or solid state, so changes in enthalpy are usually dominated by the internal energy U.

Entropy S is a complex concept in thermodynamics, often associated with how exchanges of heat at a given temperature change the enthalpy and efficiency of gas-based mechanical systems, such as internal combustion engines and gas turbines. In the context of material phases in processing, the relevant interpretation of entropy is as a measure of the atomic or molecular 'disorder' in the system, and how this and the enthalpy change as the phases change. For example, for a solid crystal to melt, heat must be supplied (at constant temperature) to overcome the intrinsic 'latent heat' of melting. We will see later that this is an enthalpy change, with a corresponding increase in the entropy. The rise in entropy is associated with the greater disorder in the liquid state, compared to the regular packed state of the solid. Similarly there is an increase in entropy when one element dissolves in another, forming a solid solution. Changes in entropy of a crystalline solid are relatively small (due to the regular atomic packing) but play a key role in determining the phases present. Entropy is important, too, in polymers, since the molecules have freedom to adopt different configurations.

So, for any composition and temperature we can (at least in principle) calculate the free energy of alternative material states (liquid solution, solid solution, mixtures of phases, etc.). Out of the unlimited number of possible states, that with the lowest free energy is the state at *thermodynamic equilibrium*. Higher energy states may form, but there is then the possibility of energy being released by a change towards equilibrium. Only at equilibrium is no change physically possible. In practical alloy processing it is often easy enough to reach equilibrium or near-equilibrium states; hence the importance of phase diagrams, which map the equilibrium phases expected for a given alloy.

Introduction to Guided Learning Unit 4:
Phase diagrams and phase transformations (GL4.1-GL4.4)

Phase diagrams and phase transformations are big topics in process metallurgy. In the following subsections, we give a concise summary of phase diagrams and the underlying science of phase transformations. *Guided Learning Unit 4* goes into depth on how to interpret phase diagrams and relate them to microstructural evolution in common processes. The rest of this chapter will be intelligible without diverting into the unit, but a more complete grasp is available by working through the *GLU4* text and exercises. It is recommended that *GLU4* is visited in two installments. GL4.1-GL4.4 cover phase diagrams and how to read them: work through this now. GL4.5-GL4.8 cover common phase transformations – you will be prompted in the text when it is best to go back to these sections.

Phase diagrams: illustrative examples Real alloys can contain up to 10 or more different elements, though many are minor trace additions or impurities. The starting point for understanding all alloys is to consider a principal element mixed with one other element, making a *binary alloy* (or system). Carbon steels are based on iron-carbon, stainless steels on iron-chromium, and so on. Phase diagrams for binary alloys are plots of temperature against composition, usually in weight % (wt%), showing fields in which the equilibrium phase or phases are fixed. Figure 11.4 shows a simple example: the phase diagram for the binary lead-tin (Pb-Sn) system. This and more complex diagrams are built up and explained in more detail in *Guided Learning Unit 4*.

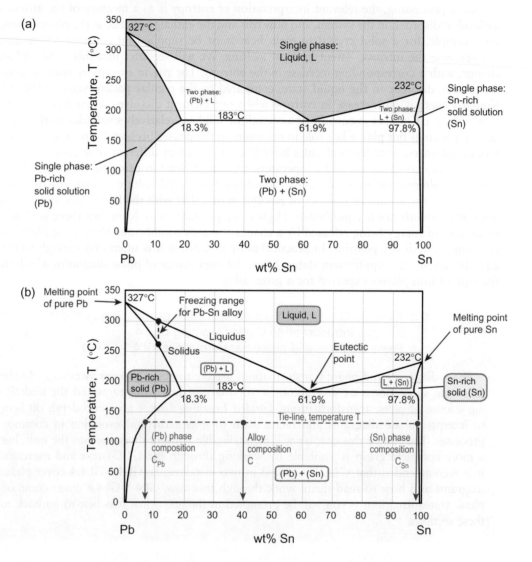

Figure 11.4 Phase diagram for the binary Pb-Sn system, showing a number of key features. In (a) the single phase fields are highlighted: liquid – red; Pb-rich solid solution (Pb) – green; Sn-rich solid solution (Sn) – yellow. These are separated by two phase fields, the relevant phases being identified by a horizontal tie-line, shown in (b) in the (Pb)-(Sn) field. Other common characteristics are labelled and described in the text.

Note the following features from Figure 11.4:

- The diagram divides up into single- and two-phase regions, separated by *phase boundaries*.
- At any point in a two-phase region, the phases present are those found at the phase boundaries at either end of a horizontal *tie-line* through that point, for its specific composition and temperature.
- The pure elements (Pb and Sn) have a unique *melting point* (at which the material can change from 100% solid to 100% liquid).
- Alloys show a *freezing range* over which solid and liquid coexist, between the boundaries known as the *liquidus* and *solidus*, so there is no longer a single melting point.
- Both Pb and Sn will dissolve one another to some extent – the regions marked *Pb-rich solid (Pb)* and *Sn-rich solid (Sn)*. Sn is more soluble in Pb than the reverse, but the maximum solubility in both cases is at the same temperature.
- At one special point, the *eutectic*, the alloy can change from 100% liquid to 100% two-phase solid at a fixed temperature – it is also the lowest temperature at which 100% liquid is possible.

The Pb-Sn system only contains solid solutions and mixtures of the two. The common appearance of *compounds* on phase diagrams is illustrated in Figure 11.5, part of the aluminium-copper (Al-Cu) system. The compound $CuAl_2$ contains one Cu atom for every two Al atoms: 33 atom % (at%) Cu. Converted to percentage by weight (see *Guided Learning Unit 4*), the compound is Al-53 wt% Cu. The phase diagram shows a thin vertical single-phase field for $CuAl_2$. The small spread of composition shows that the stoichiometry need not be exactly in the 1:2 atomic ratio of Cu:Al, but some excess Al can be accommodated in the $CuAl_2$

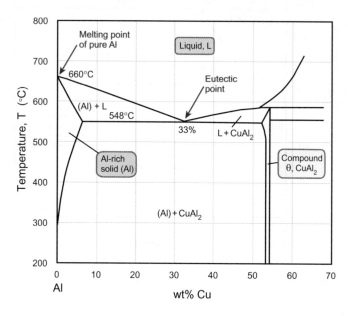

Figure 11.5 Phase diagram for part of the binary Al-Cu system, showing the compound $CuAl_2$ at 53 wt% Cu.

crystal structure. Most compounds show some spread in their composition, but some are strictly stoichiometric – the single-phase field then appears as a vertical line on the diagram. Note that between pure Al and $CuAl_2$, the form of the phase diagram is similar to the binary system in Figure 11.4, with a eutectic point at 33 wt% Cu.

Figure 11.6 shows the important Fe-C phase diagram. It, too, has a compound (iron carbide, Fe_3C) and a eutectic point (at 4.3 wt% C and 1147°C), but in other respects it is a bit more complicated, with a number of additional features (explained more fully in *Guided Learning Unit 4*):

- Pure iron shows *three* phases in the solid state: as the temperature increases iron changes from α-iron (BCC lattice) to γ-iron (FCC lattice) before reverting to BCC δ-iron and then melting.
- The single-phase *austenite* field (solid solution of carbon in FCC γ-iron) has a *eutectoid point* at its base – at this point the single-phase austenite at a composition of 0.8 wt% C can transform on cooling to a mixture of *ferrite* (a solid solution of carbon in BCC α-iron) and *cementite* (the compound Fe_3C).

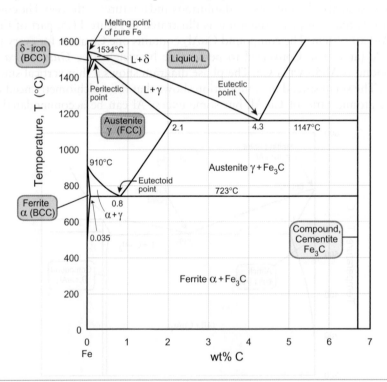

Figure 11.6 Phase diagram for part of the binary Fe-C system, up to the compound iron carbide (cementite) Fe_3C at 6.7 wt% C. The single phases are liquid and three solid solutions of carbon in iron: α-iron (ferrite), γ-iron (austenite), and δ-iron. The Fe-C system contains a eutectic point, and the austenite field closes at bottom and top, respectively, with a eutectoid point and a peritectic point.

• The austenite field closes at the top in a *peritectic point*, at which single-phase austenite with this composition transforms on heating to a mixture of liquid and BCC δ-iron.

As discussed in *Guided Learning Unit 4* (GL4.8), an important practical consequence of this diagram relates to heat treatments that exploit the difference in solubility of carbon in FCC and BCC iron.

Thermodynamics and kinetics of phase transformations We noted earlier that the equilibrium state was that with the lowest free energy at a given composition and temperature. The boundaries on phase diagrams show where the free energy of the two different states on either side of the boundary are the same. So what happens if we start at equilibrium in one field of the diagram and change the temperature until we cross a boundary?

The free energy of any state varies with temperature – Figure 11.7 shows this schematically for the case of the liquid and solid states of a pure metal. The two curves cross at the melting point, T_m, where liquid and solid can co-exist in equilibrium. Above the melting point, liquid is the state of lower free energy; below it, solid is the lower. If we cool a liquid below its melting point, its free energy as a liquid is then greater than that as a solid – the system can release energy if it solidifies. This is the change in free

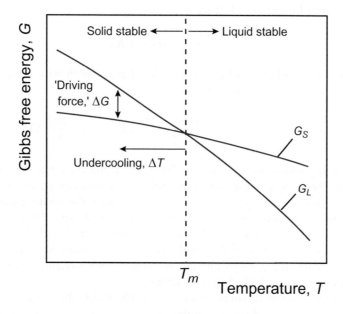

Figure 11.7 Schematic variation of free energy G with temperature for pure liquid (G_L) and solid (G_S). Above the melting point T_m, $G_L < G_S$ and liquid is the equilibrium state; below T_m, solid is stable, and $G_S < G_L$. The driving force ΔG is the difference between the two curves when, for example, liquid is undercooled by an amount ΔT below its melting point, as shown.

energy ΔG, shown in the figure, that is commonly called the *driving force*, even though it is not actually a force at all. Note from the figure that on heating a solid above its melting point, there would then be a driving force to melt to a liquid. In both cases the driving force increases with distance away from the equilibrium temperature T_m. The temperature differences ΔT below or above T_m are known as the *undercooling* or the *superheating*, respectively.

It is straightforward to show how the driving force varies with the undercooling ΔT. Consider liquid and solid at the melting point T_m, where the free energies are equal:

$$G_S = G_L \quad \text{so} \quad \left(H - T_m S\right)_S = \left(H - T_m S\right)_L \tag{11.2}$$

Hence the *changes* in enthalpy and entropy, ΔH and ΔS, between solid and liquid are related by

$$\Delta S = \frac{\Delta H}{T_m} \tag{11.3}$$

The enthalpy change ΔH ($=H_L - H_S$) may be familiar already as the 'latent heat of melting' – the heat that must be applied to change solid to liquid at its melting point. The entropy increases in proportion, as in Equation (11.3), giving no net change in free energy G.

Now if we undercool the liquid to a temperature T below the melting point, the driving force is given from Equation (11.2) as

$$\Delta G = \Delta H - T\,\Delta S \tag{11.4}$$

For modest undercooling, neither ΔH nor ΔS change much with temperature, so we can substitute from Equation (11.3), giving

$$\Delta G = \Delta H - T\left(\frac{\Delta H}{T_m}\right) = \frac{\Delta H}{T_m}\left(T_m - T\right) = \frac{\Delta H}{T_m}\Delta T \tag{11.5}$$

So to a first approximation the driving force for a phase change on cooling increases in proportion to the undercooling (or conversely the superheating, on heating). Solid-state phase changes follow exactly the same principles, with the melting point replaced with the equilibrium temperature between the two states, and with driving forces that are of the order of around one-third those for solidification. There is still 'latent heat' released at constant temperature when a solid crystal changes into another solid phase (or phases), but it is smaller. Similarly the entropy change associated with a change between two crystals is lower, as both still have a high degree of order.

As noted earlier, microstructures also evolve *without* the phases changing. An example is *precipitate coarsening*, in which a population of precipitates of various sizes evolves such that the average size increases. Small ones dissolve back into the surrounding matrix, and their atoms diffuse over and attach to the larger ones. We can still approach this using thermodynamics, considering free energy before and after the change. But in this case it is not a change of phase driving the process, but the energy associated with the *interface* between the precipitates and the surrounding matrix. This surface energy between two different crystals is associated with the atomic misfit between the phases across the boundary between them. It is analogous to the 'surface tension' that tries to pull water droplets to be spherical. The total interfacial energy of the system is the surface energy per unit area, γ, times the total area of interface. Hence coarsening occurs because the total surface area falls if the precipitates form a smaller number of larger particles.

Another important example in metals processing is recrystallisation. Deformed metals are packed full of dislocations, with an associated energy per unit length (introduced in Chapter 5). Recrystallisation wipes out the deformed grains, replacing them with new grains of much reduced dislocation density (see later) – the driving force is the change in stored dislocation energy.

Returning now to the solidification example, Equation (11.5) shows that the more we cool down, the more energy is available to drive solidification. We might therefore expect solidification (or any other phase change) to go faster and faster the more we cool a liquid below the equilibrium temperature. But this is not the case: having thermodynamics on our side is a necessary condition, but it is not sufficient – we also need a *kinetic mechanism* by which the microstructural change takes place. Atoms need to relocate, which in the vast majority of cases means they must diffuse (even if they only need to hop a fraction of the atomic spacing). This brings in a further dependence on temperature, and (crucially) a dependence on the time available – both things that are strongly influenced by the manufacturing process (as in the bike frame example in Section 11.3).

The kinetics of diffusion were discussed in Chapter 8, in relation to material performance at elevated temperatures. Diffusion is at the heart of material processing, too. Recall that there is a strong exponential temperature dependence of the rate of diffusion, known as Arrhenius' law:

$$\text{Diffusion rate} \propto \exp(-Q/RT) \qquad (11.6)$$

The rate of a phase transformation therefore depends on both the driving force and the diffusion rate. Their temperature dependencies are in opposition – the more we cool, the higher the driving force, but the lower the diffusion rate. How do they trade off? Figure 11.8 shows what happens. At low undercooling, diffusion is rapid but the driving force is small, falling to zero at the equilibrium temperature – the net transformation rate is low, limited by the driving force. At the other extreme (approaching 0 K) the driving force is maximised, but diffusion is now so slow that nothing can happen. Somewhere in between, a maximum transformation rate is found where both thermodynamics and kinetics are favourable. This is an important and counter-intuitive outcome – diffusion-controlled changes go fastest at some particular undercooling, not when they are hottest.

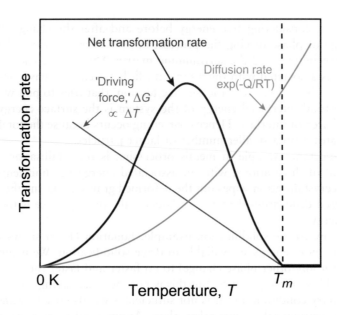

Figure 11.8 Schematic variation of transformation rate with temperature, showing the competition between driving force and diffusion rate, leading to a maximum rate at some intermediate undercooling.

Nucleation and growth There is a further refinement we should know about in materials processing. When a liquid solidifies, solid first has to appear from somewhere, after which the interface between solid and liquid can migrate to enable atoms to switch from one phase to the other at the boundary. We call these two stages *nucleation* and *growth* (Figure 11.9).

Nucleation presents another subtlety of thermodynamics. At first glance, there is a driving force, so surely off it goes? But the first nuclei to form generate an interface that wasn't

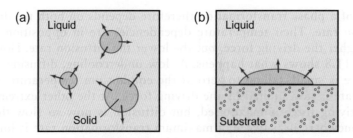

Figure 11.9 Nucleation and growth, illustrated for solidification. (a) The solid nucleus forms spontaneously as a sphere within the liquid (homogeneous nucleation); (b) the solid nucleus forms as a spherical cap on a solid substrate, such as the mould wall (heterogeneous nucleation).

there before (between the liquid and solid). As noted earlier, interfaces between phases have a surface energy (due to the atomic mismatch). So some energy has to be provided to create this interface, and the source of this energy is the free energy of the transformation. The free energy released depends on the *volume* of the new phase formed, whereas the energy penalty at the interface depends on the *area* of its surface. For small particles of the new phase the surface area energy represents a significant barrier to transformation; for larger particles the volumetric energy release dominates. As a consequence of this trade-off there is a *critical radius* for nucleation. Only above this size do nuclei become stable and grow – smaller nuclei are unstable and melt back into the liquid.

Nucleation and growth have their own characteristic rate variations with temperature, but both depend on the same factors: the driving force and the diffusion rate. So both tend to zero at the equilibrium temperature and at 0 K – hence the net transformation rate, combining nucleation and growth, still peaks somewhere in between. Figure 11.8 therefore represents the temperature dependence of the overall transformation rate.

The spontaneous formation of new phases, such as solid crystals within the bulk of a liquid, is strictly known as *homogeneous nucleation* [Figure 11.9(a)]. The figure also shows [in (b)] an alternative way to start a transformation: the new solid phase attaches itself to some other interface in the system, here the solid surface of the mould holding the liquid. This is called *heterogeneous nucleation*. It turns out that the amount of undercooling needed for this route is significantly smaller than for homogeneous nucleation, so in practice heterogeneous behaviour starts first and dominates solidification. Qualitatively we can see why. Figure 11.9 shows the same volume of new solid formed by both nucleation mechanisms. Three different interfaces are involved in the heterogeneous case (between liquid, solid, and the substrate material), with associated surface energies between each pair. The net effect is that the solid-liquid interface achieves a much larger radius than in the homogeneous case, using the same number of atoms. Growth at this interface will proceed as if it is part of a much larger nucleus. Hence we can exceed the critical radius with far fewer atoms in the solid, by forming a heterogeneous nucleus. Metals processing often exploits heterogeneous nucleation to influence the length scale of the new microstructure being formed (particularly grains). Examples are provided later.

Time-temperature-transformation diagrams and the critical cooling rate   Foundry engineers worry about how long it takes for a casting to solidify, rather than the rate at which the solid-liquid interface is moving. But it is clear that these are directly related – inversely in fact: the faster the transformation rate, the sooner it will both start and finish. First we flip Figure 11.8 on its side, so that it shows temperature against rate of transformation (Figure 11.10, left); then we can infer the figure shown to the right, in which the time taken to reach a given fraction transformed is a minimum where the rate is a maximum, and the time tends to infinity for temperatures at which the rate falls to zero. The characteristic shape leads to the name of 'C-curve' for these plots of diffusion-controlled phase transformations. A set of C-curves depicting the onset and progress of a transformation to its completion is called a *time-temperature-transformation (TTT) diagram*. For some heat treatments (notably steels) these are as central to processing as the phase diagram.

TTT diagrams are widely used to study diffusion-controlled phase transformations. The C-curve shape reflects the transformation rate as a function of temperature, assuming that we can somehow cool rapidly from above the transformation temperature and then hold isothermally at a fixed undercooling. This is fine in the laboratory with small samples, but real

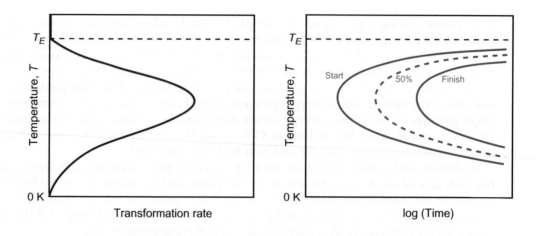

Figure 11.10 Temperature against the rate of transformation *(left)*, and the corresponding form of temperature-time-transformation (TTT) diagram *(right)* – maximum rate leads to minimum transformation time. 'C-curves' are shown for the start, 50% completion and finish of the transformation.

processing involves continuous cooling from above the transformation temperature to room temperature. In this scenario we reach for a variant set of curves called *continuous cooling transformation (CCT) diagrams*. Without going into details, the essential C-shape is retained in continuous cooling, with the curves shifted to longer times and lower temperatures.

Figure 11.11 shows the valid temperature histories for each type of diagram. From this we extract one core idea: in continuous cooling there is a *critical cooling rate* (CCR), which

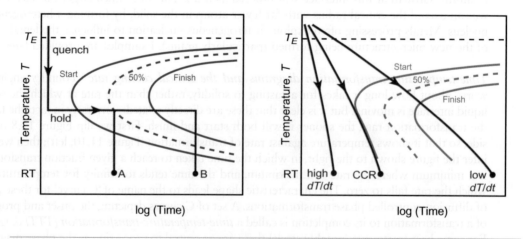

Figure 11.11 (a) A TTT diagram, for quench-and-hold isothermal treatments to room temperature *(RT)*; at B the transformation is 100% complete, while at A a 'split transformation' has occurred, with only 50% transformed; (b) a CCT diagram, broadly the same shape, but for continuous cooling histories. The critical cooling rate *(CCR)* is the minimum rate that avoids the start of any diffusion-controlled transformation.

just avoids the onset of the diffusional transformation. If the cooling rate dT/dt exceeds this value, then (in general) the initial state at high temperature is retained, at a temperature when it is no longer the equilibrium state, but is unable to transform because the kinetics are much too sluggish. In thermodynamics terms this is a *metastable* state. Forcing components to cool at an accelerated rate is a common trick in heat treatment, to deliberately bypass equilibrium and achieve an alternative microstructure as a route to better properties than slow cooling.

Introduction to Guided Learning Unit 4:
Phase diagrams and phase transformations (GL4.5-GL4.8)

At this point, we recommend that you finish working through *Guided Learning Unit 4* (GL4.5-GL4.8). These sections illustrate the evolution of microstructure for a number of important processes (solidification and heat treatment), in relation to the relevant phase diagrams and TTT diagrams.

The rest of this chapter illustrates a range of manufacturing processes and the microstructural evolution that takes place during processing. These case studies illustrate how control of composition and cooling history determine the microstructure in industrial processes, and the challenges in achieving the desired properties.

11.5 Metal shaping processes

Casting of metals Casting is a relatively cheap shaping route, well suited to making complex three-dimensional shapes. It is used over a wide range of sizes, from large ship propellers to engine parts, machinery, and sculptures, down to toys, household fittings, and medical implants. Tweaking the chemistry of castings to improve their properties has been a bit of a black art for centuries. Nowadays it is possible to use sophisticated software tools to model everything from the choice of composition to the way to pour the metal into the mould, to control the grain size and to minimise porosity and residual stress.

In casting, a liquid above its melting point is poured into a mould, where it cools by thermal conduction. New solid forms by *nucleation*, with crystals forming spontaneously in the melt, or on the walls of the mould, or on foreign particles in the melt itself [Figure 11.12(a)]. Solidification is complete when crystals growing in opposing directions impinge on one another, forming grain boundaries — each original nucleus is the origin of a grain in the solid [Figure 11.12(b)]. A fine grain size is important for optimal properties and is therefore produced by stimulating heterogeneous nucleation — fine-scale solid particles, called *inoculants*, are added to the melt (e.g. TiB_2 is added to aluminium castings). At the other extreme, we sometimes wish to grow a single crystal, with no grain boundaries at all (e.g. to give creep resistance in cast Ni superalloy turbine blades for jet engines). Here solidification is started from a small single crystal 'seed,' and the liquid is cooled slowly from that end so that no new nuclei appear and the starter nucleus grows along the entire length of the component. Examples of common cast microstructures are illustrated later and in *Guided Learning Unit 4*.

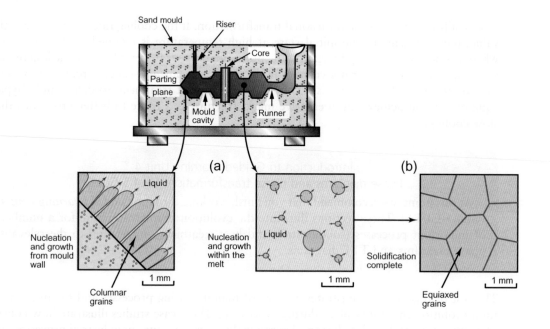

Figure 11.12 Solidification in metal casting: (a) nucleation and growth of solid crystals on the mould wall and within the melt; (b) impingement of growing solid crystals forms the grains and grain boundaries.

The rates of nucleation and growth (and thus grain size and shape) depend on the imposed cooling history, and this is governed by heat flow. This is a multiple-parameter problem: it depends on the thermal properties of the metal and the mould, the contact between the two, the initial liquid temperature, the release of latent heat on solidification, and the size and shape of the casting. This is a good example of the material-process-design coupling discussed in Section 11.2 requiring technical evaluation: the geometry of the component combines with the operation and equipment design of the process, and with the choice of alloy to dictate the cooling rate; the microstructural response to that cooling rate (and thus properties) depends on the alloy chemistry.

A further complication in casting is that the liquid has a uniform concentration of solute but the solid does not. The solubility of impurities and solutes in the solid is less than that in the liquid, which leads to concentration gradients in the solid, particularly enriching the last solid to form (i.e. the grain boundaries). This *segregation* is explained in *Guided Learning Unit 4*. It can lead to problems with embrittlement or corrosion sensitivity at the boundaries. In the case of impurity segregation, it is impractical and expensive to try and remove them from the melt. The trick is to render them harmless by giving them something else to react with, to form a solid compound distributed throughout the casting. For example, manganese is added to carbon steels in order to wrap up the impurity sulphur as harmless MnS, rather than letting it concentrate at the grain boundaries as brittle FeS.

The difference in solubility between liquid and solid is also a problem with respect to gaseous impurities dissolved in the melt. On solidification, most of the dissolved gas is rejected into the remaining liquid, but it must eventually be released as a separate phase, as trapped bubbles of gas, giving *porosity*. Good casting practice traps these bubbles as micro-porosity throughout, preventing the gas from accumulating into a big cavity in the middle of the component. A more expensive option is to 'out-gas' the melt under vacuum before it is poured. In steel-making, high levels of dissolved gas are inevitable – the carbon content is reduced to the required level by injecting oxygen into high-carbon molten iron, leaving excess oxygen in solution. To deal with this, reactive elements such as aluminium may be added to the melt. These react with oxygen in solution to form solid oxides, which are again trapped as harmless particles in the casting – a trick known as 'killing' the steel. It is another example of how a small alloying addition is used to fix a problem due to the presence of a trace impurity.

Deformation processing of metals   Solid-state deformation processes (rolling, forging, extrusion, drawing) exploit the plastic response of metals, in particular their *ductility* – that is, their ability to remain intact without damage when subjected to large strains and shape changes. Most processes are compressive rather than tensile, to avoid the problem of necking in tension. Forming is often conducted hot, to exploit the reduced yield stress, effectively shaping the components by creep at high strain-rates. Wrought metal products (i.e. shaped by deformation) are as ubiquitous as castings: beams, columns, and tubes for buildings and offshore structures; bodies, panels, and engine parts for every type of vehicle; machinery, tools, pipework, wire, food and drink containers – the list goes on. Deformation makes the required shape, but it also drives the internal changes in microstructure. The component temperature, strain-rate, and strain during shaping and the subsequent thermal history have multiple influences on microstructure and properties (outlined below).

First, the temperature determines the phases present (see Section 11.3, and *Guided Learning Unit 4*). This affects the material strength during forming but also has implications for what happens during cooling afterwards. For example, hot rolling of carbon steels takes place in the *austenite* field, at temperatures of 900°C or more (Figure 11.6). On cooling, rolled carbon steels form a mixture of grains of *ferrite* and *pearlite* (the latter being a two-phase mixture of iron carbide and ferrite), as described in *Guided Learning Unit 4*. The grain size and propor-tions of ferrite and pearlite (and hence properties) depend on the steel composition and the cooling rate. As usual, this thermal history depends on process and design variables – the size and shape of the component, the initial temperature, and how the product is handled after shaping (e.g. thin strip is coiled up, and thus remains hot for longer).

We have seen how casting determines the initial grain size in the solid state. Perhaps surprisingly, the grain structure can be manipulated further in solid-state deformation pro-cessing. In the first instance, when the component changes shape, the grains inside follow suit to accommodate the overall strain – Figure 11.13(a) shows how rolling 'pancakes' the grains. But it is the microstructure *within* the grains that matters: work hardening occurs, increasing the dislocation density. If this is the final shaping step, this is exploited to strengthen the final product, for example, deep drawing the walls of an aluminium beverage can or drawing a copper wire. For earlier forming stages, however, work hardening is a problem as it pushes up the forming loads and limits the ductility. It is therefore necessary to *anneal* the metal (soften with heat), and this heat treatment offers the opportunity to control the grain structure, too.

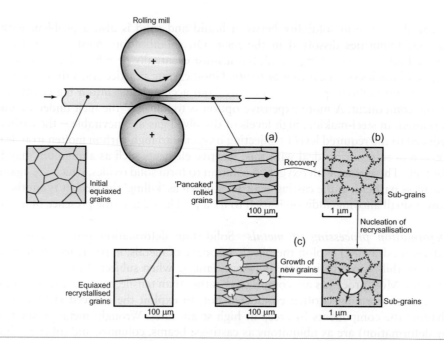

Figure 11.13 Grain structure evolution in the solid state by deformation and annealing. (a) Grains follow the shape change in the component. (b) The recovery mechanism: dislocations re-arrange as sub-grains (note the magnified scale). (c) The recrystallisation mechanism: new grains form by migration of boundaries from a few sub-grain nuclei, replacing the deformed microstructure.

Annealing involves *recovery* and *recrystallisation*, which are solid-state changes in the dislocation and grain structure at elevated temperature. This may require a separate heat treatment after forming, but they can also take place concurrently during hot deformation. In *recovery*, dislocations interact and rearrange into organised patterns forming *subgrains* – tiny regions within grains with a small change in lattice orientation at their boundaries [Figure 11.13(b)]. Extended heating leads to *recrystallisation* [Figure 11.13(c)]. New grains form from a few sub-grains that exceed a critical size, growing by migration of the grain boundaries (by atoms hopping by diffusion from the old to the new grains at the interface). This sweeps up and eliminates most of the stored dislocations, removing the work hardening completely, thereby reducing the yield stress and restoring ductility. As noted earlier, recrystallisation has a driving force (the stored dislocation energy) and a kinetic mechanism (boundary migration), though it is *not* a phase transformation. In common with solidification, however, the resulting grain size depends on the number of nuclei, here the number of sub-grains that exceed a critical size and whose boundaries migrate during recrystallisation, forming the new grains.

The scope for manipulating the grain structure during forming is therefore a major distinction between the cast and wrought alloys. On the plus side, the grain sizes in wrought alloys are much smaller (typically 10–100 μm). But it is one of the most coupled, multiple-variable problems to control, dependent on alloy composition and initial microstructure, deformation rate, process temperatures, component geometry, and strain – achieving a uniform grain structure in a component is virtually impossible. And there is yet more complexity – forming leads to a statistical distribution of orientations of the crystallographic planes within the grains, with some being strongly favoured. This grain *texture* leads to anisotropic strength and formability, particularly in rolled sheet – it's not as easy to make a circular beverage can as you might think.

11.6 Heat treatment and alloying of metals

Precipitation hardening Wrought metal products, and some castings and powder processed parts, are often subjected to heat treatment. This serves various purposes. We've already seen annealing to soften an alloy before further forming. Normalising is another – slow cooling from high temperature as the final step in making a component, the objectives being to minimise the residual stress, and to produce a microstructure of lower strength but high toughness. Many alloys also undergo heat treatments to enhance strength by *precipitation hardening*. Maximising strength without loss of toughness is essential for many heavy-duty applications of steels (e.g. gears, cranks, cutting tools). It is also essential in the low-density aluminium alloys used widely for aircraft and other transport applications.

In a typical heat treatment, the shaped component is heated to high temperature, cooled at a controlled rate, and (usually) reheated to an intermediate temperature. This exploits the solid-state phase changes that occur with temperature, the use of a quench to avoid transformations that are ineffective for hardening, and finally low-temperature precipitation for maximum strength. Figure 11.14 illustrates a common sequence:

- *Solution heat-treat* at elevated temperature, to form a solid solution.
- *Quench* to room temperature at a cooling rate dT/dt above the critical cooling rate (defined in Figure 11.11); this retains the high-temperature solid solution state at room temperature, avoiding the diffusion-controlled transformations to equilibrium low-hardening phases.
- Precipitate fine-scale hardening phases in the subsequent reheat.

Precipitation hardening is a generic mechanism in many alloy systems, but two heat treatments of this type are particularly prevalent: the 'quench and temper' of steels and 'age hardening' of aluminium alloys. These are discussed in more detail in Guided Learning Unit 4. Although both follow the general picture illustrated in Figure 11.14, there are subtle microstructural details in each case.

First consider age hardenable Al alloys (as in the bike frame example, Figure 11.3). The effectiveness of age hardening in aluminium alloys was illustrated in Chapter 6 on a fracture toughness–strength property chart (see Figure 6.17). On quenching, the supersaturated solid solution has only modest strength. On a practical note, the quench is usually an integral part of the shaping step (e.g. extrusion also provides the solution heat treatment). During ageing the yield strength rises to a peak, significantly harder than that produced by coarse equilibrium

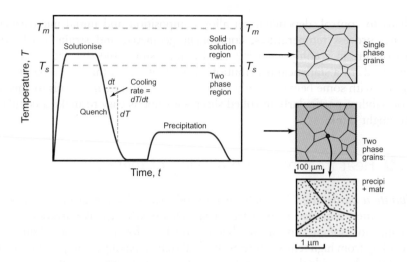

Figure 11.14 Schematic thermal profile for heat treatments that produce precipitation hardening, illustrating microstructural changes being induced: a solid solution at high temperature, and precipitation of hardening particles (shown at an enlarged scale) at a lower temperature.

precipitates on slow cooling. But this is not only because of the fine scale of the precipitates formed – age hardening exploits the formation of *metastable* phases rather than the equilibrium phases. Not all Al alloys are able to do this, but there are multiple classes of heat-treatable Al alloys that vary in their critical cooling rate, age-hardening response, weldability, susceptibility to corrosion, and so on – again, many things to consider in the technical evaluation step.

Consider now the quench and temper treatment for carbon steels. The first stage is to solutionise in the FCC austenite field ('austenitisation'), dispersing the carbon into solid solution. Quenching to room temperature traps the carbon in supersaturated solid solution. But here there is also the FCC to BCC phase change in iron to be accommodated. In spite of the quench this still takes place, by an unusual diffusionless mechanism (see *Guided Learning Unit 4*). The result is a metastable phase called *martensite*. The high supersaturation of carbon causes severe distortion of the BCC iron lattice, giving martensite a high yield strength but very low fracture toughness. In this state the component is too brittle to place in service, and is even susceptible to cracking during the quench due to induced thermal stresses. It is essential to restore the toughness, and this is achieved by tempering at an intermediate temperature. This leads to precipitation of the equilibrium phase, iron carbide. The yield strength falls as toughness is restored, but the final yield strength is substantially higher than that of the ferrite-pearlite microstructure formed on slow cooling.

The effectiveness of quenching and tempering may again be illustrated via a fracture toughness–strength property chart, as in Figure 11.15, and annotated with selected micrographs. First the effect of carbon content on the properties in the normalised condition can be seen: from 0 to 0.8 wt% carbon the strength increases steadily as the proportion of pearlite to ferrite increases. Then taking a medium carbon steel (0.4 wt% C) as an example, we see how quenching produces the unusual microstructure and properties of martensite, en route to the final tempered condition (which can be fine-tuned by choice of temper temperature, as shown in the figure). Note that the precipitation during tempering is too fine-scale to be resolved in an optical micrograph – the appearance remains similar to martensite. Quenching and tempering are further enhanced by making *alloy steels* – carbon steels with modest alloying additions – as discussed in the following case studies.

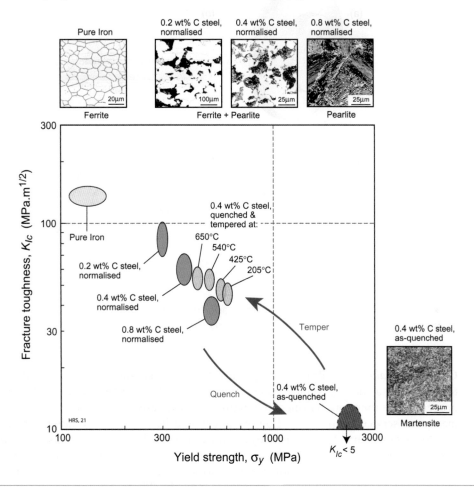

Figure 11.15 Fracture toughness–yield strength property chart for plain carbon steels, showing the effect of carbon content in the normalised condition and the effect of quenching and tempering for medium carbon steel. Selected micrographs show the corresponding microstructures. (Images courtesy: ASM Micrograph Center, ASM International, 2005).

Case studies: ferrous alloys For structural and mechanical applications, steels and other alloys based on iron dominate. They are intrinsically stiff, strong, and tough and mostly low cost. High density is a drawback for transport applications, allowing competition from light alloys, wood, and composites. Figure 11.16 illustrates the diversity of applications for ferrous alloys. These reflect the many classes of alloy – different chemistries combined with different process histories. Examining each in turn highlights the key material property, composition, and processing factors at work.

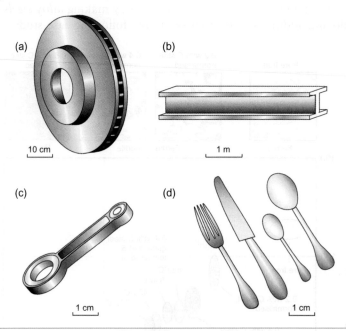

Figure 11.16 A selection of products made from alloys based on iron: (a) cast iron: brake disk; (b) low carbon steel: I-beam; (c) low alloy steel: connecting rod; (d) stainless steel: cutlery.

Cast iron: brake disk Brake disks [Figure 11.16(a)] use sliding friction on the brake pads to decelerate a moving vehicle, generating a lot of heat in the process. The key material properties are therefore hardness (strength) for wear resistance, good toughness, and a high maximum service temperature. Low weight would be nice for a rotating part in a vehicle, but this is secondary. All ferrous alloys fit the bill – so why cast iron? This is because casting is the cheapest way to make the complicated hollow shape of the disk, and the as-cast microstructure usually does not require further heat treatment to provide the hardness required. The carbon content of cast irons is typically 2% to 4% (by weight) giving a lot of iron carbide, the hard compound in ferrous alloys giving excellent precipitation hardening. Cast irons can also be transformation hardened at the surface to enhance their wear resistance (e.g. by laser hardening).

Cast irons illustrate another way in which minor alloying additions are used to make significant microstructural modifications, with consequent property changes. Figure 11.17(a) shows the microstructure of grey cast iron, with thin flakes of graphite as one phase in a two-phase matrix of ferrite and iron carbide. In this form, cast iron has excellent *machinability* – that is, it can be machined quickly with minimal or no lubricant – an example of a property that is important during processing, rather than in service. The graphite flakes are brittle and crack-like, making it easy for chips to break off during machining. In service, however, this brittleness may be a problem. Cast iron can be toughened using an alloying trick called *poisoning*. The addition of a small amount of cerium or magnesium has the effect of radically changing the morphology (shape) of graphite in iron – it spherodises instead of forming the usual crack-like flakes, and the fracture toughness and failure strength of the cast iron are enhanced (illustrated later). The surrounding matrix remains ferrite and iron carbide, depending on the alloy composition and cooling history. Figure 11.17(b) shows such a 'nodular' cast iron containing spheroidal graphite, in this case with a surrounding matrix of ferrite. This poisoning technique is also used to enhance toughness in Al-Si casting alloys – small sodium additions lead to the brittle Si forming fine-scale rounded clumps, rather than the needles characteristic of eutectic solidification (see *Guided Learning Unit 4*).

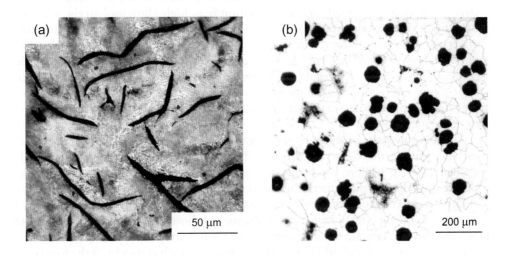

Figure 11.17   Alternative microstructures in cast iron containing 3.5 wt% carbon: (a) grey cast iron, showing flake graphite; (b) nodular cast iron (at a lower magnification), showing spherodised graphite.

Plain carbon steel: I-beams, cars, and cans   If we had to single out one universally dominant material, we might choose mild steel – iron containing 0.1% to 0.2% carbon. Almost all structural sections [Figure 11.16(b)], automotive alloys, and steel packaging (beer and food cans) are made of mild steel (or a variant enhanced with a few other alloying additions). All of these applications are wrought – the alloy is deformed extensively to shape. The microstructure consists of ferrite and a small amount of pearlite. This provides the excellent ductility needed both for processing and in service. The fracture toughness is high, and is combined with the inherent strength of iron, which may be increased by deformation processing (work hardening). Ductility, toughness, and decent strength are vital in a structural material – we would rather a bridge sagged a little rather than broke in two, and in a car crash our lives depend on the energy absorption of the front of the vehicle.

Alloy steels and heat treatment: cranks, tools, and gears The moving and contacting parts of machinery are subjected to very demanding conditions: high bulk stresses to transmit loads in bending and torsion (as in a crank or a drive shaft), high contact stresses where they slide or roll over one another (as in gears), and often reciprocating loads promoting fatigue failure (as in connecting rods [Figure 11.16(c)]). Strength with good toughness is everything. Plain carbon steels up to 0.8 wt% C give effective precipitation hardening, particularly when quenched and tempered (Figure 11.15). But the density of iron is a problem. Make the steel stronger, and you can use less of it and save weight. This is particularly true for fast-moving parts in engines because the support structure can also be made lighter if the inertial loading is reduced (so-called 'secondary weight savings'). By now the solution should not be a surprise: yet more alloying, and suitable processing.

A bewildering list of additions to carbon steel can be used to improve the strength – Mn, Ni, Cr, V, Mo, and W are just the most important! Some contribute directly to the strength, giving a solid solution contribution (e.g. high-alloy tool steels, with up to 20% tungsten). More subtle though is the way quite modest additions (<5% in low alloy steels) affect the alloy's response to the quench and temper heat treatment (discussed earlier for plain carbon steels). The problem with quenching plain carbon steels is that the critical cooling rate is high. As a result, only thin sections can be quenched to form martensite – the cooling rate within the component being physically limited by heat flow, governed by the component thickness, and the inherently low thermal conductivity of steel (compared to other metals).

Alloying solves the problem by slowing the diffusional transformations to ferrite and iron carbide, shifting the C-curves of Figure 11.11 to much longer times. This enables slower quench rates to achieve the target martensitic microstructure. Hence bigger components can be quenched to this supersaturated state and thus tempered. The technical term for the ability of a steel to form martensite is *hardenability* – the higher the hardenability, the lower the critical cooling rate. In high alloy steels, such as those used for cutting tools, the effect is so pronounced that cooling in air is sufficient to form martensite. An additional benefit of using alloy steels is that a bit of additional precipitation strength comes from the formation of alloy carbides (rather than iron carbide). Compounds such as tungsten carbide survive to higher temperatures than iron carbide without dissolving back into the surrounding iron – a useful source of high-temperature precipitation strength (e.g. for cutting tools).

For practical design purposes, it is more helpful to relate the response of an alloy steel to the *size* of component that can successfully be quenched and tempered, rather than a critical cooling rate. An industrial test for hardenability is the *Jominy end-quench*: a bar of standard dimensions is austenitised and then suspended vertically and water-quenched on the bottom end [Figure 11.18(a)]. This gives a continuous variation in cooling rate with distance from the end [Figure 11.18(b)], with the rate being defined as the average rate between two reference temperatures. As martensite is so much harder than any other state in carbon steels, measuring the hardness variation along the bar is sufficient to indicate the extent of martensite formation. Typical hardness profiles for a plain carbon and an alloy steel (of similar carbon content) are shown in Figure 11.18(c), for both as-quenched and tempered conditions. The alloy steel retains a high hardness for a much greater distance than the plain carbon steel – that is, martensite can form at a much lower cooling rate and can subsequently be tempered, meaning it has high hardenability.

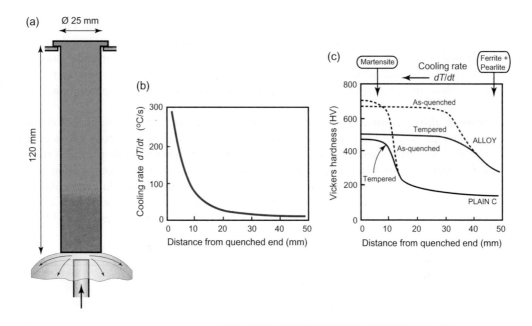

Figure 11.18 (a) Jominy end-quench test; (b) the corresponding variation of cooling rate along the bar. (c) Hardness profiles indicate the distance over which a substantial fraction of very hard martensite forms in the as-quenched condition, and thus the distance that can achieve a target hardness in the tempered condition.

The last design step is to relate cooling rates in real components to Jominy data, which is simply a matter of heat flow analysis. For a given quenching medium (e.g. into water or oil, or just cooling in the air), the cooling rate at the interior point of slowest cooling in the component can be calculated (using the same reference temperatures to define the cooling rate as in the Jominy test). This cooling rate correlates with a specific Jominy distance, so the hardness at this Jominy distance is that expected in the centre of the component in a given alloy. Everywhere else the hardness will be equal or greater than this because, by definition, the cooling rate is higher in all other locations – indeed a higher surface hardness is often desirable, for wear resistance.

So in its full complexity, heat treatment of steels presents another example in which the properties depend on coupling between material, process, and design detail – the alloy composition determines the hardenability, and the quenching process and component size determine the cooling rate, and hence the microstructural outcome and final properties.

Stainless steel: cutlery The English city of Sheffield made its name on stainless steel cutlery [Figure 11.16(d)], and still manufactures engineering steels. The addition of substantial amounts of chromium (up to 20% by weight) imparts excellent corrosion resistance to iron, avoiding one of its more obvious failings: rust. It works because the chromium reacts more strongly with the surrounding oxygen, protecting the iron from attack (see Chapter 9). Nickel is also usually added for other reasons – one being to preserve the material's toughness at the very low cryogenic temperatures needed for the stainless steel pressure vessels used to store liquefied gases. Another is that both chromium and nickel provide solid solution hardening – this and work hardening (during rolling and forging) being the usual routes to strength in stainless steels, though some Fe-Cr-C alloys are also heat treated for precipitation strength.

Ferrous alloys: summary In all these applications, strength and toughness dominate the property profile, and in summary we can now map all classes of ferrous alloys on the chart of these two properties, as in Figure 11.19. This paints a remarkable picture – starting with soft, tough pure iron (top left), we can manipulate it, by changing composition and process history, to produce more or less any combination of strength and toughness we like, increasing the strength by a factor of 20 or more. Every final material (with the exception of as-quenched martensite) is above the typical fracture toughness threshold (>15 MPa.m$^{1/2}$)

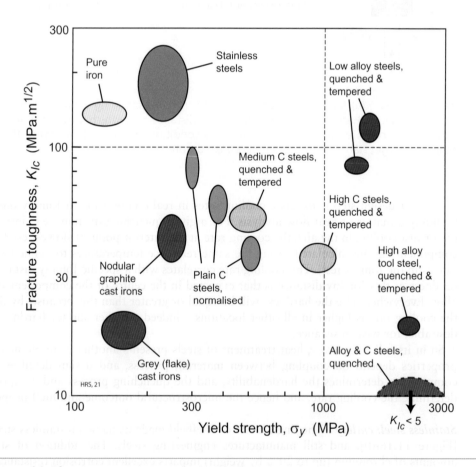

Figure 11.19 Fracture toughness–yield strength property chart for ferrous alloys. Composition and heat treatment enable a wide range of combinations of these properties.

for structural or mechanical application. Cast irons primarily offer low cost and lower pouring temperatures than pure iron, for a modest increase in strength, but note the effectiveness of poisoning in restoring toughness by forming nodular rather than flake graphite (Figure 11.17). Plain carbon steels in the normalised condition are standard for structural use, unless the corrosion resistance of stainless steels is required. The chart shows the benefit of quench-and-temper treatments in raising the strength without damaging the toughness, particularly in alloy steels. This figure explains why ferrous alloys are so important: no other material is so versatile, with literally hundreds of different steels and ferrous alloys available commercially.

11.7 Joining, surface treatment, and additive manufacturing of metals

Joining of metals   Fusion welding of metals involves heating and cooling. This may cause phase transformations and other microstructural evolution in: (1) the 'weld metal,' which is like a localised continuous miniature casting, where melting and resolidification occur; and (2) the solid region around the weld, where microstructural changes can occur as in a heat treatment process, but with thermal cycles taking seconds rather than hours. The same goes for hot, solid-state welding, with the weld metal replaced by a hot deformation zone. The solid region surrounding a weld where property changes occur is referred to as the 'heat-affected zone' (Figure 11.20). Once again the thermal history couples the material, the way the process is operated (e.g. power and speed), and design details such as joint thickness. Thermal welding can have major consequences for joint performance – depending on the alloy and the welding process, the material may be softened, or embrittled, or lose its corrosion resistance. Figure 11.20 illustrates the contrasting hardness profiles after welding, as observed in low carbon and low alloy steels, and heat-treatable aluminium alloys.

Hardness profiles are the simplest first indicator of property changes across a welded joint. Low carbon steels are weldable, as on cooling the material reverts to ferrite and pearlite – the weld and HAZ have similar hardness to the original material. Low alloy steels form martensite more easily, as the critical cooling rate is lower, increasing the risk of *weld embrittlement* (the high hardness of martensite being a characteristic indicator). Oil rigs and bridges have collapsed without warning as a result of this behaviour, usually because the welding process was conducted incorrectly, leading to cooling rates that were too high. Tempering a steel weld to eliminate martensite is expensive – better to avoid it in the first place by choice of an appropriate steel and welding procedure. In contrast, heat-treatable aluminium alloys are age hardened to their peak strength – on welding there is only one way for the hardness to go, and that is down. Over-ageing due to the weld thermal cycle can lead to a permanent loss of 50% or more of the strength. In weldable Al alloys, the metastable precipitates dissolve instead, enabling some subsequent recovery of strength by room temperature ('natural') ageing.

Welding metallurgy is a big field of study – welds are often the 'Achilles heel' in design, being critical locations that determine failure and allowable stresses. The material properties may be damaged in some way, as earlier, but welding can also generate residual stresses, stress concentrations, cracks, and crevices, potentially causing fatigue and corrosion problems. Adhesive technologies have the advantage that they don't deform or heat the components being joined, leaving the microstructure and properties intact. But they are not trouble free – they still concentrate stress and contain defects, requiring good design and process control.

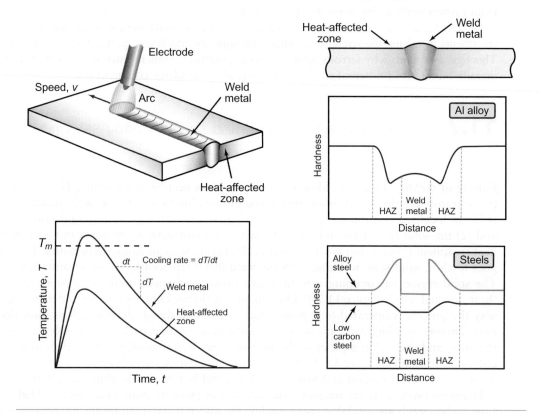

Figure 11.20 A weld cross-section with corresponding thermal histories in the weld metal and heat-affected zone. To the right are typical hardness profiles induced across welds in heat-treatable aluminium alloys, low carbon steel, and low alloy steel.

Surface treatment of metals Surface engineering exploits many different mechanisms and processes to change the surface microstructure and properties. Some simply add a new coating material, with its own microstructure and properties, leaving the substrate unchanged – the only problem is then making sure that they stick. Others induce near-surface phase transformations by local heating and cooling – *transformation hardening*. Figure 11.21 shows an example – laser hardening of steels. The laser acts as an intense heat source traversing the surface, inducing a rapid thermal cycle in a thin surface layer. Heating a depth of order 1 mm into the austenite field is readily achieved, with conduction into the cold material underneath providing the quench needed to form martensite. The high hardness of martensite gives excellent wear resistance, and the brittleness is not a problem when it is confined to a thin layer on a tough substrate.

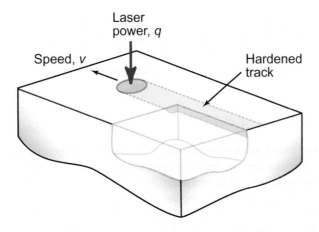

Figure 11.21 Laser hardening: a surface treatment process that modifies microstructure. The traversing laser beam induces a rapid thermal cycle, causing phase changes on both heating and cooling. The track below the path of the laser has a different final microstructure of high hardness.

Direct diffusion of atoms into the surface is also feasible, but only for interstitial solute atoms at high temperature for diffusion to occur over a distance of any significance. This mechanism is used to increase the carbon content (and thus hardness) of steels – the surface treatment known as *carburising* (see Chapter 8). Even better is a hybrid surface treatment: first carburise, and then transformation harden too, giving very hard high carbon martensite.

Additive manufacturing of metals AM processes for metals use localised moving heat sources, such as a laser, electron beam, or welding torch, to sequentially melt a powder or a wire that fuses as a solid deposit onto a part, building it up layer by layer. The pros and cons of AM processes for metals are explored in *Guided Learning Unit 3*, with the principal limitations being their low economic production volumes and poor surface finish. Nonetheless, AM is proving attractive in a number of fields, notably aerospace, automotive, and medical engineering, particularly for complex shapes and bespoke parts in high-strength alloys (Ti and Ni alloys, stainless, and tool steels).

As for all metals processing, the resulting microstructure and properties of an AM component depend on details of how the process is operated, the part geometry, and the alloy – the usual coupling of heat flow, temperature history, and material response to that history. The local melting of powder/wire by a traversing heat source means that AM resembles both laser surface cladding and fusion welding, and the microstructures and defects produced are indeed very similar. But there are specific challenges in AM due to the sequential layering process. This leads to a periodic surface ripple perpendicular to the layers, while unmolten powder tends to stick to the outer surface of the deposit – both reasons for the high surface roughness associated with AM. Sequential layer deposition also results in many interfaces between each new deposit and the material below. In metals this tends to lead to more trapped oxide and porosity, with poorer mechanical properties. On the other hand, the rapid cooling in AM can produce fine-scale microstructures with enhanced properties – the outcome will be sensitive to the exact composition and imposed cooling history.

11.8 Powder and glass processing

Ceramics, glasses, and powder metallurgy Compared to metals, ceramics and glasses are relatively niche engineering materials. Ceramic applications exploit their high hardness (cutting and grinding tools), high-temperature resistance (furnace liners), and high electrical resistivity (spark plugs, electronics, power line insulators). Ceramic products can only be shaped by filling a mould with loose powder and compacting it at high temperature (Figure 11.22). Glasses are used mainly for optical properties, both in sheet and fibre form, and for containers such as bottles – applications that also exploit the rather easier processing of glass, by moulding and drawing of (amorphous) viscous liquid. But ceramics and glasses all have the problem of brittleness. No matter what other advantages they have, they are

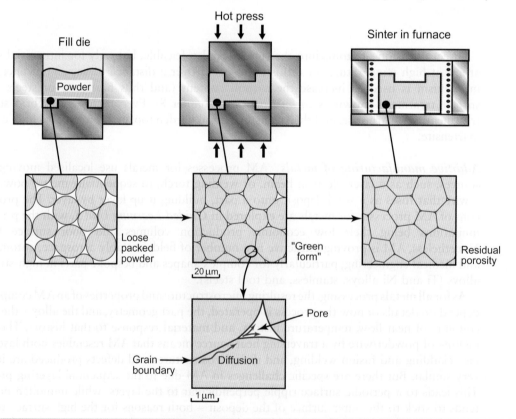

Figure 11.22 Mechanism of powder compaction in hot pressing and sintering.

susceptible to fracture, which may be traced back in many cases to the inherent defects built in during processing. Defect control is therefore a key factor in ceramic and glass processing. The main microstructural evolution during compaction is the shrinkage of porosity, with the boundaries between particles becoming grain boundaries. Powder compaction mechanisms are closely related to those of diffusional flow and creep (see Chapter 8). In purely thermal compaction (sintering), atoms diffuse along the particle boundaries to fill in the pores, as shown in Figure 11.22. Compaction is accelerated by imposing external pressure as well as temperature – 'hot isostatic pressing' (or HIPing) – giving particle deformation by creep.

The pores and microcracks in ceramics largely reflect the size of powders used in the first place, but good-quality die design and mould filling procedures are then needed to control the porosity distribution. Uniform compaction is surprisingly difficult to achieve, even in simple shapes (due to the large contraction from loose packed powder to finished near-fully-dense product).

Ceramic compositions tend to be largely discrete, each being a specific compound (alumina, silicon carbide, etc.). Some ceramic-metal composites are produced, such as tungsten carbide and cobalt, to make a hard, tough material for dies and cutting tools. In contrast, glass compositions can be tuned continuously over a wide range to achieve particular combinations of colour and other optical properties (e.g. the graded compositions used in optical fibres). More recent developments with glass processing are associated with coating technology – examples are scratch and chip resistance (in windscreens) or self-cleaning windows in high-rise buildings. Window coatings also have an increasingly important role in the sustainability of buildings, reducing heating and cooling energy demands by limiting the influx of excessive solar heating in the summer, while in the winter letting in the light but limiting the radiated heat losses.

Powder metallurgy uses similar compaction techniques, usually for metals with high melting point or high strength, or for thin sections (for which casting and deformation are difficult), or where a homogeneous microstructure is important (e.g. tool steels, refractory or magnet materials). It is also used to produce special products for which this process offers other key benefits – such as controlled amounts of porosity (useful in bearings to provide oil-traps to maintain lubrication) or to inlay small-scale features of one alloy into another (such as designer logos in various consumer products). Once compacted, the response to heat treatments follow the same underlying metallurgy of phase transformations as discussed previously (e.g. powder processed steel components may well be quenched and tempered for optimum strength and toughness).

11.9 Polymer and composite processing

Polymers and elastomers Polymers are versatile materials in many respects – easy to mould into complex shapes at low cost, easy to join together by snap fitting, and easy to colour. Bulk polymers provide a spread of values for Young's modulus, strength, and fracture toughness, though all are relatively low. Polymer chemistry, and the use of additives and fillers, give some freedom to adapt these properties, though the ability to make large changes in strength is much more limited than for metals. But bulk polymers offer scope for changing elastic properties, which metals do not. Polymers find wide application in packaging, for diverse household and workplace applications (computer and phone casings, toys, pens, kitchen appliances, and containers), and notably in transport: typically 15% of a car's weight is made up of thermoplastics. Design with polymers is often a question of ensuring sufficient stiffness, through shape detailing (discussed in *Guided Learning Unit 3*) and choice of polymer and fillers, but low cost and aesthetics are other key design drivers.

Chapter 3 introduced the structure of polymers, and how chemistry is used to produce the thousands of variants (or 'grades') available today. Long-chain molecules are produced by polymerisation of a basic monomer unit of carbon, bonded with hydrogen, chlorine, oxygen, and so on. Each monomer has a discrete chemistry – C_2H_4 makes polyethylene, for example. Dissimilar monomers only polymerise together in selected combinations, to make *copolymers* (with two monomers) or *terpolymers* (with three). For example, ABS is a terpolymer, combining monomers of acrylonitrile, butadiene, and styrene (hence the name). Many grades of polypropylene (PP) are copolymers with ethylene, to improve toughness at low temperatures. *Polymer blending* extends the range further, giving a mixture of more than one type of chain molecule (e.g. PBT-PC). This, too, is rare – in contrast to the ease of forming liquid and solid solutions atomically in metals – because the thermodynamics of mixing of dissimilar polymer molecules works against molecular inter-diffusion (something that has implications for thermoplastic recycling, when different polymer products can be difficult to separate before remelting). Recycling is also complicated by the fact that bulk polymers usually contain additives, dyes, and fillers, for all sorts of reasons: as flame retardants, colouring agents, plasticisers, or for protection against UV degradation, as well as glass fibres or ceramic particles to enhance mechanical or thermal properties. Processing primarily plays a role, in a given polymer or elastomer, via manipulation of the molecular configuration. As described in Chapters 3-5, some thermoplastics will partially crystallise, and thermosets and elastomers are cross-linked. These are influenced by the polymer chemistry but also (as for metals) by the deformation history and rate of cooling after shaping.

Polymer moulding   In spite of the wide variety in polymer chemistries, most processes work with most polymers. Thermoplastics are commonly moulded as viscous liquids, so what matters are the factors that change the melt viscosity. Particulate and fibre fillers increase the viscosity, so the ease of moulding will be influenced by the filler volume fraction. But there are some variations even with virgin polymer, which always comes with a statistical distribution in the length of the molecular chains – viscosity is influenced by the average chain length and its standard deviation. Large, thin parts can only be easily injection moulded using grades with a low 'melt flow index' (MFI), i.e. consisting of relatively short and light chains. The downside of a short chain length is principally a reduction in toughness. Other common thermoplastic moulding processes, such as extrusion, have no difficulty with long-chain, high-MFI polymers – and we get better toughness as a bonus.

For semi-crystalline thermoplastics, crystallinity is an important factor to consider in processing. For polypropylene and polyethylene, crystallinity is essential – without it their stiffness and strength would be too poor for use at room temperatures. Crystallinity leads to increased shrinkage – as much as 5% by volume compared to the amorphous state. Allowance must be made for this when designing a mould, or process operating conditions adjusted, for example, by maintaining the pressure in injection moulding after filling, to supply more polymer to compensate for the shrinkage. The thermodynamic and kinetic factors controlling the evolution of crystallinity in relation to temperature and time are broadly the same as for the phase transformations discussed earlier in the chapter – indeed, on cooling a semi-crystalline thermoplastic from the melt, the extent of crystallinity follows a set of C-curves (as in Figure 11.11). Hence variations in cooling rate within a single moulding, or from batch to batch, can lead to variability in crystallinity, and thus variable shape accuracy and final properties. The sensitivity of crystalline fraction to process history explains why amorphous thermoplastics are often preferred for high-precision parts.

In contrast to injection moulding, processes like extrusion, thermoforming, and blow moulding are more analogous to deformation processing than to casting – the polymer is deformed or stretched to shape. Process conditions again play a significant role in determining the final molecular arrangement, and thus properties, because the strain imposed during polymer deformation may partially align the molecules, and if cooling is sufficiently rapid this alignment may be 'frozen in.' Figure 11.23 shows the blow moulding of a drink bottle, from a preformed closed tube called a *parison*. In some variants the parison is stretched longitudinally with an internal rod prior to blow moulding. The stretching process of blow moulding aligns the molecules in the bottle wall, enhancing stiffness and strength. The degree

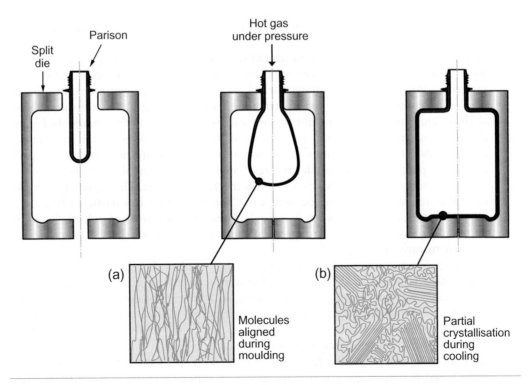

Figure 11.23 Evolution of molecular microstructure in polymer moulding: (a) alignment of molecules during viscous flow in shaping; (b) partial crystallisation during cooling.

of stretch is even customised to give greater enhancement in the circumferential direction, which carries the greater tensile stress under pressure. Partial crystallisation during cooling may also enhance stiffness and strength in semi-crystalline thermoplastics.

Additive manufacturing of polymers   Polymer AM processes were assessed in relation to conventional processes in *Guided Learning Unit 3*. As for metals, the range of processes and their applicability across an increasing range of polymers is steadily evolving. The key limitations are the same – only modest production runs are economic, and the surface finish is poor. Polymer AM is primarily used in concept design, for visualisation models and shape prototyping, including making scale models (e.g. of buildings for presentations). If the polymer acts as a binder with a metal or ceramic, AM can also be used for pre-forming complex powder metal parts, and the manufacture of sand moulds for conventional casting. A valuable area of application is to make cheap bespoke parts, such as small batches of shaped containers, or components like handles and hinges needed to repair and keep old products in service. Examples such as these have been a lifeline in health care in developing economies for medical consumables and hospital diagnostic equipment. As for bulk shaping, polymers in AM show less microstructural sensitivity than metals, and the resulting finish and properties are less of an issue, as strength is rarely critical in AM applications of polymers. Polymer AM often implies a 'jetting' process, which tends to produce disc-shaped 'splats' of solid, as the droplets flatten when they hit the surface, giving a greater area of interface (and associated porosity) parallel to the layer build direction. This causes some anisotropy in strength and ductility.

Polymer fibres   The most dramatic impact of processing on polymer properties is in making *fibres*. These are drawn from the melt, or for even better results, cold drawn. Now the molecular alignment is much more marked, so that the fibre properties exploit the covalent bonding along the chain. We can illustrate the resulting strength and stiffness on a property chart. To emphasise the lightweight performance with respect to other materials, Figure 11.24 shows the specific strength σ_y/ρ and specific stiffness E/ρ. Bulk polymers, in spite of all the chemical trickery, span only about a decade in strength and stiffness, and are outperformed by metals, particularly on specific stiffness. Standard polymer fibres of nylon or polyester are stiffer and over 10 times stronger than the bulk polymer. They quickly overtake most natural fibres, such as cotton and hemp, but not silk, which is a particularly good natural fibre. However, aramid fibres (such as Kevlar), and special grades of polyethylene fibres, are up to 100 times stiffer and stronger than bulk polymer, with specific properties close or superior to carbon and glass fibres.

Fibre properties are all very impressive – the problem though is getting the fibres into a form that is useful to the engineer. One option is composites – aramid fibres are used in the same way as carbon and glass fibres, in a matrix of epoxy resin. As discussed in Chapter 4, the resulting composites shown in Figure 11.24 lie between those of the matrix and fibre, where they compete strongly with the best metals, such as low-alloy steels. Alternatively, we use polymer fibres in exactly the way that natural fibres have been used for millennia: twisted, tangled, or woven to make rope, cables, felt, fabrics, and textiles. Fibre-based materials are a whole discipline of materials engineering in themselves, with many important applications. The geometric structure of ropes and textiles is essential to the way they work – it gives them flexibility in bending, combined with high stiffness and

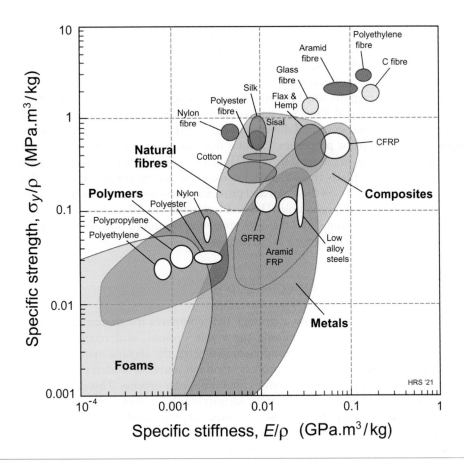

Figure 11.24 Strength-specific stiffness property chart. Bulk polymers are stiffened and strengthened by drawing into fibres, competing with natural, carbon, and glass fibres. Most products use the fibres in a composite, such as CFRP, or woven into a fabric.

strength in tension. The drape of clothing exploits the differences in stiffness parallel to and at 45° to the weave (the 'bias'). Ropes can be stretched, coiled, and knotted elastically due to their twisted and woven architecture. The loss of alignment of the fibres in making a rope or weave does mean that their load-carrying capacity is never quite as good as that of the fibres themselves in their pristine form. Nonetheless, it is clear from the property chart that cables based on the best polyethylene fibres, for example, can potentially compete with conventional steel cabling.

Making composites   As noted previously, most polymers are in fact particulate composites – bulk polymer melts to which filler powders have been added before moulding to shape. We opt to classify these as polymer variants, largely because the scale of the additives is very fine, almost at the scale of the polymer's own molecular microstructure. Fibre-reinforced polymer composites – CFRP, GFRP, and KFRP (reinforced with the aramid fibre Kevlar) – contain chopped or continuous fibres on a much coarser length-scale than the molecular, and require dedicated manufacturing techniques. Their potential performance in lightweight design has been illustrated in earlier chapters, but can these ideal properties be achieved routinely in practice? The answer depends on the complexity of the component geometry and the impact of joining. Composites are straightforward to produce in flat panels. 'Pre-preg' layers of partly cured resin containing unidirectional fibres are stacked and cured under pressure in an autoclave, usually with the layers aligned in several directions (e.g. 0°, 90°, and ±45°) to give reasonably uniform 'quasi-isotropic' in-plane properties. Circular tubes are also easy to make by filament winding. More complex shapes are more difficult to laminate, particularly if the section changes in thickness. Joints are particularly troublesome. Fibre composites cannot be welded, and drilling holes for bolts and rivets damages the fibres and provides stress concentrations that can initiate cracks and delaminations around the hole. The best technique is adhesive joining, and the increasing use of composites has stimulated the development of improved adhesives. Careful joint design and manufacture are necessary to avoid premature failure in or around the joint.

11.10 Summary and conclusions

Selecting a process must follow a strategy that compares the design requirements with the capabilities of processes, determining the viable options, and recognising the significant coupling between details of the component geometry, the specific material composition, and the processing conditions. Generic process attributes and cost models can capture some of the simple technical, quality, and economic capabilities of a process. This enables processes to be screened out and ranked approximately, identifying the classes of process to consider in full technical detail. A systematic approach to initial screening and cost modelling is presented in *Guided Learning Unit 3*, which also outlines the key design issues for technical evaluation. As in earlier chapters of this book, control of microstructure and properties emerges as a recurring theme, in which processing plays a central role. Shaping processes such as casting, forming, and moulding do more than make a shape – simultaneously they impose a thermal history and often a deformation history that govern the evolution of microstructure and properties – in all classes of materials. Surface treatments may locally change the microstructure and properties, and joining components together can change things further – joints often being the design-limiting location or the source of disaster. Therefore processing offers many opportunities for innovation in design, together with responsibility for careful control. The theme of this chapter is the influence of processing on achieving the target properties in design of products with each class of materials.

Metals show the greatest versatility – not surprising given the many different elements and alloy compositions available, the diversity of microstructures that form in the solid state, and their inherent castability, formability, and weldability. One class alone (ferrous alloys) covers a wide domain of the key structural properties of strength and toughness. Other alloy systems use the same principles of modifying chemistry, forming, and heat treatment to develop different microstructures and property profiles. Many of the microstructural changes presented are governed by the theory of phase diagrams and phase transformations introduced in this chapter and covered in greater depth in *Guided Learning Unit 4*.

Ceramics, glasses, and polymers have some scope for variation of composition and process. Blending, additives and fillers, cross-linking, and crystallinity provide modest ranges in the properties of polymers, including the elastic modulus. Most effective is the production of fibres, taking bulk polymers into completely new territory, well beyond the performance of natural fibres. Making use of these fibre properties in practical products needs imaginative hybrid construction and processing as composites, ropes, textiles, and fabrics.

11.11 Further reading

Ashby, M. F. (2017). *Materials selection in mechanical design* (5th ed.). Butterworth-Heinemann. ISBN 978-0-08-100599-6. (The source of the process selection methodology, giving a summary description of the key processes in each class).

Ashby, M. F., & Jones, D. R. H. (2013). *Engineering materials II* (4th ed.). Butterworth-Heinemann. ISBN 978-0080966687. (Popular treatment of material classes, and how processing affects microstructure and properties).

ASM Handbook Series (1971–2004). Heat treatment (vol. 4); Surface engineering (vol. 5); Welding, brazing and soldering (vol. 6); Powder metal technologies (vol. 7); Forming and forging (vol. 14); Casting (vol. 15); Machining (vol. 16). ASM International. (A comprehensive set of handbooks on processing, occasionally updated, and now available online at www.products.asminternational.org/hbk/index.jsp).

Kalpakjian, S., & Schmid, S. R. (2013). *Manufacturing engineering and technology* (7th ed.). Pearson Education. ISBN-13: 978-0133128741. (Like its companion below, a long-established manufacturing reference).

Kalpakjian, S., & Schmid, S. R. (2016). *Manufacturing processes for engineering materials* (6th ed.). Pearson Education. ISBN-13: 978-0134290553. (A comprehensive and widely used text on material processing).

Tempelman, E., Shercliff, H. R., & Ninaber van Eyben, B. (2014). *Manufacturing and design – Understanding the principles of how things are made* (1st ed.). Butterworth-Heinemann. ISBN 978-0-08-099922-7. (The source of the 'manufacturing triangle' in this chapter, and the approach to the technical evaluation of process-material-design interactions).

Thompson, R. (2015). *Manufacturing processes for design professionals.* Thames and Hudson. ISBN-13: 978-0500513750. (A structured reference book with remarkable diverse illustrations, for industrial and product designers, architects, and engineers alike).

11.12 Exercises

Exercise E11.1 Define the following terms in relation to phase diagrams: free energy, thermodynamic equilibrium, solidus, liquidus, freezing range, eutectic point, compound.

Exercise E11.2 Define the following terms in relation to phase transformations: driving force, latent heat, Arrhenius' law, nucleation and growth, TTT diagram, critical cooling rate.

Exercise E11.3 Briefly explain the meaning of the following terms in relation to casting processes: homogeneous nucleation, heterogeneous nucleation, inoculants, segregation, aluminium-killed steel, poisoning.

Exercise E11.4 Briefly explain the meaning of the following terms in relation to deformation processing: annealing, recovery, recrystallisation.

Exercise E11.5 Blacksmiths and other traditional manual metal-workers have hammered wrought ferrous alloys into shape for centuries. Use your knowledge of deformation processing to suggest reasons for forming these alloys both when they are red-hot and when they are cold.

Exercise E11.6 Briefly explain the meaning of the following terms in relation to bulk and surface heat treatment of metals: age hardening, martensite, tempering, hardenability, transformation hardening, carburising.

Exercise E11.7 Exercises E11.3, E11.4, and E11.6 cover many of the reasons that alloying additions are made in metals processing. See if you can identify even more reasons for alloying, with examples.

Exercise E11.8 The table shows typical data for strength and fracture toughness of a selection of copper alloys, both cast and wrought. Sketch a property chart (on log scales) and plot the data. Use the chart to answer the following:

(a) How do the cast and wrought alloys compare on fracture toughness, at comparable strength?
(b) Rank the strengthening mechanisms (as indicated in the table) in order of effectiveness.
(c) Do the trends observed in (a) and (b) follow a similar pattern to aluminium alloys (see Figure 6.17)?

Alloy	Process route	Main strengthening mechanisms	Yield strength (MPa)	Fracture toughness (MPa.m$^{1/2}$)
Pure Cu	Cast	None	35	105
Pure Cu	Hot rolled	Work	80	82
Bronze (10% Sn)	Cast	Solid solution	200	55
Brass (30% Zn)	Cast	Solid solution	90	80
Brass (30% Zn)	Wrought + annealed	Solid solution	100	75
Brass (30% Zn)	Wrought	Solid solution + Work	400	35
Cu – 2% Be	Wrought + heat treated	Precipitation	1000	17

Exercise E11.9   The table shows the yield strength for a number of pure metals in the annealed condition, together with typical data for the strongest wrought alloys available based on these metals. By what factor is the strength increased in each alloy system, relative to that of the pure metal on which it is based? Which shows the greatest absolute increase in strength?

Base element	Yield strength (MPa) (pure metal)	Strongest wrought alloy	Yield strength (MPa) (alloy)
Iron	120	Fe-Mo-Co-Cr-W-C quenched and tempered tool steel	2750
Aluminium	25	Al-Zn-Mg-Cu age-hardened	625
Copper	35	Cu-Be-Co-Ni age-hardened	1250
Magnesium	65	Mg-Al-Zn-Nd age-hardened	435
Nickel	70	Ni-Be cold worked	1590

Exercise E11.10   The cooling rate dT/dt (in °C s^{-1}) at the centre of a long bar of steel is predicted to vary varies with the diameter D (in mm) according to the equation:

$$dT/dt \approx 12{,}600 \times D^{-1.9}$$

A steel shaft 40 mm in diameter is to be hardened by austenitising at 850°C followed by quenching into cold oil and tempering. The centre of the bar must be 100% martensite after quenching, and five low alloy steels A to E are available. The critical cooling rates (CCR) for formation of 100% martensite, and the cost of each steel, are:

Steel	CCR (°C s^{-1})	Cost (£/kg)
A	110	0.50
B	5.5	0.90
C	4.5	1.25
D	42	0.60
E	2	1.00

(a) Find the cooling rate at the centre of the 40 mm bar.
(b) Select the best steel for this application from those listed in the table, justifying your choice.

Exercise E11.11   In plain carbon and alloy steels, *hardenability* and *weldability* are considered to be opposites. Explain why this is the case. For a steel to be suitable for laser transformation hardening, would you look for high or low hardenability? Why?

Exercise E11.12   Additive manufacturing (AM) competes with conventional processes for low production runs in specific applications. With reference to typical AM processes for metals and for polymers, explain why the mechanical properties tend to be inferior in components produced by AM.

Exercise E11.13   The ceramic alumina Al_2O_3 is hot or cold pressed and sintered with a variety of binders to facilitate densification. These binders have a much lower Young's modulus than the alumina, and there is usually some residual porosity, giving nominal compositions from 85% to 100% alumina. The table shows the resulting density, Young's modulus, and compressive strength for a range of aluminas. Plot a graph showing the variation of these properties with alumina composition, accounting for the trends in each.

Composition (% Al_2O_3)	Density (kg/m^3)	Young's modulus (GPa)	Compressive strength (GPa)
85.0	3400	220	1.8
88.0	3470	250	2.0
90.0	3520	276	2.5
94.0	3670	314	2.3
97.0	3800	355	2.3
99.5 (fine grain)	3850	375	6.3
100.0	3960	400	2.6

Exercise E11.14   Briefly explain the meaning of the following terms in relation to polymer processing: copolymer, polymer blend, crystallisation, fibres. Why is it possible to weld and recycle thermoplastics, but not a thermoset or an elastomer?

Exercise E11.15   The excellent specific properties of natural and artificial fibres were highlighted in Figure 11.24. It is also of interest to explore other property combinations that measure performance (e.g. the maximum elastic stored energy). The table summarises typical data for various fibres, together with some bulk polymers, and steel, for comparison. Calculate the following performance indices for the materials given: (a) maximum elastic stored energy (per unit volume), σ_f^2/E; (b) maximum elastic stored energy (per unit mass), $\sigma_f^2/E\rho$.

Which material appears best on each criterion? Which criterion would be more important for climbing ropes? Why is it not practical for the properties of fibres to be exploited to their maximum potential?

	Hemp	Spider-web silk	Bulk nylon	Aramid fibre	Polyester fibre	PE fibre	Nylon fibre	Alloy steel wire
Young's modulus (GPa)	8	11.0	2.5	124	13	2.85	3.9	210
Strength (MPa)	300	500	63	3930	784	1150	616	1330
Density (kg/m^3)	1490	1310	1090	1450	1390	950	1140	7800

Chapter 12
Materials, environment, and sustainability

The 'Three Capitals': Natural, Manufactured, and Human.

Chapter contents

https://doi.org/10.1016/B978-0-08-102399-0.00012-1

12.1 Introduction and synopsis

The practice of engineering consumes vast quantities of materials and relies on a continuous supply of them. We start by surveying this consumption, emphasising the materials used in the greatest quantities. Increasing population and living standards cause this consumption rate to grow – something it cannot do forever. Finding ways to use materials more efficiently is a prerequisite for a sustainable future.

There is a more immediate problem: present-day material usage already imposes stress on the environment in which we live. The environment has some capacity to cope with this, so that a certain level of impact can be absorbed without lasting damage. It is clear, however, that current human activities exceed this threshold with increasing frequency, diminishing the quality of the world in which we now live and threatening the well-being of future generations. *Design for the environment* is generally interpreted as the effort to adjust our present product design efforts to correct known, measurable environmental degradation; the timescale of this thinking is 10 years or so, an average product's expected life. *Sustainable design* is more than this: it involves the more complex balance of environmental concerns with those of economics, legality, and social acceptability over a much longer timescale. This chapter explores these and the ways in which materials are involved.

12.2 Material production, material consumption, and growth

Material consumption   Speaking globally, we produce roughly 10 billion (10^{10}) tonnes of engineering materials each year. This production is responsible for about 16% of global energy consumption and carbon release to the atmosphere (cement production alone is responsible for almost 5%). Figure 12.1 gives a perspective: it is a bar-chart of the production of a range of materials, including those used in the greatest quantities. It has some interesting messages. On the extreme left, for calibration, are hydrocarbon fuels – oil and coal – of which we currently use 10 billion tonnes per year. Next, moving to the right, are metals. The scale is logarithmic, making it appear that the consumption of steel (the first metal) is only a little greater than that of aluminium (the next); in reality, the consumption of steel exceeds by a factor of 10 that of all other metals combined.

Polymers come next: 50 years ago their consumption was tiny; today the combined consumption by weight of the commodity polymers, polyethylene (PE), polyvinyl chloride (PVC), polypropylene (PP) polystyrene, and polyethylene-terephthalate (PET) approaches that of steel – by volume it is greater.

These may be big, but the *really* big ones are the materials of the construction industry. Steel is one of these, but the consumption of wood for construction purposes exceeds that of steel even when measured in tonnes per year (as in the diagram), and since it is a factor of 10 lighter, if measured in m^3/year, wood eclipses steel. Concrete, the material of bridges, dams, airports, and mega-cities, is bigger still. Fibres, too, are important: think of clothing, carpets, blankets, sheets, tents, sails, parachutes. Until the late 19th century all were

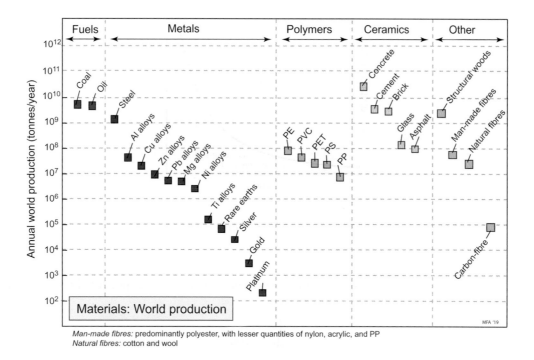

Figure 12.1 The annual consumption of hydrocarbons *(left-hand column)* and of engineering materials *(the other columns)*.

natural in origin – hemp, linen, cotton, wool, silk. Over the course of the 20th century, synthetic chemistry overtook nature as a provider of fibres and remains so today. Last in this column is carbon, the fibre used to reinforce high-performance composites. Just 30 years ago this material would not have crept onto the bottom of this chart. Today its production is approaching that of some light alloys and is growing fast.

The data on this figure describe the big players; collectively they account for some 98% of *all* material production, by weight. Actual material consumption is even greater than this because recycling feeds material back into use, adding to that produced from virgin raw materials – a point we return to in a moment.

The growth of consumption Demand for materials is growing exponentially with time (Figure 12.2), simply because both population and living standards grow exponentially. One consequence of this is dramatised by the following statement: at a global growth rate of just 3% per year we will mine, process, and dispose of more 'stuff' in the next 23 years than we have since the start of the industrial revolution 250 years ago. You can prove this for yourself in the Exercises in this chapter, using the analysis below.

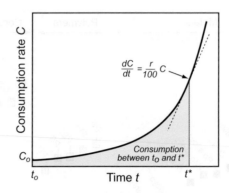

Figure 12.2 Exponential growth in consumption.

If the current rate of consumption in tonnes per year is C, then exponential growth means that

$$\frac{dC}{dt} = \frac{r}{100} C \tag{12.1}$$

where t is time and r is the percentage fractional rate of growth per year. Integrating over time gives

$$C = C_o \exp \left\{ \frac{r(t - t_o)}{100} \right\} \tag{12.2}$$

where C_o is the consumption rate at time $t = t_o$. The *doubling-time t_D* of consumption rate is given by setting $C/C_o = 2$ to give

$$t_D = \frac{100}{r} \log_e(2) \approx \frac{70}{r} \tag{12.3}$$

Steel production is currently growing at 3.7% per year over the last decade, doubling in 19 years. The market for carbon fibre is currently growing at 8% per year; it doubles in just 9 years.

Example 12.1 Calculating growth rates

The world production P of silver in 2010 was 23,800 tonnes. By 2018 it had grown to 27,300 tonnes. Assuming exponential growth, what was the annual growth rate of production of silver in ths period?

Answer. Exponential growth of production P tonnes per year is described by

$$P = P_o \exp\left\{\frac{r(t-t_0)}{100}\right\}$$

where P_o is the rate of production at time $t = t_o$. Setting $P = 27,300$ and $P_o = 23,800$ tonnes per year, and $(t - t_o) = 8$ years, then solving for the growth rate r gives

$$r = \frac{100}{(t-t_o)} \ln\left(\frac{P}{P_o}\right) = 1.7\% \text{ per year}$$

Example 12.2 Cumulative growth

A total of 23.6 million cars were sold in China in 2015, with sales at the time increasing by 7.3% per year. The total number on the road at the end of 2015 was 203 million. If this growth rate had continued, how many cars would have been on China's roads by the end of 2022, assuming that the number removed from the roads in this time interval can be neglected?

Answer. Starting with Equation (12.2):

$$C = C_o \exp\left\{\frac{r(t-t_0)}{100}\right\}$$

The cumulative number Q of cars entering service in the time interval $(t-t_o)$ is found by integrating this equation over time

$$Q = \int_{t_o}^{t} C\, dt = \frac{100\,C_o}{r}\left\{\exp\left\{\frac{r\,(t-t_o)}{100}\right\} - 1\right\}$$

Substituting for $C_o = 23.6 \times 10^6$, $r = 7.3\%$ per year, and $(t-t_0) = 7$ years (i.e. 2016 to 2022), the additional number of cars by the end of 2022 is $Q = 216 \times 10^6$, giving a total of 419×10^6 overall (more than double the 2015 figure).

[In reality, the growth rate fell sharply after 2016, and actually became negative. The total number of cars in China in 2022 was 315 million, slightly more than in the USA, but far fewer *per capita* – on this measure China ranked 90th in the world].

The picture, then, is one of a global economy ever more dependent on a continuous supply of materials and the energy it takes to make them. This raises two concerns:

- Concern for the health of the environment, the life-support system for all living things
- Concern for security of access to natural resources, the support system for economic well-being and growth

We begin with the environment.

12.3 Natural capital and the materials life cycle

All life as we know it evolved on earth. The continued existence of life on earth depends on the resources that nature provides: clean air, fresh water, productive land and oceans, bio-diversity, and, of course, the resources of minerals and energy from which materials are made. We refer to these collectively as *Natural Capital*. The word *capital* is deliberately chosen: a capital is an asset. You can draw it down to use it; you can build it up by conservation or saving; and you can exchange it for other goods and services. Natural Capital is represented in this chapter by the icon shown in Figure 12.3.

Natural Capital
Clean atmosphere, productive land, oceans and biosphere, fresh water, material and energy resources

Figure 12.3   Natural capital.

We draw down natural capital to create materials and products. Some – clean air, fresh water, vegetable and animal stocks – are renewable, provided they are properly managed. Others – minerals and hydrocarbon deposits, for example – are not renewable (once used or burned they are gone) and are finite in extent, although for many the natural deposits are so large that serious depletion has not, until recently, been seen as a problem. As we have seen, the rate at which this draw-down is taking place is accelerating, with consequences that are now cause for concern. To understand this we must examine the life cycle of materials and products.

The material and product life cycle (**Figure 12.4**)   Ores and feedstocks, drawn from natural capital, are processed to give materials; these are manufactured into products that are used; and at the end of their lives, a fraction is reused or recycled, while the rest is incinerated or

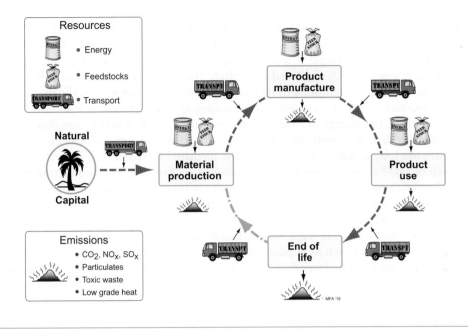

Figure 12.4 The material and product life cycle.

committed to landfill. Energy and materials are consumed at each point in this cycle (we shall call them 'phases'), with an associated penalty of CO_2, SO_x, NO_x, and other emissions – low grade heat and gaseous, liquid, and solid waste. Resource consumption, emissions, and their consequences are assessed by *life-cycle analysis* (LCA). A rigorous LCA examines the life cycle of a product and assesses in detail the associated resource consumption and environmental impact of each phase of life, summing them to give a life assessment. To do this you need information about the life history of the product at a level of precision that is only available after it has been manufactured and used: this type of LCA is a product-assessment tool, rather than a design tool. This has led to the development of more approximate "streamlined" life-assessment methods that seek to combine acceptable cost with sufficient accuracy to guide decision making during the design process – the choice of materials being one of these decisions. We shall follow this route, focusing on the *embodied energy* and *carbon footprint* associated with the life of a product. First we need to define what these words mean.

12.4 Embodied energy and carbon footprint of materials

Materials are made from naturally occurring ores and feedstocks. Making them requires energy and releases emissions, both on a very large scale. Nations now seek ways to minimise the environmental impact of material production, so this has to be quantified. Two eco-metrics – embodied energy and carbon footprint – characterise material production.

Embodied energy The embodied energy, H_m, of a material is the energy that must be committed to create 1 kg of usable stock – 1 kg of steel stock, or of PET pellets, or of cement powder, for example – measured in MJ/kg. The *carbon footprint*, $CO_{2,e}$, is the associated release of CO_2, adjusted to include the carbon-equivalent of other associated emissions, in kg/kg. It is tempting to try to estimate embodied energy via the thermodynamics of the processes involved – extracting aluminium from its oxide, for instance, requires the provision of the free energy of oxidation to liberate it. While it is true that this much energy must be provided, it is only the beginning. The thermodyamic efficiencies of industrial processes are low, seldom reaching 50%. The feedstocks used in the extraction or synthesis themselves carry embodied energy, and transport is involved. The production plant has to be lit, heated, and serviced; and as it was built (at least in part) for the purpose of making the material or product, there is an 'energy mortgage' – a share of the energy consumed in building the plant in the first place.

The upper part of Figure 12.5 shows, much simplified, the inputs to a PET production facility: oil derivatives such as naptha and other feedstock, direct power (which, if electric, is generated in part from hydrocarbon fuels with an efficiency of about 34%), and the energy of transporting the feedstock to the facility. The plant has an hourly output of usable PET

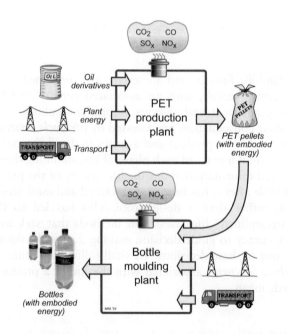

Figure 12.5 The flows of material and energy in PET and bottle production.

granules. The embodied energy of the PET, $(H_m)_{PET}$, with usual units of MJ/kg, is then given by

$$(H_m)_{PET} = \frac{\Sigma \, \text{Energies entering plant per hour}}{\text{Mass of PET granules produced per hour}}$$

The carbon footprint is defined in a similar way.

Figures 12.6 and 12.7 show the embodied energy and carbon footprint per kg for common materials. When compared per unit mass, metals, particularly steels, appear as attractive choices, demanding much less energy than polymers. But when compared on a volume basis, the ranking changes, and polymers lie lower than metals. The light alloys based on aluminium, magnesium, and titanium are particularly demanding, with energies and carbon emissions that are high by either measure.

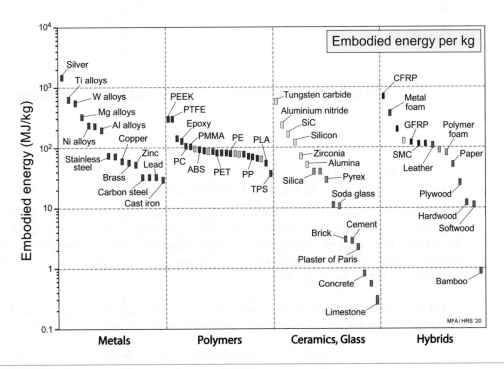

Figure 12.6 Bar chart of embodied energy of basic materials by weight.

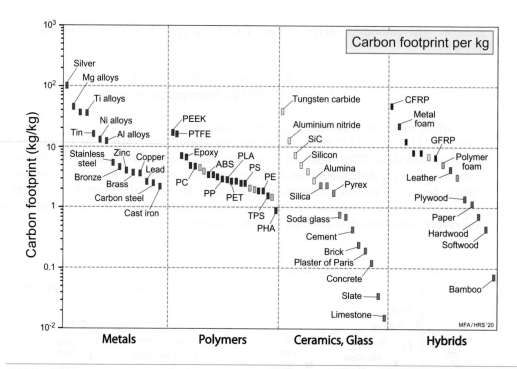

Figure 12.7 Bar chart of carbon footprint of basic materials by weight.

Example 12.3 Material substitution

The casing of a power tool is made of 0.75 mm thick aluminium alloy sheet. It is proposed to replace it with a polypropylene (PP) moulding of the same overall size, but 3 mm thick. Does the replacement PP moulding reduce the embodied energy and carbon footprint? (Data for density, embodied energy. and carbon footprint may be found in Appendix A, Tables A2 and A8).

Answer. The relevant data from Appendix A are shown in the following table. The last two columns list embodied energy and carbon footprint per unit volume.

Material	Density ρ (kg/m^3)	Embodied energy H_m (MJ/kg)	Carbon footprint CO_2 (kg/kg)	$H_m \rho$ (GJ/m^3)	$CO_2 \rho$ (Mg/m^3)
Aluminium alloy	2800	190	12	530	34
Polypropylene (PP)	900	69	2.9	62	2.6

The ratio of embodied energy per unit volume of aluminium is 8 times that of PP; its carbon footprint is 13 times larger. The PP casing is 4 times thicker than the aluminium one and thus requires 4 times the volume of material, but it still has lower embodied energy and carbon footprint than the aluminium casing.

Processing energy The processing energy, H_p, associated with a material is the energy, in MJ, used to shape, join, and finish 1 kg of the material to create a component or product. Thus polymers, typically, are moulded or extruded; metals are cast, forged, or machined; ceramics are shaped by powder methods. A characteristic energy and carbon footprint per kg is associated with each of these. Continuing with the PET example, the granules now become the input (after transportation) to a facility for blow-moulding PET bottles for water, as shown in the lower part of Figure 12.5. There is no need to list the inputs again – they are broadly the same, the PET bringing with it its embodied energy $(H_m)_{PET}$. The output of the analysis is the energy committed per bottle produced, and the carbon footprint is assessed in a similar way.

There are many more steps before the bottle reaches a consumer and is drunk: collection, filtration, and monitoring of the water, transportation of water and bottles to a bottling plant, labelling, delivery to a central warehouse, distribution to retailers, and refrigeration prior to sale. All have energy inputs, which, when totalled, give an energy commitment and carbon footprint, even for a product as simple as a plastic bottle of cold water.

At the end of its first life as a bottle the material is still there. What happens to it then?

Recycling: ideals and realities We buy, use, and discard paper, packaging, cans, bottles, furniture, smartphones, computers, cars, even buildings. Why not retrieve the materials they contain and use them again? It's not that simple.

First, some facts. There are (simplifying again) two sorts of 'scrap,' by which we mean material with recycling potential. In-house scrap is the off-cuts, ends, and bits left in a material production facility when the usable material is shipped out. Here ideals can be realised: almost 100% is recycled, meaning that it goes back into the primary production loop, even though some more energy is used in doing so. But once a material is released into the outside world, the picture changes. It is processed to make parts that may be small, numerous, and widely dispersed; it is assembled into products that contain other materials; it may be painted, printed, or plated; and its subsequent use contaminates it further. To reuse the material, it must be collected (not always easy), separated from other materials, identified, decontaminated, chopped, and processed. Imperfect separation causes problems: even a little copper or tin damages the properties of steel; residual iron embrittles aluminium; heavy metals (lead, cadmium, mercury) are unacceptable in many alloys; PVC contamination renders PET unusable; and dyes, water, and almost any alien plastic renders a polymer unacceptable for its original purpose, meaning that it can only be used in less demanding applications (a fate known as 'down-cycling').

Despite these difficulties, recycling can be economic, both in material and energy terms. This is particularly so for metals: the energy commitment per kg for recycled aluminium is about one-tenth of that for virgin material; that for steel is about one-third. Some inevitable contamination is countered by addition of virgin material to dilute it. Metal recycling is economic and helps save energy and reduce emissions.

Example 12.4 Material substitution

Consider again the aluminium casing of the power tool (as in Example 12.3). How does the comparison between PP and aluminium change if 100% recycled aluminium is used for the casing instead? (Table A9 contains the relevant data: the embodied energy for recycled aluminium is 33 MJ/kg, and the carbon footprint is 2.6 kg/kg).

Answer. Using the density of aluminium ($= 2800$ kg/m^3), the embodied energy and carbon footprint for 100% recycled material per unit volume, calculated as in Example 12.3, are reduced to 92 GJ/m^3 and 7.3 Mg/m^3, respectively. The energy is now closer to the value for virgin PP (72 GJ/m^3), and the carbon footprint is now about 2.5 times that of PP (2.8 Mg/m^3). But since the aluminium wall thickness is 4 times less than the PP equivalent, the aluminium casing now has the lower profile in both embodied energy and carbon footprint.

The recycling picture for plastics is less rosy. Figure 12.8 illustrates this for PET. In the upper part of the figure, bottles are collected and delivered to the recycling plant as mixed plastic – predominately PET, but with PE and PP bottles, too. Some energy is saved, but not a lot – typically 50%.

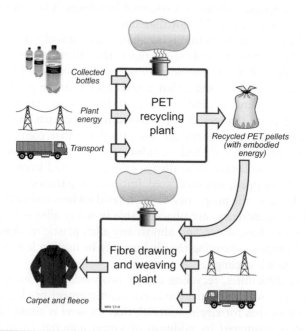

Figure 12.8 The flows of material and energy during the recycling of plastic bottles to recover PET.

Recycling of PET, then, can offer some energy savings. But is it economic? Time, in manufacture, is money. Collection, inspection, separation, and drying are slow processes, all adding cost. The quality of recycled material also may be less good than the original – plastics are rarely recycled in a closed loop, but go into secondary products (as shown in the lower part of Figure 12.8). If the recycled stuff were as good as new it would command the same price; in reality it commands little more than half. Using today's technology, the cost of recycling plastics is high, and the price they command is low, which is not a happy combination.

The consequences of this are brought out by Figure 12.9. It shows the fraction of current supply that derives from recycling. For paper and cardboard it is high, as it is for many metals: most of the lead and cast iron, and approaching half the steel, aluminium, and copper we use today has been used at least once before. For plastics the only significant success is PET with a recycle fraction of about 32%, up from 18% a decade ago, but for the rest the contribution is small, or (for many) it is zero. Oil price inflation and restrictive legislation could change all this, but for the moment that is reality.

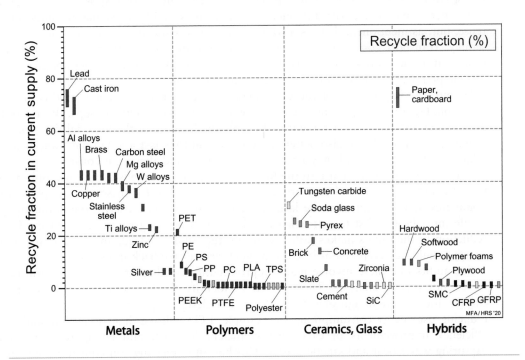

Figure 12.9 The fractional contribution of recycled materials to current consumption. For metals and paper/cardboard, the contribution is large; for polymers, it is small.

Before we go further, a word of warning. Some material properties (e.g. density, modulus, specific heat) can be measured with precision, so the values listed in handbooks can be accepted as accurate and reproducible. Eco-properties are not like that. Assessing embodied energies, carbon footprints, and recycle fractions is an imprecise science. The values depend on local conditions – copper mined and refined in Peru has an embodied energy that differs from that produced in the United States because the ores and technologies of the two countries differ, and even within the United States there are differences between different producers. The numbers in Appendix A, Tables A8 and A9 must be viewed as approximate, with a precision of $\pm 10\%$, at best.

Eco-audits of products: streamlined life-cycle analysis With this background, we can explore the way products consume energy in each of the four life-phases of Figure 12.4. The procedure is to tabulate the main components of the product together with their material and mass (the bill of materials [BoM]). The embodied energy and processing energy associated with the product are estimated by multiplying each mass by the appropriate energies H_m and H_p, and adding them. The use energy is calculated from the power consumption and the 'duty cycle' (i.e. the proportion of the time that it is in use), and the source from which the power is drawn, allowing for suitable efficiencies in converting energy from one form to another (e.g., electricity to heat). To this should be added the energy associated with maintenance and service over product life. The energy of disposal is more difficult: some energy may be recovered by incineration, some saved by recycling, reconditioning, or reuse, but, as mentioned, there is also an energy cost associated with collection and disassembly. Transport costs can be estimated from the distance of transport and the energy/ tonne.km of the transport mode used – typical data for transport energies are given later in this chapter. Carbon footprint is assessed in a similar way.

Despite the imprecision of the data mentioned earlier, the outcome of this analysis is revealing. Figure 12.10 presents the evidence for a range of product groups.[1] Products 1 and 2 are material-intensive; for these it is the embodied energy of the material that dominates. Buildings (Product 3) are also material-intensive, but they last a long time – and if they are inhabited, they are lit and may be heated or cooled, and these can consume more energy than the materials that went into the building (that shown here is a car park, unheated but lit). In contrast, Products 4 to 6 are use energy-intensive – for these, the use phase totally dominates. If you want to make large reductions in life-cycle energy, the dominant phase of life must be the first target; when the differences are as great as those in the figure, a reduction in the others makes little impact on the total – indeed, an increase in embodied energy would be acceptable if it led to a much greater payback in use energy.

Eco-audits like these involve many approximations – they are order-of-magnitude estimates, not precise analyses. Their role is to identify the big contributors to energy and carbon, focusing re-design objectives. Product re-design is proving increasingly effective in reducing use-energy consumption: for example, LED lighting, light-weighting and energy-efficiency improvements of cars, energy-efficient housing, and energy-recovery systems capturing waste heat and kinetic energy. This has an interesting consequence: the embodied energies of the materials of products (which is often increased by these changes) are emerging as an increasing focus of attention.

---

[1] These eco-audits are discussed in greater depth in the companion text: Ashby, M.F. (2021), - see 12.10 Further Reading.

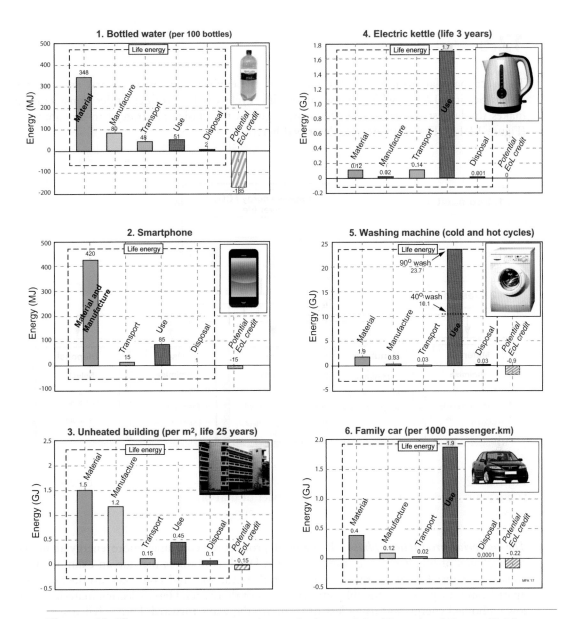

Figure 12.10 The energy consumed at each phase of the life cycle of Figure 12.4 for a range of products. The columns show the embodied energy of the materials, that for manufacture, transport, use, and disposal over the stated life. Carbon emissions follow an almost identical pattern.

12.5 Materials and eco-design

To select materials to minimise the impact on the environment we must first ask – as we did in Section 12.4 – which phase of the life cycle of the product under consideration makes the largest contribution? The answer guides the choice of strategy to ameliorate it (Figure 12.11). The strategies are described here.

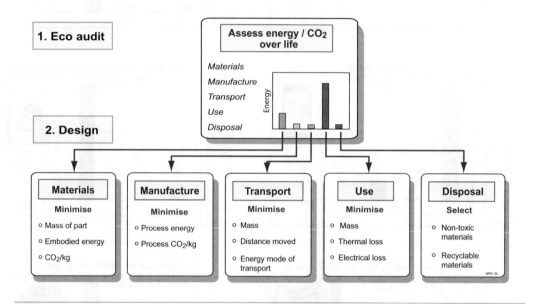

Figure 12.11 Eco-design starts with an analysis of the life phase(s) to be targeted. Subsequent actions aim at minimising the impact of that phase.

The material production phase If material production consumes more energy than the other phases of life, it becomes the first target. Drink containers provide an example: they consume materials and energy during material extraction and container production, but apart from transport and possible refrigeration, not thereafter. Here, selecting materials with low embodied energy and using less of them are the ways forward. Large civil structures – buildings, bridges, roads – are material-intensive. For this reason architects and civil engineers concern themselves with embodied energy and (in the use phase) the thermal efficiency of their structures.

Example 12.5 Relative contributions to embodied energy of products

A catalytic converter for a car has a core of alumina ceramic honeycomb weighing 0.7 kg, carrying 4 g of platinum. The active components are contained in a stainless steel casing weighing 1.2 kg. Which component contributes most to the embodied energy of the product?

Component	Mass (kg)	Embodied energy (MJ/kg)
Alumina core	0.7	52
Platinum catalyst	0.004	2.9×10^5
Stainless steel casing	1.2	73

Answer. Multiplying the mass by the embodied energy per kg yields the following values:

Component	Energy contribution to product (MJ)
Alumina core	36
Platinum catalyst	1160
Stainless steel casing	88

The platinum catalyst accounts for only 0.2% of the mass of the converter but 90% of its embodied energy.

The product-manufacture phase   The energy required to shape a material is usually much less than that to create it in the first place. Certainly it is important to save energy in manufacture, but higher priority often attaches to the local impact of emissions and toxic waste during manufacture, depending on the location.

The product-use phase   The eco-impact of the use phase of energy-using products has nothing to do with the embodied energy of the materials themselves – indeed, minimising this may frequently have the opposite effect on use-energy. Use-energy depends on mechanical, thermal, and electrical efficiencies; it is minimised by maximising these. Fuel efficiency in transport systems (measured, say, by MJ/km) correlates closely with the mass of the vehicle itself; the objective then becomes that of minimising mass. Energy efficiency in heating systems is achieved by minimising the heat flux out of the system where we don't want it, while for refrigeration we aim to minimise the heat flux into the system. Energy efficiency in electrical generation, transmission, and conversion is maximised by minimising the various losses in the system, such as ohmic heating in the conductors; here the objective is to minimise electrical resistance while meeting necessary constraints on strength, cost, etc. Selection to meet these objectives is exactly what earlier chapters of this book were about.

Example 12.6 Carbon footprint and cost of use phase

A small car weighs 1300 kg; it covers 175,000 miles over its life at an average fuel economy of 35 miles per US gallon. The car carries a spare wheel and tyre weighing 13 kg, which the owner has never had occasion to use. If fuel consumption scales linearly with vehicle weight, what has been the carbon and cost penalty of carrying the spare wheel over the vehicle life? (Assume gasoline costs around $3.5 per gallon in the United States. Burning 1 US gallon of gasoline releases 11 kg of carbon dioxide).

Answer. The car burns 175,000/35 = 5000 gallons of petrol costing $17,500 over its life and releases 56,540 kg of carbon dioxide. Increasing its weight by 1% (the contribution of the spare wheel) thus carries the penalty of 565 kg of carbon dioxide and costs about $175 over the vehicle life. In Europe the carbon penalty is the same, but the cost is about twice as much.

The product disposal phase   Disposal in landfill contaminates water and soil, and wastes materials. Increasingly, legislation dictates disposal procedures, take-back, and recycle requirements, and – through landfill taxes and subsidised recycling – deploys market forces to influence the end-of-life choice.

Transport   Globalisation means that materials and products are transported over large distances before they reach the consumer. This is often seen as misguided – better, it is argued, to use indigenous materials and to manufacture locally. Economists argue otherwise – the lower labour costs in emerging economies more than offset the cost of transporting materials and products between them and developed countries where they are sold and used. Table 12.1 gives an idea of the energy and carbon footprint of various transport modes.

Table 12.1   Approximate energy and carbon footprint for transport.

Transport type and fuel	Energy (MJ/tonne.km)	Carbon footprint (kg CO_2/tonne.km)
Shipping - Diesel	0.18	0.014
Rail - Diesel	0.35	0.027
Large truck (8 axle) - Diesel	0.71	0.055
Small truck (2 axle) - Diesel	1.5	0.12
Family car - Diesel	1.4–2.0	0.11–0.15
Family car – Gasoline/Petrol	2.2–3.0	0.15–0.2
Family car – Hybrid	1.4	0.10
Aircraft, long haul - Kerosene	6.5	0.54
Aircraft, short haul - Kerosene	13	1.1

Example 12.7 Estimating energies and carbon footprints for transport

Television sets are manufactured in Shanghai, China, and transported to Le Havre, France. What is the carbon footprint per set (shipping mass 21 kg/unit) if sent by sea freight? By air freight? (Search online for estimates of the relevant distances).

Answer. The distance from Shanghai to Le Havre by sea (via the Suez Canal) is 19,200 km. The carbon footprint of this transport leg is $0.021 \times 0.014 \times 19{,}200 = 5.6$ kg CO_2.

The distance from Shanghai to Le Havre by air over Russia is 9330 km. The carbon footprint of this transport leg is $0.021 \times 0.54 \times 9330 = 106$ kg CO_2, about 19 times more than by sea.

12.6 Materials dependence

Critical (strategic) materials New technologies create new material demands. Wind and solar power, advanced batteries, energy-efficient lighting, electric vehicles, aerospace, and IT are large in scale and rely more heavily on the less-abundant elements, many in the lower part of the Periodic Table of Elements. Demand has grown rapidly in the last 2 decades for lithium (Li), indium (In), tellurium (Te), gallium (Ga), antimony (Sb), beryllium (Be), high-purity quartz, and rare earth elements (REE), in particular, and supply has not always kept pace. For some elements this is because they are very rare; for others, geopolitical issues can result in export quotas, or conflict zones in or near supplying countries can disrupt supply.

Elements that are essential to a nation for economic or security reasons, and for which supply is uncertain, are classified as 'critical' (Table 12.2). Nations respond by creating stockpiles, seeking strategic appliances with supplier countries to get privileged access, and promoting

Table 12.2   Examples of critical elements and their applications

Critical element	Applications
Lithium (Li)	Batteries, Al-Li alloys
Indium (In)	Flat panel displays, touch screens, solders
Tellurium (Te)	Photovoltaics
Gallium (Ga)	LEDs, smartphones, photovoltaics
Beryllium (Be)	Cu-Be alloys for aerospace; satellites
Rare earths (notably Nd, Eu, Ds, Y, La, Ce, Te)	Magnets, lasers
Platinum group metals (Pt, Pd, Ir, Os, Rh, Ru)	Catalysts, auto exhaust emission control

Source: Ashby M.F. (2021) - see 12.10 Further Reading.

recycling schemes to recover stock that has already been used. But this supply risk does not always translate into business risk. The requirements of a given manufacturer depend on the concentration of the element in their products and the scale of production; criticality becomes a business risk when the element provides key functionality that cannot be achieved in any other way. Businesses minimise risk by identifying alternative suppliers and exploring substitutes for the element, should supply shortage drive prices to an unacceptable level.

Restricted (hazardous) substances Business risk arises in another way. There is increasing awareness that some elements and an increasing number of compounds can cause damage to human health and the bio-sphere. Many of these are substances used in extracting and processing materials (e.g. hexavalent chromium in chrome plating) or are added to materials to give better functionality (e.g. fire retardants and plasticisers in plastics). Nations restrict the use of these substances – the severity of restriction depends on the quantity used and the level of damage they might cause. In the United States, such substances are documented in a series of additions to the Toxic Substance Control Act (TSCA) of the Environmental Protection Agency. In Europe they are published as REACH[2] directives; early warning of new additions is provided in the wonderfully named SIN (Substitute It Now) list.

Dealing with risk Dependence on critical elements or restricted substances exposes a manufacturer to business risk and loss of market. The risk is not static: changing relationships between nations can rupture supply chains, and the list of restricted substances increases every year. Increasingly, manufacturers protect themselves by screening during product development, exploring substitutes before they are needed, and, if they can afford it, engaging in R&D to create lower-risk alternatives. Beyond that, they adopt measures to use materials more efficiently (reducing demand) and to retain ownership of the products they make and the materials they use (allowing reuse). This circular approach to material conservation has increasing appeal.

Circular materials economics[3] Since the industrial revolution, large-scale mining and global trade have allowed material costs to decline[4] in relative terms. At the same time labour costs have risen, stimulating the development of manufacturing methods that minimise the use of labour (mass production, robotic manufacture, etc.). The low cost of materials encouraged industry to adopt a linear materials economy, summarised as: *take – make – use – dispose* (Figure 12.12).

Some 80% of material usage still follows such a path, but there is increasing awareness that this cannot go on. The global population is increasing, and all strive for a higher standard of living. The global consumption of materials, at present about 77 billion

---

[2] REACH = Registration, Evaluation, Authorisation, and Restriction of Chemicals

[3] https://ellenmacarthurfoundation.org/

[4] World Bank (2013) http://blogs.worldbank.org/prospects/category/tags/historical-commodity-prices

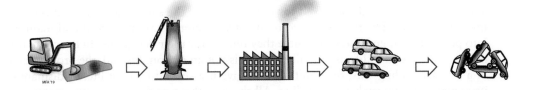

Figure 12.12   The linear materials economy: take – make – use – dispose.

tonnes per year, is expected to rise to 100 billion by 2030[5]. The stress on the global eco-sphere caused by industrial development is already a cause for concern. One way forward is to establish a more circular materials economy[6,7] (Figure 12.13).

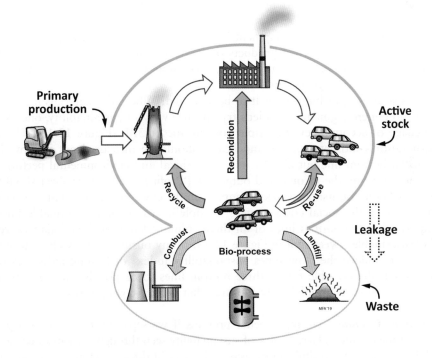

Figure 12.13   The circular materials economy. The aim is to retain materials in the Active stock box by reuse, reconditioning, and recycling, minimising leakage into the Waste box.

[5] http://www.foe.co.uk/sites/default/files/downloads/overconsumption.pdf

[6] Ellen MacArthur Foundation (2014), www.ellenmacarthurfoundation.org

[7] Webster, K., Bleriot, J., & Johnson, C. (2013). *A new dynamic: Effective business in a circular economy.* Ellen MacArthur Foundation. ISBN 978-0-9927784-1-5.

Materials in a circular economy are seen not as a disposable commodity but as a valued asset to be tracked and conserved for reuse, in rather the same way that financial capital is invested, recovered as revenues, and re-invested. Materials are produced and manufactured into products that enter service, where they remain for their design life. In the circular model, disposal at the end of life as landfill or waste is not an option. Instead, the product is re-used in a less demanding way, re-conditioned to give it a second lease of life, or dismantled into its component materials for recycling. All three of these options retain the materials of the product as *active stock* (the upper green box in the figure). It may be impractical to recycle perishable materials; those that are bio-degradable can be composted, returning them to the bio-sphere; those that are combustible can be incinerated with energy recovery; only the remaining residue goes to landfill. The more material that can be retained within the green Active stock box (the upper part of Figure 12.13), the less needs to be added through primary production. It is a step toward a sustainable materials future.

12.7 Materials and sustainable development

What is a 'sustainable development'? Here is a short answer: a sustainable development is one that provides needed products or services in ways that reduce the drain on natural resources, is legal, economically viable, acceptable to all stakeholders, and equitable both within and between generations.

This sounds right but how is it to be achieved? Where do materials fit in? The definition gives no concrete guidance. So let's try another view of sustainability, one expressed in the language of accountancy: the triple bottom line or 3BL (Figure 12.14). The idea is that a corporation's ultimate success and health should be measured not just by the traditional financial bottom line but also by its social/ethical and environmental performance. Instead of just reporting the income and outgoings ('Profit'), the balance sheet should also include two further accounts: one tracking impact on the environmental balance sheet ('Planet') and one tracking the social consequences ('People'). In this view, sustainable business practice requires that the bottom lines of all three columns show positive balances, represented by the 'Sustainable' sweet spot in Figure 12.14. Many businesses now claim to implement 3BL reporting; indeed the Dow Jones Sustainability Index[8] of leading industries is based on it. But is it really possible for all three bottom lines to be positive at the same time? Do the terms 'Planet' and 'People' truly capture what we are trying to say?

The three Capitals   We make better progress if we separate the circles of Figure 12.14 and expand their content. Here a view of sustainability seen through the lens of economics can help. Global or national 'wealth' can be seen as the sum of three components: the *net natural capital*, the *net manufactured capital*, and the *net human capital* (Figure 12.15). They are defined as such:

- *Natural Capital* – clean atmosphere, fresh water, fertile land, productive oceans, accessible minerals and energy

---

[8] Dow Jones Sustainability Index (2012), http://www.sustainability-index.com/

Figure 12.14 The triple bottom line: planet, profit, people.

- *Manufactured Capital* – industrial capacity, institutions, roads, built environment, financial wealth (gross domestic product [GDP])
- *Human Capital* – health, education, skills, technical expertise, accumulated knowledge, culture, heritage, and (ultimately) happiness

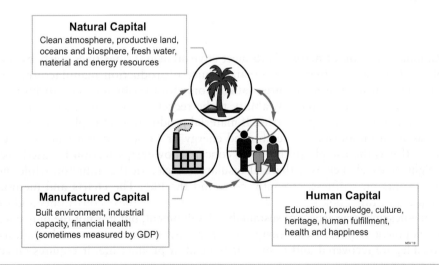

Figure 12.15 The three capitals. They underpin society as we know it today.

We encountered the first of these earlier when exploring eco-design. Like natural capital, manufactured and human capitals can be regarded as assets on which we draw and into which contributions can be made. They are mutually supportive provided they remain in balance. Each depends on the health of the others, as suggested by the grey arrows. Natural capital is drawn down to create manufactured capital. The wealth generated in this way underwrites the education and welfare systems that support and expand human capital, although it can also diminish it by encouraging unfair labour practices and social inequality. It is the understanding and foresight enabled by education that guides human interventions to support natural capital, and it is the resources generated by manufactured capital that pay for it. Sustainable development is not one thing, it is many. While the starting point may be an environmental one, sustainability extends far beyond eco-design to embrace economics, legislation, and social and ethical issues. Some brief examples will bring out the difficulties.

Advocates of *bio-fuels and bio-polymers* do so because bio-derived products diminish dependence on fossil hydrocarbons, but the land and water required to grow them is no longer available for the cultivation of food. *Carbon taxes* are designed to stimulate a low-carbon economy, but they increase the price of energy and hence of materials and products. *Design for recycling* is intended to meet the demand for materials with less drain on natural resources, but it constrains the use of light-weight composites because most cannot be recycled. The motivation for *ethical sourcing of raw materials* (sourcing them only from nations with acceptable records of human rights) is that of social responsibility, but a side effect is to suppress jobs where they may be most needed. Many proposals for sustainable technology aim to support the three capitals of Figure 12.15, but they often support only one. So sustainability is a multifaceted ideal, but this has great practical value: it sets targets to be worked towards, even if perfect fulfilment is not possible.

12.8 Summary and conclusions

Rational selection of materials to meet environmental objectives starts by identifying the phase of product-life that causes greatest concern: production, manufacture, use, or disposal. Dealing with all of these requires data not only for the obvious eco-attributes (energy, CO_2 and other emissions, toxicity, ability to be recycled, etc.) but also data for mechanical, thermal, and electrical properties. If material production is the phase of concern, selection is based on minimising the embodied energy or associated emissions (e.g. CO_2 production). If it is the use-phase that is of concern, however, selection is based instead on low weight or excellence as a thermal insulator or electrical conductor, while meeting other constraints on stiffness, strength, cost, and the like. The charts and methods in earlier chapters of this book give guidance on how to meet these constraints and objectives.

Eco-design is one aspect of sustainable development, but it is not the only one. Sustainable development requires clean energy and responsibly sourced materials that, as far as it is possible, are recovered and re-used at the end of product life. It requires an economy that generates sufficient wealth to provide for daily needs and for investment in education, health, and industrial infrastructure. If it is really to be sustainable, it must be equitable, not merely catering to the wishes of a sector of society but benefitting society as a whole. Rational assessment of a proposed sustainable development requires good data and the willingness to debate the implications of the facts for all three key capitals: natural, manufactured, and human.

You can find out much more about materials, the environment, and sustainability in two companion texts: *Materials and the Environment* and *Materials and Sustainable Development*, listed in Further Reading at the end of this chapter.

12.9 Appendix: some useful quantities

Energy contents of fuels

Coal, lignite	15–19 MJ/kg
Coal, anthracite	31–34 MJ/kg
Oil	11.69 kWh/L = 47.3 MJ/kg
Petrol (gasoline)	35 MJ/L = 45 MJ/kg
Gas	10.42 kWh/m^3
LPG	13.7 kWh/L = 46.5–49.6 MJ/kg

Conversion factors

- 1 BthU = 1.06 kJ
- 1 kWh/kg (sometimes written kW/kg/hr) = 3.6 MJ/kg
- 1 barrel of oil = 42 US gallons = 159 L =138 kg = 6210 MJ
- At \$50 per barrel, \$1 buys 124 MJ
- To convert energy from kJ/mol to kJ/kg multiply by 1000/(Atomic weight in kg/kmol).

12.10 Further reading

Allwood, J. M., & Cullen, J. M. (2012). *Sustainable Materials: With Both Eyes Open*. UIT. ISBN 978-1-906860-05-9. (The authors present a carefully researched and documented case for increased material efficiency).

Ashby, M. F. (2021). *Materials and the Environment – Eco-informed Material Choice* (3rd ed.). Butterworth Heinemann Ltd. ISBN 978-0-12-821521-0. (A teaching text that provides the resources – background, methods, data – to enable environmental issues relating to materials to be explored in depth).

Ashby, M. F. (2022). *Materials and Sustainable Development* (2nd ed.). Butterworth-Heinemann Ltd. ISBN 9780323983617. (An introduction to the concept of sustainable development with case studies and exercises).

Baker, Susan (2013). *Sustainable development*. Routledge. ISBN 978-1-134460-07-6. (A concise and readable explanation of sustainable development taking a global and national perspective).

Brundlandt, D. (1987). *Report of the World Commission on the Environment and Development.* Oxford University Press. ISBN 0-19-282080-X. (A much-quoted report that introduced the need and potential difficulties of ensuring a sustainable future).

Dasgupta, P. (2010). Nature's role in sustaining economic development. *Phil. Trans. Roy. Soc. B, 365,* 5–11. (A concise exposition of the ideas of economic, human and natural capital).

Ellen MacArthur Foundation. (2017). www.ellenmacarthurfoundation.org (The Ellen MacArthur Foundation has become an enabler of thinking and communication on the circular economy).

Fuad-Luke, A. (2002). *The eco-design handbook.* Thames and Hudson. ISBN 0-500-28343-5. (A remarkable source-book of examples, ideas, and materials of eco-design).

Graedel, T. E. (1998). *Streamlined Life-cycle Assessment.* Prentice-Hall. ISBN 0-13-607425-1. (An introduction to LCA methods and ways of streamlining them).

Kyoto Protocol. (1997). United Nations, Framework Convention on Climate Change. Document FCCC/CP1997/7/ADD.1. http://cop5.unfccc.deO14040/14041/14042/14043, Environmental management - life cycle assessment (and subsections), Geneva, Switzerland. (The first international consensus on combating climate change - leading to the currently annual 'Conference of the Parties' [COP]).

MacKay, D. J. C. (2008). *Sustainable energy – Without the Hot Air.* UIT Press. www.withouthotair.com/. ISBN 978-0-9544529-3-3. (MacKay brings a welcome dose of common sense into the discussion of energy sources and use. Fresh air replacing hot air, with the most engaging style).

McDonough, W., & Braungart, M. (2002). *Cradle to Cradle, Remaking the Way We Make Things.* North Point Press. ISBN-13 978-0-86547-587-8. (The book that popularised thinking about a circular materials economy).

Mulder, K., Ferrer, D., & Van Lente, H. (2011). *What is Sustainable Technology.* Greenleaf Publishing. ISBN 978-1-906093-50-1. (A set of invited essays on aspects of sustainability, with a perceptive introduction and summing up).

Schultz, K., & Bradley, D. (2016). Critical Mineral Resources for the 21st Century. USGS. https://minerals.usgs.gov/east//critical/index.html. (An analysis of critical elements from a US perspective).

12.11 Exercises

Exercise E12.1   What is meant by *embodied energy per kilogram* of a metal? Why does it differ from the free energy of formation of the oxide, carbonate, or sulphide from which it is extracted?

Exercise E12.2   Iron is made by the reduction of iron oxide, Fe_2O_3, with carbon; aluminium by the electro-chemical reduction of bauxite, basically Al_2O_3. The enthalpy of oxidation of iron 7.4 MJ/kg, that of aluminium to its oxide is 30.4 MJ/kg. Compare these with the embodied energies of cast iron and of aluminium, retrieved from Appendix A, Table A8. What conclusions do you draw?

Exercise E12.3   Copper, atomic weight 59 kg/kmol, can be extracted from natural copper oxide, CuO. The free energy of formation of CuO is 134 kJ/mol. This much energy *must* be provided to liberate 1 mol of copper from CuO. How does this compare with the reported embodied energy of copper, 59 MJ/kg (Appendix A, Table A8)?

Exercise E12.4 Aluminium, molecular weight 27 kJ/kmol, is made by the electrolysis of bauxite, essentially Al_2O_3, releasing 2 mols of aluminium per mol of bauxite. The free energy of formation of Al_2O_3 is 1640 kJ/mol; this much energy *must* be provided to liberate the 2 mols of aluminium from Al_2O_3. How does this compare with the embodied energy of aluminium, 190 MJ/kg (Appendix A, Table A8)?

Exercise E12.5 Window frames are made from extruded aluminium. It is argued that making them instead from extruded PVC would be more environmentally friendly (meaning that less embodied energy is involved). If the section shape and thickness of the aluminium and the PVC windows are the same, and both are made from virgin material, is the claim justified? You will find embodied energies and densities for the two materials in Appendix A, Tables A2 and A8.

Exercise E12.6 A range of office furniture includes a chunky hardwood table weighing 25 kg and a much lighter table with a 2 kg virgin aluminium frame and a 3 kg glass top. Which of the two tables has the lower embodied energy? Use data from Appendix A, Table A8 to find out.

Exercise E12.7 Rank the three common commodity materials low carbon steel, aluminium alloy, and polypropylene by carbon footprint per unit weight (kg/kg), CO_2, and carbon footprint per unit volume (kg/m^3), $CO_2.\rho$, where ρ is the density, using data from Appendix A, Tables A2 and A8 for density and carbon footprint.

Exercise E12.8 A bicycle design has an aluminium alloy frame weighing 11 kg; a medium carbon steel wheel-set of total weight 1.4 kg; and tyres, grips, and saddle, all butyl rubber, weighing 0.9 kg. What, approximately, is the carbon footprint of the materials of the bike? How does it compare with the carbon footprint of 1 L of gasoline (2.9 kg/L)? You will find eco-data for the materials in Appendix A, Table A8.

Exercise E12.9 What is meant by the process energy per kilogram for casting a metal? Why does it differ from the latent heat of melting of the metal?

Exercise E12.10 How does the carbon footprint of polypropylene compare with that of zinc alloy:

(a) by weight?
(b) by volume?
Use the data from Appendix A, Tables A2 and A8 to find out.

Exercise E12.11 Global sales of electric vehicles (EVs) in 2018 were estimated to be 1.7 million units and growing exponentially. Sales were expected to double by 2021. What was the expected annual growth rate?

Exercise E12.12 Use the internet to research rare earth elements. What are they? Why are they important? Why is there concern about their availability?

Exercise E12.13 The world production of rare earth elements rose from 60,000 tonnes in 1994 to 138,000 tonnes in 2018 and was growing exponentially. What was the annual growth rate? How long would it have taken for production to double?

Exercise E12.14 Most electric vehicles (EVs) use lithium-ion batteries. An EV using today's technology requires, on average, about 8 kg of lithium. In 2018 EV production was 1.7 million, growing at 23% per year. In the same year, the global production of lithium was 43,000 tonnes. If lithium production remains unchanged, how long will it be before the demand for lithium from EVs exceeds current production?

Exercise E12.15 'Lithium production, driven by electric vehicle production, is expected to rise by 23% over the next two years.' (Bloomberg Technology, 28 Jul 2019). What is the annual growth rate of lithium production? What is the doubling time of production?

Exercise E12.16 Global gold mine production, P_o, in 2018 was 3332 tonnes of gold, up 2% from the previous year. If this growth rate continues, what is the expected production in 2030?

Exercise E12.17 The global production of platinum in 2017 was 224 tonnes, of which a quarter was derived from recycling.* That may not sound like much, but at the current price of US$32/g it is worth about US $6.9 billion ($6.3 \times 10^9$). The catalytic converter of a car requires about 2 g of platinum-group metals, of which 75% is platinum. Car manufacture in 2017 was approximately 74 million vehicles. If all have catalytic converters, what fraction of the world production of platinum was absorbed by the auto industry in 2017?

Exercise E12.18 Global water consumption tripled in the 50 years to 2020. What was the annual growth rate, r %, in consumption C assuming exponential growth? By what factor would water consumption be predicted to increase between 2020 and 2050?

Exercise E12.19 The price of cobalt, copper, and nickel has fluctuated wildly in the past decade. Those of aluminium, magnesium, and iron have remained much more stable. Why? Research this by examining uses (which metals are used in high value-added products?) and the locations of the principal mining areas.

* http://www.platinum.matthey.com/documents/new-item/pgm%20market%20reports/pgm_market_report_may_2017.pdf

Exercise E12.20 Why is the recycling of metals more successful than that of polymers?

Exercise E12.21 Which phase of life would you expect to be the most energy-intensive (in the sense of consuming fossil fuel) for the following products?

- Toaster
- Two-car garage
- Bicycle
- Motorbike
- Wind turbine
- Ski lift

Indicate, in each case, your reasoning in one sentence.

Exercise E12.22 Vehicle tyres create a major waste problem. Use the internet to research ways in which the materials contained in car tyres can be used, either in the form of the tyre or in some decomposition of it.

Exercise E12.23 What are the US CAFÉ rules relating to car fuel economy? Use the internet to find out and report your findings in 10 sentences or less.

Exercise E12.24 The European Union's ELV (End of Life Vehicles) Directive on the disposal of vehicles at the end of life dictates what fraction of the weight of the vehicle must be recycled and what fraction is permitted to go to landfill. What are the current fractions? Use the internet to find out.

Exercise E12.25 Estimate the energy to cast copper by assuming it to be equal to the energy required to heat copper from room temperature (20°C) to its melting temperature T_m, and then melt it, requiring the latent heat of melting, L. Compare this with the actual casting energy. The table lists the necessary data. What conclusions do you draw?

Material	Specific heat, C_p (J/kg.K)	Melting temp, T_m (°C)	$C_p (T_m - T_o)$ (kJ/kg)	Latent heat, L (kJ/kg)	Casting energy (kJ/kg)
Copper	380	1030	384	210	9100

Exercise E12.26 The aluminium window frame of Exercise E12.5 is, in reality, made not of virgin aluminium but of 100% recycled aluminium. Recycled PVC is not available, so the PVC window continues to use virgin material. Which frame now has the lower embodied energy? Embodied energies for recycled materials are listed in Appendix A, Table A9.

Exercise E12.27 It is found that the quality of the window frame of Exercise E12.26, made from 100% recycled aluminium, is poor because of the pickup of impurities. It is decided to use aluminium with a 'typical' recycled content of 44% instead. The PVC window is still made from virgin material. Which frame now has the lower embodied energy? The data you will need are given in the previous related exercises.

Exercise E12.28 A chemical engineering reactor consists of a stainless steel chamber and associated pipework weighing 3.5 tonnes, supported on a mild steel frame weighing 800 kg. The chamber contains 20 kg of loosely packed alumina spheres coated with 200 g of palladium, the catalyst for the reaction. Appendix A, Table A8 lists embodied energies for the first three materials; that for palladium is 1.9×10^5 MJ/kg.

Material	Embodied energy (MJ/kg)
Stainless steel	81
Mild steel	33
Alumina	53
Palladium	41,500

Exercise E12.29 You are about to buy a wedding ring, but you and your partner also have environmental consciences and wish to assess the embodied energies for two options. You would like to buy a 24 carat (100%) gold ring weighing 10 g; your partner prefers one made of 50-50 platinum-rhodium alloy weighing 15 g. How different are the embodied energies of the two rings? (The table lists data for precious metals).

Material	Embodied energy (MJ/kg)
Gold	250,000
Platinum	290,000
Rhodium	610,000

To put the difference in context, you propose a 'light-bulb test': assuming a conversion efficiency of 0.38 for primary energy to electrical energy, for how many hours could you run a 10 W LED lamp on the energy saved if you settle for the ring with the lower embodied energy? (Remember 1 kWh = 3.6 MJ).

Exercise E12.30 The embodied energy of a mobile phone is about 400 MJ. The charger, when charging, consumes about 1 W. The phone is used for 2 years, during which it is charged overnight (8 hours) every night. Is the use energy larger or smaller than the energy to make the phone in the first place? What conclusions can you draw from your result? (In making the comparison, remember that electrical power is generated from primary fuel with an efficiency of about 38%).

Exercise E12.31 Cast iron scrap is collected in Europe and shipped 19,000 km to China where it is recycled. The energy to recycle cast iron is 5.2 MJ/kg. How much does the transport stage add to the total energy for recycling by this route? Is it a significant increase? Transport energy and carbon footprint are listed in Table 12.1.

Exercise E12.32　Bicycles, weighing 15 kg, are manufactured in South Korea and shipped 10,000 km to the US West Coast. On unloading they are transported by large (8 axle) trucks to the point of sale, Chicago, Illinois, a distance of 2900 km. What is the transport energy per bicycle? To meet Christmas demand, a batch of the bicycles is air-freighted from South Korea directly to Chicago, a distance by air of 10,500 km. What is the transport energy then? Transport energy and carbon footprint are listed in Table 12.1. The bikes are made almost entirely out of aluminium. How do these transport energies compare with the total embodied energy of the bike? Take the embodied energy of aluminium to be 190 MJ/kg.

Exercise E12.33　You are Assistant Professor of History at a German university and an active member of the Green Party. You have just been appointed to a Chair of History at the University of Sydney, Australia, 24,000 km away. You have a large library of books, which you cherish. The total weight of the books is 1200 kg. You could send them by sea freight, but that would take months and with a risk of damage. Federal Express could air-freight them, and they would get there before you, safe and sound. Weigh the environmental consequences of these two options in terms of carbon release to atmosphere. Carbon footprints for freight-transport are listed in Table 12.1.

Exercise E12.34　In 2008 there were 65 million cars on Chinese roads. By the end of 2018 this had increased to 240 million.

(a) If growth was exponential, what was the average growth rate of car numbers in China?
(b) If this growth continues, how many cars will be on China's roads in 2030?
(c) Conduct a sanity check. What is the population of China today? How fast is it growing? Is the car population calculated in (a) really going to continue until 2030? Use the internet to find the data you need.

Exercise E12.35　An analyst in 2023 predicts that the future growth rate in electric car sales will be approximately $= 20\%$ per year worldwide, and that the cars will have an average life of 13 years. The rare-earth metal *neodymium* is an ingredient of the high-field permanent magnets used at present for their motors. Assuming a fraction $f = 80\%$ of the neodymium in the current motors can be recovered at end of life, and that the amount of neodymium used per vehicle remains essentially unchanged, what fraction of future neodymium demand for car motors could be met?

Exercise E12.36　Global annual sales of smartphones in recent years have exceeded 1.5 billion units. The average smartphone weights 150 g and has a composition corresponding to that for electronic waste listed in the table. How much gold has been used annually for the production of smartphones in recent years?

Typical compositions electronic waste

Electronic waste	kg / tonne
Plastics	300
Glass and ceramics	300
Copper	200
Iron	80
Tin	40
Silver	2
Gold	0.6
Palladium	0.1
Other	77.3

Exercise E12.37 (a) From 2008 – 2014, smartphone sales doubled roughly every 2 years. What was the annual % growth rate in sales at this time?

(b) From 2014 – 2017, growth in smartphone sales had slowed to approximately $r = 4\%$ per year, with global smartphone sales in 2017 being 1.54 billion. If sales had continued to grow at the same rate, what would the anticipated sales have been in 2023? Search online to find out what has actually happened to annual smartphone sales.

Exercise E12.38 The statement is frequently made that "at a global growth rate of just 3% per year we will mine, process and dispose of more 'stuff' in the next 25 years than in the entire history of human engineering." Assuming that consumption started with the dawn of the industrial revolution, in 1750, is the statement true?

Exercise E12.39 Show that the index for selecting materials for a strong panel, loaded in bending, with the minimum embodied energy content is

$$M = \frac{\sigma_y^{1/2}}{H_m \, \rho}$$

where H_m is the embodied energy/kg of the material, σ_y is its yield strength, and ρ is its density. The panel area is fixed, and only the thickness may be varied.

Exercise E12.40 Behind the idea of sustainable development lies the concept of three essential capitals. What are they, and what are their characteristics?

Guided Learning Unit 1
Simple ideas of crystallography

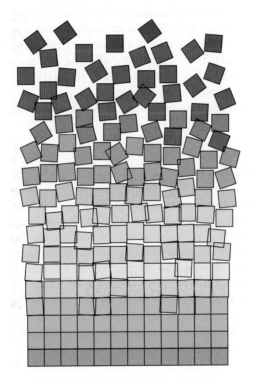

Chapter contents

https://doi.org/10.1016/B978-0-08-102399-0.00020-0

Introduction and synopsis

Though they lacked the means to prove it, the ancient Greeks suspected that solids were made of discrete atoms that were packed in a regular, orderly lattice to give crystals. Today, with the 21st-century techniques of X-ray and electron diffraction and lattice-resolution microscopy available to us, we know that all solids are indeed made of atoms and that most (but not all) are crystalline. The common engineering metals and ceramics are made of many small crystals, or *grains*, stuck together at *grain boundaries* to make *polycrystals*. Many properties of crystalline materials – its strength, stiffness, toughness, conductivity, and so forth – are strongly influenced by their atomic packing.

Crystallography is a geometric language for describing the three-dimensional arrangement of atoms or molecules in crystals. Simple crystalline and non-crystalline structures were introduced in Chapter 3. Here we go a little deeper, exploring a wider range of structures, interstitial space (important in understanding alloying and the heat treatment of steel), and ways of describing planes and directions in crystals (crucial in the production of single crystal turbine blades, beverage cans, and microchips).

Exercises are provided throughout, so do these as you go along; solutions are provided online.

GL1.1 Crystal structures

Three-dimensional crystal lattices may be characterised by the repeating geometry of the atomic arrangement they contain, particularly exploiting their high degree of symmetry. Most atomic bonding gives a well-characterised equilibrium spacing between neighbouring atoms, so for the purposes of understanding crystal packing, the atoms may be treated as hard spheres in contact. Spheres can be packed to fill space in various different ways – in fact, there are 14 distinguishable three-dimensional lattices. If you are a crystallographer or mineral scientist you need to know about all of them, but engineering materials, for the most part, are based on three simple structures, and we concentrate on these.

Each lattice is characterised by a unit containing a small number of atoms, which repeats itself in three dimensions, called the *unit cell*.

> **Definition:** The *unit cell* of a crystal structure is the unit of the structure, chosen so that it packs to fill space, and which, translated and stacked regularly, builds up the entire structure. The *primitive unit cell* is the smallest such cell.

Figure GL1.1 shows the most important unit cells. The first lattice is the *triclinic* unit cell, which is the most general – the edge lengths (or *lattice constants*) a, b, and c are all different, and none of the angles is 90°. The other 13 space (or *Bravais*[1]) lattices are all special cases of this one – for example, for all the cubic unit cells, $a = b = c$ and $\alpha = \beta = \gamma = 90°$.

The great majority of the 92 stable elements are metallic, and of these, the majority (68 in all) have one of the last three simple structures in Figure GL1.1: close-packed hexagonal (CPH, or sometimes HCP), face-centred cubic (FCC), and body-centred cubic (BCC). Both CPH and FCC lattices are close-packed – the spheres fill as much space as possible – and can be constructed by stacking close-packed planes of spheres. Let's investigate this as an exercise.

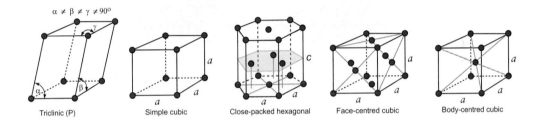

Figure GL1.1 The general triclinic unit cell, and the four most important lattices for engineering materials: simple cubic (SC), close-packed hexagonal (CPH), face-centred cubic (FCC), and body-centred cubic (BCC).

[1] Auguste Bravais (1811–1863), French crystallographer, botanist, and physicist, sought to explain the shapes of mineral crystals by analysing the figures formed by points distributed regularly in space. Others had tried this and had concluded that the number of such 'lattices' distinguishable by their symmetry was finite and small, but it took Bravais to get it right, demonstrating that there are exactly 14.

Exercise

(NB solutions for this and all Exercises are provided online – details in Preface).

GL1.1   Figure GL1.2 shows a close-packed layer of spheres, with a partial second layer stacked on top. Identify two alternative ways to stack the third layer, then draw them on the two views shown. Look at the alignment between the atoms in the first and third layers. What do you observe?

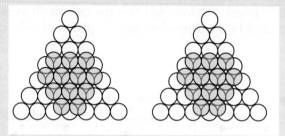

Figure GL1.2   Two layers of close-packed planes.

The close-packed hexagonal structure   The CPH structure is usually described by a hexagonal unit cell, with an atom at each corner, one at the centre of the hexagonal faces, and three in the middle layer. In this packing, the close-packed layers are clearly seen, and alternate close-packed layers are aligned (as observed for one case in Exercise GL1.1), giving an ABAB... sequence (Figure GL1.3). Unit cells are often drawn with the atoms reduced in size, to show clearly where their centres lie with respect to one another, but in reality atoms will touch in certain *close-packed directions*.

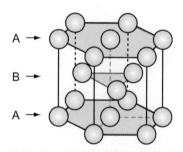

Figure GL1.3   The CPH structure of packed spheres showing the ABA stacking.

Of the metallic elements, 30 have the CPH structure, including:

Material	Typical uses of element and its alloys
Zinc	Die-castings, plating
Magnesium	Lightweight structures
Titanium	Light, strong components for airframes and engines; biomedical and chemical engineering
Cobalt	High-temperature superalloys, bone-replacement implants
Beryllium	A light, stiff metal; its use is limited by expense and potential toxicity

CPH metals have the following characteristics.

- They are reasonably ductile, allowing them to be hot forged, but in a more limited way than FCC metals (so they are also commonly cast).
- Their structure makes them more anisotropic than FCC or BCC metals.

The face-centred cubic structure The FCC structure is described by a cubic unit cell with one atom at each corner and one at the centre of each face. Atoms touch along the diagonals of the cube faces, and the centre atoms in any pair of adjoining faces also touch one another. The close-packed layers are not so obvious in FCC, but they can be seen by looking down any diagonal of the cube. Figure GL1.4 shows that the close-packed layers are now stacked in an ABCABC... sequence (the other case in Exercise GL1.1).

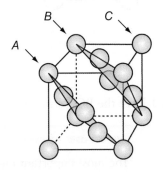

Figure GL1.4 The FCC structure of packed spheres showing the ABC stacking.

Among the metallic elements, 17 have the FCC structure, including:

Material	Typical uses of element and its alloys
Aluminium	Airframes, and bodies of trains, trucks, cars; beverage cans
Nickel	Turbine blades and disks
Copper	Conductors, bearings
Lead	Batteries, roofing, cladding of buildings
Austenitic stainless steels (based on iron)	Stainless cookware, chemical and nuclear engineering, cryogenic engineering
Silver, gold, platinum	Jewellery, coinage, electrical contacts

FCC metals have the following characteristics:

- They are very ductile when pure, work hardening rapidly but softening again when annealed, allowing them to be shaped by deformation processing (e.g. rolled, forged, or drawn).
- They are generally tough and resistant to crack propagation (as measured by their fracture toughness, K_{1c}).
- They retain their ductility and toughness to absolute zero, something very few other structures allow.

The body-centred cubic structure The BCC structure is described by a cubic unit cell with one atom at each corner and one in the middle of the cube (Figure GL1.5). The atoms touch along the internal diagonals of the cube, and this structure is not made up of close-packed planes.

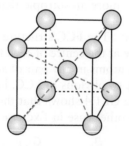

Figure GL1.5 The BCC structure of non-close-packed spheres.

Of the metallic elements, 21 have the BCC structure (most are rare earths), including:

Material	Typical uses of element and its alloys
Iron, mild steel	The most important metal of engineering: construction, cars, cans, and more
Carbon steels, low alloy steels	Engine parts, tools, pipelines, power generation
Tungsten	Lamp filaments
Chromium	Electroplated coatings

BCC metals have the following characteristics:

- They are ductile, particularly when hot, allowing them to be shaped by deformation processing (e.g. rolled, forged, or drawn).
- They are generally tough and resistant to crack propagation (as measured by their fracture toughness, K_{1c}) at and above room temperature.

- They become brittle below a 'ductile-brittle transition temperature,' limiting their use at low temperatures.
- They can generally be hardened with interstitial solutes.

Exercises

GL1.2   (a) For the CPH structure, show that the ratio of the lattice constants $c{:}a$ is equal to 1.633.
(b) The atomic radius of CPH magnesium is 0.1605 nm. Find the lattice constants c and a for Mg.

GL1.3   Find the ratio of the lattice constant (the size of the cube) to the atomic radius for: (a) FCC; (b) BCC.

Atomic packing fraction and theoretical density The density of a crystalline material depends on the mass of the atoms and the way in which they are packed. We can find the theoretical density of the crystal directly from the mass and volume of the unit cell. An important first step is to be clear how many atoms are fully contained within a given unit cell (since atoms on the boundaries of the unit cell are shared with neighbouring cells). This also allows us to find the *atomic packing fraction* (i.e. the proportion of space that is occupied by the atoms, represented as solid spheres).

Example GL1.1

Find the atomic packing fraction for the FCC lattice.

Answer. The volume of the FCC unit cell, with a lattice constant a, is equal to a^3. This can be expressed in terms of the atomic radius, using the result from Exercise GL1.3(a): $a = 2\sqrt{2}\,R$. Hence the unit cell volume is $16\sqrt{2}\,R^3$. Now consider the FCC unit cell in Figure GL1.4. Atoms on the corners are shared between eight unit cells, while those on the faces are shared between two. The number of atoms within a single unit cell is thus:

$$8 \times 1/8\,(\text{corners}) + 6 \times 1/2\,(\text{faces}) = 4\text{ atoms}$$

Hence the atomic packing fraction for FCC $= \dfrac{4 \times \dfrac{4}{3}\pi R^3}{16\sqrt{2}R^3} = 0.74\ (74\%)$

Exercise

GL1.4   Use the same method to find the atomic packing fractions for: (a) CPH; (b) BCC. Compare these with the value for FCC, and comment on the results.

The theoretical density is found as follows. The mass of 1 atom is given by A/N_A, where A is the atomic mass (in kg) of 1 mol of the element, and $N_A = 6.022 \times 10^{23}$ is Avogadro's[2] number (the number of atoms per mol). So if the number of atoms per unit cell is n, and its volume is V_c, the theoretical density is:

$$\rho = \frac{nA}{V_c N_A}$$

Since the atomic mass and the crystal packing are both physically well-defined, this explains why the densities of metals (and ceramics) have narrow ranges, and there is no scope to modify the density of a solid (e.g. by processing a metal differently).

Example GL1.2

The diameter of an atom of nickel is 0.2492 nm, and the atomic mass of Ni is 58.71 kg/kmol (note the units: kg *per kmol*). Calculate the theoretical density of FCC nickel.

Answer. From Exercise GL1.3(a), $a = D\sqrt{2}$, where the atomic diameter is D. Hence the lattice constant for Ni is $\sqrt{2} \times 0.2492$ nm $= 0.3524$ nm. Hence the volume of the unit cell $V_c = (0.3524 \times 10^{-9})^3$ m³.

The mass of a Ni atom $= A/N_A = 58.71 \times 10^{-3}/(6.022 \times 10^{23}) = 9.749 \times 10^{-26}$ kg. As the number of atoms per unit cell $n = 4$, the theoretical density of nickel is:

$$\rho_{Ni} = 4 \times 9.749 \times 10^{-26}/(0.3524 \times 10^{-9})^3 = 8911 \text{ kg/m}^3$$

Exercises

GL1.5 Gold has an FCC structure, a density of 19,300 kg/m³, and an atomic mass of 196.967 kg/kmol. Estimate the dimensions of the unit cell and the atomic diameter of gold.

GL1.6 (a) Determine the theoretical density for BCC iron, Fe (for which the atomic mass is 55.847 kg/kmol and the lattice constant is 0.2866 nm, at room temperature).
(b) A sample of pure iron is heated from room temperature, 20°C, to 910°C. The sample expands uniformly, such that the strain ε in all directions is $\varepsilon = \alpha \Delta T$, where α is the thermal expansion coefficient, and ΔT is the temperature change. Find the expected value for the lattice constant at 910°C, and thus the % decrease in the density compared to the room temperature value.
(c) At 910°C, pure iron undergoes an important phase transformation to FCC (more on this in *Guided Learning Unit 4*). Use the atomic packing fractions of BCC and FCC to determine the % increase in density during the transformation of the Fe sample, and thus find the % reduction in the linear dimensions of the sample when this occurs.

GL1.7 α-titanium has a CPH structure, a density of 4540 kg/m³, and an atomic mass of 47.9 kg/kmol. Calculate the dimensions of the unit cell and the atomic diameter of titanium.

---

[2] Amedeo Avogadro (1776–1856), Italian professor at the University of Turin, though originally a church lawyer and schoolteacher. The hypothesis and associated constant named after him stem from his essay on the molecular theory of gases, postulating that there are a fixed number of molecules in a given volume of an ideal gas, at constant temperature and pressure. Sacked from his chair for revolutionary activities against the king, the university's official stance was that he was on leave to concentrate on research, but this particular sabbatical lasted for 10 years before he was reinstated.

GL1.2 Interstitial space

Interstitial space is the unit of space between the atoms or molecules. The FCC and CPH structures both contain 'interstitial holes' (or 'interstices') of two sorts: *tetrahedral* and *octahedral*; BCC effectively only contains tetrahedral holes. Four examples are shown in Figure GL1.6.

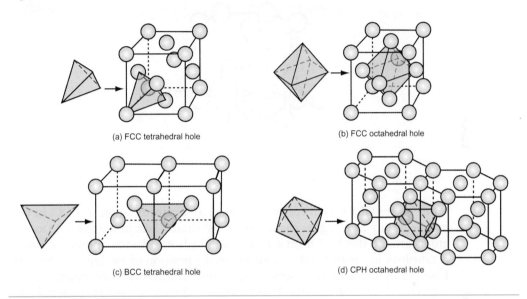

(a) FCC tetrahedral hole

(b) FCC octahedral hole

(c) BCC tetrahedral hole

(d) CPH octahedral hole

Figure GL1.6 (a, b) Two types of interstitial holes in the FCC structure, (c) tetrahedral hole in the BCC structure, and (d) octahedral hole in the CPH structure.

These holes are important because foreign atoms, if small enough, can fit into them. For both FCC and CPH structures, the tetrahedral hole can accommodate, without strain, a sphere with a radius of 0.22 of that of the host. The octahedral holes are larger – you can find out how much in Exercise GL1.8. In reality atoms are not completely hard, so foreign atoms that are somewhat larger than the holes can be squeezed into the interstitial spaces.

This is particularly important for carbon steel. At room temperature, carbon steel is BCC iron with carbon in some of the interstitial holes. BCC tetrahedral holes can hold a sphere with a radius 0.29 times that of the host without distortion. Carbon goes into these holes, but, because it is too big, it distorts the structure. It is this distortion that gives carbon steels much of their strength.

We encounter interstitial holes in another context later: they provide a way of understanding the structures of many oxides, carbides, and nitrides.

Exercises

GL1.8   Calculate the diameter of the largest sphere that will fit into the octahedral hole in the FCC structure, relative to the diameter of the host spheres.

GL1.9   Identify a tetrahedral hole in the CPH lattice of Figure GL1.7.

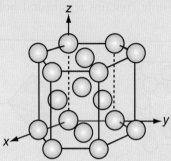

Figure GL1.7   The CPH structure (this view is rotated by 60° compared to Figure GL1.3, to aid visualisation).

GL1.10   From Figure GL1.6(c), there appears to be an octahedral hole in BCC, in the middle of any face (e.g. at the centre of the shared face in the figure). Do the atoms around this hole form a regular octahedron? Identify another identical octahedron in the BCC lattice and locate the position of the hole on the unit cell. Find the size of the largest sphere that could fit into this space, relative to the diameter of the host spheres. Hence explain why carbon sits in the tetrahedral holes in BCC iron.

GL1.11   Derive the result given in the text, that the largest sphere that fits in a tetrahedral hole in FCC or CPH lattices has a diameter equal to 0.22 times that of the host spheres.

GL1.3 Describing planes

The properties of crystals depend on *direction*. The elastic modulus of hexagonal titanium (CPH), for example, is greater along the hexagonal axis than normal to it. This difference can be used in engineering design. In making titanium turbine blades, for example, it is helpful to align the hexagonal axis along a turbine blade to use the extra stiffness. Silicon (which also has a cubic structure) oxidises in a more uniform way on the cube faces than on other planes, so silicon 'wafers' are cut with their faces parallel to a cube face. For these and many other reasons, a way of describing planes and directions in crystals is needed. *Miller*[3] *indices* provide it.

---

[3] William Hallowes Miller (1801–1880), British mineralogist, devised his index system (the 'Millerian system') in 1839. He also discovered a mineral which he named – wait for it – Millerite.

Definition: The *Miller indices of a plane* are the *reciprocals* of the intercepts that the plane makes with the three axes defining the edges of the unit cell, reduced to the smallest integers.

Figure GL1.8 illustrates how the Miller indices of a plane are found. Draw the plane in the unit cell so that it does *not* contain the origin – if it does, displace it along one axis until it doesn't. Find where the plane intersects the axes, and measure the three intercepts, taking one unit as the cell edge-length, and take the reciprocals. (If the plane is *parallel* to an axis, its intercept is at infinity and the reciprocal is zero). Reduce the result to the small-est set of integers by multiplying or dividing through by a constant to get rid of fractions or common factors. The Miller indices of a plane are always written in *round* brackets: (100), (111), and so on. Note that the plane intercepts may be negative (e.g. $x = y = 1$, $z = -1$): the indices are then written as $(11\bar{1})$. Work through the examples in the figure to check that you get the same indices. Note that there are many planes of the 111 type; two are shown in Figure GL1.8. The complete *family* of planes is described by putting the indices in curly brackets:

$$\{111\} = (111),(11\bar{1}),(1\bar{1}1),(\bar{1}11)$$

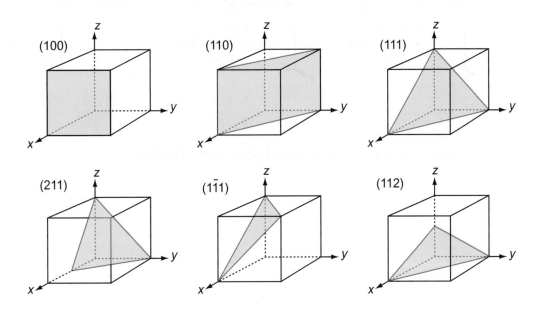

Figure GL1.8 The Miller indices of planes from four common families.

Exercises

GL1.12 What are the Miller indices of the planes shown in Figure GL1.9(a)–(c)? (Remember that, to get the indices, the plane must not pass through the origin, and you should eliminate fractions or common factors).

GL1.13 When iron is cold it cleaves (fractures in a brittle way) on the (100) planes. What will be the angle between intersecting cleavage faces?

GL1.14 Mark on the cells of Figure GL1.9(d)–(f) the following planes: $(\bar{1}11)$, (120), $(\bar{1}\bar{1}2)$.

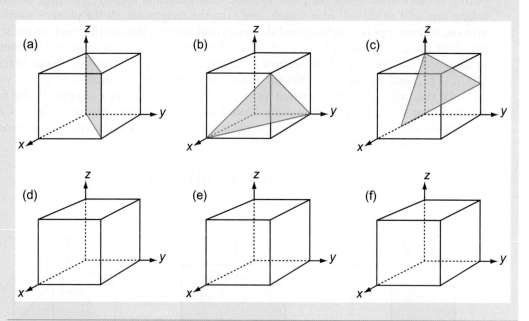

Figure GL1.9 Identifying the Miller indices of planes.

GL1.15 What is the family of close-packed planes in the FCC lattice?

GL1.4 Describing directions

> **Definition:** The *Miller indices of a direction* are the components of a vector parallel to that direction, starting from the origin, reduced to the smallest integer values (but *not* in this case taking reciprocals).

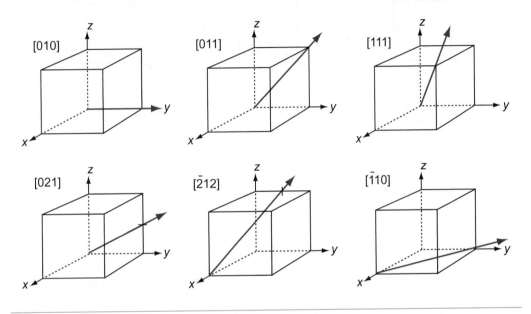

Figure GL1.10 Miller indices of directions. Always translate the vector so that it starts at the origin.

Figure GL1.10 shows how to find the Miller indices of a direction. Draw a line from the origin, parallel to the direction, extending it until it hits an edge or face of the unit cell. Read off the coordinates of the point of intersection. Get rid of any fractions by multiplying all the components by the same constant. Check that you agree with the direction indices in the figure. The notation $\bar{1}$ simply means an intersection point is −1. The Miller indices of a direction are always written in *square* brackets (e.g. [100]), and, as with planes, there are several directions of the 100 type. The complete *family* of directions is described by putting the indices in angle brackets:

$$< 100 > = [100], [010], [001]$$

Exercises

GL1.16   Identify the Miller indices of the three directions shown in the sketches of Figure GL1.11(a)–(c).

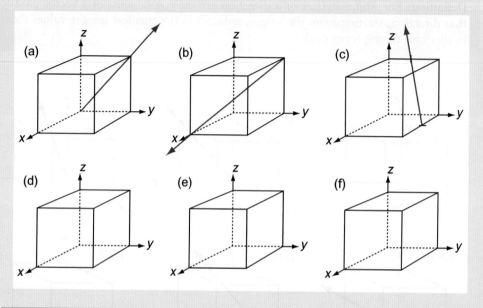

Figure GL1.11   Identifying the Miller indices of directions.

GL1.17   Mark, on the cells of Figure GL1.11(d)–(f), the following directions: $[1\bar{1}1]$, $[210]$, and $[22\bar{1}]$.

GL1.18   What is the family of close-packed directions in the FCC lattice?

GL1.5 Ceramic crystals

Technical ceramics give us the hardest, most refractory materials of engineering. Those of greatest economic importance include the following:

- Alumina, Al_2O_3 (spark plug insulators, substrates for microelectronic devices)
- Magnesia, MgO (refractories)
- Zirconia, ZrO_2 (thermal barrier coatings, ceramic cutting tools)
- Uranium dioxide, UO_2 (nuclear fuels)
- Silicon carbide, SiC (abrasives, cutting tools)
- Diamond, C (abrasives, cutting tools, dies, bearings)

The ceramic family also gives us many of the functional materials – those that are semi-conductors, are ferromagnetic, show piezoelectric behaviour, and so on. Their structures often look complicated but can usually be made comprehensible by reading them as atoms of one type arranged on a simple FCC, CPH, or BCC lattice, with the atoms of the second type (and sometimes three or more) inserted into the interstitial spaces of the first.

The diamond-cubic (DC) structure We start with the hardest ceramic of the lot – diamond – of major engineering importance for cutting tools, abrasives, polishes, and scratch-resistant coatings, and at the same time as a valued gemstone for jewellery. Silicon and germanium, the foundation of semiconductor technology, have the same structure.

Figure GL1.12 shows the unit cell. Think of it as an FCC lattice with an additional atom in four of its eight tetrahedral interstices, labelled 1, 2, 3, and 4 in the figure. The tetrahedral hole is far too small to accommodate a full-sized atom, so the others are pushed farther apart, lowering the density. The tetrahedral structure is a result of the covalent nature of the bonding in carbon, silicon, and germanium atoms – they are happy only when each has four nearest neighbours, symmetrically placed around them (see Chapter 3).

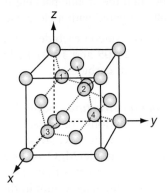

Figure GL1.12   The diamond-cubic (DC) structure.

Silicon carbide, like diamond, is very hard, and it too is widely used for abrasives and cutting tools. The structures of the two materials are closely related. Carbon lies directly above silicon in the Periodic Table of Elements. Both have the same crystal structure and are chemically similar. So it comes as no surprise that a compound of the two with the formula SiC has a structure like that of diamond, with half the carbon atoms replaced by silicon, as in Figure GL1.13 (the atoms marked 1, 2, 3, and 4 in Figure GL1.12 are the ones replaced).

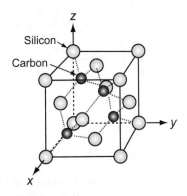

Figure GL1.13   The structure of silicon carbide.

Materials with the diamond-cubic structure	Comment
Carbon, as diamond	Cutting and grinding tools, jewellery
Silicon, germanium	Semiconductors
Silicon carbide	Abrasives, cutting tools

Exercises

GL1.19 How many atoms are there in the unit cell of the DC structure, shown in Figure GL1.12? Show that the lattice constant for DC is equal to 2.309 D, where D is the atomic diameter. Hence find the atomic packing fraction for the DC structure, and comment on the result compared with a close-packed structure.

GL1.20 The lattice constant of DC silicon carbide is $a = 0.436$ nm. The atomic masses of silicon and carbon are 28.09 and 12.01 kg/kmol, respectively. What is the theoretical density of silicon carbide?

Ceramics with the rocksalt (halite) structure Several important oxides have the formula MO, where M is a metal ion. Oxygen ions are large, usually bigger than those of the metal. When this is so, the oxygen packs in an FCC structure, with metal atoms occupying the octahedral holes in this lattice – an example, MgO, is shown in Figure GL1.14. This is known as the rocksalt (or halite) structure because it is that of sodium chloride, NaCl, with chlorine replacing oxygen, and sodium replacing magnesium in the figure. Sodium chloride (common salt) itself is a material of engineering importance. It has been used for road beds and buildings, there are plans to bury nuclear waste in rocksalt deposits; and large single crystals of rocksalt are used for the windows of high-powered lasers.

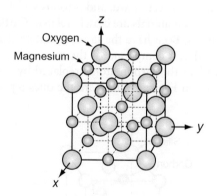

Figure GL1.14 The halite structure of MgO, typical of many simple oxides.

Materials with the rocksalt structure	Comment
Magnesia, MgO	A refractory ceramic with useful strength
Ferrous oxide, FeO	One of several oxides of iron
Nickel oxide, NiO	Ceramic superconductors
All alkali halides, including salt, NaCl	Feedstock for chemical industry, nuclear waste storage

Exercises

GL1.21 (a) Consider the halite structure shown in Figure GL1.14. If the oxygen ions form a true FCC lattice, then the oxygen ions should touch on the face diagonals. Determine the diameter of the largest metal ion that can fit into the locations shown in the figure, without distorting the lattice, relative to an oxygen diameter of unity.

(b) The diameters of oxygen and magnesium ions are 0.252 and 0.172 nm, respectively. Are the oxygen ions actually close-packed in the MgO lattice?

GL1.22 (a) Sodium chloride (NaCl) has a lattice constant $a = 0.564$ nm. Determine the density of NaCl, given that the atomic masses of sodium and chlorine are 22.989 and 35.453 kg/kmol, respectively.

(b) The ratio of the diameters of sodium and chlorine ions is 0.69. Show that the chlorine and sodium atoms are touching along the edges of the unit cell, and find the size of the gaps between the chlorine atoms on the diagonal of the face of the unit cell.

Oxides with the corundum structure   A number of oxides have the formula M_2O_3, among them alumina (Al_2O_3). The oxygen ions, the larger of the two, are arranged in a CPH-like stacking. The M ions occupy two-thirds of the octahedral holes in this lattice, one of which is shown for alumina, filled by an Al ion, in Figure GL1.15. The hole is not big enough to accommodate the aluminium ion, and the oxygen lattice is pushed slightly apart – so it is not actually close-packed, but the atom centres fit the CPH structure. It is also the case that in this instance the hexagonal unit of Figure GL1.15 is not a unit cell. A larger and more complicated unit cell is needed, so that when these are stacked to form the lattice, the metal ions also fall into a regular pattern of hexagons, occupying the correct proportion of holes. This need not concern us further here.

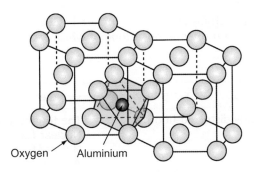

Oxygen   Aluminium

Figure GL1.15   The M atoms of the corundum structure lie in the octahedral holes of a CPH oxygen lattice, of which one is shown. Two adjacent cells are needed to show an octahedral hole.

Materials with the corundum structure	Comment
Alumina, Al_2O_3	The most widely used technical ceramic
Iron oxide, Fe_2O_3	The oxide from which iron is extracted
Chromium oxide, Cr_2O_3	The oxide that gives chromium its protective coating
Titanium oxide, Ti_2O_3	The oxide that gives titanium its protective coating

Oxides with the fluorite structure   The metal atoms in the important technical ceramic zirconium dioxide, ZrO_2, and the nuclear fuel uranium dioxide, UO_2, being far down in the Periodic Table of Elements, are large in size. Unlike the oxides described earlier, the M in MO_2 is now larger than the O. So it is the M atoms that form a FCC-like structure with the oxygen atoms in the eight tetrahedral holes – one such site in ZrO_2 is shown in Figure GL1.16.

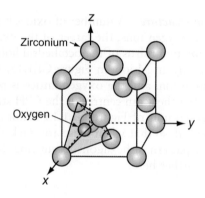

Figure GL1.16   The M atoms of fluorite-structured oxides like ZrO_2 pack in an FCC array, with oxygen in the eight tetrahedral holes (one of which is shown).

Materials with the fluorite structure	Comment
Zirconia, ZrO_2 (slightly distorted fluorite)	The toughest refractory ceramic
Urania, UO_2	Nuclear fuel
Thoria, ThO_2	Nuclear fuel
Plutonia, PuO_2	A product of nuclear fuel reprocessing, a fuel in itself

Exercise

GL1.23 Uranium dioxide UO_2 has the fluorite structure shown in Figure GL1.16. The oxygen ions are too large to fit into the locations shown in the figure, pushing the metal lattice further apart (so it is not strictly close-packed).

(a) Assuming the oxygen ions touch all the surrounding uranium ions, find the lattice constant for UO_2, given that the ionic diameters of uranium and oxygen are 0.222 and 0.252 nm, respectively.

(b) Determine the theoretical density of UO_2, given that the atomic masses of uranium and oxygen are 238.05 and 16.00 kg/kmol, respectively.

GL1.6 Polymer crystals

Many polymers crystallise – if their chains are sufficiently regular. The long chains line up and pack to give an ordered, repeating structure, just like any other crystal. The low symmetry of the individual molecules means that the choice of lattice is limited. Figure GL1.17 is a typical example; it is polyethylene. It looks complicated, but can be considered as a body-centred rectangular unit cell, shown dashed in the figure, with each grey marker in the unit cell being replaced by a C_2H_4 unit oriented as shown, with the lower C atom placed at each marker centre.

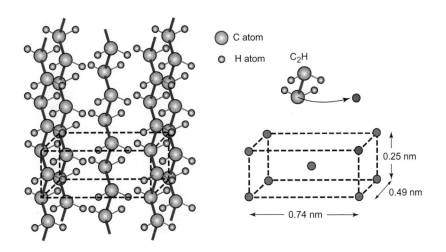

Figure GL1.17 The structure of a crystalline polyethylene.

Few engineering polymers are completely crystalline, but many have as much as 90% crystallinity. Among those of engineering importance are:

Material (crystallinity)	Typical uses
Polyethylene, PE (65–90%)	Bags, tubes, bottles
Nylon, PA (65%)	High-quality parts, gears, catches
Polypropylene, PP (75%)	Mouldings, rope

Exercise

GL1.24   (a) The dimensions of the unit cell of a polyethylene crystal are shown in Figure GL1.17. Identify the number of complete C_2H_4 monomers per unit cell, and hence show that the density of crystalline polyethylene is close to 1000 kg/m^3. The atomic masses of carbon and hydrogen are 12.01 and 1.008 kg/kmol, respectively.

(b) Amorphous polyethylene has a density of 870 kg/m^3. Explain the difference in value compared to that calculated in (a). Use a simple rule of mixtures to find the density of a semi-crystalline polyethylene with 70% crystallinity.

Guided Learning Unit 2
Material selection in design

A pole vaulter – the pole stores elastic energy.
(istock.com)

Chapter contents

https://doi.org/10.1016/B978-0-08-102399-0.00021-2

GL2.1 Introduction and synopsis

The systematic design of products was introduced in Chapter 2, exploring the constraints and objectives that must be met, such as the mechanical responses to loads, or the effect of heat or other service conditions, either applied deliberately or unintentionally. To recap briefly, for mechanical and structural applications, the *functional constraints* automatically include the *stiffness* – the allowable deflection has a limit; and the *strength* – the product must not fail under load. But for a design to be optimal, some aspect of the design will also be pushed to as low or as high a value as possible – minimum mass or cost, for example, or maximum energy stored or transferred: these we call *objectives*. In reaching the optimum solution, the designer's freedom of choice – the *free variables* – are primarily certain *geometric constraints* (section shapes and sizes) and the material.

In this *Guided Learning Unit* we develop a systematic approach to material selection, first for *stiffness-limited* design (following Chapter 4) and then *strength-limited* design (following Chapter 5). This involves simple modelling of the mechanical response in a given application, to identify the key trade-off in material properties that enables the performance to be optimised – that is, we need a *material performance index*. We explain how to plot them onto *material property charts* and illustrate their use via case studies. As in all the *Guided Learning Units*, exercises are embedded within the text, to give practice in the techniques as you go along. Fully worked solutions are provided online. Once grasped, the combination of index lines and charts is widely applicable in design limited by other physical behaviour – fracture, heat transfer, or electrical conduction. Examples will be found in many chapters of the book.

GL2.2 Screening and ranking: property limits and trade-offs

Let us start with a problem of re-design – finding a replacement material for an application that already exists. A common challenge in personal transport applications, from bicycles to cars, is to save weight, to reduce the energy needed to move it along, be this provided by a person, a combustion engine, or a battery – the objective is to minimise the mass. But a constraint on all vehicles is that they need to be sufficiently stiff to support the load of their passengers, contents, and self-weight while being propelled along the road. In choosing the material it is clear we will need to consider the properties of density (governing the mass) and Young's modulus (governing the stiffness), and for this we already have the $E - \rho$ property chart introduced in Chapters 1 and 4 (Figure GL2.1). If we are looking at an existing design made of steel, for example, a simple-minded view would be to ask: which materials are stiffer and lighter than steel? This will certainly be better if we make it exactly the same size as the existing steel version. What materials does this approach suggest?

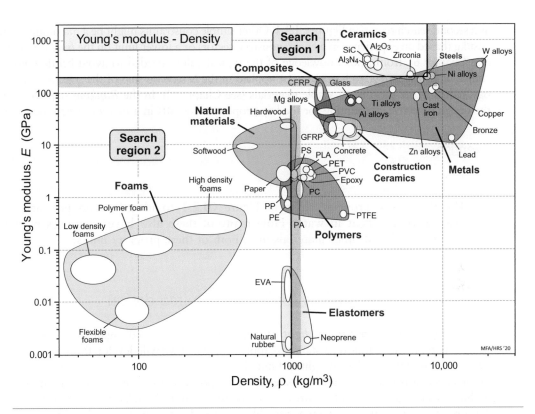

Figure GL2.1 Material property chart for Young's modulus and density. Two 'search regions' are shown for materials with (1) higher E and lower ρ than steels, (2) lower ρ than water.

Example GL2.1

Use the $E - \rho$ property chart in Figure GL2.1 to identify materials that have a higher value of E and a lower value of ρ than steels, using the rectangular 'search region 1' on the chart, located with the *Steels* bubble in its bottom right-hand corner. Do these materials sound like realistic candidates for a bicycle or parts of a car?

Answer. From the $E - \rho$ chart, the only materials meeting these criteria are various *ceramics*: silicon carbide (SiC), alumina (Al_2O_3), aluminum nitride (Al_3N_4), and zirconia. Ceramics are inherently brittle, with poor resistance to impact – hardly appropriate for a vehicle.

So approached in this way, the answer is self-evidently impractical – what have we missed? The answer is that we do not have to improve on *both* properties to reduce the weight – a heavier material may work, provided its modulus is sufficient; likewise, a less stiff material might work, provided it is light enough. Optimising material choice is therefore a matter of *property trade-offs*: we come to this shortly. But any design imposes multiple constraints on the material of which it is made, and some of these can be captured as simple property limits. For example, 'this solid component must float in water' translates into the property limit $\rho < 1000$ kg/m^3 – this is shown as a second search region in Figure GL2.1. Other simple constraints might be a maximum material cost/kg (based on experience in a given industrial sector) or a lower limit on the *fracture toughness* (see Chapter 6), from prior knowledge of the typical values needed to avoid brittle failures. So simple limits on property charts may be applied in the *Screening* step of the selection methodology, reducing the shortlist of viable materials (see Chapter 2). In the *Ranking* step we optimise the design objective, accounting for the key property trade-offs in a *material performance index*.

Exercise

GL2.1 Use the $E - \rho$ property chart in Figure GL2.1 to identify solid polymers and elastomers that float in water. How many of these also have a Young's modulus $E > 1$ GPa? What other materials meet both of these property limits?

GL2.3 Material indices for stiffness-limited design

Minimising weight in tension: a light, stiff tie Ranking materials against a design objective using a material index follows a standard approach, which we will develop through an exemplar problem: materials for a lightweight tie limited by its stiffness in tension (see the biplane example in the cover picture of Chapter 4). Its length L is specified, and it must carry a tensile force F without extending elastically by more than δ – its stiffness must be at least $S^* = F/\delta$ (Figure GL2.2). The objective is to make it as light as possible, and the designer may vary the cross-section area A (a *free variable*). These design requirements are summarised in Table GL2.1.

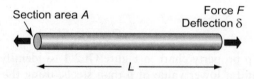

Section area A
Force F
Deflection δ
L

Figure GL2.2 A tensile tie of stiffness $S = F/\delta$, where F is the load and δ is the extension or deflection.

Table GL2.1 Design requirements for a light, stiff tensile tie

Function	• Tensile tie	
Constraints	• Stiffness S^* specified	Functional constraint
	• Length L specified	Geometric constraint
Objective	• Minimise mass m	
Free variables	• Choice of cross-section area A	
	• Choice of material	

To get a feel for the problem, we will analyse a specific problem – and then show how to derive a general material index for this design scenario and how to use it on a property chart to select from all possible materials.

Exercise

GL2.2 (a) Using Hooke's law for uniaxial tension (see Chapter 4), show that the stiffness of the tie is $S = AE/L$, and hence, for a tie of solid circular cross section, find an expression for the radius R needed to provide the target stiffness S^*.

(b) A particular tie of length $L = 0.3$ m carries a load $F = 100$ N, and the extension δ is not to exceed 0.1 mm. Use the material properties in the table to find the required radius R if the tie were to be made in each of these materials.

(c) Write an expression for the mass m of the tie and determine the mass of each of these materials. Which is the lightest?

	Young's modulus E (GPa)	Density ρ (kg/m^3)
PEEK	3.8	1300
Natural rubber	0.0015	950
Ti alloy	102	4600

What we see from this example is that the functional constraint (stiffness) determines the area required, via Young's modulus E, and this area then determines the component mass (via the density ρ). Let's therefore generalise the analysis, to identify the trade-off between E and ρ explicitly.

The first step is to seek an equation that describes the *objective* – the design requirement we wish to maximise or minimise, here the mass m of the tie:

$$m = AL\rho \qquad \text{(GL2.1)}$$

for area A, length L, and density ρ. The *geometric constraint* is that L is fixed, but we can vary the *free variable* A to meet the *functional constraint*: the section area A must be sufficient to provide the specified stiffness S^*:

$$S^* = AE/L \text{ or } AE = S^*L \text{ (constant)} \qquad \text{(GL2.2)}$$

The higher the modulus E, the smaller the area A needed to give the necessary stiffness; if E is high, a smaller A is needed. But will it be lighter? That depends on the density – so we eliminate the free variable A by rearranging the constraint Equation (GL2.2) and substituting back into the objective Equation (GL2.1), giving

$$m = S^* L^2 \left(\frac{\rho}{E} \right) \qquad \text{(GL2.3)}$$

Note that both S^* and L are specified, so the lightest tie that will provide the required stiffness S^* is that made of the material with the smallest value of ρ/E. We could define this as the material index of the problem, seeking the material with a minimum value, but it is more usual to express indices in a form for which a maximum is sought. We therefore invert Equation (GL2.3) and define the material index M_t (subscript t for tie) as:

$$M_t = \frac{E}{\rho} \qquad\qquad (GL2.4)$$

This particular property ratio is also called the *specific stiffness*. Materials with a high value of M_t are the best choice provided that they also meet any other constraints of the design, such as the need for sufficient toughness. Note that we can rank the materials that pass the screening steps in decreasing order of their value for M_t, without even needing to specify the values for the stiffness S^* or the length L.

Exercise

GL2.3   Go back to Exercise GL2.2 and evaluate the material index $M_t = E/\rho$ for the three materials, ranking them in order. Do these values scale as expected with the reciprocal of the mass? What is the ratio?

Material indices and property charts   Plugging numbers into a material index equation is a tedious process – the power of the method lies in combining them with material property charts, specifically due to their logarithmic scales. Materials with a constant value of the material index M_t for a light, stiff component in tension all satisfy

$$M_t = E/\rho = \text{constant}, C$$

that is, for a given stiffness, materials of equal mass have a constant value C of the index. Taking logs,

$$\log(E) = \log(\rho) + \log(C) \qquad\qquad (GL2.5)$$

On the $\log(E) - \log(\rho)$ property chart, this is the equation of a straight line of slope 1.

Figure GL2.3 shows a schematic of the $E-\rho$ chart showing a set of lines for the material index E/ρ plotted onto it, increasing in factors of 10. All the materials that lie on a line of constant $M_t = E/\rho$ perform equally well as a light, stiff tie; those on a higher line will all perform better – they are lighter for the same stiffness. Each time we move the line to a higher value of M_t, the shortlist is reduced until we are left with the lightest options: composites, ceramics, and the top end of the metals and natural materials. Note that we are ranking first here – screening will reduce the shortlist significantly, when we consider factors such as toughness, cost, resistance to corrosion, environmental impact, and ease of manufacturing.

Indeed, the selection line on Figure GL2.3 may need to be dropped to a lower value if everything on our initial shortlist is eliminated by screening.

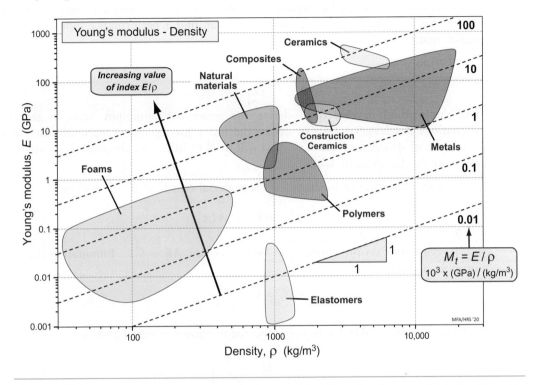

Figure GL2.3 A schematic $E–\rho$ chart showing a grid of lines for the material index $M_t = E/\rho$.

Printed property charts give a complete overview of material classes and are far easier to work with than tables of data (such as Appendix A). But the number of materials that can be shown on a chart is obviously limited (as in Figure GL2.1). Selection using them is practical in order to understand the concepts. This problem is overcome by a computer implementation of the method, which is the purpose of the Granta EduPack software – for details, see the Preface to this book.

Exercise

GL2.4 (a) Use the property tables in Appendix A to find typical values of the Young's modulus E and the density ρ for magnesium alloys and low carbon steels. Evaluate the specific stiffness, E/ρ, in each case. How similar are they?

(b) Use a graphical technique on the property chart in Figure GL2.1 to make the same comparison. What do you notice about the specific stiffness of other structural metals (Al, Ti, Ni, and other ferrous alloys)?

Minimising material cost or 'embodied energy'   What if the objective for our tensile tie is not to minimise weight but to choose the cheapest material? The material cost is only part of the cost of a shaped component – there is also the manufacturing cost, to shape, join, and finish it. But in the first instance we can find a performance index to minimise material cost.

Example GL2.2

Use the method developed to derive the index for minimum mass in tension [Equations (GL2.1)–(GL2.4)] to derive the index for minimum material cost for a tensile tie of specified stiffness S^*. You will need to introduce the material cost/kg of the material, C_m ($/kg or £/kg).

Answer. The objective function for the material cost C of the tie will be

$$C = m\,C_m = A L C_m \rho$$

The constraint on the stiffness, as before, is that $AE = S^* L$. Eliminating the free variable A gives

$$C = S^* L^2 \left(\frac{\rho C_m}{E} \right)$$

Hence the material index to maximise for minimum material cost is

$$M = \frac{E}{(\rho C_m)}$$

Note that this is the same as the index of Equation (GL2.4) with density ρ replaced by cost per unit volume (ρC_m).

So this material index combines three material properties, and we are back to plugging in numbers, unless we happen to have a property chart showing E against (ρC_m). This is precisely what the Granta EduPack software enables us to generate (and then to superimpose limit and index lines, and to perform selection from a property database).

In the emerging age of green design, we may wish to minimise environmental impact rather than cost (which would become a constraint instead). For transport applications, it is often best to minimise mass, as before, if the energy used during the vehicle lifetime dominates, and is strongly dependent on vehicle weight. But designers also need to explore which materials give a product the lowest *embodied energy* (i.e. the energy expended in producing the raw material from ores and other sources). Considerations such as these are part of the important techniques of *life cycle analysis* (see Chapter 12). Here you can explore embodied energy in some exercises.

Exercises

GL2.5   Use the same method as for minimum cost, to derive the material index for minimum embodied energy in a tensile tie, where the embodied energy/kg of the material is H_m (MJ/kg).

GL2.6   Use the property tables in Appendix A to find typical values of the cost/kg C_m and the embodied energy/kg H_m for magnesium alloys and low carbon steels. Compare the performance of these alloys on the basis of minimum cost and minimum embodied energy, for application as a stiffness-limited tensile tie.

Minimising weight in bending: a light, stiff panel   The mode of loading that most commonly dominates in engineering is not tension, but *bending*—think of floor joists, wing spars, golf clubs, bicycle forks, and shelving. As we will see, the indices for bending differ from those for tension, and this (significantly) changes the optimal choice of material. We start by modelling an elastic panel, specifying stiffness and seeking to minimise weight.

A panel is a flat slab, like a tabletop, of length L, width b, and thickness h. Consider a panel supported at its edges and loaded in bending by a central load F distributed uniformly across the width [Figure GL2.4(a)]. The stiffness in bending is explained in Chapter 4 – it depends on the 'flexural rigidity,' the product of Young's modulus E, and the 'second moment of area' I of the cross section:

$$S = \frac{F}{\delta} = \frac{C_1 EI}{L^3} \qquad \text{(GL2.6)}$$

where C_1 is a constant that only depends on the distribution of the total load and the support arrangement (see Figure 4.17) – for three-point bending the value is $C_1 = 48$. For a rectangular cross section, width b, and depth (thickness) h, the second moment of area is

$$I = \frac{bh^3}{12} \qquad \text{(GL2.7)}$$

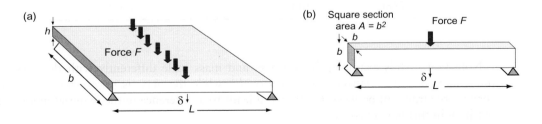

Figure GL2.4   Loading in bending: (a) panel; (b) beam of square section. In both cases the stiffness $S = F/\delta$, where F is the load and δ is the deflection.

The stiffness constraint for the panel is a limit on the stiffness, $S^* = F/\delta$, while the objective is to make it as light as possible. The length L and width b are specified – the dimension we can vary is the thickness h. Table GL2.2 summarises the design requirements. As before, we'll first explore the analysis via a numerical example and then derive and use the index for minimum mass.

Table GL2.2 Design requirements for a light, stiff panel in bending

Function	• Panel loaded in bending	
Constraints	• Stiffness S^* specified	Functional constraint
	• Length L and width b specified	Geometric constraints
Objective	• Minimise mass m	
Free variables	• Choice of panel thickness h	
	• Choice of material	

Exercise

GL2.7 (a) For the panel shown in Figure GL2.4(a), find an expression for the thickness h needed to provide the target stiffness S^*.

(b) A particular panel loaded in three-point bending has a span (length) $L = 1$ m, a width $b = 20$ cm, and carries a central load $F = 100$ N. The additional deflection δ due to this load must not exceed 1 cm. Use the material properties in the table to find the required thickness h in each of these materials.

(c) Write an expression for the mass m of the panel and determine the mass of each of these materials. Which is the lightest?

	Young's Modulus E (GPa)	Density ρ (kg/m^3)
Steel	210	7800
Al alloy	70	2800
CFRP	100	1500

Note that in this example, the stiffness and mass scale differently with thickness: the functional constraint (stiffness) depends on h^3, while the mass is proportional to h. This changes the trade-off between E and δ and leads to a difference in the material index – find out how in this next exercise.

Exercise

GL2.8 (a) Show that the material index M_p for a light, stiff panel in bending [Figure GL2.4(a)], subject to the design requirements in Table GL2.2, is

$$M_p = \left(\frac{E^{1/3}}{\rho} \right)$$

   (b) Evaluate this index for the three materials in Exercise GL2.7. Check that the index scales as expected with the reciprocal of the mass.
   (c) What do you notice about the values of M_p for the two metals, compared to their values in tension $M_t = E/\rho$ (the specific stiffness)?

Note that once again we can rank materials without needing to specify the fixed parameters in Equation (GL2.6): the load and allowable deflection, the panel length and width, or even the details of the bending configuration (which change the constant C_1). The last step is to explore the material index for bending on a material property chart, giving an overview of the competition across all materials (rather than the handful covered in the previous exercises). Charts also help to visualise the differences in selection of light stiff materials for bending rather than tension.

Exercise

GL2.9 Materials with a constant value of the index $M_p = E^{1/3}/\rho$ will have equal performance for a panel in bending (equal mass to meet the stiffness requirement). What will be the slope of a line of constant M_p on an $E - \rho$ property chart?

The material index for bending $M_p = E^{1/3}/\rho$ doesn't look very different from the previous index for tension, $M_t = E/\rho$ – a high value of E and a low value of ρ give high values for both indices. But the difference is significant and leads to different optimum choices of material. Figure GL2.5 shows both indices plotted on the $E - \rho$ property chart, located through aluminum alloys, to explore the competition with this common 'light alloy.'

For tension, we can see (as in Exercise GL2.4) that many structural alloys have very similar values to Al alloys – Mg, Ti, Ni alloys, cast iron, and all steels – as well as hardwood and glass (the last of these being unlikely for tensile applications). CFRP stands out as being lighter than all these options (assuming ceramics are eliminated due to their brittleness). But notice how the ranking changes for bending of a panel: for the metals, there is now a clear ranking of the 'light alloys' – Mg, then Al, then Ti alloys – which are all better than ferrous and Ni alloys. GFRP now competes, while CFRP still stands out but is matched by woods, and even foams enter the picture with polymers not far away – we will see shortly why other considerations may rule

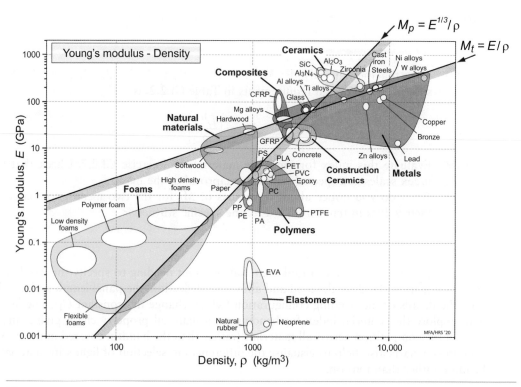

Figure GL2.5 The E–ρ chart showing lines for the material indices $M_t = E/\rho$ (tension) and $M_p = E^{1/3}/\rho$ (panel bending), both located to pass through Al alloys.

them out. Glass again competes – when we can overcome the concerns about impact resistance with clever processing, they are indeed used for 'panels': windscreens, tabletops, and shelving. So the effect of mode of loading is important – simple-minded design for bending using the specific stiffness as a guide is far from optimal. At a glance, we can understand the historical evolution of lightweight bending-dominated structures such as aircraft and sports equipment, from wood to Al alloys to fibre composites – but not steels.

Exercises

GL2.10 Panel design may also target minimum cost or embodied energy. Find a material index to maximise for each of these objectives, for a panel subject to the same functional and geometric constraints as before (Table GL2.2).

GL2.11 In Exercise GL2.6, you found values for cost/kg C_m and embodied energy/kg H_m for magnesium alloys and low carbon steels. Use these data to compare the performance of these alloys on the basis of minimum cost and minimum embodied energy, for application as a stiffness-limited panel in bending.

The constraint of size limits For both the tensile and bending cases, no limits were placed on the free variable of size. But meeting the functional constraint of stiffness meant that the size was dictated by the Young's modulus, to satisfy

$$EA = \text{constant (in tension)} \quad \text{or} \quad Eb^3 = \text{constant (in panel bending)} \qquad \text{(GL2.8)}$$

Very low modulus materials may compete in principle, but only if it is possible to use them in the very large sections required to meet the stiffness target. It is for this reason that foams and polymers are often eliminated from the shortlist. In contrast to the application of a material index, it is not possible to apply size limits without having values for the design variables – these determine the value of the "constants" in Equation (GL2.8). For tension, for example, $EA = S^*L$ [Equation (GL2.2)]. But given values for S^* and L, an *upper* limit on size can be converted into a *lower* limit on E (and similarly for bending). This is a common reason for needing to apply *property limits*, exactly as we did in Figure GL2.1, screening out materials to one side of a horizontal or vertical line. And there may be lower limits on size, too – a very stiff material need only be very thin, but this may cause practical difficulties in handling (sharp edges or wires, or thin panels that are too easy to buckle and dent).

Minimising weight: light, stiff beams Bending was introduced through the design of a panel, where the geometric constraints on the cross section are simple: fixed width, variable thickness. But bending loads are also carried by beams supported at one or both ends, and beams can be made in a wide variety of shapes: solid rectangles, cylindrical tubes, I-beams, and more. Some of these are too complicated to apply the previous method directly; however, if we constrain the shape to be *self-similar* (such that all dimensions change in proportion as we vary the overall size), the problem becomes tractable again. The following exercise derives the corresponding results for a beam of the simplest shape: a square cross section.

Exercise

GL2.12 Consider a beam of square section $b \times b$ that may vary in size (both width and depth, but these are always equal). Figure GL2.4(b) shows this square-section beam in three-point bending. The stiffness $S^* = F/\delta$ and span L are specified.
  (a) Derive the material index M_b for minimum mass of a beam in bending.
  (b) Determine the slope of a line on the $E - \rho$ property chart for which $M_b = \text{constant}$.
  (c) Add this index line to Figure GL2.5, intersecting Al alloys, and comment on the change in competition between Al and other materials, compared to the cases of tension and panel bending.
  (d) What would you need to know to set a limit on Young's modulus, if there was an upper limit on the size of the cross-section b?

Light, stiff beams: the effect of shape   The analysis of beams conveys the effect of material on mass for a given stiffness, but only if we have no freedom to change the shape. Look at any construction site, or the front forks of your bicycle, and you will see that cross sections of beams are not solid squares. Shaping the cross section enables an increase in the bending stiffness of the section (through the second moment of area I) *without* using more material and changing the cross-sectional area A (and thus mass). Figure GL2.6 shows how this is achieved, by locating material away from the 'neutral axis' of bending (defined in Chapter 4). Figure GL2.6 shows (across the top) that the stiffness of a shaped section relative to a solid square can be 10-20, or higher – this ratio is known as the *shape factor* (for stiffness). Since it is an increase in I at constant area (or mass), its effect is to increase the 'flexural rigidity' (EI). It could therefore be thought of as a way of 'enhancing the apparent modulus' – and indeed this is what an advanced analysis shows. Note that in practice, the stiffness is usually specified, and we want to use as little material as possible. So in reality, shaping is mostly used to achieve the same stiffness at lower mass, but we can still analyse the mass saving in terms of the shape factor. Techniques to quantify these gains in efficiency due to shaping, when selecting materials for beams, are beyond the scope of this book – details may be found in the Further Reading at the end of this *Guided Learning Unit*.

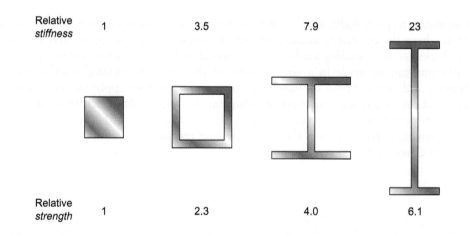

Figure GL2.6 The effect of section shape of a beam on its stiffness and strength, relative to a square section. All sections have the same cross-sectional area, and thus equal mass per unit length.

   If all materials could be made equally easily into efficient thin-walled shapes, we could neglect shape, and the problem would only depend on the Young's modulus again. But this is the key point: some classes of materials are much easier to shape than others – metals, for example, can be extruded or rolled into thin-walled I-beams and hollow profiles; wood on the other hand is cut from trees and comes in solid rectangular sections, or it can be shaped with adhesives and fasteners into thick-walled box sections and I-beams. A further factor is the limitation on how thick the walls of the section need be to avoid *buckling* under the compressive stresses acting on one side of the beam in bending: low modulus materials need to be thicker to avoid this.

In summary therefore, for beam bending there is a second hidden ranking of materials by shape factor, which must in practice be superimposed on the effect of material properties. Metals are best, followed quite closely by composites; woods and polymers are much more limited – which is why wood has been replaced so often in performance-driven design by metals and composites, and why polymer products can end up being heavier than steel, due both to their low modulus and low shape factor. Remember though that for tension and for bending of a flat panel, shape is not a factor – in tension the area can be any shape, while a panel is constrained to be a rectangular shape.

GL2.4 Case studies in stiffness-limited design

Here we present a number of case studies as extended exercises, to practice the methods covered so far and to consider multiple aspects of a design together.

Light levers for corkscrews   The design of a corkscrew was considered in Chapter 2. The lever of the corkscrew (Figure GL2.7) is loaded in bending: the force F creates a bending moment $M = FL$. The lever needs to be stiff enough that, when a cork is being extracted, the bending displacement, δ, is acceptably small. If the corkscrew is intended for travellers, it should also be light. The section is rectangular, and we will make the assumption of *self-similarity*, meaning that we are free to change the scale of the section but not the ratio of the lengths of the two sides. The material index for a self-similar light, stiff beam was derived earlier (Exercise GL2.8):

$$M = \frac{E^{1/2}}{\rho} \tag{GL2.9}$$

There are other obvious constraints. Corkscrews get dropped and must survive impacts of other kinds, so brittle materials like glass or ceramic are unacceptable. Standard consumer items generally avoid the use of expensive materials, so we might exclude materials with a cost over (£ or $)10/kg. The corkscrew will be washed and must therefore resist exposure to water. There will also be an upper limit on the size of cross-section, to fit the user's hand

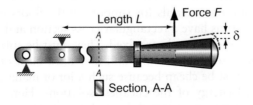

Figure GL2.7   A corkscrew lever must be adequately stiff and, for travelling, be as light as possible.

and the characteristic length of the lever (determined by the force needed to remove the cork and the force a user can comfortably apply).

Exercise

GL2.13   (a)   Draw up a table of design requirements for the corkscrew, as in previous examples in the text, summarising the function, constraints, objective, and free variables.

(b)   Use the $E - \rho$ property chart and the index in Equation (GL2.9) to identify a shortlist of materials that perform well. (*Hint*: re-use the figure from Exercise GL2.12.)

(c)   Apply the constraints on impact resistance and cost/kg to refine the shortlist. Refer to Table A2 in Appendix A for data on material prices.

(d)   Check if the materials remaining on the shortlist are included as preferred materials for a fresh water environment (see Table A10 in Appendix A, which lists materials with 'excellent' resistance). If not, use your experience of similar products to assess whether corrosion is likely to be a problem.

(e)   Comment on whether the upper limit on section size is likely to influence the shortlist.

(f)   Can you think of any other factors for the designer to consider in the selection?

Low-cost structural materials for buildings   The most expensive thing that most people ever buy is the house they live in. Roughly half the cost of a house is the cost of the materials of which it is made, and these are used in large quantities (family house: around 200 tonnes; large apartment block: around 20,000 tonnes). The materials are used in three ways: structurally to hold the building up; as cladding, to keep the weather out; and as 'internals,' to insulate against heat and sound and to provide comfort and decoration. Over the lifetime of the building, heating and cooling a building contributes significantly to the lifetime environmental impact and carbon emissions, so architects and structural engineers increasingly need to know the embodied energy in a structure, as there is a trade-off between (1) the initial 'investment' in energy and emissions and (2) those of the building in the next 50 to 100 years. So we will consider cost and embodied energy.

Consider the selection of materials for supporting the floors in a house (Figure GL2.8). We will assume these beams have a rectangular cross section and are self-similar as before – the ratio of the sides of the rectangle are fixed. They must be stiff, so that the building does not flex too much under internal loading; and they must be strong, so that there is no risk of it collapsing. They must be cheap because such a lot of material is used, but we will also consider the embodied energy of the cheapest solutions. Here we consider the stiffness-limited design of the floor joists; strength-limited design comes next.

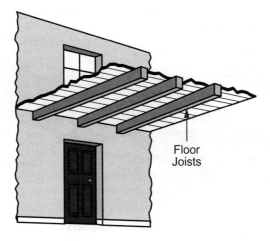

Figure GL2.8 The materials of a building are chosen to perform many different roles. Here we explore low-cost materials for the joists that support the internal loads on the floors.

Exercise

GL2.14 (a) Draw up a table of design requirements for floor beams, as in previous examples in the text, summarising the function, constraints, objective, and free variables.

(b) Derive the material index for minimum cost of a stiffness-limited beam in bending, for the case of a rectangular beam of fixed self-similar shape.

(c) Figure GL2.9 shows a chart of Young's modulus E against cost per unit volume ($\rho\, C_m$). Note that the cost/m^3 scale has been normalised – all values have been divided by a constant reference value: the cost/m^3 of *mild steel* (low carbon steel). Show that this has no influence on the application of the material index from (b). Use the property chart and the index from (b) to identify a shortlist of materials that cost the same or less than *mild steel*.

(d) Comment on the suitability of each material on your shortlist in terms of other material properties, and the ease of manufacturing long prismatic shapes (more on this in Chapter 11). Which materials could be made cheaper by shaping of the cross section?

(e) Modify the material index to the case of minimum embodied energy for the beam in bending. Use Appendix A to look up typical values for E, ρ, and H_m for the remaining materials on your list. Evaluate the new index for these materials, and comment on the results.

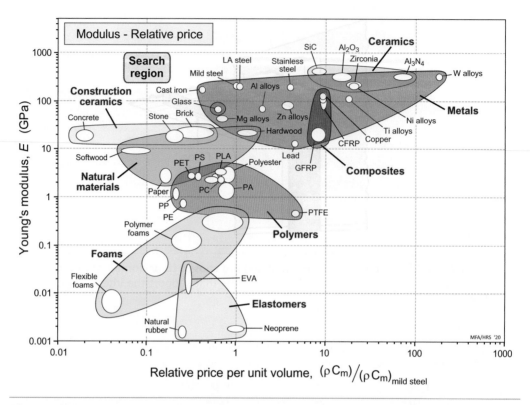

Figure GL2.9 Modulus–relative price chart, used here for the selection of materials for cheap, stiff floor beams.

GL2.5 Material indices for strength-limited design

Many structures loaded in tension or bending must, in addition to being stiff enough, be able to carry the prescribed loads without yielding. If the elastic limit is reached before the maximum allowable deflection, the design is *strength-limited*, and the functional constraint changes to $\sigma_{max} < \sigma_y$. Possible objectives remain as they were before – the structure should be as light as possible, or avoid failure at minimum cost or environmental impact. With this change, the procedure is the same as before, so once we have established how to handle the constraint, deriving the various material indices is straightforward.

Minimising weight: a light, strong tie-rod Consider again the cylindrical tensile tie of given length L carrying a tensile force F as in Figure GL2.2, with the new constraint that it must not yield, but remain elastic. The objective is to minimise its mass. The length L is specified but the cross-section area A can be varied as Table GL2.3 summarises. As for stiffness, the first calculation is for a specific case, then this is generalised to derive the material index.

Table GL2.3   Design requirements for a light, strong tensile tie

Function	• Tensile tie	
Constraints	• Support load F without yielding	Functional constraint
	• Length L specified	Geometric constraint
Objective	• Minimise mass m	
Free variables	• Choice of cross-section area, A	
	• Choice of material	

Exercise

GL2.15   (a)   Derive a simple expression for the radius R of a circular tensile tie needed to support a tensile load F such that the stress is below the yield stress (or elastic limit) σ_y.

(b)   A circular tie of length $L = 0.3$ m carries a load $F = 1000$ N. Use the material properties in the table to find the value of R needed if the tie were to be made in each of these materials.

(c)   Determine the mass m of each of these materials. Which is the lightest?

	Yield strength (or elastic limit) σ_y (MPa)	Density ρ (kg/m³)
PEEK	79	1300
Natural rubber	24	950
Ti alloy	870	4600

As before, the functional constraint determines the area needed (via the strength), and this in turn determines the mass (via the density). Derivation of the material index for the requirements in Table GL2.3 is simple. The objective, mass m of the tie is

$$m = AL\rho \qquad \text{(GL2.10)}$$

The functional constraint, for a cross-sectional area A, is

$$\sigma = \frac{F}{A} < \sigma_y \quad \text{or} \quad A\sigma_y = F \text{ (constant)} \qquad \text{(GL2.11)}$$

Eliminating the free variable A in Equation (GL2.10) gives

$$m \geq FL\left(\frac{\rho}{\sigma_y}\right) \qquad \text{(GL2.12)}$$

So the lightest tie that will carry F safely is that made of the material with the smallest value of ρ/σ_y. Following the same convention as before, we invert this, and seek materials with the largest values of

$$M_t = \frac{\sigma_y}{\rho} \tag{GL2.13}$$

This material index for light, strong components loaded in tension is called the *specific strength*. Plotting index lines on charts will be considered together with the results for bending, derived next.

Minimising weight: light, strong panels and beams   Strength in bending is explained in Chapter 5. Bending produces a linear variation of stress across the section, with the maximum stress σ_{max} at the surface at the greatest distance y_m from the neutral axis, and this stress scales with the bending moment M according to

$$\sigma_{\max} = \frac{My_m}{I} = \frac{M}{Z_e} \tag{GL2.14}$$

where I is the second moment of area as before ($= bh^3/12$ for a rectangular section) and the 'elastic section modulus', $Z_e = I/y_m$. The position of the worst and therefore design-limiting moment M depends on the distribution of the load and configuration of the supports but is proportional to the load F – for three-point bending, as in Figure GL2.4, $M = FL/4$. The functional constraint in bending is therefore used to eliminate the free dimensional variable by rearranging Equation (GL2.14) with stress $\sigma_{max} = \sigma_y$. Let's find the material indices for bending, starting with the design requirements for a panel (Table GL2.4).

Table GL2.4   Design requirements for a light, strong panel in bending

Function	• Panel in bending	
Constraints	• Must support bending load F without yielding	Functional constraint
	• Width b and span L specified	Geometric constraint
Objective	• Minimise mass m	
Free variables	• Choice of thickness h	
	• Choice of material	

Exercises

GL2.16   (a) Show that the material index M_p for a light, strong panel in bending, subject to the design requirements in Table GL2.4, is

$$M_p = \frac{\sigma_y^{1/2}}{\rho}$$

(b) Evaluate this index and the corresponding expression for tension [Equation (GL2.13)] for the three materials in the table. Comment on the competition between the materials in both modes of loading.

	Yield stress σ_y (MPa)	Density ρ (kg/m^3)
Low alloy steel	870	7800
Al alloy	350	2800
CFRP	760	1500

GL2.17 Consider a beam of square section $b \times b$ in which the width and depth may vary (but are always equal). The load F and the span L in three-point bending are specified. Derive the material index M_b for minimum mass of a strength-limited beam in bending.

Light, strong beams: the effect of shape Earlier it was noted that shaping the section of a beam increases its bending stiffness (through the second moment of area I) at constant mass. Equation (GL2.14) shows that, for a given bending moment, the maximum stress depends not on I but on $Z_e = I/y_m$ (the elastic section modulus). So shaping the section also increases the strength at constant mass, but the effect is weaker than for stiffness because part of the increase in I comes from increasing the depth of the beam, and thus y_m. Figure GL2.6 shows (across the bottom) the typical increase in the moment to cause yield for shaped sections, relative to a solid square – this ratio is another *shape factor* (for strength). As before, it is usually the allowable bending moment that is specified, and the aim is to use as little material as possible and to minimise the mass – for a full analysis, see Further Reading. For stiffness shape factors, we noted that manufacturing and buckling limit the efficiency that can be achieved in different material classes. The same applies to strength – metals and composites improve the most when shaped, woods and polymers fall behind.

Minimising material cost or 'embodied energy.' To change the objective from minimum mass to minimum cost or embodied energy, the indices change exactly as before. In all cases, ρ is simply replaced by (ρC_m) or (ρH_m), where C_m and H_m and the cost/kg and embodied energy/kg of the material, respectively.

Exercise

GL2.18 Modify the material index for the strength-limited panel in bending (Exercise GL2.16) for the cases of minimum cost and minimum embodied energy. Use Appendix A to look up typical values for C_m and H_m for the materials in Exercise GL2.16, and evaluate the new indices. Comment on changes in the ranking using each index.

The constraint of size limits Upper and lower limits on the dimensions of the cross-section will place limits on material yield stress, in the same way as for Young's modulus in stiffness-limited design. For example, meeting the functional constraint of remaining in the elastic regime gave these relationships between size and yield stress:

$$\sigma_y A = \text{constant (in tension),} \quad \text{or} \quad \sigma_y h^2 = \text{constant (in panel bending)} \quad \text{(GL2.15)}$$

As for stiffness, applying a size limit on the viable yield stress requires values for the design variables to plug into the right-hand sides of Equation (GL2.15). Then *upper* limits on size translate into *lower* limits on σ_y (and vice versa), which can be applied to eliminate materials outside these limits on a yield stress property chart – examples below.

Material indices and property charts The strength-density property chart was introduced in Chapter 5. Now we can use this chart for strength-limited design at minimum mass. The three relevant material indices are

$$M_t = \frac{\sigma_y}{\rho} \text{ (tensile tie),} \quad M_b = \frac{\sigma_y^{2/3}}{\rho} \text{ (bending beam),}$$

$$\text{and } M_p = \frac{\sigma_y^{1/2}}{\rho} \text{ (bending panel)} \qquad \text{(GL2.16)}$$

Exercise

GL2.19   On a strength-density property chart, what are the slopes of the lines representing constant values of each of the indices for lightweight, strength-limited design.

Figure GL2.10 shows the strength-density chart, with a particular value of σ_y/ρ identified, passing through the high-strength end of the elongated bubble for aluminium alloys.

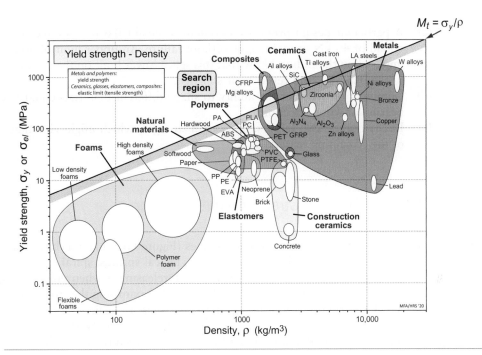

Figure GL2.10 The strength-density chart showing the index σ_y/ρ, located through the top of the bubble for Al alloys.

Exercises

GL2.20 Add the index line for lightweight strong panels in bending to Figure GL2.10, also intersecting the top of the Al alloy bubble, and comment on the change in competition between Al and other materials, compared to tensile loading. Do any classes of materials compete with Al alloys on strength, when they did not on stiffness (see Exercise GL2.12)?

GL2.21 You are to design a set of traffic-control columns that rise up from beneath the road to provide a barrier to cars. To minimise the power requirements for lifting the columns, they must be as light as possible. They have a specified height L and may be assumed to be built in at the ground. A solid cylindrical cross section has been chosen, but the radius R may be varied. In service the columns are vertical and must not fail when a given load W is applied horizontally at mid-height. The maximum stress in a bollard is given by $\sigma_{max} = 2WL/\pi R^3$ (if you wish, derive this for yourself, using the beam bending formulae in Chapter 5).
(a) Which of the material indices for strength-limited design is relevant here, and why?
(b) Use the strength-density property chart in Figure GL2.10 to identify materials that will be the same mass or lighter as the strongest *low alloy (LA) steels*, excluding brittle materials.

(c) A particular design proposes a column of height 0.6 m, required to resist a load of 40 kN at mid-height without failure. An upper limit of 50 mm is imposed on the radius. Show that this sets a lower limit on the failure stress of the material, and revise your shortlist.

(d) Comment on whether the ability to shape different materials could have a bearing on your selection.

(e) Briefly outline two factors, other than strength, which might eliminate any of the remaining materials.

GL2.6 Case study in strength-limited design

Materials for springs Springs come in many shapes (Figure GL2.11) and have many purposes: axial springs (e.g. a rubber band), leaf springs, helical springs, spiral springs, torsion bars. All depend on storing elastic energy when loaded, releasing it when unloaded again – Table GL2.5 summarises the requirements, for the case of maximising the stored energy per unit volume. The elastic energy stored in a material carrying a tensile stress σ was derived in Chapter 4 – it is given by $\sigma^2/2E$ (per unit volume). If the spring yields or fractures it ceases to fulfill its function, so the maximum value of σ must not exceed the elastic limit σ_{el}, when the stored energy is $\sigma_{el}^2/2E$ per unit volume. Dropping the factor of 2 in the index (as it is the same for all materials), the best material for a spring limited by its volume is therefore that with the greatest value of

$$M = \frac{\sigma_{el}^2}{E} \qquad\qquad (GL2.17)$$

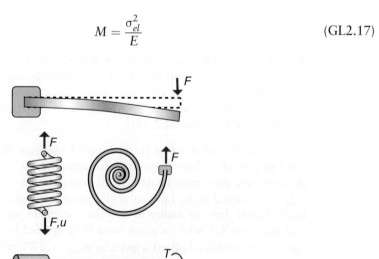

Figure GL2.11 Springs: leaf, helical, spiral, and torsion-bar. Springs store elastic energy. The best material for a spring, regardless of its shape or the way it is loaded, is that of a material with a large value of σ_{el}^2/E.

Table GL2.5   Design requirements for springs

Function	• Elastic spring
Constraint	• No failure: maximum stress in spring below elastic limit
Objective	• Maximum stored elastic energy per unit volume
Free variable	• Choice of material

Though it is less obvious, the index is the same for springs loaded in bending and torsion – leaf springs, torsional springs, and coil springs – the best choice for one is the best choice for all. Essentially it is because the stored elastic energy per unit volume is always the stress × strain, so if the stress varies with position, then the stored energy in each little volume is still $\sigma^2/2E$, and integrating this over the volume only introduces a numerical correction that is the same for all materials.

The choice of materials for springs of minimum volume is shown in Figure GL2.12. A line of slope 2 links materials with equal values of σ_{el}^2/E, with the highest values of M now toward the

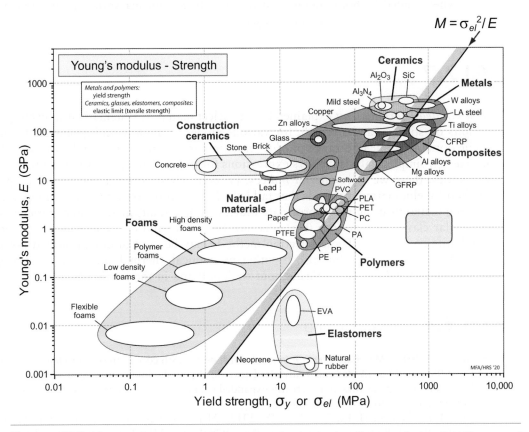

Figure GL2.12   The Young's modulus–strength chart, showing the index line for stored energy of an elastic spring limited by its volume (σ_{el}^2/E).

bottom right. The best choices include the highest strength alloy steels (known as *spring steels*), *titanium alloys* (good but expensive), and *CFRP* (now used for truck springs). Not far behind are *Al and Mg alloys* (lightweight) and *tungsten alloys* (heavy), and *GFRP* (a cheaper option than CFRP). Materials with much lower modulus and strength compete directly: the polymer PA – *nylon* – is used for springs in children's toys and, of course, *elastomers* (rubber bands, including those used for bunjee jumping). It is interesting to note how there is close competition across almost every class of materials: light and heavy metals, polymers, elastomers, and composites.

Exercise

GL2.22   Certain springs need to be as light as possible to minimise inertial loads (e.g. valve springs in high-performance automobile engines) because they move with the valves. Derive the index for light springs to guide material choice (i.e. to maximise elastic stored energy per unit mass). For the top three structural materials identified from Figure GL2.12 as efficient volume-limited springs, estimate a value for the highest strength variant from Figure GL2.12, and look up typical values for the other properties needed in Appendix A. Hence find the spring material that performs best when limited by both volume and weight. Identify further spring applications for which this material is commonly used.

GL2.7 Summary and conclusions

This *Guided Learning Unit* explained how to select materials systematically, meeting *constraints* on stiffness and strength, for various modes of loading and geometric freedom, while optimising one of a range of *objectives*: minimising mass, material cost, or embodied energy. Simple analysis delivers expressions for the objective, containing only material properties: these we call *material indices*. Property charts offer a convenient graphical method for eliminating materials that fall below target property limits (*screening*) and for *ranking* materials using the appropriate indices. Simple limits arise from experience with particular classes of materials (e.g. must not be brittle) and as a consequence of setting size limits on a design. These selection methods may be used in design limited not only by stiffness and strength but also by other mechanical, thermal, and functional properties. Examples will be found throughout the book and in the Further Reading below.

GL2.8 Further reading

Ashby, M. F. (2017). *Materials selection in mechanical design* (5th ed.). Butterworth-Heinemann. ISBN 978-0-08-100599-6. (A dedicated advanced text on the methodology presented here, with a full discussion of shape factors and a greatly expanded catalogue of solutions to mechanical/structural problems, applied to material selection).

Ashby, M. F., Shercliff, H. R., & Cebon, D. (2019). *Materials: Engineering, science, processing and design* (4th ed.). Butterworth-Heinemann. ISBN 978-0-08-102376-1. (A more comprehensive version of this text, giving greater depth on shape factors, and a wider range of structural analysis and case studies in material selection).

GL2.9 Final exercises

GL2.23 The effect of imposing an upper limit on a variable dimension of a design was discussed in the text. For example, for a stiffness-limited panel with an upper limit on thickness, a limit may be found on Young's modulus – but only if we have values for the required stiffness, span, etc. Without these values, however, we can still compare the *relative* sizes we would need in competing materials in order to meet the functional requirement, in terms of the properties of these materials.

From the analysis of stiffness-limited designs in the text, use the stiffness constraint to identify the relationships between the free variable and the Young's modulus, for the three modes of loading considered. Hence derive expressions for the ratio of the sizes needed in two materials, in terms of their Young's moduli E_1 and E_2 for the following design scenarios and free variables: (a) a cylindrical tensile tie (radius ratio), (b) a square section beam in bending (side length ratio), (c) a panel in bending (thickness ratio). Find a value of these ratios, for the comparison of *nylon* (PA) (E_1) and *CFRP* (E_2), using typical values for E in Appendix A. Comment on the practical feasibility of using polymers for stiffness-limited applications.

GL2.24 A sailing enthusiast is seeking materials for lightweight panels to use in a sea-going yacht. The panels are of rectangular cross-section and will be loaded in bending, as shown in the figure. The span L and width b of the panels are fixed, but the depth d may vary (up to a maximum value). The required stiffness is specified as a maximum allowable deflection δ under the expected central load W. The central deflection for the simply supported span shown is given by:

$$\delta = \frac{W L^3}{48 E I} \quad \text{where} \quad I = \frac{b d^3}{12}$$

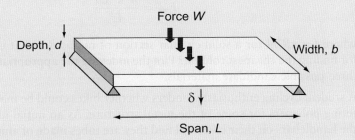

The designer is interested in two scenarios: (i) minimum mass; and (ii) minimum material cost.

(a) Show that the stiffness and geometric constraints lead to the relationship $E d^3 = \text{constant}$, where E is Young's modulus. Explain why the design specification leads to a minimum allowable value for E, and find an expression for this minimum value.

(b) Write down an expression for the first objective to be minimised (mass) and use the stiffness constraint to eliminate the free variable (thickness). Hence define the material performance index to *maximise* for a minimum mass design.

(c) Use the Young's modulus–density property chart (Figure GL2.5) to identify a shortlist of candidate materials with high values of this index (excluding ceramics and glasses; why is this?).

(d) A particular panel design specifies a span of $L = 1$ m and width $b = 20$ cm, with a load $W = 12$ N, and a maximum allowable deflection $\delta = 3$ mm. An upper limit of 10 mm is imposed on the thickness. Evaluate the resulting limit that this imposes on Young's modulus, and use the property chart to update the shortlist of materials.

(e) Refer to Table A10 in Appendix A to check whether any of the remaining materials are not recommended for use in water, sea water, or a marine atmosphere (which may not eliminate them from consideration, but require additional protection – more on this in Chapter 9).

(f) Modify your material performance index to describe the alternative objective of minimum material cost. Use the material property chart in Figure GL2.9 to identify a shortlist of materials, applying the same additional constraints as before. Are there any new candidate materials on this shortlist compared to those for minimum weight design?

GL2.25 A material is required for a cheap column with a solid circular cross section that must support a load F^* without buckling. It is to have a length L. Write down an equation for the material cost of the column in terms of its dimensions, the price per kg of the material, C_m, and the material density ρ. The cross-section area A is a free variable – eliminate it using the constraint that the load must not exceed the buckling load (from Chapter 4):

$$F_{crit} = \frac{n^2 \pi^2 \, E I}{L^2}$$

where $I = \pi R^4/4$ for a solid circular section of radius R. What is the material index for finding the cheapest column? Plot the index on the appropriate chart and identify three possible candidate materials.

GL2.26 A student cycling enthusiast wonders whether bikes could be made lighter by substituting polymers for some of the metal structure. As an initial idea, they investigate the handlebars on their bike, and find they are tubes made of aluminium of diameter 20 mm, with a wall thickness of 1.5 mm. They have identified a glass-filled nylon as a relatively stiff thermoplastic with a Young's modulus $E = 4.5$ GPa, which can easily be extruded and formed to a handlebar shape. The handlebar length cannot be changed, and to fit the existing handlebar mount, brake levers, and so on, the outer diameter is also fixed, but the wall thickness can vary.

(Recall from Chapter 4 that the second moment of area for a solid cylinder of radius R is $I = \pi R^4/4$, and that for a hollow cylinder, it can be found by subtracting I for the inner radius from I for the outer radius.)

(a) Find the flexural rigidity (EI) of the aluminium handlebar, taking $E = 70$ GPa.
(b) Show that this stiffness cannot be matched with the glass-filled nylon, even if the cross section is solid.

GL2.27 The question of lightweighting bicycles using a material such as glass-filled nylon in place of aluminium was introduced in Exercise GL2.26. Parts of the frame, such as the tubing and forks, have more freedom than the handlebars to vary the cross-section size as well as the shape. We can investigate the potential of polymers more generally using material indices, assuming that bending is the limiting mode of loading. Data for the competing materials are aluminium: $E = 70$ GPa, $\rho = 2700$ kg/m^3; glass-filled nylon: $E = 4.5$ GPa, $\rho = 1260$ kg/m^3.

(a) Compare the values of material index for bending with a self-similar fixed shape, $E^{1/2}/\rho$, for aluminium and the glass-filled nylon.
(b) Could efficient shaping of the polymer increase the competition with aluminium?

GL2.28 Derive an index for selecting materials for a panel of given span and width that meets a constraint on bending strength at the design load and is as *thin* as possible. How would you search for suitable materials on a property chart?

GL2.29 Explain how you would conduct a similar analysis to that in Exercise GL2.23, for the case of *strength-limited design*, to relate the sizes and yield strengths for two possible materials, for the same three scenarios and free variables: (a) a cylindrical tensile tie (radius ratio); (b) a square section beam in bending (side length ratio); (c) a panel in bending (thickness ratio). Evaluate the size ratios, for the comparison of *CFRP* and *nylon* (PA) (using typical values for σ_y in Appendix A). Are polymers a more practical proposition for strength-limited rather than stiffness-limited applications?

GL2.30 A low-temperature furnace operating at 150°C uses solid shelves of rectangular cross section for supporting components during heat treatment. The shelves are simply supported at the edges with the components located toward the centre, where the temperature is uniform. The designer wishes to minimise the mass, for ease of auto-mated removal of the shelves, but failure of the shelves in bending must be avoided. The width b and length L of the shelves are fixed, but the thickness d can vary, up to a specified limit.

(a) For a specified load W applied at mid-span of the shelves, the maximum stress in bending is given by:

$$\sigma_{max} = \frac{3WL}{2bd^2}$$

Show that the performance index to be maximised for minimum mass is $M = \sigma_f^{1/2}/\rho$, where σ_f is the material strength and ρ is the density. Explain why there is also a lower limit on the material strength.

(b) Use a strength-density property chart to identify a shortlist of candidate materials (excluding ceramics and glasses; why is this?). Eliminate materials that would not meet the thickness limit, for which a minimum strength of 200 MPa is suggested. Check whether the remaining materials can operate at the required temperature – to do this, look at Figure 8.7, which indicates the usual upper limit on the service temperature of materials.

GL2.31   A helical (coil) spring, as in the figure, has a specified number of turns n and radius R. The thickness of the wire d is a free variable. The aim is to make the lightest spring possible. Two designs are considered: one in which the *stiffness* of the spring S is fixed, and one in which the *failure load F* is fixed.

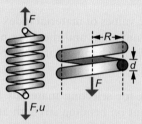

(a) Use Equations (5.27) and (5.28) for helical springs in Chapter 5 to determine the material indices for the two designs. Assume $G = 3/8\ E$.

(b) Using the material properties in the table (for high-strength variants of each material), establish which of these materials is best for each case.

(c) Compare the results of (b) with the ranking based on the performance index for maximum elastic energy storage per unit volume for springs [Equation (GL2.17)]:

$$M = \frac{\sigma_{el}^2}{E}.$$

Material	Yield stress σ_y (MPa)	Density ρ (kg/m^3)	Young's Modulus E (GPa)
Low alloy steel	1800	7800	205
Ti Alloy	1100	4600	110
CFRP	1000	1500	100

GL2.32   A helical spring of specified stiffness S has n turns and a fixed radius R, and is made from a wire of diameter d, which may be varied. You will need Equations (5.27) and (5.28) for helical spring stiffness and failure load from Chapter 5.

(a) Find an expression for the mass of the spring, and hence derive a performance index to be maximised, for each of two possible design objectives for the spring: minimum mass and minimum cost.

(b) Rank the materials in the table in order of increasing mass and increasing cost.

(c) A particular spring has $n = 6$ turns, a radius $R = 0.5$ cm, and a required stiffness $S = 100$ N/m. Find the required wire diameter for the lightest material and for the cheapest material, and comment on whether these are reasonable values for the given radius.

(d) If the maximum extension required is 1 cm, check whether this can be achieved in each case without exceeding the elastic limit, and modify your material choice if not.

	Young's modulus E (GPa)	Yield stress σ_y (MPa)	Density ρ (kg/m³)	Material cost C_m (£/kg)
Low alloy steel	205	1800	7800	0.8
Mg alloy	44	400	1800	3.0
Nylon	1.4	70	1100	2

In the final exercise, constraints are imposed on both stiffness and strength. It is not then possible to use the material indices because the actual limiting dimensions of the free variable need to be compared, choosing the greater value in each material (to ensure that both criteria can be met). But in other respects the analysis is the same, eliminating the free variable to optimise the objective and checking on the effect of size limits on viable property values.

GL2.33   A number of large display boards are required for a conference on the sustainability of materials. Each board consists of a rectangular panel of material whose length L and width b are fixed, and whose thickness d may vary. The panels are most heavily loaded by their *self-weight* when being carried horizontally, simply supported along their short edges of width b. As the boards are to be used as a design case study, the objective is to minimise their environmental impact. The central deflection δ and the maximum bending stress σ_{max} due to its self-weight are given respectively by

$$\delta = \frac{5\rho g b^4}{32 E d^2} \quad \text{and} \quad \sigma_{max} = \frac{3\rho g b^2}{4d}.$$

where ρ is the density, E is Young's modulus, and $g = 9.8$ ms^{-2} is the acceleration due to gravity.

(a) The length and width of one set of panels are 3 m and 2 m, respectively, and the thickness must be between 5 and 20 mm. For each of the materials in the table, find the minimum thickness required to meet both a stiffness constraint that the maximum allowable deflection is 20 cm, and a strength constraint that the maximum stress must not exceed half the failure strength, σ_f. Eliminate any materials that fail to meet the target range of thickness.

(b) The organisers have imposed upper limits of £30 per panel and 35 kg per panel. Eliminate any of the materials identified in part (b) that fail to meet either constraint. For the remaining shortlist, evaluate the embodied energies, and identify the material with the lowest value.

(c) All the materials are imported, and travel a distance of 1000 km. Evaluate the transportation energies for truck freight, for which the energy consumption is 0.94 MJ/tonne.km. Does the ranking of the shortlisted materials change, based on the total embodied and transportation energy? (More on this in Chapter 12.)

Material	Density ρ (kg/m³)	Young's modulus E (GPa)	Failure strength σ_f (MPa)	Cost/kg C_m (£/kg)	Embodied energy/kg H_m (MJ/kg)
Al foam	360	0.9	1.3	9	370
Biocomposite	530	3	10	0.8	7
Pine	600	10	40	0.8	12
Rigid polymer foam	320	0.3	6	8	81
Fibreboard	750	4	10	0.3	18

Guided Learning Unit 3
Process selection in design

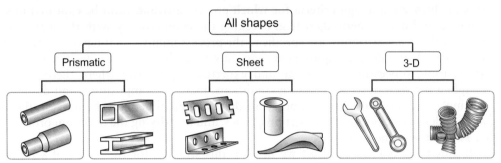

Classification of component shape.

Contents

https://doi.org/10.1016/B978-0-08-102399-0.00022-4

GL3.1 Introduction and synopsis

The strategy for selecting manufacturing process was introduced in Chapter 11 (see Figure 11.1): identification of design requirements, screening of options using process attributes, and ranking by cost (for shaping processes), followed by technical evaluation. The manufacturing triangle (see Figure 11.2) highlighted the dominant criteria to be considered, sub-divided into technical, quality, and economic. Initial screening is limited to straightforward design criteria for which process attributes can be captured in simple numeric and non-numeric data (e.g. checking the compatibility with the material to be processed). Other attributes are specific to the process family – shaping, joining, or surface treatment. Many design requirements can only be addressed in the technical evaluation stage, using processing knowledge and expertise – primarily where there is complex coupling between process operating conditions, material composition, and design detail. Examples are the freedom in shape detail that is compatible with the process, the avoidance of defects, achieving the target properties, and identifying the optimum process speed (a key factor in the cost).

In this *Guided Learning Unit* we develop this strategy, for the stages of screening on process attributes and ranking shaping processes with a simple cost model. As for materials, the data for processes are presented graphically as far as possible, as *process attribute charts*. Worked examples and embedded exercises focus on shaping processes, but the same principles apply for joining and surface treatment (discussed briefly). Looking at everyday objects is a powerful learning tool to appreciate aspects of process selection in design; pulling them apart is even better, given access to a workshop. The exercises include suggestions for hands-on learning about materials processing. Solutions to all the exercises are provided online – it is recommended that you check your answers as you go along, to consolidate your learning.

GL3.2 Selection of shaping processes: attributes for screening

The goal in defining process attributes is to identify the characteristics that discriminate between processes, to enable their screening against the design requirements. Each of the three process families – shaping, joining, and surface treatment – has its own set of characteristic attributes. As we shall see, the viable ranges of many attributes largely reflect physical limits of a given process. We start with the technical and quality attributes for shaping processes – joining and surface treatment come later. One technical attribute applies to all process families – compatibility with material – so we examine this first.

Material–process compatibility Figure GL3.1 shows a material–process compatibility matrix for shaping processes, including some finishing processes. Compatible combinations are marked by dots, the colour of which identifies the material family. Its use for screening is straightforward – specify the material and some processes are immediately eliminated.

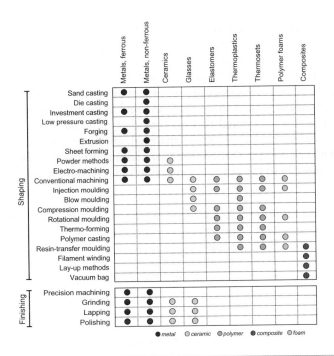

Figure GL3.1 The material–process compatibility matrix, for shaping and finishing processes.

Example GL3.1

Figure GL3.1 shows that each material class – metals, polymers, ceramics, etc. – largely has its own dedicated set of shaping processes. Use your knowledge of processing (or conduct some research) to identify which of the compatible processes operates on material in the liquid state and which in the solid state. What property of a liquid means that metals and polymers have distinct processing methods? Why is casting not compatible with ceramics? Why are many solid-state processes conducted at elevated temperature?

Answer. Liquid processes are metal casting and all polymer and glass moulding processes; the rest are solid-state. The key property of a liquid in relation to processing is its *viscosity*. This is low for metals, making them 'runny' so that they flow easily into a mould under gravity; polymers and glasses form viscous liquids, needing pressure to force them into a mould. Ceramics have high melting points – there is nothing available to contain them in the liquid state! Solid metals and ceramics are often shaped at elevated temperature so that they deform plastically by creep (see Chapter 8), requiring much lower machine loads and giving greater ductility and speed in shaping bulk solids and compacting powders.

Coupling of material selection and process selection   The choices of material and of process are clearly linked. The compatibility matrix shows that the choice of material isolates a sub-domain of processes. The process choice feeds back to material selection. If, say, *aluminium alloys* were chosen in the initial material selection stage, and *casting* in the initial process selection stage, then only the subset of aluminium alloys that are suitable for casting can be used. If the component is then to be *age-hardened* (a heat-treatment process for strength), material choice is further limited to *heat-treatable aluminium casting alloys*. So we must continually iterate between process and material, checking that inconsistent combinations are avoided. The key point is that neither material nor process choice can be carried too far without giving thought to the other. Once past initial screening, it is essentially a co-selection procedure.

Shape-process compatibility   Alongside materials at the top of the manufacturing triangle we also find *part size/shape* – the second key attribute of a shaping process being the types of shapes it can make. A simple high-level classification is illustrated in Figure GL3.2, with three generic classes of shape. The merit of this approach is that many processes clearly map onto specific shapes – identifying the shape class offers immediate discrimination in choosing process. Figure GL3.3 shows the matrix of viable combinations.

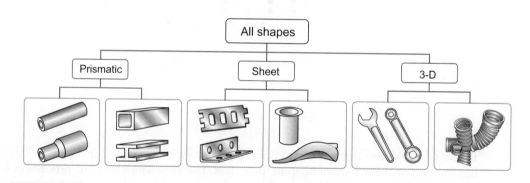

Figure GL3.2   The classification of shape.

Note that prismatic shapes, made by extrusion for example, have a special feature: they can be made in continuous lengths. The other shapes are made by discrete 'batch' processes. This distinction has a bearing on the cost of manufacture: batch processes require many more cycles of setting up than continuous processes. Continuous processes are well-suited to long, prismatic products such as railway track and I-beams. Extrusion is a particularly versatile continuous process because complex prismatic profiles, incorporating internal channels and longitudinal features such as ribs and stiffeners, can be manufactured in one step. Note that prismatic shaping processes are also used to make standard stock material such as tubes, plate, and sheet. Look out, too, for three-dimensional parts that are prismatic in one cross-sectional direction and made in large quantities – these have often been extruded or rolled, cut to length, and possibly formed further, rather than shaped by a batch net-shape process. A lot of foodstuffs are produced this way, too.

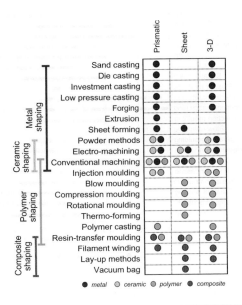

Figure GL3.3 The shape-process compatibility matrix.

Exercise

GL3.1 Look at a selection of metal and polymer products around your home, and classify some of their main components into the shape categories of Figure GL3.2, looking out for parts that were cut and formed from prismatic or sheet stock material.

Mass and section thickness There are limits to the size of component that a process can make. Size can be measured by volume or by mass, but since the ranges of both cover many orders of magnitude, whereas densities span only a factor of about 10, it doesn't make much difference which we use here – big things are heavy, whatever they are made of. Figure GL3.4 shows the 'normal' operational limits, meaning the mass range over which the process works without undue technical difficulty. All can be stretched to smaller or larger extremes, but with a probable cost penalty, in part because the equipment is no longer standard. During screening therefore it is important to recognise 'near misses' – processes that fall within a factor of 2 or so of a limit, but that could be reconsidered. The colour coding for material compatibility has been retained, with more than one colour where the process spans more than one material family. Most processes span a mass range of about a factor of 1000 or so.

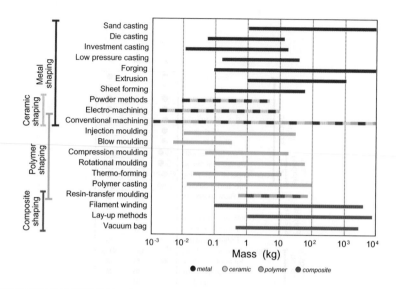

Figure GL3.4   The process – mass range chart, for shaping processes. The colours refer to the material, as in previous charts – multiple colours indicate compatibility with more than one material family.

Figure GL3.5 shows a second chart giving the ranges of section thickness that each shaping process is capable of producing. It is the lower end of the ranges – the minimum section thickness – where the physics of the process imposes limits.

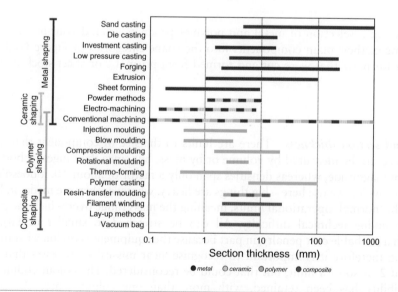

Figure GL3.5   The process – section thickness chart, for shaping processes, colour-coded by material as in previous charts.

Exercises

GL3.2 Screening using process attributes aims to eliminate processes that are unable to meet a design requirement. Use Figures GL3.4 and GL3.5 to examine the competition between metal shaping processes for the following:

(a) Component mass around 0.01, 1, and 100 kg
(b) Section thickness around 0.5, 5, and 50 mm

Repeat the comparisons for polymer shaping.
What do you conclude about how discriminating these attributes are in selecting process?

GL3.3 The figure shows a hot liquid metal or polymer entering a channel in a cold mould, as in casting or injection moulding. What physical behaviour do you think leads to a lower limit on the thickness *t* of the channel?

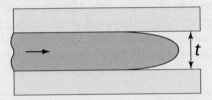

Tolerance and roughness Turning next to the *quality* corner of the manufacturing triangle (see Figure 11.2), two straightforward process attributes are the *tolerance, T,* and the *surface roughness, R.* These quantify the dimensional precision and the surface finish of a component, respectively. The tolerance T on a dimension y is specified as $y = 100 \pm 0.1$ mm, or as $y = 50^{+0.01}_{-0.001}$ mm (indicating that there is more freedom to oversize than to undersize). Tolerances are important to ensure that parts fit together properly during product assembly or that the product works properly – for instance, that moving parts are not too tight or too slack.

The surface quality is often specified as well, though not necessarily over the entire surface. It has a strong influence on product aesthetics, but there are technical issues, too – for example, for the faces of flanges that must mate to form a seal, or for sliders running in grooves. It is also important for resistance to fatigue crack initiation under cyclic loading (see Chapter 6). The roughness, R, is a measure of the irregularities on a surface (Figure GL3.6) – mathematically it is defined as the 'root-mean-square (RMS)' amplitude of the wavy bumps and grooves in the surface. Modern equipment uses laser profilometry to map roughness over a region of a component, using interference from laser light reflected off the irregular surface. Numerically the roughness can be as low as 0.01 μm for the polished surface of a mirror, but is more typically 0.5 to 10 μm for most shaping and machining processes. In design it is specified as an upper limit (e.g. $R < 10$ μm), and it is unlikely to matter if the surface quality delivered is better than is strictly needed.

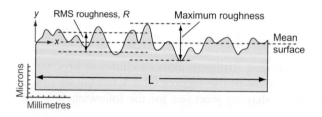

Figure GL3.6 A section through a surface showing its irregular profile (highly exaggerated in the vertical direction). The irregularity is quantified by the roughness, *R*.

Figures GL3.7 and GL3.8 show the characteristic ranges of tolerance and roughness of which processes are capable, again retaining the colour coding by material family. Data for finishing processes are added below the shaping processes.

Metal castings produce a finish determined by the roughness of the moulds – sand castings are rough, while die castings are smooth. No shaping processes for metals, however, do better than $T = 0.1$ mm and $R = 0.5$ μm. Machining or other finishing processes are commonly used after metal casting or deformation processing to bring the tolerance or finish up to the desired level, creating a *process chain*. Ceramics need to be finished with great care, on account of their brittleness, but can be manufactured to high precision and smoothness.

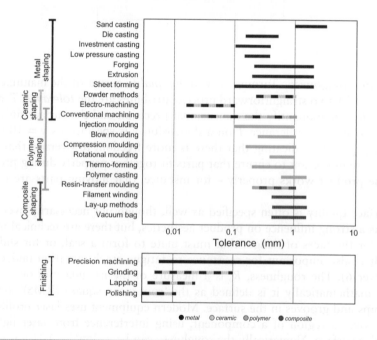

Figure GL3.7 The process-tolerance bar chart, for shaping processes and some finishing processes, enabling process chains to be selected.

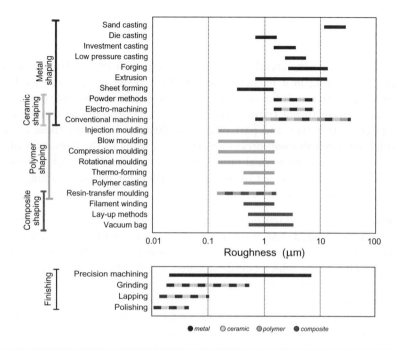

Figure GL3.8 The process-roughness bar chart, for shaping processes and some finishing processes, enabling process chains to be selected.

But precision and finish carry a cost – it is an expensive mistake to over-specify precision or finish, and we should only set demanding targets over the parts of a component where it is really necessary. Moulded polymers, too, inherit the finish of the moulds, which are precision machined from much harder metals that retain their dimensions and finish, so polymer components rarely need finishing. Tolerances in polymers are no better than for metals – in spite of the much lower processing temperatures, internal stresses left by moulding cause distortion, and polymers are prone to creep in service.

Exercise

GL3.4 (a) Use Figures GL3.7 and GL3.8 to examine the competition between shaping processes, for each of metals and polymers, for the quality attributes of (i) dimensional tolerance, (ii) surface roughness.

What do you conclude about how discriminating these attributes are in selecting process, compared to the technical attributes considered in Exercise GL3.2?

(b) With reference to the figures, explain why finishing is rarely needed for polymer components.

GL3.3 Estimating cost for shaping processes

Accurately estimating commercial process costs is a specialised job, largely based on in-house experience of operating in a particular sector. So here we simply aim to illustrate how a number of generic factors influence the costs of competing process routes. The *economic corner* of the manufacturing triangle (see Figure 11.2) prompts us to consider production rate, investment, batch size, and part price. We illustrate here how these are connected via a cost model for batch shaping processes. This also leads to the definition of the *economic batch size* – an indicative numerical range of production run for which processes are usually economic. This completes the set of attributes needed for initial screening – examples and exercises of screening follow in Section GL3.4.

The cost model   The manufacture of a component consumes resources (Figure GL3.9), each of which has an associated cost. We capture these inputs in the simplest form of cost model through the parameters defined in Table GL3.1. Some values are characteristic *cost-related attributes* of each process, which could be stored in a database, while others are *user-specific inputs*, such as local labour and energy costs. In brief, the cost of producing a component of a given mass combines: (i) the cost of the materials used in its manufacture; (ii) a share of the cost of dedicated *tooling*, which is dependent on the number of parts being made (the batch size); and (iii) a share of the hourly *overhead* costs of operation, which is dependent on the production rate. Let's build up these terms into a model.

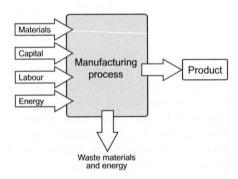

Figure GL3.9   The inputs to a manufacturing process.

Table GL3.1   Symbols, definitions, and units in the cost model

Resource		Symbol	Unit
Materials	Cost of stock material	C_m	$/kg
Capital	Cost of tooling	C_t	$
	Cost of equipment	C_c	$
Time-dependent costs	Overhead rate (including labour, administration, energy, rent, etc.)	C_{oh}	$/h

Consider the manufacture of a component weighing m kg, made of a material costing C_m \$/kg. In practice, processes rarely use exactly the right amount of material but generate some scrap (in mould feeding channels or machining swarf). Some is lost altogether, while most is recycled but may lose value – in this case an allowance can be made by increasing the mass of material to a value that is effectively consumed, higher than the mass of the component. There may also be a cost of 'consumables,' such as lubricants. Here we will assume these corrections are small, and estimate the material contribution per part, $C_{material}$, to be:

$$C_{material} = mC_m \qquad\qquad (GL3.1)$$

Table GL3.1 shows two capital costs, and we deal with these separately. First, the cost C_t of a set of tooling – dies, moulds, fixtures, and jigs – is what is called a *dedicated cost*: one that must be wholly assigned to the production run of this single component. It is written off against the numerical size n of the production run. Again, practical details will dictate that tooling wears out and may need to be replaced multiple times to complete the target production run. To keep things simple, we will assume that the given tooling cost is written off purely over our batch size n – numerical adjustments could easily be made to account for multiple sets of tooling. Thus the tooling cost per part, $C_{tooling}$, takes the form:

$$C_{tooling} = \frac{C_t}{n} \qquad\qquad (GL3.2)$$

The *capital cost of equipment*, C_c, by contrast, is rarely dedicated. A given piece of equipment can be used to make many different components by installing different die-sets or tooling. It is usual to convert the capital cost of *non-dedicated* equipment into an equivalent hourly cost, by dividing it by a *capital write-off time*, t_w, being the number of production hours for which the machine operates in its lifetime, which will be many years. This quantity C_c/t_w, is then an effective hourly cost of the equipment – a bit like a rental cost that you charge to yourself. As there are other generic hourly costs of production, we will roll this part of the capital cost in with the other time-dependent factors in Table GL3.1 – labour, energy, and so on, giving an overall effective overhead rate, $\dot{C}_{oh}$. The cost per part of overheads (including equipment write-off rate), $C_{overhead}$, is the hourly overhead rate divided by the production rate/hour $\dot{n}$:

$$C_{overhead} = \frac{\dot{C}_{oh}}{\dot{n}} \qquad\qquad (GL3.3)$$

The total cost of shaping per part, C_{part}, is the sum of these three terms:

$$C_{part} = C_{material} + C_{tooling} + C_{overhead} = mC_m + \frac{C_t}{n} + \frac{\dot{C}_{oh}}{\dot{n}} \qquad\qquad (GL3.4)$$

Values for C_m are available in databases – typical values are given in Appendix A. Process costs and production rates can cover quite wide ranges and may be proprietary information – representative values are stored in the Ansys/Granta Edupack software (see Preface), and typical values of C_t, $\dot{C}_{oh}$, and $\dot{n}$ will be provided in the following examples and exercises. Given the mass m of the component, we can then evaluate the variation of cost per part with batch size – let's explore this graphically.

Figure GL3.10 plots Equation (GL3.4), showing the cost per part, C_{part}, against batch size, n, for a small component shaped by three alternative processes: sand casting, die casting, and low-pressure casting. Note that the cost axis is shown as a cost per part divided by the material contribution, mC_m. At small batch sizes the unit cost is dominated by the dedicated costs of tooling, C_t/n. As the batch size n increases, this contribution falls until it becomes negligible, and the curve runs out to a constant value that is the sum of the material cost and the time-dependent overhead cost, $\dot{C}_{oh}/\dot{n}$. Competing processes differ in tooling cost C_t, overhead rate $\dot{C}_{oh}$ (including a share of equipment cost), and production rate $\dot{n}$. Sand-casting equipment is cheap, but the process is slow; die-casting equipment costs much more but runs much faster. Mould costs for low-pressure die casting are greater than for sand casting; those for high-pressure die casting are higher still. The combination of all these factors for each process causes the $C_{part} - n$ curves to cross, as shown in Figure GL3.10.

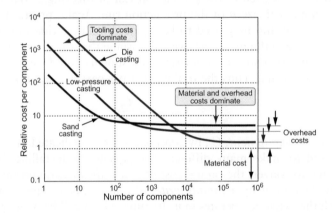

Figure GL3.10 The cost of casting a small component as a function of the batch size, for three competing processes.

Exercises

GL3.5 The figure shows the relative cost per part, as in Figure GL3.10, for a reference process A, and three variants B-D in which one parameter in the cost equation has been changed.

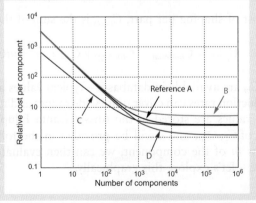

(a) Identify which of the curves B-D corresponds to decreasing the value of: (i) the tooling cost; (ii) the overhead rate; (iii) the production rate, justifying your choices.

(b) On the same graph (i.e. relative to the *initial* material cost), add the cost curve for a material for which the cost per kg is 10 times higher, with everything else unchanged.

GL3.6 A small PE bucket is to be manufactured by injection moulding or rotational moulding. The designer wishes to estimate the manufacturing cost for different batch sizes, using the cost Equation (GL3.4). Use the data in the table to determine the cheaper process for batch sizes of 1000 and 50,000.

Process	mC_m (£)	C_t (£)	$\dot{C}_{oh}$ (£/hr)	$\dot{n}$ (parts/hr)
Injection moulding	0.25	5000	20	120
Rotational moulding	0.20	1000	20	40

Economic batch size The crossover of the cost-batch size curves means that the process that is cheapest depends on the batch size. This suggests the idea of an *economic batch size* – a range for which each process is likely to be the most competitive. If the data are available, Equation (GL3.4) allows the cost of competing processes to be compared and ranked; if not, the economic batch size provides an indicative screening step. Figure GL3.11 is a chart of this attribute, for the same shaping processes as the earlier charts, with the same colour coding by material class. Processes such as investment casting of metals and lay-up methods for composites have low tooling costs but are slow, which favours small batch sizes. The reverse is true for die casting of metals and injection moulding of polymers: they are fast, but the tooling is expensive, requiring large batch sizes to be economic.

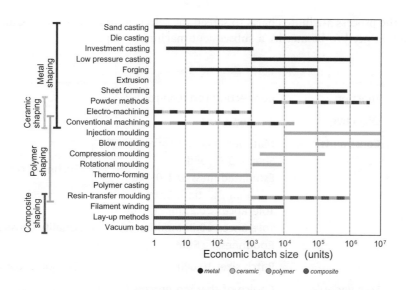

Figure GL3.11 The process–economic batch size bar chart for shaping processes.

GL3.4 Case studies: selection of shaping processes

Now we can apply the selection methodology to a range of case studies – first a worked example, then exercises for you to complete. In each case, the design requirements provide a target list of process attributes for screening, with the first example also using the cost model.

Example GL3.2 Shaping a steel connecting rod

The figure shows a schematic connecting rod ('con-rod') to be made of a medium carbon steel, with the design requirements summarised in the table. Note the higher dimensional precision on the internal bearing surfaces of the boreholes and the low surface finish required over the entire component due to the risk of fatigue crack initiation for a con-rod, as it carries high cyclic loads.

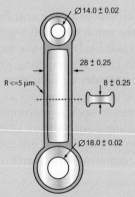

Use the process attribute charts to screen the selection of process, noting any requirements that need to be met by a subsequent machining step.

Design requirements for shaping a steel connecting rod

Function	• Connecting rod (con-rod)
Objective	• Minimise cost
Constraints	• Material: medium carbon steel ⎫
	• Shape: 3-D solid ⎪
	• Estimated mass: 0.3–0.4 kg ⎬ Technical constraints
	• Minimum section: 8 mm ⎭
	• Tolerance: <0.25 mm (surface) ⎫
	• Tolerance: <0.02 mm (bores) ⎬ Quality constraints
	• Roughness: <5 μm ⎭
	• Batch size: 10,000 Economic constraint
Free variable	• Choice of shaping process

Answer. First, we screen on material and shape compatibility (ferrous metal, 3D shape) – the relevant parts of Figures GL3.1 and GL3.2 (i.e. metal shaping) are annotated below. These requirements eliminate die casting, low-pressure casting, extrusion, and sheet forming, but many options remain (including all the surface finishing processes).

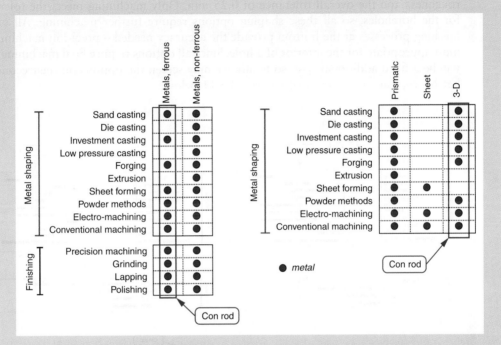

Next we consider mass and section thickness – Figures GL3.4 and GL3.5 (for metal shaping – extracts below). Of the remaining processes, sand casting looks marginal for this component mass (by a factor of ~2), but we'll keep it in for now; all processes meet the thickness requirement.

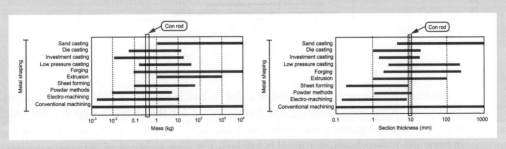

The quality attribute charts below – extracts of Figures GL3.7 and GL3.8 – narrow the choice considerably. Sand casting is disqualified by both tolerance and roughness constraints – it will require machining over the entire outer surface. The remaining processes – *investment casting, forging, powder methods,* and *machining* – all achieve the target roughness and the overall tolerance of 0.25 mm. Only machining meets the tolerance for the boreholes, so all these shaping options require further machining. All of the finishing processes at the bottom provide the accuracy needed – precision machining is most appropriate for the interior of a hole. Since all options require final machining, this will be a fixed additional cost, so finally we may screen the options on their economic batch size, using the extract of Figure GL3.11 below.

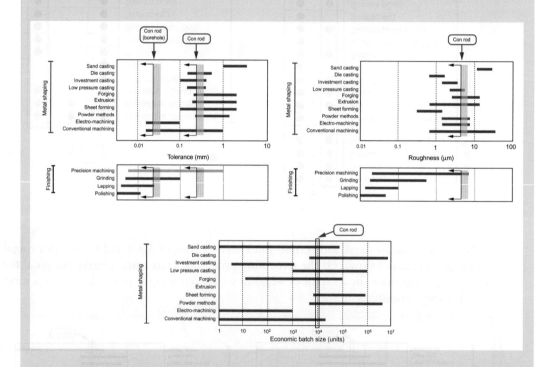

Investment casting is not economic for a batch size of 10,000, and conventional machining (from solid) is marginal. Sand casting for the shaping step is economic, but requires machining over the entire surface, again suggesting higher cost than other options (as well as being marginal for mass). The final shortlist of processes is therefore: *forging* and *powder methods* (both with local machining of the boreholes).

This case study reveals that there is close competition for the manufacture of carbon steel components in the standard mid-range size and shape of a connecting rod. Surface finish and accuracy present some problems, but finishing by machining is not difficult for carbon steel. The final choice is determined by cost considerations – the next exercise.

Exercise

GL3.7 Costs may be estimated using Equation (GL3.4) for the shortlisted processes of forging and powder methods, together with the option of sand casting, which was considered borderline in the initial screening. Data are provided in the table. For convenience, we assume that the same amount of material is used in each process, so the costs are again normalised to the material cost, mC_m. Plot the cost per part against the batch size, identifying the cheapest option for the target batch size $n = 10,000$, and comment on the competition between the processes with batch size.

Parameters for the cost model	Sand casting	Forging	Powder methods
* Material, mC_m	1	1	1
* Dedicated tooling cost, C_t	200	1000	7000
* Basic overhead, $\dot{C}_{oh}$ (per hour)	80	90	100
Production rate, $\dot{n}$ (per hour)	2	30	10

* Note that costs have been normalised to mC_m

For a con-rod, the subsequent technical evaluation of the chosen process will be dominated by issues such as avoiding cracking, minimising residual stress, and developing the required microstructure and properties – all critical factors in achieving the required fatigue performance. These are all sensitive to the choice of operating conditions such as processing temperature, deformation rate, and lubrication, together with details of the design of the tooling. The microstructure and properties will also be sensitive to the composition of the steel and its heat treatment after shaping – this aspect of technical evaluation is discussed further in Chapter 11.

Exercises

GL3.8 The figure shows the ceramic insulator within the body of an automotive spark plug. The design constraints affecting choice of process are summarised in the table below. The insulator must seal tightly within the assembly, setting the limits on precision and surface finish. Spark plugs are made in large numbers – the projected batch size is 100,000.

Use the process attribute charts to identify the processing options.

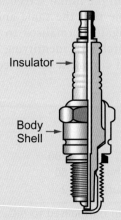

Design requirements for shaping a ceramic spark plug insulator

Function	• Spark plug insulator	
Objective	• Minimise cost	
Constraints	• Material alumina	
	• Shape: 3D hollow	
	• Estimated mass: 0.04–0.06 kg	Technical constraints
	• Minimum section: 1.5 mm	
	• Tolerance: $<\pm 0.3$ mm	Quality constraints
	• Roughness: $< 10\,\mu$m	
	• Batch size: 100,000	Economic constraint
Free variable	• Choice of shaping process	

GL3.9 The figure shows a PE kitchen chopping board, with a stainless steel handle. The design constraints affecting choice of process are summarised in the table below, with a total batch size of 10,000. The chopping area is to be made rough with a pattern of protruding dimples (shown magnified in the figure). This is a safety feature, to limit slipping of the food items being cut. Fitting the handle is considered separately after the exercise.

Use the process attribute charts to identify the processing options. How do you think the rough pattern of dimples could be produced on the chopping area?

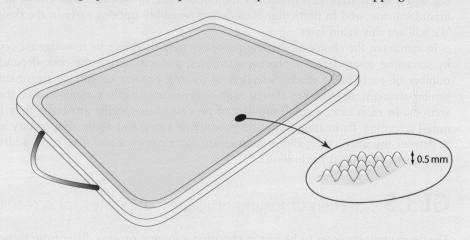

Design requirements for shaping a thermoplastic kitchen chopping board

Function	• Kitchen chopping board	
Objective	• Minimise cost	
Constraints	• Material: polyethylene, PE (thermoplastic)	
	• Shape: solid 3D	
	• Estimated mass (PE): 800 g	Technical constraints
	• Minimum section 8 mm	
	• Tolerance: 0.1 mm	
	• Roughness: 1 μm (outside chopping area), 0.5-mm dimpled surface over chopping area	Quality constraints
	• Batch size: 2000 in each of five colours	Economic constraint
Free variable	• Choice of shaping process	

These exercises demonstrate the limited options for shaping ceramics and the greater versatility for moulding polymers, exploited for many household items such as the chopping board. PE is a good material choice, being cheap and easy to clean – essential for kitchenware – and readily custom-coloured using food-safe additives. The overall shape here is not challenging – making the drainage groove near the edge is simply achieved with a matching ridge inside the mould (an option for making the patterned rough surface, too). At first glance we might think of making the board, and then asking how to attach the stainless steel handle. Polymer moulding has a particular advantage in its ability to 'mould in' metal components mounted securely in the mould cavity, combining shaping and joining into a single step. This illustrates the importance of technical know-how in design and manufacturing, and in particular considering assembly options early in the design process. We will see this again later.

In summary, the choice of shaping processes may therefore be investigated systematically by screening using data for process attributes, and estimating the cost, depending on the number of parts to be made. Selection of joining process or surface treatment follows a similar strategy, and the key design requirements are briefly summarised in the following sections. In each case, a small number of process class-specific attributes may be used for initial screening. Estimating costs and production rates, and addressing quality aspects such as defects, microstructure, and properties are more complex and need to be addressed in the subsequent 'Technical Evaluation.'

GL3.5 Selection of joining processes: attributes for screening

The design requirements to be met in choosing a joining process differ from those for choosing a shaping process. Even material compatibility is more involved than for shaping, in that joints are often made between dissimilar materials.

Material compatibility A simple classification of joining processes is shown in Figure GL3.12, indicating the material classes that are routinely compatible with them. Note that

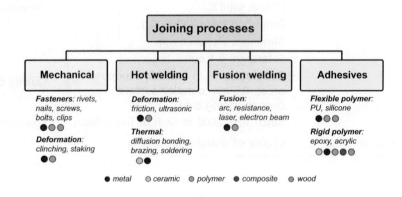

Figure GL3.12 The classification of joining processes, showing the material classes (including wood) for which they are most commonly applied.

wood is now included – for shaping it was self-evident that this almost exclusively implied machining. Mechanical joining spans a diverse range of processes, including fasteners and snap fittings, as well as a variety of related techniques that exploit mechanical interlocking and friction to hold things together. Mechanical joints are difficult in ceramics and composites due to their inherent susceptibility to damage and cracking, but they can work with special precautions. Adhesives on the other hand are universal, and particularly important for fibre composites and wood. Welding is subdivided here, to distinguish between those that do not melt the part being joined ('hot welding') and those that do ('fusion welding'). The latter is the common interpretation of the term 'welding' – using a localised moving heat source to melt a controlled volume of material at the interface between the components. Most solid-state welding uses high temperature plasticity, often generating heat by friction. Brazing and soldering are commonly thought of as welding processes, but they use a low melting metal to join the parts, which do not melt themselves – we could in principle classify this as using a 'metallic adhesive'.

When the joint is between dissimilar materials, further considerations arise, both in manufacture and in service. The process must obviously be compatible with both materials – but this is no guarantee that the process works with dissimilar materials. For example, fusion welding is not viable for metals from different families – apart from the aspect of needing similar melting points, the random alloy composition generated by melting two alloys together is highly unlikely to have acceptable properties – so we need a non-fusion solution. Even then, service conditions may present limitations. Dissimilar metals in joints may form a galvanic corrosion couple, if close to water or a conducting solution (see Chapter 9). This is one reason for including *service conditions* in the manufacturing triangle (see Figure 11.2).

Exercise

GL3.10   Look at a bicycle and classify a selection of different joints into Mechanical, Welding, or Adhesives. Identify joints between dissimilar materials.

Joint geometry and mode of loading   Three important joining-specific design considerations are the geometry of the joint, the thickness of the parts, and the way the joint will be loaded – further aspects of *design detail* and *service conditions* in the manufacturing triangle. Figure GL3.13 illustrates standard joint geometries and modes of loading. The choice of geometry depends on the materials involved, the shapes and thicknesses of the parts, and the mode of loading – and all these influence the choice of process: it is a highly coupled problem. Adhesive joints support shear but are poor in peeling. Adhesives need a good working area – lap joints are fine, butt joints are not. Rivets and staples, too, are well adapted for lap joints in shear, but not tension. Part thickness also plays a role – riveting can be used with thin, equal sections but not with thick; adhesives cope with almost any thickness. Welding is highly adaptable, but matching choice of process to material, geometry, and loading quickly leads to technical considerations – the geometries and thicknesses that can be welded are governed by the physics of plastic deformation and/or heat flow, which are clearly dependent on the material and the process operating conditions.

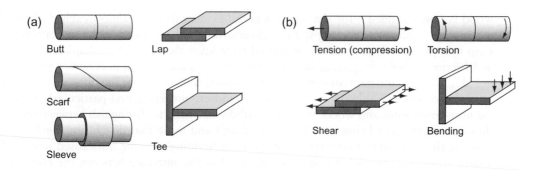

Figure GL3.13   (a) Standard joint geometries, and (b) modes of loading.

Secondary functionalities   Most joints carry load, but that is not their only function. Joints may enable or prevent the conduction of heat or electricity, or they may operate at an elevated temperature, or may serve as a seal (to contain or exclude gases or liquids). Manufacturing conditions also impose practical constraints on joint design. Electron beam welding sets a size limit on an assembly, as it must fit within a vacuum chamber. On-site joining operations require portable equipment, possibly with portable power supplies (e.g. hot-plate welding of polymer pipelines). Portability, on the other hand, effectively means there is no upper limit on the size of the assembly – bridges, buildings, shipbuilding, and aircraft make this self-evident.

An increasingly important consideration is the requirement for economic disassembly during and at the end of the product life, to facilitate repair and material reuse and to minimise contamination during recycling. This favours threaded fasteners that can be undone, or adhesives that can be loosened with solvents or heat.

Exercise

GL3.11   Look at the products around your house (or your garage, or garden shed), identifying as many different types of joints as you can. Note the class of process used, the material(s) being joined, and the geometry of the joint. Select a few examples, then draw up a list of design requirements that the joint (and therefore the process) needs to satisfy.

In summary, for selection of joining process, after an initial check on compatibility with material and joint geometry, we quickly turn to technical evaluation to understand the physical limitations of each material-process combination, accounting for multiple influences of the component shapes and joint geometry. The quality corner of the manufacturing triangle for joining is particularly demanding – avoidance of cracks, control of thermal and mechanical stresses, and (for welding in particular) complex evolution of microstructure and properties. Further examples feature in Section GL3.7 and Chapter 11, but the conclusion is clear: think about joining implications early in the design process.

GL3.6 Selection of surface treatment processes: attributes for screening

The surfaces of products are subject to different design requirements than the bulk of the material beneath. It is the surface region that is exposed to the environment and experiences the localised stresses due to contact between components, and product aesthetics are strongly determined by the surface finish and appearance. Process selection in design therefore includes consideration of the need for surface treatment, and the potential costs and benefits. Figure GL3.14 categorises the classes of surface treatment process.

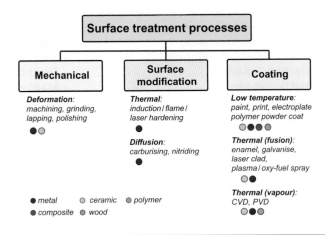

Figure GL3.14   The classification of surface treatment processes, showing the material classes (including wood) for which they are most commonly applied.

Material compatibility   Selection of shaping process introduced the value of mechanical surface treatments to correct dimensional precision and roughness, particularly for metals and ceramics. Surface treatments, however, cover a much wider range of functions (see below). Screening processes first on material compatibility is straightforward, largely determined by the physics of how the processes work. This is the basis of the process classes and their sub-divisions in Figure GL3.14: removal of thin layers by mechanical deformation, modification of the surface microstructure (by heat and/or by diffusing in additional atoms), and applying a coating (by painting, dipping, deposition, or spraying – all leaving a solid layer bonded to the surface, formed from solid particles, or a molten layer, or directly from a vapour). Note the dominance of metals across all process classes in Figure GL3.14 – the majority of the applications are in fact for steels, for reasons we now explore.

Purpose of the treatment   After material compatibility, the most discriminating surface treatment-specific attribute proves to be the purpose or function of the treatment (again governed by *service conditions* in the technical corner of the manufacturing triangle (see Figure 11.2). Table GL3.2 illustrates the diversity of functions that surface treatments can provide. Some protect, some enhance performance, while others are primarily aesthetic. All add cost, but the added value can be large. Protecting a component surface can extend product life and increases the interval between maintenance cycles. Coatings on cutting tools enable faster cutting speeds and greater productivity. Surface hardening processes may enable the substrate alloy to be replaced with a cheaper material (e.g. using a plain carbon steel with a hard carburised surface or a coating of hard titanium nitride [TiN], instead of using a more expensive alloy steel). This explains why steels dominate in surface treatment – they are the workhorse structural alloys for applications susceptible to wear, fatigue, and corrosion. Surface coatings for aesthetic purposes (colour, reflectivity) and corrosion protection apply to most material classes but are largely unnecessary for polymers – these are inherently resistant to chemical attack, and it is usually easier and cheaper to simply colour the polymer itself (with dye or a filler).

Table GL3.2   The design functions provided by surface treatments

• Correction of dimensions (tolerance)	• Thermal conduction
• Correction of surface finish (roughness)	• Thermal insulation
• Corrosion protection (liquids)	• Electrical insulation
• Corrosion protection (gases)	• Magnetic properties
• Wear resistance	• Decoration
• Friction control	• Colour
• Fatigue resistance	• Reflectivity

Exercise

GL3.12   Look at the products around your house (or your garage, or garden shed), identifying as many different types of *surface treatments* as you can, and noting the *class of material* they have been applied to, and the *function(s)* of the surface treatment.

Secondary compatibilities   Material compatibility and design function are the dominant considerations in selecting a finishing process, but there are others. Some surface treatments leave the dimensions, precision, and roughness of the surface unchanged. Others deposit a relatively thick layer with a rough surface, which may require another finishing treatment. Component geometry also influences the choice. 'Line-of-sight' deposition processes coat only the surface at which they are directed, leaving inaccessible areas uncoated. And again there are recycling implications – coatings on products contribute significantly to contamination and downgrading of materials at the end of product life.

GL3.7 Technical evaluation

The initial screening stages for all process classes, and the cost model and case studies for shaping, have shown how far we can take process selection using structured data and process attributes. For each process class, it is possible to address a number of factors on the checklist of the manufacturing triangle from Figure 11.2, but not all. Throughout the chapter it was noted that there are complex issues requiring technical evaluation. These commonly fall under one or more of these categories:

- Part shape and design details (collectively called *form freedom*) – both the constraints imposed by how the process works, and opportunities to exploit potential benefits (including *design for joining*)
- Common *defects*, and best practice in how to avoid them
- Limits on *production rate*
- The resulting *microstructure* and *properties*

All of these are issues for which the outcome is closely integrated with the process physics – i.e. how it works, and the coupling between the component geometry, material behaviour, and process operating conditions. Examples of the first three are briefly illustrated later to give a flavour of the issues – a full discussion is beyond the scope of this text: see *Further Reading* for more information. Microstructure and properties, however, are core themes of this book – so we devote Chapter 11 to this topic, in the context of processing.

Form freedom   Many shaping processes involve filling a shaped cavity within a mould, by pouring or forcing hot liquids into the cavity and leaving them to solidify – metal casting, and polymer moulding, for example. Powder methods and forging use shaped dies to force solid materials to compact or deform plastically to a required shape. In all cases, we need to be able to remove the moulds or dies to extract the shaped component without damaging it. This has implications for designing the detail of the shape to make it easy to process – Figure GL3.15 shows examples.

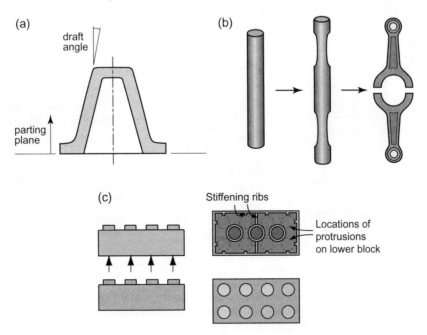

Figure GL3.15   Form freedom in shape design to facilitate: (a) mould or die separation in casting or forging; (b) material redistribution in multiple-stage forging; (c) efficient internal stiffening in polymer injection moulding.

As far as possible, we use two-part moulds that meet at a flat surface called the *parting plane* – this keeps the costs down, as the machinery is simpler and cheaper if movement (and loads) are in a single direction, also making the turnaround faster between making each part. So knowledge of how the process works is essential to incorporate features such as tapered 'draft angles' [Figure GL3.15(a)], and managing the position of re-entrant shapes. Solid-state shaping starts from a simple pre-form shape such as a cylinder. If the cross section of the part varies considerably in thickness, multiple deformation steps are needed, each with its own set of tooling, to redistribute the material [Figure GL3.15(b)]. Injection moulding of polymers is particularly versatile in enabling shape detail to be incorporated in a single process step. Clever internal shaping is used to customise the stiffness of a part – Figure GL3.15(c) shows the thin internal ribs of material that enable building blocks to have just the right stiffness to be joined and separated by a child.

Avoidance of defects   Casting and moulding involve liquid material flow into a cavity and subsequent cooling and thermal contraction – both are sources of *defects: porosity* and *cracking*, or poor dimensional precision due to impeded flow and shrinkage. Smooth complete mould filling is promoted by curved component shapes and careful choice of channel thickness (Figure GL3.16). From Chapters 5 and 6 we know that sharp corners are *stress concentrations* that may lead to cracking – note that components are commonly subjected to stress while they are being made, not just in service. Any process involving hot material cooling down is likely to involve thermal stresses (see Chapter 7) – technical process knowledge and modelling are needed to avoid cracking and lost production. The same goes for heat treatment of shaped components to modify their microstructure – quenching steels from over 800°C into cold water, for example (see Chapter 11).

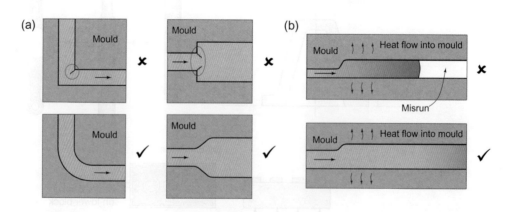

Figure GL3.16   Avoidance of defects in casting or injection moulding by attention to detail in shape: (a) curved corners, to give smooth metal flow, and avoid cracking at stress concentrations *(shown circled)*; (b) sufficient channel thickness, to avoid a 'misrun' (i.e. incomplete filling due to premature solidification).

Design for joining Processing is multi-stage, from primary shaping to joining and finishing treatments. These should not be considered in isolation – some of the complexity (and opportunities) revealed by technical evaluation involve through-process couplings between manufacturing stages. The ability to successfully join and assemble products is intimately connected to prior shaping processes – indeed there is a whole branch of manufacturing that addresses this, known as *design for assembly*. Figure GL3.17 shows a couple of illustrative examples to make this point.

First, extrusion offers substantial form freedom in the detail of the prismatic cross-sectional shape. This allows a structural shape to be made, but incorporating features that facilitate the joining of these shapes – Figure GL3.17(a) shows edge detail that enables multiple simple extrusions to be slotted together and joined by seam welding to assemble a large area of stiff platform (e.g. the deck on a ship or an oil rig).

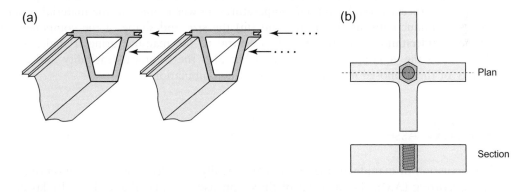

Figure GL3.17 Examples of design for joining: (a) edge detail in extrusion to facilitate assembly of a platform from simple sections; (b) threaded metal inserts moulded into a polymer injection moulding, to enable a tight bolted joint to be made.

Injection moulding of polymers is also prominent in design for assembly. The enormous form freedom has allowed polymers to replace many components in consumer goods and vehicles – these were previously made in multiple parts in metals but are now made with far fewer plastic mouldings that are quick and easy to assemble. Here the low stiffness and large failure strain of polymers is an advantage, enabling the incorporation of snap fittings into the design of mouldings, where the loading is relatively light (closures for storage boxes, for instance). Where tight joints are needed, however, mechanical fasteners such as threaded screws and bolts may be required, and then the polymer is too weak or brittle to take the metal fastener. But injection moulding readily enables metal 'inserts' to be moulded in, allowing parts to be screwed together, transferring the load while avoiding a polymer thread. Figure GL3.17(b) illustrates this versatility in designing for joining with polymers – note the hexagonal shape for the insert, to prevent its rotation during tightening of the threaded fastener.

Production rate The cost model for batch processes included a production rate in parts/hour. Continuous processes such as extrusion also have an inherent output rate, expressed as a length/hour. All processes also have a characteristic 'yield' – the fraction of production that reaches the target quality, the rest being scrapped (and mostly recycled internally). We also need to recognise that all production involves set-up time as well as actual processing – the fraction of the time spent on set-up varies widely depending on the scale and complexity of the equipment and the degree of automation. It is not surprising therefore that the generic data for production rate in the cost model can vary so widely for a given process.

When it comes to the processing step itself, it is easier to capture the process speed, by modelling the underlying physics of heat flow, liquid flow, deformation, and microstructure evolution. A large proportion of processes involve heating and cooling – process cycle times are frequently governed by inherent heat conduction. This enables everything from sophisticated software to simple semi-empirical rules to be used to scale process time or speed with: (i) the key process variables (initial temperatures, power input); (ii) the material properties (thermal conductivity, specific heat); and (iii) the size and shape of the component. This physical understanding of the process is central to the manufacturing triangle, not only determining the economics of production but also the technical feasibility and resulting quality of the proposed material-process-design combination.

GL3.8 Assessment of new processes: additive manufacturing

No discussion of manufacturing and design is really complete without mention of *additive manufacturing* (AM): the build-up of three-dimensional (3D) shapes layer by layer, using some form of deposition technique. 3D printers are one example, in which a polymer wire is fed through a heated chamber and extruded as required as the feed head moves in two dimensions. Other polymer techniques involve injecting adhesive into powders or selectively curing a UV-sensitive polymer. AM with metals resembles surface treatment or welding, depending on the deposition method and heat source (laser, electron beam, or welding arc) – either a pre-deposited powder is melted by the moving heat source, or powder or wire is fed directly into the focused hot spot on the surface.

What AM methods all have in common is the avoidance of dies and moulds, and almost unlimited geometric ('form') freedom and customisation – Figure GL3.18 illustrates these benefits, also showing that process steps can be eliminated. On the downside, they are inherently slow and only suited to small batch sizes, while the quality is usually inferior compared to conventional processing (in terms of precision, surface finish, porosity, and properties).

It remains early days for many AM methods in production, but it is informative to explore how AM processes currently compete with conventional processes, using our methodology of initial screening on process attributes.

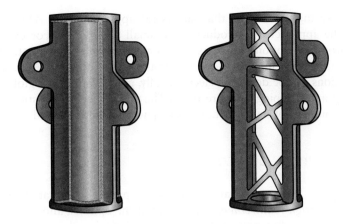

Figure GL3.18 Exploiting form freedom in additive manufacturing to reduce the part count and eliminate a welding step, while lightweighting the component.

Exercise

GL3.13   At present, the dominant processes for industrial AM are *powder bed fusion* (for metals, using lasers) and *photo-polymerisation* (for polymers). Indicative data for these processes are shown in the table, covering the same process attributes as before.

Add the data to copies of the process attribute charts, then comment on the overlap between these processes and casting and moulding processes (for metals and polymers, respectively). Which are most discriminating: the technical, quality, or economic attributes?

Attribute data for selected additive manufacturing processes

Attribute	Powder bed fusion (metals)	Photo-polymerisation (polymers)
Material compatibility	Ferrous (carbon, tool, stainless-steels)	Thermoplastic (PMMA)
	Non-ferrous (Al, Co, Cu, Ni, Ti alloys)	Thermoset (epoxy)
Shape compatibility	3D, sheet, prismatic	3D, sheet, prismatic
Mass range (kg)	0.01–20	0.1–10
Minimum section thickness (mm)	0.8–100	0.5–100
Tolerance (mm)	0.2–0.8	0.1–2
Roughness (μm)	8–125	100–125
Economic batch size	1–10	1–10

This exercise makes it clear that successful applications of AM need to exploit the strengths of the approach for production of small batch sizes. For example, AM enables on-demand production of one-off parts with simple portable equipment. This is attractive, as the machines are closely integrated with computer-aided design (CAD), so all the manufacturing instructions can be delivered electronically, giving immediate global reach and avoiding stockpiling and shipping of spare parts. The positive characteristics of AM are proving attractive in a number of fields, notably aerospace, automotive, and medical engineering.

GL3.9 Summary and conclusions

Selecting a process follows a strategy with some parallels to that for selecting materials: identification of *design requirements*, *screening* of the options, and *ranking* of the processes that survive screening, but now ending with a *technical evaluation*. For process selection, the requirements convert directly into the attributes of the process to be used in the screening step. Process attributes capture their technical capabilities and some aspects of product quality. All process families require screening for compatibility with material to which they are to be applied, but after this the attributes and constraints are specific to the family. For shaping, component size, geometry, and precision and finish are most discriminating; for joining, it is joint geometry and mode of loading; for surface treatments, it is the purpose for which the treatment is applied (wear, corrosion, aesthetics, etc.). Ranking is based on cost if this can be modelled, as it can for batch shaping processes. Cost models combine characteristic process attributes such as tooling, capital costs and the production rate, with user-specific parameters such as overhead rates and required batch size. The economic batch size provides a crude alternative to the cost model – a preliminary indication of the batch size for which the process is commonly found to be economic.

The design requirements frequently point to important aspects of processing that fall beyond a simple screening process. The correct choice of shape detail to facilitate processing, the avoidance of defects, achieving the required material properties, and determining the process speed (and thus economics) – all of these require technical expertise, and often involve coupling between the process operating conditions, the material composition, and design detail. This highlights the importance of the technical evaluation stage in making the final choice of processes, refining the design itself, and the material and process specifications, in parallel. Emerging technologies, such as additive manufacturing, need to be evaluated alongside conventional processing – the chapter concluded with a preliminary review of AM techniques, following the same selection methodology.

The interactions between manufacturing processes and the resulting microstructure and properties emerged repeatedly as a theme in this chapter and require deeper discussion – this is the topic of Chapter 11.

GL3.10 Further reading

Ashby, M. F. (2017). *Materials selection in mechanical design* (5th ed.). Butterworth-Heinemann. ISBN 978-0-08-100599-6. (The source of the process selection methodology, giving a summary description of the key processes in each class).

ASM Handbook Series. (1971–2004). *Heat treatment* (vol. 4); *Surface engineering* (vol. 5); *Welding, brazing and soldering* (vol. 6); *Powder metal technologies* (vol. 7); *Forming and forging* (vol. 14); *Casting* (vol. 15); *Machining* (vol. 16). ASM International. (A comprehensive set of handbooks on processing, occasionally updated, and now available online at www.asminternational.org/hbk/index.jsp).

Kalpakjian, S., & Schmid, S. R. (2013). *Manufacturing engineering and technology* (7th ed.). Pearson Education. ISBN-13: 978-0133128741. (Like its companion below, a long-established manufacturing reference).

Kalpakjian, S., & Schmid, S. R. (2016). *Manufacturing processes for engineering materials* (6th ed.). Pearson Education. ISBN-13: 978-0134290553. (A comprehensive and widely used text on material processing).

Tempelman, E., Shercliff, H. R., & Ninaber van Eyben, B. (2014). *Manufacturing and design – understanding the principles of how things are made* (1st ed.). Butterworth-Heinemann. ISBN 978-0-08-099922-7. (The source of the 'manufacturing triangle,' and the approach outlined in this chapter to the technical evaluation of process-material-design interactions).

Thompson, R. (2015). *Manufacturing processes for design professionals*. Thames and Hudson. ISBN-13: 978-0500513750. (A structured reference book with remarkable diverse illustrations, for industrial and product designers, architects, and engineers alike).

GL3.11 Final exercises

GL3.14 A manufacturing process is to be selected for an aluminium alloy piston. Its weight is between 0.8 and 1 kg, and its minimum thickness is 4 to 6 mm. The design specifies a precision of 0.5 mm and a surface finish in the range 2 to 5 µm. It is expected that the batch size will be around 1000 pistons. Use the process attribute charts to identify a subset of possible manufacturing routes, taking account of these requirements. Would the selection change if the batch size increased to 10,000?

GL3.15 Before the advent of the devices and streaming services for music and video, the medium of choice was the compact disc (CD)/digital video disc (DVD), and they are not yet entirely obsolete. The huge numbers produced suggests the physical disc itself must be quick and cheap to produce, but with unusually high precision being required due to the nature of the data storage/retrieval mechanism. Reading a digital disc involves tracking the reflections of a laser from microscopic pits at the interface between two layers on the disk. These pits are typically 0.5 µm in size and 1.5 µm apart, and so the tolerance (and roughness) must be rather better than 0.1 µm for the disk to work. Standard discs are made from polycarbonate, with a thickness of 1.2 mm, and a mass of around 15 g.

Use the process attributes charts to identify shaping processes that can meet the design requirements for material, shape, mass, thickness, and high-volume production. Can the candidate processes routinely achieve the tolerance and finish needed? See if you can find out how they are made in practice.

GL3.16 A process is required for the production of 10,000 medium carbon steel engine crankshafts. The crankshaft is a complex three-dimensional shape about 0.3 m in length with a minimum diameter of 0.06 m. At the position of the bearings, the surface must be very hard, with a surface finish better than 2 µm and a dimensional accuracy of ±0.1 mm. Elsewhere the surface finish can be 100 µm, and the tolerance is 0.5 mm. Use the process attribute charts to identify a suitable fabrication route for the crankshaft.

GL3.17 A heat-treatable aluminium alloy automotive steering arm has a mass of 200 g, a minimum thickness of 6 mm, and a required tolerance of 0.25 mm. Production runs over 50,000 are expected. Discuss the implications for the manufacture of the component by forging, using the process attribute charts. Suggest an alternative processing route, commenting on how this might influence the choice of Al alloy and final properties.

GL3.18 In a cost analysis for casting a small aluminium alloy component, costs were assigned to tooling and overheads (including capital) in the way shown in the table. The costs are in units of the material cost of one component. Use thse simple cost model presented in Equation (GL3.4) to identify the cheapest process for a batch size of: (i) 100 units; (ii) 10^6 units.

Process	Sand casting	Investment casting	Pressure die	Gravity die
Material, $C_{material}$	1	1	1	1
Overhead rate (including capital), $\dot{C}_{oh}$ (hr^{-1})	500	500	500	500
Tooling, C_t	50	11,500	25,000	7500
Production rate $\dot{n}$ (hr^{-1})	20	10	100	40

GL3.19 A flanged cylinder is to be manufactured by casting. The figure shows two suggested cross sections.

(a) For either geometry, where would you need to place the parting plane for this casting?

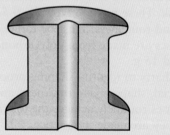 

(b) Which of the two designs is preferable, and why?

GL3.20 Look closely at a number of polymer bottles with threaded caps. The pre-form is likely to have been injection moulded, and the bottle then blow moulded. Look for a parting line showing the location of the parting plane when the bottle was made. How was the thread manufactured on the bottle? Could the thread inside the cap be made the same way, and if so, how was the mould removed? Can you identify geometric features of the bottle and cap that add functionality, and have been produced directly by detailing the mould shape? Can you find similar features on any glass bottles?

GL3.21 A restorer of musical instruments needs to replace five keys on an antique clarinet. All five have different shapes with complex curved surfaces, and these have been 3D-scanned into a CAD package from an equivalent instrument. The keys are made of a Cu alloy and weigh between 10 and 20 g. The minimum section thickness is 3 mm, and the dimensional tolerance is 0.2 mm. The keys will be electroplated and polished to give a shiny, silver finish, so the roughness after shaping is less important: 20 μm is the suggested target. Use the process attribute charts, supplemented with the data for *powder bed fusion* from Exercise GL3.13, to assess the design requirements for the clarinet keys for both powder bed fusion and the current process for making clarinet keys: investment casting. Does AM compete for this application?

GL3.22 A student wishes to produce 10 identical prototype polymer housings for an electronic device and has access to a photo-polymerisation system. The target shape is a hollow two-part shell about 3 to 4 mm thick. Before proceeding with the design, the student wishes to check whether there are other polymer shaping processes available that should be considered. Which other polymer shaping processes may be economic for a batch size as low as 10? Are they also compatible with the overall shape and section thickness? Compare the relative performance of the conventional and AM processes for tolerance and roughness, based on the current capabilities of photo-polymerisation summarised in the attribute charts, supplemented with the AM data in Exercise GL3.13. The student wishes to incorporate shaped features into the housing, for mounting the device securely inside, and to facilitate closure of the housing with screws. Would this be easier with AM or with the competing conventional processes?

Guided Learning Unit 4
Phase diagrams and phase transformations

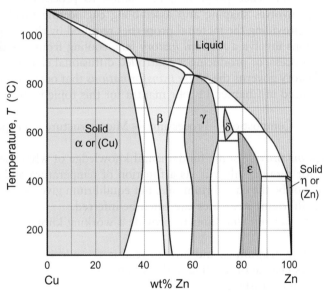

Chapter contents

https://doi.org/10.1016/B978-0-08-102399-0.00023-6

Introduction and synopsis

Phases and phase transformations are familiar in everyday life: rain and snow are the products of phase transformations in the sky succumbing to gravity – the word *precipitation* is used in phase transformation theory, and this is no coincidence. We make roads safer in the ice and snow by spreading salt – lowering the melting point of the water by changing its composition and causing a phase change. Bubbles rising in a glass of beer signify gases dissolved in the beer forming a separate phase, while in a boiling pan of water the bubbles of steam are formed by the water itself changing phase from liquid to vapour.

Phase diagrams and phase transformations are central to understanding microstructure evolution (and hence properties) in relation to processing, particularly for metals. Manufacturing involves shaping and assembling engineering products and devices, while simultaneously providing the material properties required by the design. Most material processing operations involve a *thermal history* (e.g. cooling from a high-temperature shaping or deposition process, or producing a controlled diffusional change in a solid product). The temperature history governs the *phase transformations* that occur, generating the microstructure. As introduced in Chapter 11, our mantra for this topic is thus:

$$\text{Compostion} + \text{Processing} \rightarrow \text{Microstructure} + \text{Properties}$$

Phase diagrams provide some fundamental knowledge of what the *equilibrium* structure of a metallic (or ceramic) alloy is, as a function of temperature and composition. The real structure may not be the equilibrium one, but equilibrium gives a starting point from which other (non-equilibrium) structures can often be inferred.

This *Guided Learning Unit* aims to provide a working knowledge of the following:

- What a phase diagram is, and how to read it
- How phases change on heating and cooling
- The resulting microstructures

On completion of this *Guided Learning Unit* you should be able to:

1. Identify and quantify the equilibrium phases at any point on binary phase diagrams.
2. Predict the microstructures that may result from simple processing histories relating to solidification and heat treatment of engineering alloys.
3. Understand how equilibrium and non-equilibrium microstructure evolution in alloy heat treatment relate to time-temperature-transformation (TTT) diagrams.

Key definitions are highlighted as they appear, and exercises are provided throughout for each topic, with a further set at the end for practice. Full solutions are provided online – use these to check your answers as you go along, to consolidate your learning (but not as a shortcut!).

GL4.1 contains some essential terminology and definitions.

GL4.2-GL4.4 show first how to interpret simple phase diagrams, leading to examples of more complex cases.

GL4.5-GL4.7 introduce phase transformations and show how phase diagrams are used to predict microstructure evolution during slow cooling (e.g. in solidification, and cooling in the solid state to room temperature).

GL4.8 extends phase transformation theory to non-equilibrium cooling in heat treatment of steels and other alloys, relating this to TTT diagrams.

This *Guided Learning Unit* fits best with Chapter 11 in two installments: GL4.1-GL4.4, dealing with phase diagrams, and GL4.5-GL4.8, covering phase transformations and microstructure evolution.

GL4.1 Key terminology

Alloys and components

> **Definition:** A *metallic alloy* is a mixture of a metal with other metals or non-metals. Ceramics, too, can be mixed to form *ceramic alloys*. The *components* are the chemical elements/compounds that make up alloys.

Examples of alloys and their components:

- *Brass*: a mixture of copper (Cu) and zinc (Zn)
- *Carbon steel*: based on iron (Fe) and carbon (C)
- *Spinel*: a ceramic alloy made of magnesia (MgO) and alumina (Al_2O_3)

Components are given capital letters: A, B, C or the element/compound symbols Cu, Zn, C, MgO, etc.

> **Definition:** A *binary alloy* contains two components. A *ternary alloy* contains three; a *quaternary alloy*, four; and so on.

Composition Alloys are defined by the components and their compositions, in weight or atom %.

> **Definition:** The composition in *weight* % of component A:
>
> $$C_A = \frac{\text{Weight of component A}}{\sum \text{Weights of components}} \times 100$$
>
> The composition in *atom* (or *mol*) % of component A:
>
> $$X_A = \frac{\text{Number of atoms (or mols) of component A}}{\sum \text{Number of atoms (or mols) of component}} \times 100$$

To convert between weight and mols:

$$(\text{Weight in grams})/(\text{Atomic or molecular wt in grams/mol}) = \text{Number of mols}$$

$$(\text{Number of mols}) \times (\text{Atomic or molecular wt in grams/mol}) = \text{Weight in grams (g)}$$

Phases   For pure substances, the idea of a phase is familiar: ice, water, and steam are the solid, liquid, and gaseous states of pure H_2O – each is a distinct phase. Processing of metallic alloys leads to microstructures in which the component elements are distributed in a number of ways. In the liquid state for metals, more or less everything dissolves completely. In the solid state, things are more complex – for example, in a binary alloy the solid microstructure usually takes one of three forms (discussed later):

- A single solid solution
- Two separate solid solutions
- A chemical compound, with a separate solid solution

Recall that a *solid solution* is a solid in which one (or more) elements are 'dissolved' in another so that they are homogeneously dispersed, at an atomic scale.

A region of a material that has a homogeneous atomic structure is called a *phase*. A phase can be identified as a cluster of as few as 10 or so atoms, but it is usually much more. In the three types of solid microstructures listed earlier, each solid solution or chemical compound would be an identifiable phase.

> **Definition:** All parts of an alloy microstructure with the same atomic structure are a single *phase*.

Exercises (reminder: answers online)

GL4.1   A 1.5-kg sample of α-brass contains 0.45 kg of Zn, and the rest is Cu. The atomic weight of copper is 63.5 and zinc 65.4. Write down the composition of copper in α-brass, in wt%, C_{Cu}. Find the compositions of copper and zinc in the α-brass, in at%, X_{Cu}, and X_{Zn}.

GL4.2   An alloy consists of X_A at% of **A** with an atomic weight a_A, and X_B at% of B with an atomic weight a_B. Derive an equation for the composition of the alloy C_A in wt% of A. By symmetry, write down the equation for the composition of the alloy C_B in wt% of B.

Constitution, equilibrium, and thermodynamics

> **Definition:** The *constitution* of an alloy is described by the following:
> (a) Phases present
> (b) Weight fraction of each phase
> (c) Composition of each phase

At *thermodynamic equilibrium*, the constitution is stable: there is no further tendency for it to change. The independent *state variables* determining the constitution are temperature, pressure, and composition. Hence the *equilibrium constitution* is defined at constant temperature and pressure for a given alloy composition.

Thermodynamics controls the phases in which mixtures of elements can exist as a function of the state variables – this is discussed further in Section 11.4. The key parameter for a given composition, temperature T and pressure p, is the *Gibbs free energy*, G, defined as [Equation (11.1)]:

$$G = U + pV - TS = H - TS$$

where U is the *internal energy*, V is the volume, H is the enthalpy ($U + pV$), and S is the entropy. The internal energy U is the sum of the atomic vibration and the bond energies between the atoms. For the liquid and solid states most relevant to materials processing, U dominates the enthalpy (pV is small), so $H \approx U$. Entropy S is a measure of the disorder in the system – when an alloy solidifies there is a decrease in entropy because the solid has a regular lattice, whereas the liquid does not.

Each possible state – liquid solution, solid solution, mixtures of phases, and so on – has an associated free energy, and that with the lowest free energy is the state at thermodynamic equilibrium.

> **Definition:** The *equilibrium constitution* is the state of lowest Gibbs free energy G, for a given composition, temperature, and pressure. An alloy in this state shows no tendency to change – it is thermodynamically stable.

Phase diagrams Pressure has a limited influence on material processing, as this primarily involves the liquid and solid states. We therefore consider processing outcomes to be controlled by the remaining two state variables: temperature and composition. Maps with these state variables as axes are called *phase diagrams* – for binary alloys they are two dimensional.

> **Definition:** A *phase diagram* (or equilibrium diagram) is a diagram with T and composition as axes, showing the equilibrium constitution.

The phase diagram of a binary alloy made of components A and B, for all combinations of T and X_B, defines the *A–B system*. Commercial alloys may contain 10 or more elements, but in all cases there is one principal element (copper alloys, aluminium alloys, and so on) plus other elements that are added deliberately or are impurities. The starting point for understanding the essential behaviour of any alloy is therefore to consider the binary systems for the principal component element and its key alloying element.

GL4.2 Simple phase diagrams, and how to read them

Melting point, liquidus, and solidus   Consider first a pure material A heated from the solid state. The *melting temperature* is the unique temperature at which the phase change to the liquid state occurs, and solid and liquid can co-exist [see Figure GL4.1(a)]. Similarly, liquid changes to gas or vapour at a unique (higher) temperature, the boiling point. In practical phase diagrams boiling is of little interest, so the diagram is usually limited to liquid and solid states.

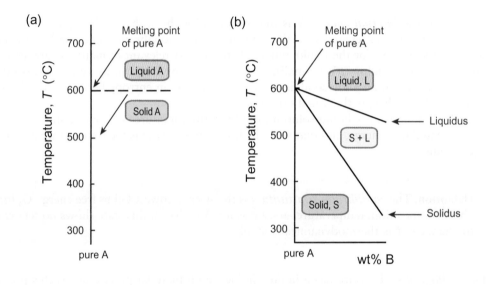

Figure GL4.1   (a) One-dimensional phase diagram for a pure substance: a temperature scale showing the phase boundary between solid and liquid – the melting point; (b) the A-rich end of a binary A-B phase diagram, illustrating partition of the melting point between solidus and liquidus boundaries.

Now if we consider the binary A–B system and add a second axis for composition, the behaviour illustrated in Figure GL4.1(b) is commonly observed: the upper limit of 100% solid and lower limit of 100% liquid separate, and there is not a unique melting point – this is known as *partition*. In the region between the two *phase boundaries* on the figure, liquid (L) and solid (S) are stable together, in proportions that depend on the temperature (see later). These boundaries have special names:

Definition: The phase boundary that limits the bottom of the *liquid* field is called the *liquidus line*; the line giving the upper limit of the single-phase *solid* field is called the *solidus line*.

Had we considered pure B first, it too would have a unique melting point and shown partition as we added some A. So what happens as we cover all possible compositions between pure A and pure B? One possible outcome is illustrated in Figure GL4.2, the phase diagram for the Cu-Ni system (the base alloy for some coinage, explaining the terminology 'coppers' in the United Kingdom, and 'nickels' in the United States). This *isomorphous phase diagram* is the simplest example: *isomorphous* meaning 'single-structured.' Here the solid state is a solid solution for all compositions, all the way from pure A to pure B. Since the atomic structure of this solid solution is the same at all compositions (with only the proportions of Cu and Ni atoms varying), it is a single phase.

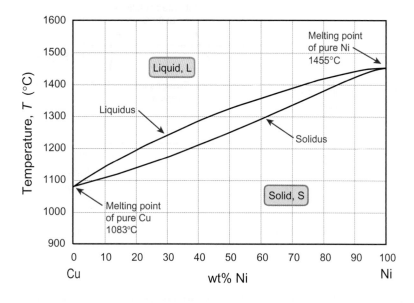

Figure GL4.2 Isomorphous phase diagram for the Cu-Ni system.

This behaviour in the solid state is actually rare – in virtually every other atomic mixture, there is a limit to the amount of an element that can be dissolved in another. We will explore this *solubility limit* with an everyday example: a cup of tea.

Solubility limits   For the purposes of illustration, think of tea as hot water, and sugar as a molecular component. Add one spoonful of sugar to hot tea, and it dissolves – the sugar disperses into solution. Keep spooning it in, and solid sugar is seen sitting in the bottom of the cup. The tea has reached its *solubility limit* at this temperature and has become *saturated*: it will not dissolve any more sugar. This saturated tea now co-exists with a second phase: solid sugar as a saturated solution, as the sugar absorbs as much tea as it can. This is characteristic of mixtures of two equilibrium phases, both of which are solutions – both phases are saturated (i.e. both are as impure as possible).

If we add sugar to cold tea, less sugar will dissolve – the saturation limit (in wt% sugar) rises with temperature. Harder to observe, but equally true, the sugar absorbs more water as it is heated. Plotting this behaviour over a range of temperatures leads to the partial phase diagram shown in Figure GL4.3. The boundaries between the single- and two-phase regions are known as *solvus boundaries*. What happens at a given temperature and composition is determined by the thermodynamics of mixing sugar and water. Minimising free energy dictates whether there are one or two phases, and in the two-phase regions fixes the proportions and compositions of the phases.

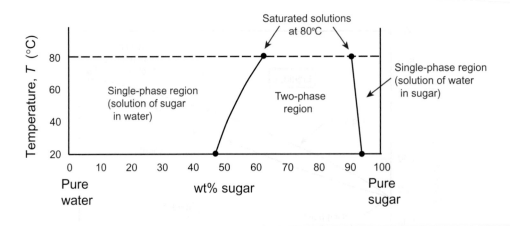

Figure GL4.3   Schematic phase diagram for sugar and tea (water) – the saturation levels of both solutions increase with temperature.

Building a simple binary diagram, and the eutectic point   The picture in Figure GL4.3 is also found in many metallic mixtures. As an exemplar binary alloy we will consider the Pb-Sn system, the basis for many years of solders used for electronic joints (before falling out of favour due to environmental and health concerns). Figure GL4.4 shows a partially complete Pb-Sn phase diagram. Note the nomenclature adopted: (Pb) for a Pb-rich single solid phase, and (Sn) for Sn-rich single solid phase. The lower part replicates the sugar-tea behaviour: the solubility of Pb in Sn (and of Sn in Pb) increases with temperature. Note that the solubilies of these elements in one another is low, especially Pb in Sn on the right of the diagram. The upper part of the diagram shows the partition behaviour from the melting points of the pure elements, as in Figure GL4.1(b).

To complete the diagram, first consider where the falling solidus boundaries meet the rising solvus boundaries. Figure GL4.4 suggests that the points where they intersect will be the points of maximum solubility (highest saturation) in the single-phase solids, and this will close the regions or *fields* which they occupy. For good thermodynamic reasons, not elaborated here, peak saturation of Sn in (Pb) occurs at the same temperature as that for Pb in (Sn). It is essentially a corollary of the fact that in a two-phase region, both phases are

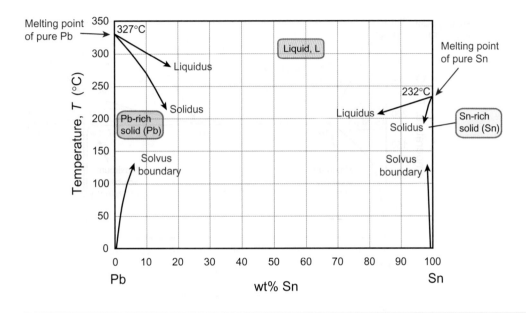

Figure GL4.4 Partial-phase diagram for the Pb-Sn system, showing the limiting behaviour at high and low temperatures.

as impure as possible. Below this temperature we have a mixture of solid solutions, with a horizontal boundary linking the two points of maximum solubility, as shown in Figure GL4.5(a), closing the two-phase field, (Sn) + (Pb).

Finally, what happens to the two liquidus boundaries? Again, thermodynamics dictates that these meet the horizontal line at a single point, and the liquid field closes in a shallow 'V' [Figure GL4.5(a)]. This point on the diagram is important and is known as a *eutectic point*. At this special temperature and composition, the alloy can change directly from liquid to two-phase solid at a fixed temperature. This also closes the two (liquid + solid) regions above the horizontal line to either side of the eutectic point. The single-phase fields are highlighted on the complete phase diagram in Figure GL4.5(b).

> **Definition:** The lower limit of the single-phase liquid field formed by the intersection of two liquidus lines is called the *eutectic point*.

For the Pb-Sn system the eutectic point is at the composition $C_{Sn} = 61.9\,wt\%$, and temperature $T = 183°C$. Eutectics will be re-visited later in relation to the microstructures that form when a eutectic composition solidifies. For the time being, note that the eutectic composition gives the lowest temperature for which 100% liquid is stable. For this reason, casting, brazing, and soldering alloys are often eutectic or near-eutectic in composition.

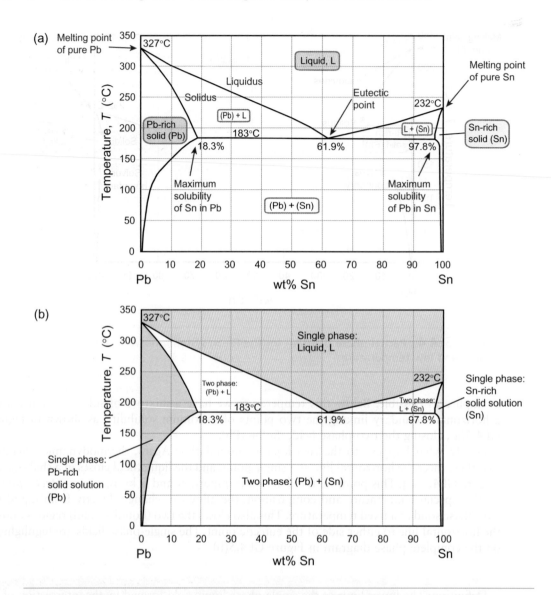

Figure GL4.5 The completed phase diagram for the Pb-Sn system, showing (a) the eutectic point and closing the liquid field; (b) the single-phase fields *(shaded)*, separated by two-phase fields.

Reading a binary phase diagram: phase compositions The state variables (temperature and composition) define a point on the phase diagram: the *constitution point*. The first thing to establish at a constitution point is the number of phases present, one or two (in a binary system). Single-phase regions are always labelled, either with the notation in Figure GL4.5, with (Pb) for Pb-rich solid and so on, or with a Greek character (α, β, γ, etc). When a phase diagram is traversed at a given temperature, crossing phase boundaries takes us from a single-phase field to a two-phase field (or vice versa) [Figure GL4.5(b)]. In the two-phase regions, a horizontal line through the constitution point ending at the adjacent phase boundaries identifies the phases present: they are the single phases beyond those boundaries. This line is called a *tie-line* (Figure GL4.6).

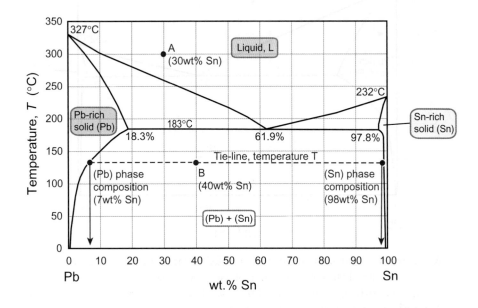

Figure GL4.6 Phase diagram for the Pb-Sn system, illustrating constitution points in single- and two-phase fields, and the tie-line defining the phases and compositions in the two-phase field.

At a constitution point in a single-phase region, the *phase composition* is simply the composition of the alloy itself (e.g. point A in Figure GL4.6). In two-phase regions (e.g. point B in Figure GL4.6), the phase compositions are given by the values on the phase boundaries at the ends of the tie-line through the constitution point. Recall that these are the saturation limits of the single-phase fields on the other sides of the boundaries.

> **Definition:** In a single-phase region, phase and alloy compositions coincide. In a two-phase region the phase compositions lie on the phase boundaries at either end of a horizontal tie-line through the constitution point.

Consider points A and B on the Pb-Sn phase diagram in Figure GL4.6. Constitution point A (temperature 300°C, composition Pb-30 wt% Sn) lies in the single-phase liquid field; the phase composition is also Pb-30 wt% Sn. Constitution point B (temperature 130°C, alloy composition Pb-40 wt% Sn) lies in a two-phase field with two solid phases identified from the ends of the tie-line: (Pb) and (Sn); the phase compositions are Pb-7 wt% Sn and Pb-98 wt% Sn, respectively.

Exercises

GL4.3 Use the Pb-Sn diagram in the figure to answer the following questions.

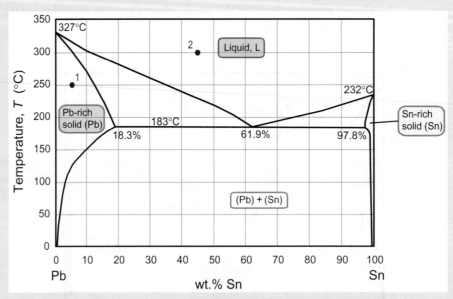

(a) What are the values of the state variables (composition and temperature) at constitution points 1 and 2, and what phases are present?

(b) Mark the constitution points for Pb-70 wt% Sn and Pb-95 wt% Sn alloys at 210°C. What phases are present in each case?

(c) The alloy at constitution point 1 is cooled very slowly to room temperature, maintaining equilibrium. Identify the temperature of the phase boundary at which a change in the phases occurs. What phase(s) is(are) present below the phase boundary?

(d) The alloy at constitution point 2 is cooled slowly to room temperature. Identify the temperatures at which phase changes occur, and the phase(s) before and after each change.

GL4.4 Use the Pb-Sn diagram in Exercise GL4.3 to answer the following questions.

(a) The constitution point for a Pb-25 wt% Sn alloy at 250°C lies in a two-phase field. Construct a tie-line on the figure and read off the two phases and their compositions.

(b) The alloy is slowly cooled. Identify the phases and their compositions (i) at 200°C, (ii) at 150°C.

(c) Indicate with arrows on the figure the lines along which the compositions of the two phases move during slow cooling from 250°C to 200°C. The overall composition of the alloy stays the same, of course. How can this be maintained as the compositions of the phases change?

Reading a binary phase diagram: proportions of phases In a two-phase field at constant temperature, the compositions of the phases are fixed at the saturation limits – the values on the boundaries at the ends of the tie-line. So different compositions at this temperature will contain different *proportions* (or *weight fractions* W) of each phase, in such a way as to conserve the overall fractions of the two elements. The proportions of each phase (by weight) in a two-phase region can be found from the phase diagram using the *lever rule* – its derivation is given as an exercise later. Consider Pb-20 wt% Sn at 250°C in Figure GL4.7, a constitution point in the two-phase field: liquid plus Pb-rich solid. To find the proportions of each phase, first construct a tie-line through the constitution point and read off the compositions of the phases:

Pb-rich solid with $C_{Sn}^{SOL} = 12\ wt\%$; liquid with $C_{Sn}^{LIQ} = 34\ wt\%$

The tie-line is of length ℓ, and the lengths of the segments to either side of the constitution point are a and b, respectively. For the example alloy of composition $C_{Sn} = 20\%$:

$$\ell = C_{Sn}^{LIQ} - C_{Sn}^{SOL} = 34 - 12 = 22\%$$

$$a = C_{Sn} - C_{Sn}^{SOL} = 20 - 12 = 8\%$$

$$b = C_{Sn}^{LIQ} - C_{Sn} = 34 - 20 = 14\%$$

The weight fractions of liquid and solid in the alloy are $W^{LIQ} = a/\ell$ and $W^{SOL} = b/\ell$. Hence:

$$W^{LIQ} = 8/22 = 36\%$$

$$W^{SOL} = 14/22 = 64\%$$

This illustrates why the name is the lever rule – it is analogous to balancing two weights on either side of a pivot, with the shorter distance being that to the greater weight (for moment equilibrium).

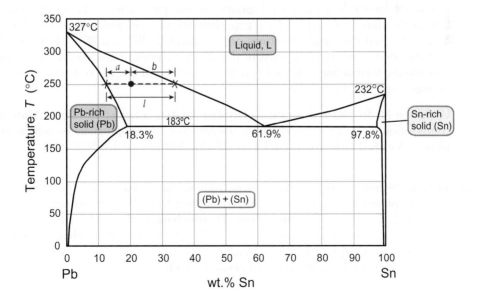

Figure GL4.7 Phase diagram for the Pb-Sn system, illustrating the lever rule for finding the weight fractions of the phases in a two-phase field.

Note the following:

- $W^{SOL} + W^{LIQ} = a/\ell + b/\ell = (a + b)/\ell = 1$ (as expected, the two fractions sum to unity)
- At the left-hand end of the tie-line, $W^{SOL} = 1$ ($a = 0, b = \ell$)
- At the right-hand end of the tie-line, $W^{LIQ} = 1$ ($a = \ell, b = 0$)

To summarise: to find the weight fractions of the phases in *any* two-phase region (liquid-solid, or two solid phases):

- Construct the tie-line through the constitution point.
- Read off the three compositions (for the alloy, and the two ends of the tie-line).
- Apply the lever rule.

Alternatively, and more approximately, the lengths can be measured directly from the phase diagram. But note that the proportions of the phases only vary linearly with composition along the tie-line if the diagram has a *linear weight % scale*. Some phase diagrams have linear *atom %* scales and may then show a non-linear weight % scale along the top of the diagram (examples later). In this case, the lever rule for weight fraction cannot be applied by direct measurement, but the equations for weight fractions can still be applied by finding the three compositions (in wt%) and evaluating a, b, and ℓ as in the previous example. Note that the concept of the atom fraction of a phase is not particularly useful, so the lever rule is not generally applied to linear atom % scales.

Exercises

GL4.5   Derive the lever rule for a general mixture of two phases, α and β. Let the composition of the alloy be C (wt% of alloying element), the compositions of the phases be C_α and C_β, and the weight fractions of the phases be W^α and W^β. (Hints: first find an expression conserving the mass of the alloying element between the alloy and the two phases, then define a, b, and ℓ in this notation and use the overall conservation of mass expressed in $W^\alpha + W^\beta = 1$.)

GL4.6   Using the Pb-Sn phase diagram in Exercise GL4.3, consider the Pb-Sn alloy with composition $C_{Sn} = 25$ wt%. What are the approximate proportions by weight of the phases identified in Exercise GL4.4, at 250°C, 200°C, and 150°C?

Intermediate phases   Many systems show *intermediate phases*: compounds that form between components. Examples are $CuAl_2$, Al_3Ni, and Fe_3C. If both components are metallic, they are called intermetallic compounds. Thermodynamically, compounds form because the particular combination of components is able to form as a single phase with a specific lattice of lower free energy than, say, a mixture of two phases. Example lattices are given in *Guided Learning Unit 1: Simple ideas of crystallography*.

The atomic % of components in a compound is called its *stoichiometry*. Compounds are written in the form noted earlier, A_xB_y, where x and y are integers. The at% of the components in an intermediate compound can easily be stated by inspection, $x/(x + y)$ and $y/(x + y)$; for example, Fe_3C contains 25 at% C. In general, the integer values x and y are small because the number of atoms that define the repeating unit of the crystal lattice is also small. Compounds therefore usually appear on phase diagrams with at% scales at simple integer ratios: 25%, 33%, 50%, and so on. In principle they therefore plot as a vertical line representing the single phase (an example follows).

As a single phase of fixed composition, intermediate phases have unique melting points (like pure components). The higher degree of thermodynamic stability means that compounds often have higher melting points. The liquid field for compositions on either side often shows a falling liquidus line, with eutectics forming between the compound and a solid solution, or between two compounds if the system shows more than one. Figure GL4.8(a) shows the unusual (and very untypical) silver-strontium phase diagram (notice the *at% scale*). This is not exactly a well-known engineering alloy, but it illustrates the 'ideal' behaviour of compounds on phase diagrams. It has four intermetallic compounds (the vertical lines) and looks like five separate phase diagrams back-to-back: the Ag-Ag_5Sr diagram, the Ag_5Sr-Ag_5Sr_3 diagram, and so on. The liquidus boundary falls from each single-phase melting point, forming five eutectics. Note that the solidus lines are all vertical – they coincide with the lines representing the single-phase compound. In all the adjoining two-phase fields, the compositions of the solid phases are fixed and do not vary with temperature.

Diagrams of this type give the impression that two two-phase fields meet at a vertical boundary, which violates the fundamental thermodynamics. Traversing a diagram at constant temperature must show one-phase, two-phase, one-phase, and so on, as boundaries are crossed, with tie-lines in the two-phase fields ending at single-phase boundaries. This remains the case here – there *is* a single-phase field in between the two-phase fields, it is one of the compounds – but the field has essentially collapsed to a single vertical line.

A more typical example with engineering significance is illustrated in Figure GL4.8(b), showing part of the Al-Cu diagram, on a wt% scale (with the corresponding at% scale across the top). This diagram is the basis of the important 'age-hardening' Al-Cu alloys, used widely in aerospace. A compound forms at 33 *at%* Cu: it is therefore $CuAl_2$ (given the name θ-phase, to signify that it is a single phase). The liquidus boundaries fall to a eutectic point at $33 wt\%$ Cu and 548°C. In contrast to the silver-strontium diagram, the θ field is not a single vertical line, but a tall thin region with a small spread in composition. In other words, $CuAl_2$ can tolerate a small amount of excess Al while remaining a single phase – some of the Cu atoms are replaced by Al, and the stoichiometry may not be exactly 1:2. We can think of it as a solid solution of Al in $CuAl_2$. Most practical compounds show some tendency to form a solid solution over a small range of composition close to stoichiometric, giving a thin single-phase field rather than a vertical line. In some cases the spread of composition is so great that it ceases to be meaningful to distinguish it as a compound at all, and simply to consider it as a solid solution.

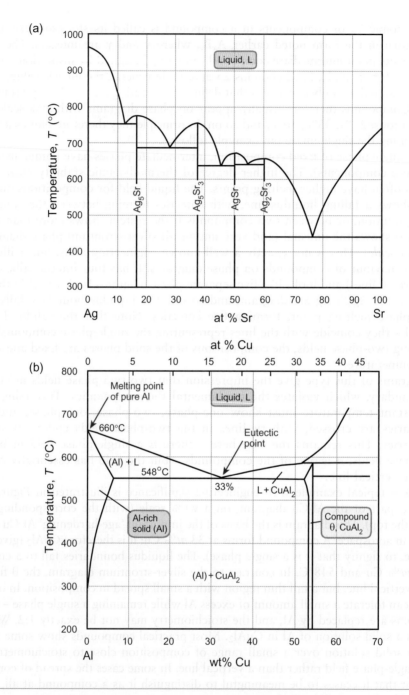

Figure GL4.8   Phase diagrams showing intermediate compounds: (a) the silver-strontium Ag-Sr system; (b) part of the aluminium-copper Al-Cu system.

Exercises

GL4.7 Use Figure GL4.8(a) to answer the following:
  (a) For an Ag-90 at% Sr alloy at 600°C:
      (i) Plot the constitution point on the phase diagram.
      (ii) Identify the phases present, and find their compositions in at%.
      (iii) The temperature is slowly reduced to 500°C. Will the phase compositions and proportions change?
  (b) For an Ag-30 at% Sr alloy at 600°C:
      (i) Plot the constitution point on the phase diagram.
      (ii) Identify the phases present, and find their compositions in at%.
      (iii) Will the proportions change if the temperature is reduced to 500°C? Why is this?
  (c) The atomic weight of Ag is 107.9 and that of Sr is 87.6. Calculate the compositions of the four intermetallic compounds in the Ag-Sr system in *weight%*.

GL4.8 Use Figure GL4.8(b) to answer the following. For an Al-4 wt% Cu alloy:
  (a) Calculate the composition in at% Cu (atomic masses of Al and Cu: 26.98 and 63.54, respectively).
  (b) At 550°C, identify the phase(s) present, and find its composition (in wt%) and proportion by weight.
  (c) Repeat for 250°C.

GL4.3 The iron-carbon diagram

The *iron-carbon* phase diagram is important in engineering, providing the basis for understanding all *cast irons* and *carbon steels* and their heat treatment. First, we consider pure iron. The low temperature form of iron is called *ferrite* (or α-iron), with a BCC lattice (body-centred cubic, defined in *Guided Learning Unit 1*). On heating, pure iron changes to *austenite* (or γ-iron) at 910°C, and switches to a face-centred cubic (FCC) lattice. Pure austenite is stable up to 1391°C, when it changes back to BCC δ-iron, before melting at 1534°C.

A key characteristic of the iron-carbon system is the different extent to which FCC and BCC iron dissolve carbon in interstitial solid solution, forming single phases. In *Guided Learning Unit 1*, it was shown that the *interstitial holes* are larger in FCC than in BCC. This leads to low solubility of carbon in BCC ferrite and δ-iron, and much higher solubility in FCC austenite. Note that the same names (ferrite, austenite, and δ) are applied equally to the different states of pure iron and to the solid solutions they form with carbon. The nomenclature 'iron-rich solid (Fe)' is not much help to us, as there are three distinct variants.

Figure GL4.9 shows the iron-carbon diagram, up to 7 wt% carbon. Just below this upper limit lies the compound *iron carbide* Fe_3C (also given the name *cementite*), with 25 at% carbon. This part of the Fe-C system covers the main carbon steels and cast irons, as indicated. This diagram is more complicated than those shown previously and needs a bit of breaking down.

First consider the picture below 1000°C, up to 2.0 wt% carbon. This shows the low solubility of carbon in ferrite, with a maximum of 0.035 wt% at 723°C. Below the transformation temperature of ferrite to austenite (910°C) the picture resembles the partition behaviour

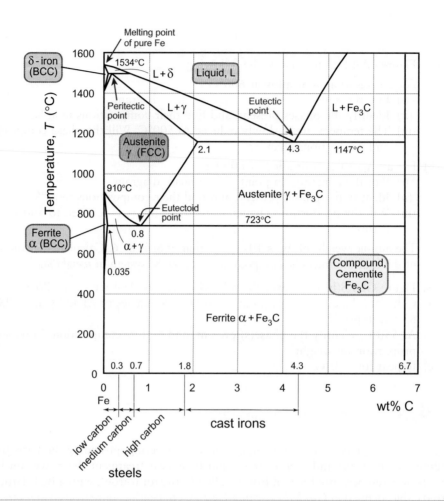

Figure GL4.9 The iron-carbon phase diagram up to 7 wt% carbon, the region covering carbon steels and cast irons.

seen below the melting point of a pure element, with two-phase boundaries falling from this temperature and a two-phase region in between. But in this case the upper phase is a solid solution (austenite), rather than a liquid. At the temperature of maximum C solubility in ferrite (723°C), the lower limit of the austenite field also forms a 'V', giving the minimum temperature at which austenite forms as a single phase, at a composition of 0.8 wt% C.

This feature on a phase diagram is called a *eutectoid point* and is particularly important in the context of carbon steels (as illustrated later when we consider their microstructures and heat treatments). Note the similarity in shape to a eutectic, with the key difference that the phase above the 'V' is a single solid phase (as opposed to a single liquid phase in a eutectic).

Definition: The lower limit of a single-phase solid field formed by two falling phase boundaries intersecting in a 'V' is called a *eutectoid point*.

Now following the rising boundary of the austenite field to the right and above the eutectoid point, we reach a point of maximum solubility at 2.1 wt% C, and the diagram to the right shows exactly the eutectic structure seen earlier. The eutectic temperature coincides with the temperature of maximum solubility of C in austenite, with falling solidus and liquidus lines enclosing the two-phase liquid + austenite region.

Completing the austenite field at the top introduces another new feature, a *peritectic point*. The austenite field closes in an inverted 'V' at the peritectic point (i.e. the maximum temperature at which this single phase forms). This temperature coincides with the temperature at which δ-iron has its maximum solubility, giving a horizontal line through the peritectic point. Above the line is a two-phase field, of which one is a liquid – here liquid + δ-iron. The phases in the other two-phase fields are readily identified from tie-lines, δ + austenite (γ) to the left of the peritectic, and γ + liquid to the right.

To summarise the key nomenclature of the iron-carbon system, the single phases are:

Ferrite: α-iron (BCC) with up to 0.035 wt% C dissolved in solid solution
Austenite: γ-iron (FCC) with up to 2.1 wt% C dissolved in solid solution
δ-iron: (BCC) with up to 0.08 wt% C dissolved in solid solution
Cementite: Fe_3C, a compound with 6.7 wt% C, at the right-hand edge of the diagram

The system has a eutectic point at 4.3 wt% C, a eutectoid point at 0.8 wt% C, and a peritectic point at 0.2 wt% C.

In passing, it should perhaps be noted that this iron-carbon diagram is not strictly an equilibrium diagram. Iron carbide is in fact a metastable state – true equilibrium is reached in thermodynamic terms in two-phase mixtures of iron and carbon. However, in most circumstances iron carbide forms readily in preference to carbon as a separate phase – it is an example of the *mechanism* of phase transformations taking priority over the thermodynamics and free energy. This pseudo-equilibrium phase diagram, including iron carbide, is therefore most commonly used. The difference does become apparent in the solidification of cast irons, alloys containing 2–4 wt% C (i.e. approaching the eutectic composition). In this case, carbon can form as an equilibrium phase in the microstructure.

Exercise

GL4.9   Use Figure GL4.9 to answer the following:
(a) For a Fe-0.4 wt% C alloy at 900°C and 600°C:
(i)   Plot the constitution points on the phase diagram.
(ii) Identify the phases present, and find their compositions in wt%.
(iii) If the temperature is slowly reduced from 900°C to 600°C, at what temperatures are phase boundaries crossed? Identify the phases present after each boundary is crossed.
(b) How does slow cooling from 900°C to 600°C differ for a Fe-0.8 wt% C alloy?

GL4.4: Interpreting more complex phase diagrams

The phase diagrams covered so far illustrate almost all the features found in binary systems. This section completes the story and shows how even very complicated diagrams can be interpreted if the rules are applied carefully.

Eutectics, eutectoids, peritectics, and peritectoids   Three of these features were seen on the iron-carbon diagram. Here we define the fourth – a *peritectoid point* – and show all four together for clarity.

A peritectoid point is similar in appearance to a peritectic, being an inverted 'V' corresponding to an upper limit of formation of a single solid phase. But the difference is that the two-phase field above is formed of two solid phases (whereas in a peritectic, one is liquid). So, to help remember which is which:

- *eu–* means a normal 'V' meeting a horizontal line; *peri–* means an inverted 'V' meeting a horizontal line.
- *–tectic* means a liquid phase is involved; *–tectoid* means all phases are solid.

Figure GL4.10 shows all four for comparison. Single solid phases are denoted by Greek letters, liquid by L. Note that each point involves *three* phases: a single phase inside the 'V' and two different phases across the horizontal line.

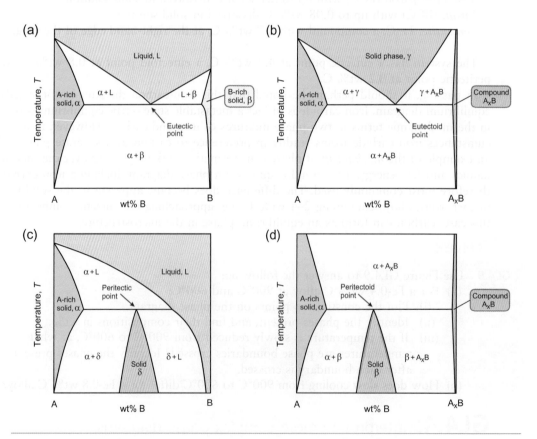

Figure GL4.10   Schematic views of (a) eutectic point, (b) eutectoid point, (c) peritectic point, (d) peritectoid point.

Compared with eutectics and eutectoids, peritectics and peritectoids are of much less engineering significance. An unusual exception is the growth of single crystals of the new high-temperature superconductors based on yttrium-barium-copper oxide. This is conducted by very slow cooling through a peritectic transformation.

Ceramic phase diagrams Ceramics are mostly compounds of a metal with one of the elements O, C, or N. They form with specific stoichiometry to satisfy the electronic balance between the elements (e.g. alumina Al_2O_3). In some cases, we are interested in mixtures of ceramics (or ceramic alloys). An example was mentioned earlier: *spinel* is made of magnesia (MgO) and alumina (Al_2O_3). Earth scientists in particular need phase diagrams for ceramics to help interpret natural minerals and microstructures.

Phase diagrams for ceramics work in the same way as for metal systems, with the elements replaced by the pure compounds. Figure GL4.11 shows the silica-alumina (SiO_2-Al_2O_3) system. It forms an intermediate single phase, known as *mullite*. Note that the top of this single-phase field closes in a peritectic point – above this point, the two-phase field is liquid + Al_2O_3.

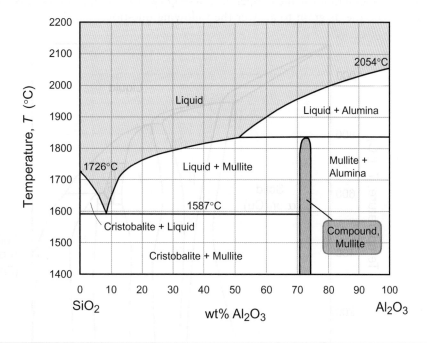

Figure GL4.11 Phase diagram for the binary ceramic silica-alumina (SiO_2-Al_2O_3) system.

Exercises

GL4.10 The Al-Si and Cr-Ni phase diagrams both have a eutectic point. Mark the eutectic point on each figure and find the eutectic temperature and composition in wt% in each system. (Note the at% scale in the Cr-Ni diagram; the atomic masses of Cr and Ni are 52.00 and 58.71, respectively.)

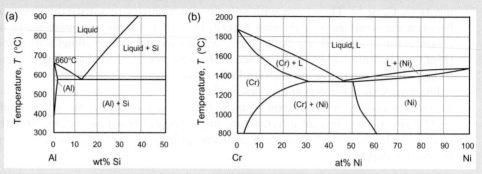

GL4.11 The phase diagram for the copper-zinc system (which includes brasses) is shown below. Use the diagram to answer the following questions.

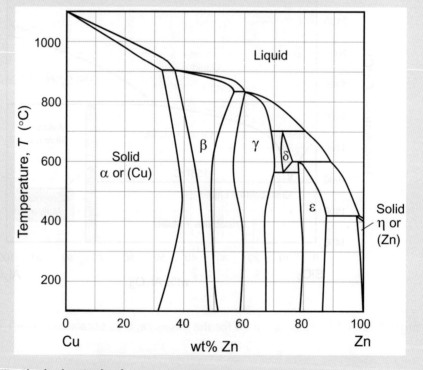

(a) (i) Shade the single-phase regions.
(ii) Highlight the eutectoid point and five peritectic points in the copper-zinc system, and write down their compositions and temperatures.

(b) The two common commercial brasses are 70/30 brass: C_{Cu} = 70 wt%, and 60/40 brass: C_{Cu} = 60 wt%. Locate their constitution points on the diagram at 200°C.

   (i) What distinguishes the two alloys?

   (ii) What roughly is the melting point of 70/30 brass?

   (iii) What are the phases in 60/40 brass at 200°C? Find their compositions and proportions.

GL4.12 Use the phase diagram for the SiO_2-Al_2O_3 system in Figure GL4.11 to answer the following:

(a) The intermediate compound mullite may be considered as having the formula SiO_2 $(Al_2O_3)_x$. Find the approximate value of x. The atomic masses of Si, Al, and O are 28.1, 26.9, and 16.0, respectively.

(b) Use the lever rule to find the equilibrium constitution of a 50 wt% Al_2O_3 alloy at 1700°C. Is it valid to measure directly from the diagram in this case? Why?

GL4.5 Phase transformations and microstructural evolution

After completing GL4.1–GL4.4 you should be able to:

- Recognise the important features of binary phase diagrams, identifying single- and two-phase fields.
- Find the proportions and compositions of the phases, at a given temperature and alloy composition.

GL4.5–GL4.8 link this understanding of phase diagrams to microstructure evolution in important industrial processes. Almost all manufacturing processes involve some combination of heating and cooling (see Chapter 11). For example:

- Casting is filling a mould with molten metal, which solidifies and cools, giving complex temperature histories due to heat conduction from the casting through the mould, with the release of latent heat.
- Hot forming shapes metal billets, again giving complex temperature histories due to conduction into the tooling and surrounding air.
- Welding often causes a thermal cycle of heating followed by cooling.
- A final stage of manufacture is often a separate heat treatment, of the whole component or its surface.

When the temperature varies in a process, the equilibrium condition of the material keeps changing (e.g. as boundaries on the phase diagram are crossed) – hence *phase transformations* take place. These transformations determine which phases are present after processing and how they are distributed among one another (i.e. the final *microstructure*). This in turn controls the material properties. So, for a complete understanding of properties and processing, we need to know more about the microstructure than is given by a phase diagram. As we will see, controlling properties relies on managing not just which phases are present but also their *morphology* (i.e. size and shape), particularly for structure-sensitive properties such as strength.

Observing phases and phase transformations   Phase diagrams tell us the equilibrium phases, their compositions, and proportions. But how is this information produced, and how do we measure whether a real process is actually following equilibrium? Many techniques are available to quantify the phases present in a microstructure, with the appropriate method to use depending on what samples are available and the different length scales of the microstructural features (seen in overview in Chapter 3).

Phase diagrams have been produced experimentally using many techniques. Traditional methods use some form of 'remote sensing' – measuring a physical characteristic that depends in a quantifiable way on the phases present, as temperature varies. Examples are *precision dilatometry* (length or density measurement), *electrical resistivity* (exploiting changes in electron mean free path), and *calorimetry* (measuring latent heat release and absorption). Direct imaging has evolved from *optical microscopy*, through *X-ray diffraction* to *scanning and transmission electron microscopy*, to *atom force microscopy* (now capable of atom-scale resolution of the spatial distribution of elements within a phase). Increasingly, phase data are determined by thermodynamic computation, from first principles. Many software packages and databases are available to compute sub-domains in alloy phase diagrams containing many different elements. They are nonetheless dependent on good experimental data for calibration of the computations.

Key concepts in phase transformations   Phase diagrams give important information needed to predict the phase transformations and final microstructure that result from a given thermal history. The real microstructure may not be at equilibrium, but phase diagrams give a starting point from which other (non-equilibrium) microstructures can often be inferred.

The fundamental thermodynamics and kinetics of phase transformations are described in Chapter 11. The key concepts in phase transformations are as follows:

- Phase transformations are driven by the resulting change in *free energy* (defined in GL4.1), also known as the *driving force*.
- At a phase boundary, the free energies of the states on either side of the boundary are equal – the driving force $\Delta G = 0$.
- Phase changes almost always involve diffusion as the kinetic mechanism by which atomic rearrangement occurs.
- Phase transformations occur via a two-stage process of *nucleation* and *growth*, in which nucleation may be spontaneous (homogeneous), or take place on some kind of interface (heterogeneous).
- *TTT diagrams* capture the extent of an isothermal transformation as a function of hold temperature and time, giving characteristic C-curves for diffusion-controlled phase transformations.
- In continuous cooling, there is a *critical cooling rate* that will just avoid the onset of the diffusional transformations.

To relate these concepts to phase diagrams and cooling in real industrial processes, we begin with examples of slow cooling, in which it is reasonable to assume that the phases present evolve to follow equilibrium: GL4.7 considers solidification, and GL4.7 covers solid-state phase changes. Finally, GL4.8 looks at examples of heat treatments in which we deliberately cool quickly to bypass equilibrium, manipulating the phases formed and their morphology.

GL4.6 Equilibrium solidification

Latent heat release on solidification Figure GL4.12(a) shows the phases found in pure iron (i.e. the phase diagram becomes a single temperature axis). If we cool iron slowly from above its boiling point, the temperature as a function of time shows two significant shelves in the cooling curve, called *arrest points*, at the boiling and melting points (at atmospheric pressure, iron boils at 2860°C and melts at 1536°C) – see Figure GL4.12(b). At each temperature there is a *phase change*: vapour-to-liquid at the boiling point and liquid-to-solid at the melting point. The arrest points on cooling are due to the release of *latent heat* (of vapourisation and melting, respectively). On heating, the reverse occurs: an arrest in the temperature rise, while heat is absorbed to melt or boil the material.

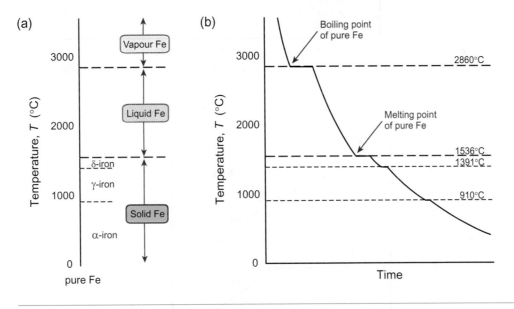

Figure GL4.12 (a) One-dimensional phase diagram for pure iron; (b) corresponding cooling curve for condensing then solidifying iron from vapour to liquid to solid.

Pure iron solidifies initially to BCC δ-iron but undergoes further solid-state phase transformations on cooling, first to FCC γ-iron at 1391°C and then back to BCC α-iron at 914°C. These transformations *also* release latent heat, but the amount is much smaller, as indicated by the modest arrests in the cooling history in Figure GL4.12(b). The mechanisms of these solid-state changes are considered later.

Alloys frequently solidify over a *range* of temperature (between the liquidus and solidus lines). In this case the latent heat is released progressively, as the equilibrium proportion of solid rises with falling temperature. The cooling curve therefore does not show a shelf at constant temperature, but the cooling rate is reduced by the progressive release of latent heat. This is illustrated in Figure GL4.13, for pure Cu and an alloy composition in the Cu-Ni system.

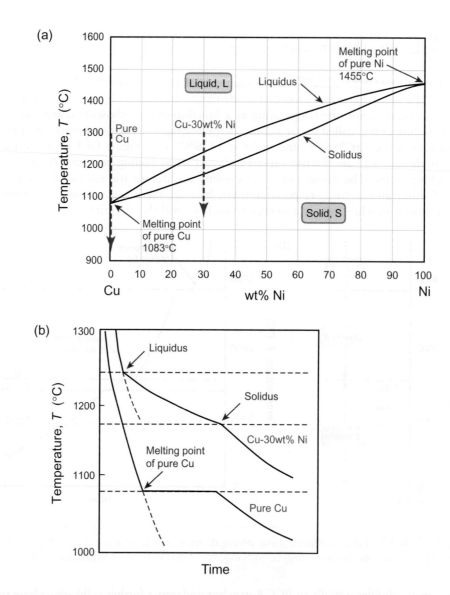

Figure GL4.13   (a) Cu-Ni phase diagram, and (b) corresponding cooling curves, for solidifying pure Cu and a Cu-30 wt% Ni alloy.

Solidification of pure metals   The mechanism of solidification is illustrated in Figure GL4.14. For *homogeneous nucleation*, embryonic solid colonies form spontaneously within the melt [Figure GL4.14(a)]. They grow stably provided they reach a critical radius. This initial barrier reflects the surface energy of the solid-liquid interface, which uses up some of the free energy released by the transformation – the probability of forming stable nuclei

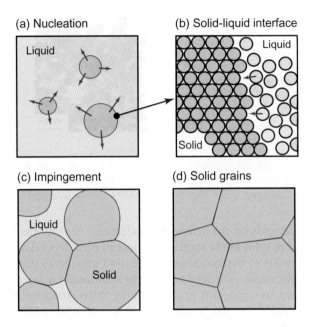

Figure GL4.14 Solidification mechanism: (a) homogeneous nucleation, (b) magnified view of atomic transfer at the solid-liquid interface, (c) growth of nuclei and the onset of impingement, (d) final solid grain structure.

increases rapidly as the liquid is undercooled below the transformation temperature. *Heterogeneous nucleation* facilitates the process, with solid nuclei forming more readily on a pre-existing solid in contact with the liquid (e.g. the walls of the mould, or high melting point particles of another solid mixed into the liquid).

Solid nuclei grow by atoms transferring at the solid-liquid interface [Figure GL4.14(b)] – the interface advancing in the opposite direction to the atomic transfer. Growth continues until *impingement* of the solid regions occurs [Figure GL4.14(c)]. Since each nucleus has its own independent crystal orientation, there is a misfit in atomic packing where they impinge. The individual crystallites remain identifiable once solidification is complete [Figure GL4.14(d)]. We call these *grains*, and the surfaces where they meet are *grain boundaries*. Grains are typically on the length-scale of 1 μm to 1 mm, and they are easily revealed by optical microscopy on polished and etched surfaces. Figure GL4.15 shows optical micrographs of pure metals after solidification, showing equiaxed grain structures. Note that some techniques generate colour contrast between grains by changing the polarisation of the reflected light [see Figure GL4.15(b)]. This emphasises the need for caution in interpreting micrographs – the image of pure Al in the figure shows a single phase – multiple colours do not necessarily indicate different phases.

So grain structure is our first example of a microstructure formed by a phase transformation. Let's emphasise a key idea that will recur from here onward. The phase diagram tells us that the liquid should transform to a single-phase solid – and that's all. It *doesn't* tell us

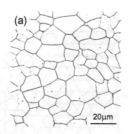

Figure GL4.15 Optical micrographs of (a) pure Fe and (b) pure Al. In (b) the etching technique produces colour contrast between different crystal orientations – but it is still all one phase. (Images courtesy: ASM Micrograph Center, ASM International, 2005.)

the grain size; this depends on the density of nucleation sites and the kinetics of the diffusive mechanism of forming solid at the interface. These in turn depend on how slowly the liquid cools and the availability of heterogeneous nucleation sites. So to describe the microstructure for *any* phase transformation, we need to bring in additional knowledge. This highlights why processing details are so important in determining microstructure and properties, and the role of processing 'tricks' to modify what happens (e.g. adding lots of fine particles to the melt to promote copious heterogeneous nucleation – the result is a much finer grain size). Process-microstructure interactions such as these are discussed in Chapter 11.

Solidification of simple binary alloys and phase reactions Alloys show a freezing range, between the liquidus and solidus lines. Consider solidification of Cu-30 wt% Ni on the phase diagram of Figure GL4.13. Solidification starts at 1240°C – nuclei form as in Figure GL4.14(c). In contrast to a pure metal, further cooling is needed for more solid to form (e.g. at 1200°C a tie-line indicates 60% solid and 40% liquid are stable), and the microstructure would appear as in Figure GL4.14(c). To reach 100% solid [Figure GL4.14(d)] we need to keep cooling to 1170°C, and the end result looks much the same as in the pure case (with the difference that the phase forming these grains is a Cu-rich solid solution, not something we can detect optically). Note that the compositions of the solid and liquid both evolve, following the ends of the tie-line.

So when an alloy is cooled, the constitution point for the alloy drops vertically on the phase diagram. The phases present change when we cross boundaries on the diagram. But the Cu-Ni example earlier showed that, if more than one phase is present, the phases may also evolve in composition on heating or cooling, *without* a change in the phases present. We call this a *phase reaction*.

> **Definition:** When any phase compositions change with temperature a *phase reaction* is taking place.

Within a single-phase field, the composition of the phase is always that of the alloy, and no phase reaction can take place on cooling. In a two-phase region, the compositions of the two phases are given by the ends of the tie-line through the constitution point. In general, as the constitution point falls vertically, the ends of the tie-line do *not* – instead they run along oblique phase boundaries. The compositions of the two phases change with temperature, and phase reactions occur. The exception would be where *both* boundaries at the end of the tie-line are vertical. This is not something associated with liquidus and solidus boundaries but is seen in the solid state, particularly with compounds [recall the unusual Ag-Sr phase diagram in Figure GL4.8(a)].

Solidification of dilute alloys   The Cu-Ni system is unusual in that the solid that forms initially is unchanged on further cooling – no transformations or reactions take place. But most binary systems show eutectics, with two-phase fields over a wide range of composition at room temperature. First we will consider solidification of a relatively dilute alloy in the Pb-Sn system, Pb-10 wt% Sn (Figure GL4.16).

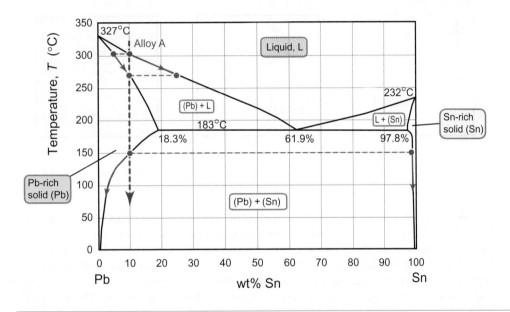

Figure GL4.16   The Pb-Sn phase diagram: solidification of a dilute alloy A.

The transformations and reactions are as follows, leading to the microstructure evolution shown in Figure GL4.17.

1. *Above 305°C*: Single-phase liquid of composition identical to that of the alloy; no phase reaction.
2. *From 305°C to 270°C*: The liquidus line is reached at 305°C; the reaction liquid→Pb-rich solid solution starts. The solid contains less tin than the liquid (see first tie-line), so the liquid becomes richer in tin and the composition of the liquid moves down the

liquidus line, as shown by the arrow. The composition of the solid in equilibrium with this liquid changes, too, becoming richer in tin also, as shown by the arrow on the solidus line: a *phase reaction* is taking place. The *proportion* of liquid changes from 100% (first tie-line) to 0% (second tie-line).

3. *From 270°C to 150°C*: Single-phase solid of composition identical to that of the alloy; no phase reaction.

4. *From 150°C to room temperature*: The Pb-rich phase becomes unstable when the phase boundary at 150°C is crossed. It breaks down into *two solid phases*, with compositions given by the ends of the third tie-line, and proportions given by the lever rule. On cooling, the compositions of the two solid phases change as shown by the arrows: each dissolves less of the other element, and a phase reaction takes place.

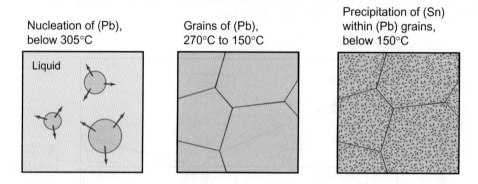

Figure GL4.17   Schematic microstructure evolution in solidification of a dilute Pb-10 wt% Sn alloy.

The formation of the Sn-rich phase within the Pb-rich grains takes place by *precipitation* from the solid solution. This mechanism, too, involves nucleation and growth, but on a much finer scale than the grain structure. Small clusters of Sn-rich solid nucleate spontaneously within the Pb-rich matrix. The fraction of this phase increases as the nuclei grow by depleting the surrounding matrix of some of its Sn. The compositions of the phases (particles and matrix) continually adjust by inter-diffusion of Pb and Sn atoms. The practicalities of solid-state precipitation are discussed further in Part 7.

Note that the phase diagram tells us the proportions of the phases, in wt%. The density of the phases is not the same, so this does not convert directly into volume % (or area fraction in a metallographically prepared cross-section). Nonetheless, provided the phase densities are not too dissimilar, the phase proportions give some idea of the proportions we expect to see in the microstructure (good enough for sketching, as in Figure GL4.17).

Exercise

GL4.13   Find the proportions and compositions of the phases formed on solidification of the Pb-10 wt% Sn alloy, at 250°C, and at room temperature (20°C).

Eutectic solidification   Next consider the solidification of the eutectic composition itself in the Pb-Sn system (Figure GL4.18, alloy B). The eutectic composition is Pb-61.9 wt% Sn. When liquid of this composition reaches the eutectic temperature (183°C), the liquid can transform to 100% solid without further cooling. This is a unique characteristic of eutectic alloys, and in this respect they resemble pure components. But the final microstructure will be very different because two solid phases form simultaneously (with proportions and compositions governed by the tie-line in the two-phase field immediately below the eutectic point). This transformation is called the *eutectic reaction*.

> **Definition:** A *eutectic reaction* is a three-phase reaction by which, on cooling, a liquid transforms into two solid phases at constant temperature: Liquid, $L \rightarrow$ Solid $\alpha +$ Solid β.

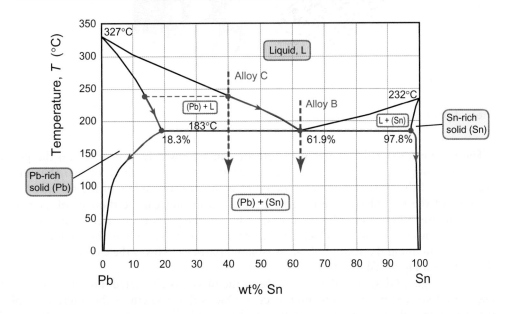

Figure GL4.18   The Pb-Sn phase diagram: solidification of eutectic alloy (B) and an off-eutectic alloy (C).

Exercise

GL4.14   Find the proportions and compositions of the solid phases formed on solidification at the eutectic point in the Pb-Sn system.

So how does this transformation take place? The phase diagram provides clues: the Pb and Sn in the liquid are uniformly mixed, but after transformation we have Pb-rich solid and Sn-rich solid. Pb and Sn must therefore inter-diffuse to generate regions in which each dominates. This must happen at an interface between liquid and two-phase solid. It will therefore be easier if the diffusion distance is kept small; that is, the two phases separate out on a small scale, such that no single atom has to diffuse too far to be able to join a growing colony of (Pb) or (Sn). Eutectics usually therefore form as intimate mixtures of the two phases, on a length-scale much smaller than a typical grain size. Figure GL4.19(a) shows a micrograph of eutectic Al-Si (the phase diagram for this system was shown in Exercise GL4.10).

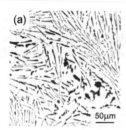 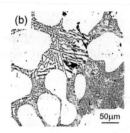

Figure GL4.19 Optical micrographs of (a) eutectic Al-Si alloy; (b) off-eutectic Al-Si alloy, showing grains of primary (Al) *(white)*. (Images courtesy: ASM Micrograph Center, ASM International, 2005.)

The proportions of the phases in a eutectic can vary widely, depending on the position of the 'V' along the eutectic tie-line. But in broad terms we can think of two characteristic dispersions of the phases. If the eutectic point is toward the middle of the tie-line, the proportions of the two phases are roughly equal, and neither can be thought of as a matrix containing the other phase. This is the case in Pb-Sn (the proportions were found in Exercise GL4.13). However, if the 'V' is located toward one end of the tie-line, then one phase forms a matrix containing the second phase as isolated particles. This is the case in Al-Si [Exercise GL4.10a and Figure GL4.19(a)].

Note in Figure GL4.19(a) that the *shape* of the phases is elongated into plates or needles. There are two reasons why this often occurs. One is that growth of the interface is planar at the atomic scale, with plates extending in the growth direction to minimise diffusion distances. The other is that the boundary between the phases is also a crystallographic boundary, and some pairs of orientations have a lower surface energy that will grow more rapidly. Forming a fine mixture of phases does mean that there is a price to be paid in surface energy between phases – the area of interface per unit volume is large. But this is a small energy penalty compared to the free energy release in the transformation as a whole. Once again we find that the detail of the transformation mechanism has an important influence on the microstructure (and hence properties).

A further phase reaction occurs on cooling of the two-phase eutectic solid (e.g. for Pb-Sn from 183°C to room temperature). As the phase boundaries on the phase diagram are not vertical, the compositions and proportions of the two solid phases evolve on cooling. Diffusion distances are automatically small within the eutectic, so both phases can become purer and adjust their proportions, by inter-diffusion.

Solidification of off-eutectic compositions Casting alloys are often off-eutectic, with a composition to one or other side of the 'V' on the phase diagram. The solidified microstructure in these cases can now be inferred: above the eutectic, partial solidification of a single-phase solid solution occurs (as seen in Cu-Ni earlier); then at the eutectic temperature the *remaining* liquid undergoes the eutectic reaction (as seen in Pb-Sn earlier).

Consider solidification of Pb-40 wt% Sn (see Figure GL4.18, alloy C). The transformations and reactions are as follows, leading to the microstructure evolution illustrated in Figure GL4.20.

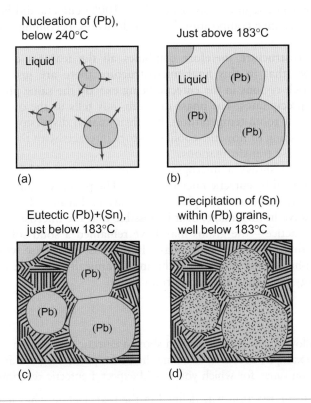

Figure GL4.20 Schematic microstructure evolution in solidification of a Pb-40 wt% Sn alloy.

1. *Above 240°C:* Single-phase liquid; no phase reactions.
2. *From 240°C to 183°C:* The liquidus is reached at 240°C, and nuclei of Pb-rich solid solution appear first [Figure GL4.20(a)]. The composition of the liquid moves along the liquidus line, that of the solid along the solidus line. This regime ends when the temperature reaches 183°C. Note that the alloy composition (40 wt% Sn) is roughly halfway between that of the new solid (18.3 wt% Sn) and the residual liquid (61.9 wt% Sn); so the alloy is about half liquid, half solid, by weight (and very roughly by volume, neglecting the difference in phase densities) [Figure GL4.20(b)].

3. *At 183°C*: The remaining liquid has reached the *eutectic point*, and this liquid undergoes solidification exactly as described before [Figure GL4.20(c)]. Note that the proportions of primary (Pb) and eutectic microstructure are the same as the solid-liquid proportions just above the eutectic temperature. Nucleation of (Sn) alongside more (Pb) will be straightforward because there are already solid grains of (Pb) present on which the eutectic can form.

4. *From 183°C to room temperature*: The two types of microstructure each evolve, independently:
   (a) The Pb-rich solid becomes unstable, and Sn-rich solid precipitates (exactly as before, in alloy A).
   (b) The eutectic region evolves exactly as 100% eutectic did before, in alloy B: both phases change composition by inter-diffusion, becoming purer.

The final microstructure therefore combines all the features discussed so far [Figure GL4.20(d)]. Note that in this final microstructure there are just *two* phases, with the Pb-rich and Sn-rich regions in the eutectic being exactly the same phases in the (Pb) grains containing (Sn) precipitates. The phase diagram only tells us the overall phase proportions (from the tie-line at room temperature). But their spatial dispersion (and hence properties) involves understanding of how the cooling history interacts with the phase diagram and the mechanisms of phase transformations.

Figure GL4.19(b) shows a micrograph of an off-eutectic Al-Si alloy. The primary Al grains are surrounded by eutectic microstructure. The primary grains have a uniform colour, appearing therefore to be a single phase. This may be because the scale of the precipitates is too fine to resolve in this image, or it is possible that the microstructure has not followed equilibrium in practice – the grains may have remained a solid solution due to some difficulty in nucleating the second solid phase. In this case the phase will be *supersaturated* in solute, and the phase is *metastable* (i.e. thermodynamically stable at room temperature at a higher free energy level than the equilibrium state).

Exercises

GL4.15   Not all alloys in the lead-tin system show a eutectic reaction on slow cooling: pure lead, for example, does not. Examine the Pb-Sn phase diagram and identify the composition range for which you would expect a eutectic reaction to be possible.

GL4.16   (a) A eutectic reaction was defined in the text. Define what happens on heating a solid of eutectic composition. Over what temperature range does this occur?
          (b) For a Pb-35 wt% Sn alloy, identify the temperatures at which melting starts and finishes on heating, and describe how the proportion of liquid evolves during melting.

GL4.17   The figure in Exercise GL4.10(a) shows the Al-Si system (the basis of aluminium casting alloys). Use this figure to answer the following:
          (a) Describe the solidification of an Al-20 wt% Si alloy, sketching the microstructure at key temperatures (e.g. onset of solidification, just above and just below the eutectic). Estimate the phase proportions in two-phase regions, and sketch the microstructures accordingly.
          (b) How would the final microstructure differ for Al-Si alloys of: (i) Al-1 wt% Si; (ii) eutectic composition?

Segregation   It was noted that when solidification starts, either on the walls of a mould or spontaneously throughout the melt, the first solid to form is purer in composition than the alloy itself (due to *partition* of the liquidus and solidus lines). For example, in Pb-10 wt% Sn (considered earlier), the first solid appearing at 305°C has a composition of 5 wt% Sn. This means that tin is *rejected* at the surface of the growing crystals, and the liquid grows richer in tin – its composition moves along the liquidus line. From 305° to 270°C the amount of primary (Pb) increases, and its equilibrium composition increases, following the solidus line. This means that lead must diffuse *out* of the solid (Pb), and tin must diffuse *in*. This diffusion takes time. If cooling is very slow, time is available and equilibrium may be maintained. But practical cooling rates are not usually slow enough to allow sufficient time for diffusion, so there is a *composition gradient* in each (Pb) grain, from the middle to the outside. This gradient is called *segregation*, and is found in almost all alloys (Figure GL4.21).

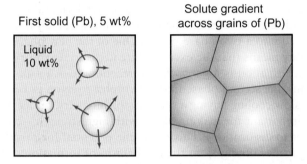

Figure GL4.21   Segregation in solidification of binary alloys.

Segregation can to some extent be 'smoothed out' after casting by *homogenisation*: holding a casting at temperature to allow some redistribution of solute by diffusion. One consequence of segregation is that the last liquid to solidify will be richer in solute than expected from the phase diagram (as the solid is purer than it should be). This may lead to the formation of eutectics at the grain boundaries (the last part to solidify), even in alloys that are not expected to show a eutectic (see Exercise GL4.15). The temperature at which melting would start on heating is thus lower than expected from the phase diagram (i.e. the eutectic temperature, not the solidus temperature).

Segregation is even more important with respect to *impurities*. No alloy is truly binary, ternary, or whatever, but will contain traces of other elements (e.g. from the ore used as the source of the metal). Impurities also dissolve more readily in the liquid than in the solid. Hence when alloys solidify, the impurities concentrate in the last bit to solidify. Even if the alloy contains only a fraction of 1% of a given impurity overall, if the solid will not dissolve it in solution, then the composition can locally be very high on the grain boundaries. This is damaging if the impurity forms brittle phases, which can be disastrous from the point of view of toughness. An example is sulphur in steel, which forms brittle iron sulphide. Rather than trying to remove the traces of impurity from the melt, the way to solve the problem is further alloying. Plain carbon steels all contain manganese, which forms MnS at higher temperatures during solidification and therefore ties up the sulphur before segregation takes place, rendering it harmless.

Another example of detrimental impurity segregation is in *fusion welding*. This is like casting on a small scale, with a melt pool under a moving heat source solidifying as it goes. Solidification finishes on the weld centre-line, and segregation concentrates impurities there. Contraction of the joint can generate high enough thermal stresses to crack the weld along the weakened grain boundaries down the weld centre: this is known as *solidification cracking*.

GL4.7 Equilibrium solid-state phase changes

Precipitation reactions   Solid-state phase transformations were introduced in GL4.6, following solidification of dilute binary alloys (Figures GL4.16 and GL4.17). The solid formed first from the liquid as grains of single-phase solid solution, but on further cooling the solvus boundary is crossed into a two-phase region. One of the new phases is present already, but the falling solubility means that the excess solute needs to be absorbed by the second phase. This type of transformation is a *precipitation* reaction: $\alpha \rightarrow \alpha + \beta$.

In many systems, the second phase in this two-phase region is a *compound*, with precipitates forming on slow cooling having a specific stoichiometry, rather than being another solid solution. This is important for making alloys with high strength – compounds are generally hard and resistant to dislocations. This offers the potential for *precipitation hardening* (see Chapter 5) – the name coming from the mechanism of formation of the second phase.

Refer to Figure GL4.8(b), the Al-rich end of the Al-Cu system. Exercise GL4.8 considered an Al-4 wt% Cu alloy at 550°C and 250°C. On cooling between these temperatures, at 490°C the solvus boundary is crossed and the θ-phase ($CuAl_2$) starts to precipitate, nucleating homogeneously within all the grains and increasing in proportion as the temperature falls. Figure GL4.22 shows the final microstructure.

Precipitates of
$CuAl_2$ in (Al)

Figure GL4.22   Schematic microstructure of a slow-cooled Al-4 wt% Cu alloy.

At modest magnification, the two-phase region may appear a uniform colour, and higher resolution is needed to distinguish the precipitates from the matrix. The scale of the microstructure is again dictated by the kinetic mechanism. By forming very large numbers of small precipitates, the average diffusion distance for solute is short, enabling the transformation to occur in the solid state. The precipitate spacing is therefore much finer than in a eutectic. And in common with a eutectic microstructure, there is an energy penalty in forming a large area per unit volume of interface between the phases, but the free energy release dominates over the surface energy.

From the point of view of precipitation hardening, it is the *spacing* of precipitates that matters. In practice the precipitates formed by slow cooling are still too far apart to be effective in obstructing dislocations. Useful strengthening comes from forming a much finer dispersion of precipitates – the means to achieve this in Al-Cu and other alloys is discussed in GL4.8.

The precipitation illustrated in Figure GL4.22 is that expected on cooling following equilibrium thermodynamics of the system, assuming no significant kinetic barrier to nucleation of precipitates. As noted earlier, it may be difficult to form a new lattice within an existing solid crystal – just having a driving force may not be sufficient to cause the phase change, leaving a metastable supersaturated solid solution on cooling. Whether or not a given system will precipitate is an alloy-specific detail, not indicated by the phase diagram.

Figure GL4.22 illustrates another feature that may be observed in precipitation reactions, on the grain boundaries. Here the precipitates are larger than within the bulk of the grains. Two factors contribute to this effect, both due to the extra 'space' associated with the boundary: first, faster nucleation can occur, as atoms can rearrange more readily into the precipitate lattice (this is a form of heterogeneous nucleation); second, the boundaries provide faster diffusion paths for solute to be drawn in to build up the precipitates. This effect doesn't always occur – again it depends on the alloy system – but it can have important practical consequences. Coarser precipitation on the grain boundaries may weaken the boundaries, with possible loss of toughness (intergranular fracture) or susceptibility to corrosive attack, forming grain boundary cracks.

Phase transformations in carbon steels   The iron-carbon system was introduced in GL4.3, including the various single-phase forms of iron-carbon solid solutions, the compound iron carbide, and the eutectoid point. Here we investigate the important phase transformations that occur in *carbon steels*; that is, Fe-C alloys with compositions between pure iron and the eutectoid composition. Figure GL4.23 shows the relevant section of the Fe-C phase diagram.

First consider cooling of pure iron, starting with FCC γ-iron at 1000°C. At 910°C, a solid-state change to BCC α-iron occurs (at constant temperature). *Nucleation* starts at the grain boundaries, where there is more space for BCC nuclei to form. The α grains grow by atoms jumping across the boundary between the two phases, adopting the crystal structure of the growing phase as they do so [shown schematically in Figure GL4.24(b)]. The interface migrates in the opposite direction to the atomic jumps. Figure GL4.24(a) and (c) show the solid-solid transformation under way, and complete at 910°C. When the growing α colonies impinge, new α-α grain boundaries are formed. Since there is typically more than one α nucleus per γ grain, the average grain size is reduced by this transformation.

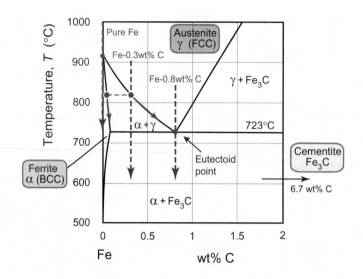

Figure GL4.23 The Fe-C phase diagram for compositions of carbon steels below 1000°C, including the eutectoid point.

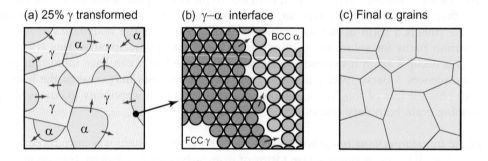

Figure GL4.24 Schematic illustration of the transformation from FCC γ-iron to BCC α-iron in pure Fe: (a) γ-α grain structure after 20% transformation; (b) atomic transfer mechanism at the phase interface; (c) final α-iron grain structure.

The eutectoid reaction The eutectoid point in the Fe-C system was introduced in GL4.3. Cooling austenite of eutectoid composition (0.8 wt% C) leads to complete transformation to ferrite and cementite, at the eutectoid temperature, 723°C (see Figure GL4.23). More generally we therefore define a eutectoid reaction as follows:

Definition: A *eutectoid reaction* is a three-phase reaction by which, on cooling, a single-phase solid transforms into two different solid phases at constant temperature: solid γ→ solid α + solid β.

The mechanism of the eutectoid reaction must transform a single solid phase into two others, both with compositions that differ from the original (given by the ends of the tie-line through the eutectoid point). In the Fe-C eutectoid reaction, austenite containing 0.8 wt% C changes into ferrite (containing almost no carbon) and cementite, with its own lattice structure (Fe_3C, containing 6.7 wt% C, or 25 at% carbon). Hence carbon atoms must diffuse to form regions of high and low composition, at the same time as the FCC austenite lattice transforms into BCC ferrite and cementite lattices. Figure GL4.25 illustrates the eutectoid transformation in carbon steels.

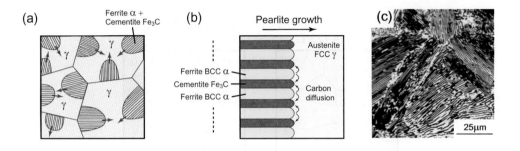

Figure GL4.25 (a) Nucleation of pearlite on austenite grain boundaries; (b) schematic view of the austenite-pearlite interface; (c) micrograph of pearlite. (Image courtesy: ASM Micrograph Center, ASM International, 2005.)

Nuclei of small plates of ferrite and cementite form at the grain boundaries of the austenite, and carbon diffusion takes place on a very local scale just ahead of the interface – the plates of ferrite and cementite grow in tandem, consuming the austenite as they go [Figure GL4.25(a) and (b)]. The resulting grains consist of alternate plates of ferrite and cementite – this microstructure has a special name: *pearlite* [Figure GL4.25(c)]. Note that pearlite is a two-phase microstructure, *not* a phase, in spite of the similarity in nomenclature. It, too, has a very large area of phase boundary, between ferrite and cementite, with associated surface energy penalty. However, kinetics once again dictate the mechanism of transformation, with C atoms on average diffusing only one plate spacing, which enables the interface between the new phases and the austenite to traverse whole grains. The length-scale of the plates in pearlite is again much finer than in a eutectic due to the solid-state diffusion. At lower magnification, pearlite may just appear as dark etching grains. This is easily mistaken for a single phase – it is not, but this is another example of a dispersion of two phases at a scale below the resolution of the microscopy technique.

Phase transformations in hypo-eutectoid steels Some commercial steels have a eutectoid composition – steel for railway track and piano strings are examples of 'pearlitic steel.' Most carbon steels are 'hypo-eutectoid,' containing less than 0.8 wt%. Mild steels contain 0.1 to 0.2 wt% C, medium carbon steels around 0.4 wt%. Here we look at how their equilibrium microstructures relate to the phase diagram.

Consider slow cooling of a medium carbon steel containing 0.3 wt% C (as indicated in Figure GL4.23). Starting with austenite at 900°C, we have a solid solution of C in FCC austenite. At 820°C, we enter the two-phase region: ferrite plus austenite. The formation of ferrite follows the same mechanism as in pure iron, nucleating on the austenite grain boundaries. The ferrite rejects carbon into the remaining austenite, the composition of which increases accordingly. Grains of ferrite grow until, just above 723°C, the proportion of ferrite to austenite is roughly 2:1 [Figure GL4.26(a)]. The remaining austenite contains 0.8 wt% C – it is at the eutectoid point. This austenite then decomposes as before into pearlite, the two-phase mixture of ferrite and cementite [Figure GL4.26(b)]. Note that, as previously, the final structure is still only two phase, but different grains have very different microstructures. The ferrite within the pearlite structure is the same stuff as the ferrite forming whole grains – it is all one phase.

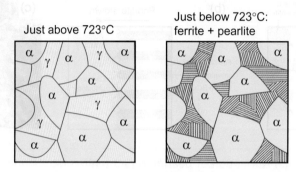

Figure GL4.26 Schematic illustration of microstructure evolution in a hypo-eutectoid steel, from ferrite + austenite to ferrite + pearlite.

Iron carbide is a hard phase, and the pearlite structure is effective in obstructing dislocation motion (due to the plate-like structure). But ferrite itself has a relatively high intrinsic strength, with very high toughness. Carbon steels that have been slow-cooled (or *normalised*) to a ferrite-pearlite microstructure provide an excellent combination of strength and toughness, widely exploited for structural and mechanical applications. As in pure iron the transformation from austenite leads to a reduction in grain size, as each old grain nucleates more than one new grain. This is used commercially for *grain refinement*, an important heat treatment as it simultaneously enhances yield strength *and* toughness.

Ferrite-pearlite microstructures are strong and tough, but we can do even better by *quenching* and *tempering* carbon steels. This heat treatment leads to the same phases, but in a different morphology, enhancing the strength without loss of toughness. We return to this in GL4.8.

Exercises

GL4.18   An Al-10 wt% Cu alloy was slow-cooled and the length-scales measured by various techniques in the final microstructures, as follows: primary grain size = 100 μm; phase spacing in the eutectic = 1 μm; precipitates within the primary grains 0.05 μm. Identify the phases involved in each of these microstructural features, and explain the differences in length-scale.

GL4.19   Pure iron cooled slowly contains 100% ferrite; the eutectoid composition contains 100% pearlite. Estimate the carbon content of the normalised hypo-eutectoid steel shown in the figure. Sketch the structure of a 0.2 wt% carbon steel after slow cooling to room temperature.

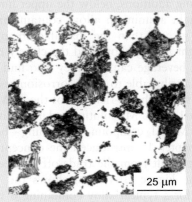

25 μm

(Image courtesy: ASM Micrograph Center, ASM International, 2005.)

GL4.20   *Hyper-eutectoid* steels contain more than 0.8 wt% C. A high carbon steel containing 1.0 wt% C is cooled slowly from 1000°C in the austenite field. Refer to the Fe-C phase diagram in Figure GL4.23 to answer the following:
   (a) At what temperature does a phase transformation begin, and what new phase then appears? Why do you think this phase tends to nucleate and grow along the austenite grain boundaries?
   (b) What happens to the compositions and proportions of the phases on cooling to just above 723°C?
   (c) What phase transformation takes place at 723°C, and does this affect the whole microstructure?
   (d) Sketch the expected final microstructure. What are the final phases at room temperature, their compositions, and proportions?

GL4.8: Non-equilibrium solid-state phase changes

The transformations considered thus far assume that the alloy is able to remain at equilibrium during cooling; that is, cooling is slow, and diffusion remains sufficiently rapid for phase fractions and compositions to adjust. In a couple of instances, though, it has been noted that this may not be the case – the kinetic mechanisms are unable to operate at practical cooling rates. For example, we noted in casting that solute gradients build up across grains (segregation); and precipitation reactions may not occur due to barriers to nucleation of the second phase, leaving a metastable supersaturated solid solution.

In this section, a number of important heat treatments are explored involving non-equilibrium responses. The common starting point in each case is to *solutionise* the alloy, dissolving the alloying additions in a high-temperature solid single phase. Then by deliberately applying cooling that avoids equilibrium, different microstructural outcomes with enhanced properties can be achieved. Further details of the resulting properties are discussed in Chapter 11.

Non-heat-treatable aluminium alloys   The *wrought* aluminium alloys are those that are shaped by deformation processing (as opposed to *casting* alloys, cast directly to shape). Both are sub-divided into non-heat-treatable and heat-treatable, as only some respond to heat treatments to modify their properties (particularly strength). Wrought Al alloys are all relatively dilute in alloying additions, below the upper limit of solid solubility in aluminium.

Virtually all wrought aluminium alloys contain magnesium – this is the single biggest application of Mg, greater than the manufacture of products in Mg alloys. The non-heat-treatable wrought Al-Mg alloys and the closely related Al-Mg-Mn alloys are familiar from the aluminium beverage can, but they also find widespread structural use, including shipbuilding (partly attributable to good corrosion resistance).

Figure GL4.27(a) shows the Al-rich end of the Al-Mg phase diagram. It looks as though any alloy up to about 14 wt% Mg solutionised in the single-phase α-field should form precipitates of the complex Mg_5Al_8 compound on cooling – but in this case kinetics dominates over thermodynamics. Chapter 11 introduced the concept of *time-temperature-transformation (TTT) diagrams* that result from the trade-off between thermodynamic driving force and the rate of diffusion. Figure GL4.27(b) shows the TTT diagram for formation of the Mg_5Al_8 compound, for a typical alloy containing 6 wt% Mg solutionised at 400°C, and then quenched and held isothermally. There is a significant barrier to the nucleation of Mg_5Al_8 – long times are needed to start this phase transformation. So the critical cooling rate is sufficiently low that a metastable Al-rich solid solution is easily retained at room temperature with only a moderate quench. It is in this state that the alloys are used commercially, exploiting solid solution hardening by the Mg, with cold deformation often used to provide additional work hardening.

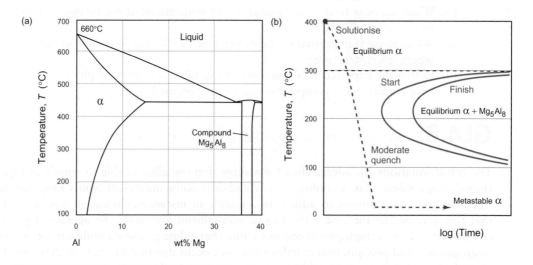

Figure GL4.27   (a) Al-rich end of the Al-Mg phase diagram; (b) schematic TTT diagram for precipitation of equilibrium Mg_5Al_8 in an Al-6 wt% Mg alloy.

Heat-treatable aluminium alloys   Commercial heat-treatable wrought aluminium alloys combine additions of magnesium with one or more of Si, Zn, Cu, and other metallic elements. But the key behaviour exploited in heat treatment is found in some binary systems, such as Al-Cu. Comparing Figure GL4.27(a) for Al-Mg alloys with Figure GL4.8(b) for Al-Cu alloys, the diagrams appear superficially the same for dilute alloys. But in Al-Cu alloys, cooling at moderate rates does follow equilibrium – the compound $CuAl_2$ readily nucleates as precipitates in a matrix of (Al) (recall Figure GL4.22), so in this system the phase diagram is followed.

However, it was noted earlier that the $CuAl_2$ precipitates formed on slow cooling are rather far apart from the perspective of interacting with dislocations, giving little strengthening. What we want is a much finer dispersion with many more nuclei forming and diffusion distances that are much shorter. This can be achieved by a two-step heat treatment: first by *quenching* to room temperature, at a cooling rate above the critical cooling rate for precipitation to occur, then by *ageing* at an intermediate temperature. The quench 'traps' the high-temperature state as a metastable supersaturated solid solution at room temperature, from which we can then precipitate on a fine scale during ageing.

As so often in metallurgy, this precipitation during ageing is more complicated than expected. The quench achieves the desired result – a supersaturated solid solution of Cu in Al. But ageing at temperatures between room temperature and about 200°C does *not* cause equilibrium $CuAl_2$ precipitates to form. Instead, metastable second phases form, and the TTT diagram for Al-Cu shows further C-curves for these phases at low temperature, as shown in Figure GL4.28(a).

This subtle difference in behaviour cannot be anticipated from the phase diagram at all, but it is fundamental to achieving the high-strength precipitation hardened alloys on which aerospace depends, as increasingly do other transportation systems (rail and road). During the ageing treatment of Al-Cu after quenching, the yield strength evolves as shown in Figure GL4.28(b).

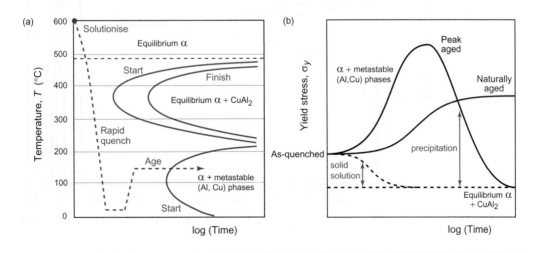

Figure GL4.28   (a) TTT diagram for precipitation of equilibrium and metastable phases in dilute Al-Cu alloys; (b) the 'ageing curve,' showing how yield strength evolves with time.

For *artificial ageing* at temperatures above room temperature, the strength rises to a peak before falling to a value below the as-quenched value. The solid solution strength in the as-quenched state is replaced with precipitation hardening, and it is this that produces the peak strength. *Natural ageing* takes place at room temperature, with the strength rising to a plateau value below the peak of the curve for artificial ageing.

The full detail of the precipitate evolution during ageing need not concern us here (see Further Reading for more information). In essence, a whole sequence of metatstable second phases form, starting with clusters of pure Cu, and progressively re-nucleating and growing through various Al-Cu phases toward the equilibrium θ phase (CuAl$_2$). At the same time, the precipitates *coarsen* (i.e. their average size and spacing increase, while their number decreases), and the way they interact with dislocations changes. In combination this all leads to the ageing curve shown in Figure GL4.28(b). Equilibrium is only eventually achieved in the final 'over-aged' condition, with coarse precipitates of CuAl$_2$. The structure is then much as it would have been after slow-cooling from the solutionising temperature. It is clear from Figure GL4.28(b) how important it is to develop the right metastable two-phase microstructure to achieve useful strength in these alloys.

Exercise

GL4.21 The figure shows the aluminium-silver (Al-Ag) phase diagram.

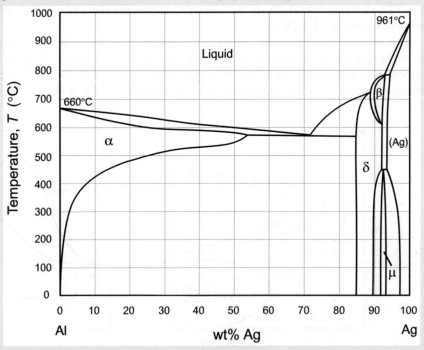

(a) An entrepreneur sees an opportunity for casting decorative artefacts in 'budget silver,' using an Al-Ag alloy. Suggest suitable compositions, explaining your reasoning.
(b) Al-Ag is known to exhibit age hardening, as illustrated earlier for Al-Cu. Identify suitable compositions for wrought heat-treatable Al-Ag alloys, and propose a possible heat-treatment sequence to achieve high strength.

Heat-treatment of carbon steels The Al-Cu example highlighted the key characteristics required for effective precipitation hardening: (a) high solubility of alloying elements at high temperature (to enable an initially uniform distribution of solute); and (b) low solubility at room temperature, with a two-phase region, including a hard compound absorbing most of the alloying additions (to enable fine-scale hard precipitates to be formed). The Fe-C system shows the same key characteristics. Figure GL4.23 shows the high solubility for carbon in iron in the austenite condition, with very low solubility in ferrite, and the prospect of forming precipitates of iron carbide.

A typical medium carbon steel can therefore be 'austenitised' (i.e. solutionised in the austenite field) at 850° to 900°C. We have seen that slow cooling in this case leads to the mixed ferrite-pearlite microstructure, itself used extensively for its excellent strength and toughness. But we can do better, still. Following the same reasoning as before, quenching at a rate above the critical cooling rate will trap the carbon in supersaturated solution. Reheating to temperatures around 500° to 600°C for 1 to 2 hours enables fine-scale precipitates of iron carbide to be formed: in steels this stage is known as *tempering*.

In general terms therefore, quenching and tempering carbon steels parallel age hardening of aluminium alloys, but there are important differences. First, the phase that precipitates is in this case the equilibrium one: iron carbide, in a matrix of ferrite. Second, there is the transformation from FCC iron to BCC iron when it is cooled from the solutionising temperature (which was not an issue in aluminium, which remains FCC throughout).

Phase transformations usually require diffusion to enable the atoms to rearrange into the new phases, but here we have an unusual phase transformation, in which the FCC iron lattice is able to transform during the quench, *without* diffusion. The resulting single phase is called *martensite*. It is metastable and is a supersaturated solid solution of carbon in iron. The iron lattice wishes to form BCC ferrite, but this has very low solubility for carbon, whereas martensite contains the full carbon content of the alloy. As a result, martensite forms with a somewhat distorted body-centred lattice (called 'body-centred tetragonal').

The transformation has a conventional driving force, as there is a difference in free energy, ΔG, between austenite and martensite. But the mechanism is quite distinct, as illustrated in Figure GL4.29(a). Small regions at austenite grain boundaries *shear* to form nuclei of the new lattice, and these propagate very rapidly across the grains forming thin lens-shaped plates of martensite [Figure GL4.29(a) and (b)]. Figure GL4.29(c) shows how this is achieved at the atomic scale purely by *straining* the FCC lattice – there is a simple geometric relationship between the atom sites in FCC and a body-centred structure, and no atom moves more than a fraction of the atomic spacing. This non-diffusive phase change is known as a *displacive transformation*.

The martensite transformation in a given steel has a characteristic temperature at which the transformation starts, and a second (lower) temperature at which it goes to completion. Since diffusion is not involved, the extent of transformation is independent of time – the start and finish of the transformation are shown on TTT diagrams by horizontal lines at the martensite start and finish temperatures [Figure GL4.30(a)].

Finally, what is the effect of quenching and tempering on properties? Since martensite contains far more carbon in solution than can normally be accommodated by BCC ferrite,

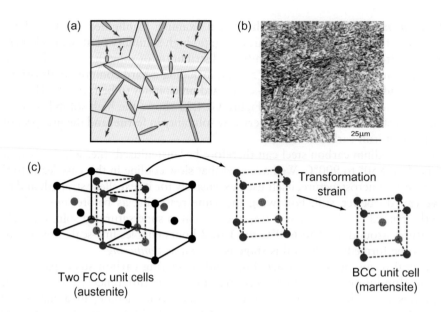

Figure GL4.29   The martensite transformation: (a) nucleation from austenite grain boundaries; (b) micrograph of martensite; (c) lattice relationship between FCC unit cells *(blue)* and the body-centred unit cell *(red)*, which strains to form the distorted structure of martensite. (b, Image courtesy: ASM Micrograph Center, ASM International, 2005.)

the distorted lattice is resistant to dislocation motion and the hardness (and yield stress) is high. But as a result the microstructure is brittle, with a low fracture toughness – similar to that of a ceramic. On tempering, fine precipitates of Fe_3C nucleate and grow, removing the carbon from solution and giving precipitation strength instead. The resulting yield strength is below that of martensite – the yield stress evolves with tempering time as shown in Figure GL4.30(b). But tempering restores the fracture toughness to acceptable levels, comparable to that of the normalised ferrite-pearlite microstructure. The net effect is a yield stress of order two to three times greater in the tempered martensite than in the normalised condition.

This outcome in terms of properties explains the widespread use of the quench-and-temper treatment for tools and machinery, but this heat treatment has some practical consequences in manufacturing steel components. Quenching red-hot steel into cold water or oil induces thermal stresses, which may result in *quench cracking* due to the inherent brittleness of the martensite. Furthermore, there is a limit to how fast the centre of a component can be made to cool, due to the time-scale of heat conduction across the section – the centre cooling rate falls as the size of component increases. There will therefore be a limiting component size that can be quenched at a rate above the critical cooling rate for the steel. For plain carbon steels (Fe-C alloys), the critical cooling rate is high: only small components can form

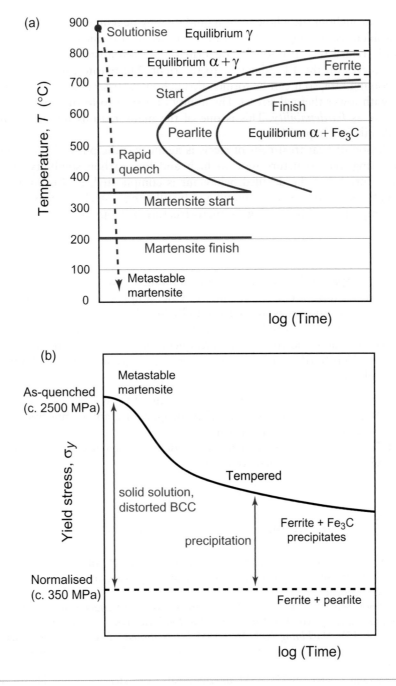

Figure GL4.30   (a) Appearance of martensite start and finish on a TTT diagram; (b) evolution of yield stress during tempering of carbon steels after quenching.

martensite throughout the whole section (and thereby be tempered afterwards). The solution to this problem is *alloying*. Additions of a few wt% of elements such as Ni, Cr, and Mo significantly retard the diffusional transformations from austenite to ferrite-pearlite, shifting the C-curves on the TTT diagram to the right. Martensite can then form at a greatly reduced cooling rate, so it can be formed throughout a larger cross-section prior to tempering (and with lower thermal stress). The ability of a steel to form martensite on quenching is known as its *hardenability*. This is one of the many reasons for producing *alloy steels* (further details in Chapter 11).

   So the lesson of heat treatment of steels is again that the phase diagram points the way to a promising microstructure (due to high and low carbon solubility in iron at different temperatures), but the detailed behaviour is complicated. It also illustrates that phase transformations can be further manipulated by alloying – steels are prime examples of our mantra: composition + processing → microstructure + properties.

Exercise

GL4.22   Three samples of high carbon steel containing 1.2 wt% carbon are to be heat treated. All three are first austenitised. One sample is cooled slowly to room temperature; another is quenched rapidly in cold water, and the third is quenched and then tempered at 600°C. The yield strength was measured in each condition, as follows: slow-cooled 800 MPa; as-quenched 3500 MPa, and quenched and tempered 1100 MPa. Refer to the Fe-C phase diagram in Figure GL4.9 to answer the following:

   (a) What is the minimum temperature required to fully austenitise the steel?
   (b) What are the phases present, their proportions, and compositions, in each condition after heat treatment?
   (c) Describe the microstructure and hardening mechanisms responsible for the yield strengths in each condition.

GL4.9 Further reading

Ashby, M. F., & Jones, D. R. H. (2013). *Engineering materials II* (4th ed.). Butterworth-Heinemann. ISBN 978-0080966687. (Popular treatment of material classes, and how processing affects microstructure and properties).

ASM Handbook Series (1971–2004). *Heat treatment* (vol. 4); *Surface engineering* (vol. 5); *Welding, brazing and soldering* (vol. 6); *Powder metal technologies* (vol. 7); *Forming and forging* (vol. 14); *Casting* (vol. 15); *Machining* (vol. 16). ASM International. (A comprehensive set of handbooks on processing, occasionally updated, and now available online at www.asminternational.org/hbk/index.jsp).

ASM Micrograph Center (2005). ASM International. (An extensive online library of micrographs documenting the composition and process history of the materials illustrated, available at www.products.asminternational.org/mgo/index.jsp).

GL4.10 Further exercises

Solutions for these exercises are also available online.

GL4.23   A special brazing alloy contains 63 wt% gold (Au) and 37 wt% nickel (Ni). The atomic weight of Au (197.0) is more than three times that of Ni (58.7). At a glance, which of the two compositions, in at%, is more likely to be correct?
(a) $X_{Au} = 0.34$, $X_{Ni} = 0.66$
(b) $X_{Au} = 0.66$, $X_{Ni} = 0.34$

GL4.24   The copper-tin system (which includes *bronzes*) is shown in the figure.

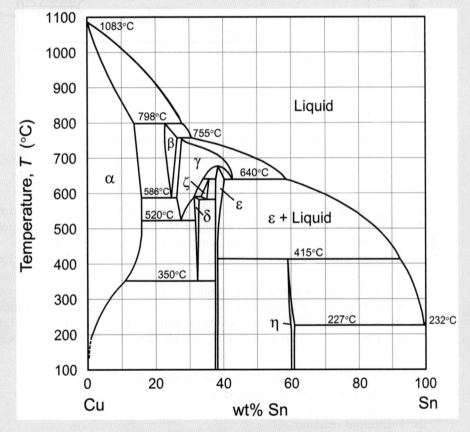

(a) Shade the single-phase regions.
(b) Highlight the four eutectoids in this system and write down their compositions and temperatures.
(c) Find the chemical formulae for the intermediate compounds ε and η (the atomic weights of Cu and Sn are 63.54 and 118.69, respectively).

GL4.25 The figure shows the phase diagram for the Al-Zn system.

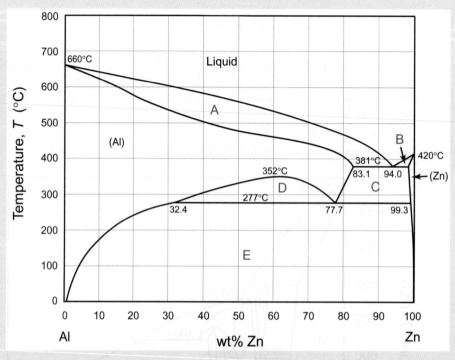

(a) Write down the phases in each of the regions labelled A, B, C, D, and E.
(b) The phase diagram contains two characteristic reaction points. For *each* point, state the type of reaction, temperature, and composition.
(c) List the hardening mechanisms potentially available to Al-Zn alloys. Explain your reasoning.
(d) An Al-Zn alloy is prepared by mixing 20 g of Al and 30 g of Zn. The alloy is melted and then allowed to cool slowly so that equilibrium is maintained. At each of the following temperatures, write down the phases present, their compositions, and the proportions of each phase: (i) 500°C, (ii) 300°C, (iii) 250°C.

GL4.26 By analogy with the definitions for eutectic and eutectoid reactions, and consideration of the phase diagrams in Part 2 of *Guided Learning Unit 4*, define:
(a) Peritectic reaction
(b) Peritectoid reaction

GL4.27 The Al-Ag phase diagram was shown in Exercise GL4.21.
(a) The Al-Ag system has one eutectic reaction. Write down the temperature of the reaction, the composition, and the phases involved.
(b) An Ag–20 wt% Al alloy is cooled slowly to room temperature from the melt. Describe and sketch the microstructural changes that occur, noting key temperatures and phase transformations.

GL4.28   The figure shows the Ti-Al phase diagram (important for the standard commercial alloy Ti-6% Al-4% V).

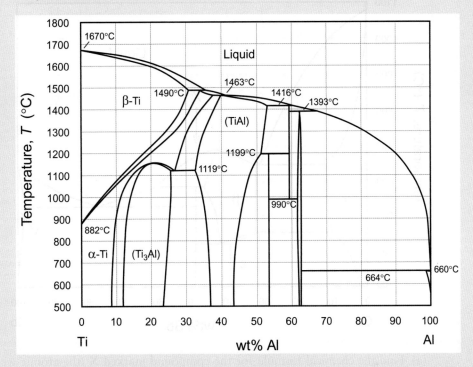

(a) Shade all single-phase fields and highlight three compounds with well-defined compositions. Find the formulae for these three compounds (the atomic weights of Ti and Al are 47.90 and 26.98, respectively). (Note that two of the single-phase fields are also shown with a compound formula in brackets – that is, these fields are solid solutions showing a spread of composition around that of each compound).

(b) Ring the five peritectic points, one peritectoid point, and two eutectoid points.

(c) On heating, over what temperature range does a Ti-6 wt% Al alloy change from α-Ti (HCP) to β-Ti (BCC)? Over what temperature range does it melt?

GL4.29   The figure shows the copper-antimony Cu-Sb phase diagram.

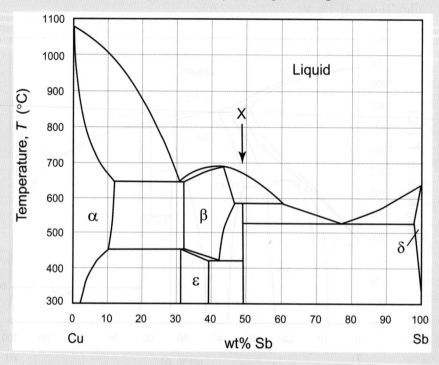

(a) Find the chemical formula for the compound marked X (the atomic weights of Cu and Sb are 63.54 and 121.75, respectively).

(b) The Cu-Sb system contains 2 eutectics, 1 eutectoid, 1 peritectic, and 1 peritectoid. Mark them all on the figure, write down the temperature and composition of each point, and identify the phases involved in each reaction, on cooling.

(c) An alloy containing 95 wt% Sb is cooled slowly to room temperature from the melt. Describe the phase changes that occur during cooling, using schematic sketches of the microstructure at key temperatures to illustrate your answer.

(d) Sketch a temperature-time curve for the 95 wt% Sb alloy over the range 650° to 450°C and account for the shape of the curve.

GL4.30 The phase diagram for the silver-copper Ag-Cu system is shown in the figure, with a linear scale of at% copper. The numbers in brackets refer to wt% copper, which is shown in the upper scale. The liquid single-phase region is indicated.

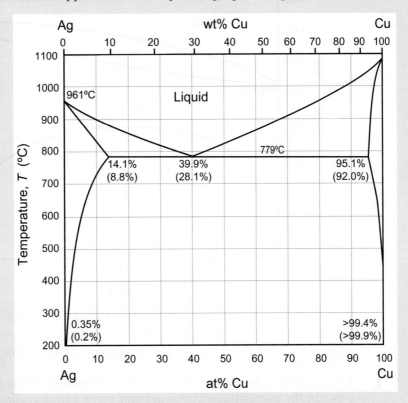

(a) On a rough sketch of the phase diagram indicate the phases that are present in each region of the diagram.

(b) An alloy containing 95 wt% Cu is heated to 1100°C and then slowly cooled to room temperature. Make labelled sketches at key temperatures to show the evolution of the microstructure, and evaluate the final proportions and compositions of the phases present.

(c) Explain how segregation can occur during the non-equilibrium cooling from the liquid state of the alloy described in (b).

(d) Describe, using the phase diagram, how you would choose an alloy composition and a heat-treatment schedule that might be suitable for producing a precipitation-hardened silver-rich alloy.

GL4.31 The figure shows the phase diagram of the Au-Ni binary system. This system shows complete solubility above 810°C and immiscibility at lower temperatures.

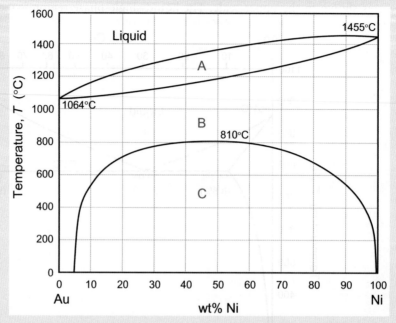

(a) Write down the phase(s) present in each of the regions labelled A, B, and C.

(b) For an Au-Ni alloy with composition 70 wt% Au, write down the phase(s) present, their compositions, and the proportions of each phase at 1200, 900, and 600°C.

(c) An Au–10 wt% Ni alloy is cooled to room temperature from the liquid state. Describe the microstructural changes that occur during cooling if equilibrium is maintained, noting key temperatures and phase transformations. Provide details of the microstructure, compositions, and relative proportions of the phases present at room temperature.

GL4.32 The figure shows schematically the microstructure of a binary Pb-Sn alloy at room temperature, after slow cooling from the liquid state. Use the Pb-Sn phase diagram in Figure GL4.18 to estimate the composition of the alloy, explaining your reasoning in terms of the evolution of the microstructure leading to that in the figure. Sketch key intermediate microstructures during the cooling history to illustrate your answer.

Appendix A
Material property data

Introductory note: read this first

Appendix A lists the applications and properties of the engineering materials that appear in the charts, examples, and exercises in the text. Material properties always have a *range* of values, often wide – particularly for *microstructure-sensitive properties* (such as yield strength). The ranges are indicated by the lengths of the bars and the dimensions of the bubbles on the charts in the main text. However, for convenience in solving problems, the tables that follow show *single, indicative 'point' values* (the average value of the true range). It is important to recognise that, in practice, properties may vary widely from those given for each material – indeed, understanding why is a central theme of this book. Some examples and exercises in the text use alternative, but equally valid, values to those in this Appendix; others refer the reader to this Appendix to look up the indicative values – facility in finding data from reliable sources is an essential skill in materials science and engineering. Each table of numerical data also starts with a schematic illustrating the property ranges of materials classes: metals, ceramics, polymers, composites, and foams.

Contents

Table A1   Material names and applications

Metals	Applications
Ferrous metals	
Cast irons	Automotive parts, engine blocks, machine tool structural parts, lathe beds
High carbon steels	Cutting tools, springs, bearings, cranks, shafts, railway track
Medium carbon steels	General mechanical engineering (tools, bearings, gears, shafts, bearings)
Low alloy ateels	Springs, tools, ball bearings, automotive parts (gears, connecting rods, etc.)
Low carbon steels	Steel structures ('mild steel') – bridges, oil rigs, ships; reinforcement for concrete; automotive parts, car body panels; galvanised sheet; packaging (cans, drums)
Stainless steels	Transport, chemical and food processing plant, nuclear plant, domestic ware (cutlery, washing machines, stoves), surgical implements, pipes, pressure vessels, liquid gas containers
Non-ferrous metals	
Aluminium alloys	
Casting alloys	Automotive parts (cylinder blocks), domestic appliances (irons)
Non-heat-treatable alloys	Electrical conductors, heat exchangers, foil, tubes, saucepans, beverage cans, lightweight ships, architectural panels
Heat-treatable alloys	Aerospace engineering, automotive bodies and panels, lightweight structures and ships
Copper alloys	Electrical conductors and wire, electronic circuit boards, heat exchangers, boilers, domestic water pipes, cookware, coinage, sculptures
Lead alloys	Roof and wall cladding, X-ray shielding, battery electrodes
Magnesium alloys	Automotive castings, wheels, general lightweight castings for transport, nuclear fuel containers; principal alloying addition to aluminium alloys
Nickel alloys	Gas turbines and jet engines, thermocouples, coinage; alloying addition to austenitic stainless steels
Titanium alloys	Aircraft turbine blades; general structural aerospace applications; biomedical implants
Zinc alloys	Die castings (automotive, domestic appliances, toys, handles); coating on galvanised steel

Ceramics	Applications
Glasses	
Borosilicate glass (Pyrex)	Ovenware, laboratory ware, headlights
Glass ceramic	Cookware, lasers, telescope mirrors
Silica glass	High-performance windows, crucibles, high temperature applications
Soda-lime glass	Windows, bottles, tubing, light bulbs, pottery glazes

Table A1   Material names and applications — cont'd

Ceramics	Applications
Technical	
Alumina	Cutting tools, spark plugs, microcircuit substrates, valves
Aluminium nitride	Microcircuit substrates and heat sinks
Boron carbide	Lightweight armour, nozzles, dies, precision tool parts
Silicon	Microcircuits, semiconductors, precision instruments, IR windows, MEMS
Silicon carbide	High temperature equipment, abrasive polishing grits, bearings, armour
Silicon nitride	Bearings, cutting tools, dies, engine parts
Tungsten carbide	Cutting tools, drills, abrasives
Non-technical	
Brick	Buildings
Concrete	General civil engineering construction
Stone	Buildings, architecture, sculpture

Polymers		Applications
Elastomers		
Butyl rubber	IIR	Tyres, seals, anti-vibration mountings, electrical insulation, tubing
Ethylene vinyl acetate	EVA	Bags, films, packaging, gloves, insulation, running shoes
Natural rubber	NR	Gloves, tyres, electrical insulation, tubing
Polychloroprene (Neoprene)	CR	Wetsuits, O-rings and seals, footware
Polyisoprene (Isoprene)	IR	Tyres, inner tubes, insulation, tubing, shoes
Polyurethane elastomers	el-PU	Packaging, hoses, adhesives, fabric coating
Silicone elastomers	SI	Electrical insulation, electronic encapsulation, medical implants
Thermoplastics		
Acrylonitrile butadiene styrene	ABS	Communication appliances, automotive interiors, luggage, toys, boats
Cellulose polymers	CA	Tool and cutlery handles, decorative trim, pens
Ionomer	I	Packaging, golf balls, blister packs, bottles
Polyamides (nylons)	PA	Gears, bearings; plumbing, packaging, bottles, fabrics, textiles, ropes
Polycarbonate	PC	Safety goggles, shields, helmets; light fittings, medical components
Polyetheretherketone	PEEK	Electrical connectors, racing car parts, fibre composites
Polyethylene	PE	Packaging, bags, squeeze tubes, toys, artificial joints
Polyethylene terephthalate	PET	Blow moulded bottles, film, sails

Continued

Table A1   Material names and applications — cont'd

Polymers		Applications
Polylactide	PLA	Biodegradable: plant pots, packaging, bags, toys, disposable cutlery
Polymethyl methacrylate	PMMA	Aircraft windows, lenses, reflectors, lights, compact discs
Polyoxymethylene (Acetal)	POM	Zips, domestic and appliance parts, handles
Polypropylene	PP	Ropes, garden furniture, pipes, kettles, electrical insulation, Astroturf
Polystyrene	PS	Toys, packaging, cutlery, CD cases
Polytetrafluoroethylene (Teflon)	PTFE	Non-stick coatings, bearings, skis, electrical insulation, tape
Polyurethane thermoplastics	tp-PU	Cushioning, seating, shoe soles, hoses, car bumpers, insulation
Polyvinylchloride	PVC	Pipes, gutters, window frames, packaging
Thermoplastic starch	TPS	Biodegradable: plant pots, packaging, bags, toys, disposable cutlery
Thermosets		
Epoxies	EP	Adhesives, fibre composites, electronic encapsulation
Phenolics	PHEN	Electrical plugs, sockets, cookware, handles, adhesives
Polyester	PEST	Furniture, boats, sports goods
Polymer foams		
Flexible polymer foam		Packaging, buoyancy, cushioning, sponges, sleeping mats
Rigid polymer foam		Thermal insulation, sandwich panels, packaging, buoyancy
Composites		
CFRP		Lightweight structural parts (aerospace, bike frames, sports goods, boat hulls and oars, springs)
GFRP		Boat hulls, automotive parts, chemical plant
Natural materials		
Bamboo		Building, scaffolding, paper, ropes, baskets, furniture
Cork		Corks and bungs, seals, floats, packaging, flooring
Leather		Shoes, clothing, bags, drive-belts
Wood		Construction, flooring, doors, furniture, packaging, sports goods

Table A2 Density and price

The densities of materials fall into three broad bands. Metals and ceramics are relatively dense. Polymers and polymer-matrix composites are less so – close to the density of water. Foams are the least dense of all – a lot of the volume is air.

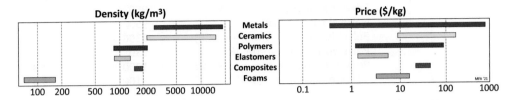

Metals		Density (kg/m³)	Price (US$/kg)
Ferrous metals	Cast iron	7100	0.36
	High carbon steel	7800	0.76
	Medium carbon steel	7800	0.76
	Low alloy steel	7800	0.84
	Low carbon steel	7800	0.76
	Stainless steel	7700	3.1
Non-ferrous metals	Aluminium alloys	2800	4.3
	Copper alloys	8300	5.6
	Lead alloys	11,000	2.9
	Magnesium alloys	1800	2.4
	Nickel alloys	8400	15
	Silver	11,000	540
	Tin	7300	18
	Titanium alloys	4600	24
	Tungsten alloys	18,000	62
	Zinc alloys	6600	3.4

Ceramics		Density (kg/m³)	Price (US$/kg)
Glasses	Borosilicate glass (Pyrex)	2200	5.4
	Silica glass	2200	7.4
	Soda-lime glass	2500	1.5
Technical ceramics	Alumina	3700	25
	Aluminium nitride	3300	130
	Silicon	2300	11
	Silicon carbide	3200	16
	Tungsten carbide	16,000	22
	Zirconia	6000	21
Non-technical ceramics	Brick	2000	1
	Cement	2000	0.11
	Concrete	2400	0.05
	Stone	2400	0.5

Continued

Table A2 Density and price — cont'd

Polymers		Density (kg/m^3)	Price (US$/kg)
Elastomers	Butyl rubber (IIR)	930	2.3
	Ethylene vinyl acetate (EVA)	950	1.8
	Natural rubber (NR)	950	1.6
	Polychloroprene (Neoprene, CR)	1300	4.7
	Polyisoprene (Isoprene, IR)	930	2.4
	Polyurethane elastomers (el-PU)	1200	2.3
	Silicone elastomers (SI)	1100	5.4
Thermoplastics	Acrylonitrile butadiene styrene (ABS)	1000	3
	Cellulose polymers (CA)	990	5.5
	Ionomer (I)	940	3.2
	Polyamides (nylons, PA)	1100	4.1
	Polycarbonate (PC)	1200	3.4
	Polyetheretherketone (PEEK)	1300	92
	Polyethylene (PE)	950	1.6
	Polyethylene terephthalate (PET)	1300	1.4
	Polylactide (PLA)	1300	3.2
	Polymethyl methacrylate (PMMA)	1200	4.3
	Polyoxymethylene (Acetal, POM)	1400	2.9
	Polypropylene (PP)	900	1.4
	Polystyrene (PS)	1000	2.2
	Polytetrafluoroethylene (Teflon, PTFE)	2200	13
	Polyurethane (tp-PU)	1200	4.7
	Polyvinylchloride (PVC)	1400	2.6
	Thermoplastic starch (TPS)	1300	2.8
Thermosets	Epoxies (EP)	1200	2.6
	Phenolics (PHEN)	1300	1.8
	Polyester (PEST)	1200	4.1
Foams	Flexible polymer foam (MD)	90	2.7
	Rigid polymer foam (MD)	110	15

Composites		Density (kg/m^3)	Price (US$/kg)
Isotropic	CFRP, epoxy matrix	1500	36
	GFRP, epoxy matrix	1900	27

Natural		Density (kg/m^3)	Price (US$/kg)
Wood-based	Paper and cardboard	940	1.1
	Softwood	510	0.88
	Hardwood	940	8.5

Table A3   Young's modulus and yield strength (elastic limit) or hardness

Metals and ceramics are both stiff and strong. Polymers are much less stiff, but their strengths overlap with those of metals. Elastomers are distinguished by their exceptionally low stiffness. Yield strength is listed for metals and polymers; for elastomers and composites (which do not show a conventional yield point), the tensile strength is given as the elastic limit. For ceramics and glasses, where compression is more relevant, the Vickers hardness is listed. Note that, for each class of metallic alloys in particular, the range of stiffness is low, while the strength can have a wide range.

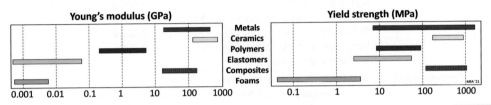

Metals		Young's modulus (GPa)	Yield strength (MPa)
Ferrous metals	Cast iron	170	390
	High carbon steel	210	630
	Medium carbon steel	210	570
	Low alloy steel	205	870
	Low carbon steel	210	300
	Stainless steel	200	540
Non-ferrous metals	Aluminium alloys	72	350
	Copper alloys	100	170
	Lead alloys	15	25
	Magnesium alloys	44	120
	Nickel alloys	210	410
	Silver	71	240
	Tin	43	10
	Titanium alloys	110	870
	Tungsten alloys	340	780
	Zinc alloys	83	160

Ceramics		Young's modulus (GPa)	Vickers hardness (HV)
Glasses	Borosilicate glass (Pyrex)	62	700
	Silica glass	71	650
	Soda-lime glass	70	460
Technical ceramics	Alumina	320	1200
	Aluminium nitride	330	1100
	Silicon	150	1000
	Silicon carbide	430	2400
	Tungsten carbide	660	2800
	Zirconia	220	1100
Non-technical ceramics	Brick	22	29
	Cement	41	5.9
	Concrete	19	6
	Stone	19	16

Continued

Table A3 Young's modulus and yield strength (elastic limit) or hardness — cont'd

Polymers		Young's modulus (GPa)	Yield strength* (MPa)
Elastomers	Butyl rubber (IIR)	0.001	4.9
	Ethylene vinyl acetate (EVA)	0.02	15
	Natural rubber (NR)	0.0016	24
	Polychloroprene (Neoprene, CR)	0.0019	17
	Polyisoprene (Isoprene, IR)	0.0024	22
	Polyurethane elastomers (el-PU)	0.0086	45
	Silicone elastomers (SI)	0.016	9
Thermoplastics	Acrylonitrile butadiene styrene (ABS)	2.4	41
	Cellulose polymers (CA)	1.7	26
	Ionomer (I)	0.29	11
	Polyamides (Nylons, PA)	1.4	50
	Polycarbonate (PC)	2.4	62
	Polyetheretherketone (PEEK)	3.8	79
	Polyethylene (PE)	0.75	23
	Polyethylene terephthalate (PET)	2.9	52
	Polylactide (PLA)	3.4	63
	Polymethyl methacrylate (PMMA)	2.7	62
	Polyoxymethylene (Acetal, POM)	2.9	59
	Polypropylene (PP)	1.2	28
	Polystyrene (PS)	2.7	34
	Polytetrafluoroethylene (Teflon, PTFE)	0.47	21
	Polyurethane (tp-PU)	1.6	46
	Polyvinylchloride (PVC)	2.6	41
	Thermoplastic starch (TPS)	0.6	19
Thermosets	Epoxies (EP)	2.4	51
	Phenolics (PHEN)	3.7	37
	Polyester (PEST)	3	36
Foams	Flexible polymer foam (MD)	0.007	0.18
	Rigid polymer foam (MD)	0.18	1.2

Composites		Young's modulus (GPa)	Yield strength* (MPa)
Isotropic	CFRP, epoxy matrix	100	760
	GFRP, epoxy matrix	20	150

Natural		Young's modulus (GPa)	Yield strength (MPa)
Wood-based	Paper and cardboard	2.8	23
	Softwood	9.3	40
	Hardwood	23	48

*See note above Table A3.

Table A4 Tensile strength and fracture toughness

Metals have both high strength and high fracture toughness K_{1c}. Composites lie lower but still have useful values of both. The low fracture toughness of ceramics and glasses makes design with them more difficult. Polymers appear to be as brittle as ceramics, but their toughness $G_{1c} = (K_{1c})^2/E$ is comparable with that of metals.

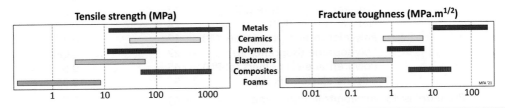

Metals		Tensile strength (MPa)	Fracture toughness (MPa.m$^{1/2}$)
Ferrous metals	Cast iron	600	35
	High carbon steel	1000	50
	Medium carbon steel	800	54
	Low alloy steel	1100	56
	Low carbon steel	450	57
	Stainless steel	820	88
Non-ferrous metals	Aluminium alloys	410	32
	Copper alloys	380	57
	Lead alloys	37	16
	Magnesium alloys	180	15
	Nickel alloys	680	94
	Silver	290	62
	Tin	14	21
	Titanium alloys	950	66
	Tungsten alloys	1000	130
	Zinc alloys	260	19

Ceramics		Tensile strength (MPa)	Fracture toughness (MPa.m$^{1/2}$)
Glasses	Borosilicate glass (Pyrex)	27	0.65
	Silica glass	84	0.69
	Soda-lime glass	33	0.59
Technical ceramics	Alumina	240	3.6
	Aluminium nitride	230	2.8
	Silicon	170	0.88
	Silicon carbide	490	4.1
	Tungsten carbide	450	2.8
	Zirconia	600	6.9
Non-technical ceramics	Brick	9.8	1.4
	Cement	2	0.4
	Concrete	1.2	0.4
	Stone	9.4	0.88

Continued

Table A4 Tensile strength and fracture toughness — cont'd

Polymers		Tensile strength (MPa)	Fracture toughness (MPa.m$^{1/2}$)
Elastomers	Butyl rubber (IIR)	4.9	0.064
	Ethylene vinyl acetate (EVA)	18	0.59
	Natural rubber (NR)	24	0.14
	Polychloroprene (Neoprene, CR)	17	0.17
	Polyisoprene (Isoprene, IR)	22	0.084
	Polyurethane elastomers (el-PU)	45	0.5
	Silicone elastomers (SI)	9	0.35
Thermoplastics	Acrylonitrile butadiene styrene (ABS)	44	2.5
	Cellulose polymers (CA)	26	1.6
	Ionomer (I)	25	2
	Polyamides (nylons, PA)	55	3.5
	Polycarbonate (PC)	67	2.2
	Polyetheretherketone (PEEK)	85	3.4
	Polyethylene (PE)	30	1.6
	Polyethylene terephthalate (PET)	57	5
	Polylactide (PLA)	57	4
	Polymethyl methacrylate (PMMA)	59	1.1
	Polyoxymethylene (Acetal, POM)	73	4
	Polypropylene (PP)	34	3.7
	Polystyrene (PS)	43	0.88
	Polytetrafluoroethylene (Teflon, PTFE)	27	1.5
	Polyurethane (tp-PU)	44	3
	Polyvinylchloride (PVC)	41	3.7
	Thermoplastic starch (TPS)	19	1
Thermosets	Epoxies (EP)	63	0.6
	Phenolics (PHEN)	46	0.98
	Polyester (PEST)	61	1.4
Foams	Flexible polymer foam (MD)	1.1	0.05
	Rigid polymer foam (MD)	1.8	0.02

Composites		Tensile strength (MPa)	Fracture toughness (MPa.m$^{1/2}$)
Isotropic	CFRP, epoxy matrix	760	11
	GFRP, epoxy matrix	150	13

Natural		Tensile strength (MPa)	Fracture toughness (MPa.m$^{1/2}$)
Wood-based	Paper and cardboard	34	2.4
	Softwood	77	3.7
	Hardwood	150	9.9

Table A5   Melting temperature or glass temperature and specific heat

Metals and ceramics have high melting points; for polymers and composites, the data are for the glass transition temperature and are much lower. Elastomers, thermosets, and composites decompose before they melt.

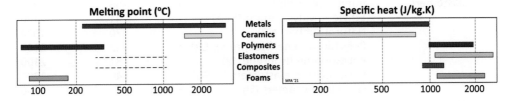

Metals		T_m or T_g (°C)	Specific heat (J/kg.K)
Ferrous metals	Cast iron	1200	480
	High carbon steel	1400	480
	Medium carbon steel	1450	480
	Low alloy steel	1500	480
	Low carbon steel	1500	480
	Stainless steel	1400	480
Non-ferrous metals	Aluminium alloys	570	940
	Copper alloys	940	380
	Lead alloys	290	130
	Magnesium alloys	540	1000
	Nickel alloys	1400	440
	Silver	960	230
	Tin	230	220
	Titanium alloys	1600	540
	Tungsten alloys	3100	130
	Zinc alloys	380	420

Ceramics		T_m or T_g (°C)	Specific heat (J/kg.K)
Glasses	Borosilicate glass (Pyrex)	520	780
	Silica glass	1200	700
	Soda-lime glass	510	900
Technical ceramics	Alumina	2100	770
	Aluminium nitride	2500	790
	Silicon	1400	690
	Silicon carbide	2300	730
	Tungsten carbide	2900	230
	Zirconia	2600	500
Non-technical ceramics	Brick	1100	800
	Cement	1100	840
	Concrete	1100	940
	Stone	1300	880

Continued

Table A5   Melting temperature or glass temperature and specific heat — cont'd

Polymers		T_m or T_g (°C)	Specific heat (J/kg.K)
Elastomers	Butyl rubber (IIR)	−68	1900
	Ethylene vinyl acetate (EVA)	−41	2100
	Natural rubber (NR)	−70	1900
	Polychloroprene (Neoprene, CR)	−42	2100
	Polyisoprene (Isoprene, IR)	−81	2100
	Polyurethane elastomers (el-PU)	−40	1700
	Silicone elastomers (SI)	−65	1100
Thermoplastics	Acrylonitrile butadiene styrene (ABS)	110	1700
	Cellulose polymers (CA)	110	1400
	Ionomer (I)	44	1800
	Polyamides (nylons, PA)	54	1500
	Polycarbonate (PC)	150	1200
	Polyetheretherketone (PEEK)	170	1500
	Polyethylene (PE)	−110	1800
	Polyethylene terephthalate (PET)	14	1200
	Polylactide (PLA)	56	1200
	Polymethyl methacrylate (PMMA)	10	1500
	Polyoxymethylene (Acetal, POM)	−12	1400
	Polypropylene (PP)	−20	1900
	Polystyrene (PS)	95	1200
	Polytetrafluoroethylene (Teflon, PTFE)	120	1000
	Polyurethane (tp-PU)	74	1600
	Polyvinylchloride (PVC)	84	1000
	Thermoplastic starch (TPS)	14	1600
Thermosets	Epoxies (EP)	130	1200
	Phenolics (PHEN)	210	1500
	Polyester (PEST)	180	1500
Foams	Flexible polymer foam (MD)	−40	2000
	Rigid polymer foam (MD)	100	1500

Composites		T_m or T_g (°C)	Specific heat (J/kg.K)
Isotropic	CFRP, epoxy matrix	130	970
	GFRP, epoxy matrix	130	1100

Natural		T_m or T_g (°C)	Specific heat (J/kg.K)
Wood-based	Paper and cardboard	-	1300
	Softwood	-	1700
	Hardwood	-	1700

Table A6   Thermal conductivity and thermal expansion

Metals and some ceramics conduct heat well. Polymers, foams, and composites insulate. Materials with high melting points (refractory metals and ceramics) have low expansion coefficient. Those with low values (notable polymers) expand greatly when heated.

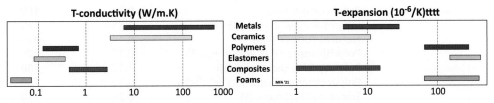

Metals		T-conductivity (W/m.K)	T-expansion (10^{-6}/K)
Ferrous metals	Cast iron	36	11
	High carbon steel	50	12
	Medium carbon steel	51	12
	Low alloy steel	43	12
	Low carbon steel	52	12
	Stainless steel	19	13
Non-ferrous metals	Aluminium alloys	160	24
	Copper alloys	120	19
	Lead alloys	31	28
	Magnesium alloys	79	26
	Nickel alloys	12	13
	Silver	420	20
	Tin	61	23
	Titanium alloys	7.3	9
	Tungsten alloys	91	4.9
	Zinc alloys	110	27

Ceramics		T-conductivity (W/m.K)	T-expansion (10^{-6}/K)
Glasses	Borosilicate glass (Pyrex)	1.1	3.6
	Silica glass	1.4	0.64
	Soda-lime glass	0.95	9.3
Technical ceramics	Alumina	25	7.2
	Aluminium nitride	170	4.7
	Silicon	140	2.5
	Silicon carbide	100	4.4
	Tungsten carbide	70	6.1
	Zirconia	2.9	11
Non-technical ceramics	Brick	0.57	9.4
	Cement	0.85	18
	Concrete	1.3	7.7
	Stone	2.1	13

Continued

Table A6   Thermal conductivity and thermal expansion — cont'd

Polymers		T-conductivity (W/m.K)	T-expansion (10^{-6}/K)
Elastomers	Butyl rubber (IIR)	0.1	180
	Ethylene vinyl acetate (EVA)	0.35	170
	Natural rubber (NR)	0.14	220
	Polychloroprene (Neoprene, CR)	0.17	220
	Polyisoprene (Isoprene, IR)	0.11	260
	Polyurethane elastomers (el-PU)	0.29	160
	Silicone elastomers (SI)	0.25	270
Thermoplastics	Acrylonitrile butadiene styrene (ABS)	0.26	95
	Cellulose polymers (CA)	0.14	120
	Ionomer (I)	0.26	230
	Polyamides (nylons, PA)	0.24	130
	Polycarbonate (PC)	0.21	120
	Polyetheretherketone (PEEK)	0.25	120
	Polyethylene (PE)	0.42	160
	Polyethylene terephthalate (PET)	0.14	120
	Polylactide (PLA)	0.14	140
	Polymethyl methacrylate (PMMA)	0.2	120
	Polyoxymethylene (Acetal, POM)	0.23	150
	Polypropylene (PP)	0.14	150
	Polystyrene (PS)	0.13	120
	Polytetrafluoroethylene (Teflon, PTFE)	0.25	140
	Polyurethane (tp-PU)	0.24	110
	Polyvinylchloride (PVC)	0.18	73
	Thermoplastic starch (TPS)	0.17	210
Thermosets	Epoxies (EP)	0.19	97
	Phenolics (PHEN)	0.15	120
	Polyester (PEST)	0.29	130
Foams	Flexible polymer foam (MD)	0.057	160
	Rigid polymer foam (MD)	0.03	37

Composites		T-conductivity (W/m.K)	T-expansion (10^{-6}/K)
Isotropic	CFRP, epoxy matrix	1.8	2
	GFRP, epoxy matrix	0.47	17

Natural		T-conductivity (W/m.K)	T-expansion (10^{-6}/K)
Wood-based	Paper and cardboard	0.24	10
	Softwood	0.26	4.7
	Hardwood	0.45	4.7

Table A7   Electrical resistivity and dielectric constant

Metals and a few ceramics conduct electricity well. All other materials are non-conductors. Some ceramics and polymers have high dielectric constants because they polarise in a field. Foams have low values because they are mostly gas (dielectric constant = 1).

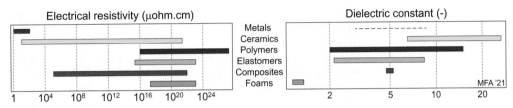

Metals		Resistivity (μohm.cm)	Dielectric constant (-)
Ferrous metals	Cast iron	52	-
	High carbon steel	18	-
	Medium carbon steel	18	-
	Low alloy steel	23	-
	Low carbon steel	17	-
	Stainless steel	75	-
Non-ferrous metals	Aluminium alloys	4.4	-
	Copper alloys	7	
	Lead alloys	24	-
	Magnesium alloys	9.1	-
	Nickel alloys	110	-
	Silver	1.7	-
	Tin	11	-
	Titanium alloys	160	-
	Tungsten alloys	11	-
	Zinc alloys	6.4	-

Ceramics		Resistivity (μohm.cm)	Dielectric constant (-)
Glasses	Borosilicate glass (Pyrex)	1.0×10^{22}	5.3
	Silica glass	1.0×10^{25}	3.8
	Soda-lime glass	2.5×10^{18}	7.1
Technical ceramics	Alumina	1.0×10^{21}	9.4
	Aluminium nitride	1.0×10^{20}	8.6
	Silicon	1.0×10^{9}	11
	Silicon carbide	3.2×10^{10}	7.5
	Tungsten carbide	45	-
	Zirconia	7.7×10^{19}	17
Non-technical ceramics	Brick	1.4×10^{14}	6.3
	Cement	6.6×10^{11}	9.8
	Concrete	5.8×10^{12}	9.8
	Stone	1.0×10^{12}	7.3

Continued

Table A7   Electrical resistivity and dielectric constant — cont'd

Polymers		Resistivity (μohm.cm)	Dielectric constant (-)
Elastomers	Butyl rubber (IIR)	3.2×10^{22}	2.3
	Ethylene vinyl acetate (EVA)	5.6×10^{21}	2.9
	Natural rubber (NR)	3.2×10^{21}	2.5
	Polychloroprene (Neoprene, CR)	1.0×10^{18}	7.3
	Polyisoprene (Isoprene, IR)	3.2×10^{15}	2.7
	Polyurethane elastomers (el-PU)	5.7×10^{17}	6.3
	Silicone elastomers (SI)	1.2×10^{20}	2.7
Thermoplastics	Acrylonitrile butadiene styrene (ABS)	9.9×10^{21}	3.0
	Cellulose polymers (CA)	9.9×10^{18}	5.1
	Ionomer (I)	9.9×10^{21}	2.3
	Polyamides (nylons, PA)	1.2×10^{20}	9.2
	Polycarbonate (PC)	3.2×10^{20}	3.2
	Polyetheretherketone (PEEK)	9.9×10^{21}	3.2
	Polyethylene (PE)	3.1×10^{23}	2.3
	Polyethylene terephthalate (PET)	9.9×10^{20}	3.6
	Polylactide (PLA)	4.3×10^{17}	3.1
	Polymethyl methacrylate (PMMA)	9.9×10^{23}	3.3
	Polyoxymethylene (Acetal, POM)	9.9×10^{20}	3.8
	Polypropylene (PP)	9.9×10^{22}	2.2
	Polystyrene (PS)	1.0×10^{26}	2.5
	Polytetrafluoroethylene (Teflon, PTFE)	9.9×10^{23}	2.1
	Polyurethane (tp-PU)	9.9×10^{18}	6.8
	Polyvinylchloride (PVC)	1.0×10^{21}	3.1
	Thermoplastic starch (TPS)	1.0×10^{17}	4.5
Thermosets	Epoxies (EP)	2.4×10^{21}	4.0
	Phenolics (PHEN)	9.9×10^{18}	4.9
	Polyester (PEST)	9.9×10^{18}	3.5
Foams	Flexible polymer foam (MD)	3.2×10^{21}	1.2
	Rigid polymer foam (MD)	1.0×10^{19}	1.1

Composites		Resistivity (μohm.cm)	Dielectric constant (-)
Isotropic	CFRP, epoxy matrix	4.0×10^{5}	-
	GFRP, epoxy matrix	6.8×10^{21}	5

Natural		Resistivity (μohm.cm)	Dielectric constant (-)
Wood-based	Paper and cardboard	1.0×10^{14}	3.9
	Softwood	1.1×10^{14}	5.6
	Hardwood	1.1×10^{14}	10

Table A8   Embodied energy, carbon footprint, and water usage

Embodied energy is the energy that must be committed to create 1 kg of usable material: 1 kg of steel stock or of cement powder, for example. The CO_2 footprint of a material is the mass of CO_2 released into the atmosphere per unit mass of material, units kg $CO_{2.eq}$/kg.[1] Water usage measures the water consumption in litres to make 1 kg of usable material. The Critical Material (Crit. Mat.) flag (X) highlights that members of this material family may contain more than 5% of a 'critical' material.

Carbon footprint, material production (kg/kg)   Water usage, material production (l/kg)

Metals		Energy (MJ/kg)	$CO_{2.eq}$ (kg/kg)	Water (l/kg)	Crit. Mat. flag
Ferrous metals	Cast iron	32	2.4	45	-
	High carbon steel	32	2.4	45	-
	Medium carbon steel	32	2.4	47	-
	Low carbon steel	31	2.3	46	-
	Low alloy steel	31	2.5	49	X
	Stainless steel	73	5.4	130	X
Non-ferrous metals	Aluminium alloys	190	12	1e3	X
	Copper alloys	59	3.6	310	-
	Lead alloys	32	2.6	480	X
	Magnesium alloys	320	45	980	X
	Nickel alloys	220	12	260	X
	Silver	1500	100	76,000	-
	Tin	230	16	11,000	X
	Titanium alloys	620	36	290	X
	Tungsten alloys	550	36	160	X
	Zinc alloys	52	4	380	-

Ceramics		Energy (MJ/kg)	$CO_{2.eq}$ (kg/kg)	Water (l/kg)	Crit. Mat. flag
Glasses	Borosilicate glass (Pyrex)	29	1.7	15	-
	Silica glass	39	2.3	1.4	-
	Soda-lime glass	11	0.76	14	-
Technical ceramics	Alumina	52	2.8	55	-
	Aluminium nitride	230	13	240	-
	Silicon	120	5	24	-
	Silicon carbide	170	7.2	58	-
	Tungsten carbide	580	38	83	X
	Zirconia	72	4	49	X
Non-technical ceramics	Brick	3	0.24	5.5	-
	Cement	2.8	0.44	37	-
	Concrete	0.82	0.12	3.4	-
	Stone	1	0.06	3.4	-

Continued

[1]The atomic weight of carbon is 12, that of oxygen is 16, so 1 kg of carbon, when burnt, produces (12 + 2 × 16) = 3.6 kg of CO_2.

Table A8   Embodied energy, carbon footprint, and water usage — cont'd

Polymers		Energy (MJ/kg)	$CO_{2.eq}$ (kg/kg)	Water (l/kg)	Crit. Mat. flag
Elastomers	Butyl rubber (IIR)	95	4.4	110	-
	Ethylene vinyl acetate (EVA)	79	2.1	2.8	-
	Natural rubber (NR)	78	2	17,000	-
	Polychloroprene (Neoprene, CR)	64	1.5	220	-
	Polyisoprene (Isoprene, IR)	86	3.9	150	-
	Polyurethane elastomers (el-PU)	82	3.2	96	-
	Silicone elastomers (SI)	120	6.5	3.30	-
Thermoplastics	Acrylonitrile butadiene styrene (ABS)	92	3.4	170	-
	Cellulose polymers (CA)	89	3.4	240	-
	Ionomer (I)	110	5.5	280	-
	Polyamides (nylons, PA)	140	7	380	-
	Polycarbonate (PC)	110	4.8	170	-
	Polyetheretherketone (PEEK)	300	17	920	-
	Polyethylene (PE)	80	1.9	58	-
	Polyethylene terephthalate (PET)	82	2.7	130	-
	Polylactide (PLA)	55	2.8	21	-
	Polymethyl methacrylate (PMMA)	100	4.9	76	-
	Polyoxymethylene (Acetal, POM)	86	3.2	250	-
	Polypropylene (PP)	69	2.9	39	-
	Polystyrene (PS)	82	2.5	150	-
	Polytetrafluoroethylene (Teflon, PTFE)	300	16	450	-
	Polyurethane (tp-PU)	82	3.2	96	-
	Polyvinylchloride (PVC)	65	2.7	190	-
	Thermoplastic starch (TPS)	37	1.6	170	-
Thermosets	Epoxies (EP)	130	6.6	28	-
	Phenolics (PHEN)	77	1.9	51	-
	Polyester (PEST)	71	2.5	200	-
Foams	Flexible polymer foam	90	3.2	170	-
	Rigid polymer foam	80	5.1	450	-

Composites		Energy (MJ/kg)	$CO_{2.eq}$ (kg/kg)	Water (l/kg)	Crit. Mat. flag
Isotropic	CFRP, epoxy matrix (isotropic)	690	48	1400	-
	GFRP, epoxy matrix (isotropic)	120	6.6	160	-

Natural		Energy (MJ/kg)	$CO_{2.eq}$ (kg/kg)	Water (l/kg)	Crit. Mat. flag
Wood-based	Paper and cardboard	51	1.2	1700	-
	Softwood	11	0.37	700	-
	Hardwood	12	0.6	700	-

Table A9   Recycle fraction, recycle energy, and recycle carbon footprint

Not all the embodied energy of a material is really 'embodied,' but some is. This retained energy means that the energy to recycle a material is usually less than that required for its first production. The recycled fraction in current supply (column 3) is an indicator of the viability of recycling. Columns 4 and 5 give an indication of the associated energy demand and CO_2 release to recycle the material into a usable form.

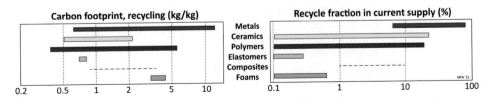

Metals		Recycle fraction (%)	R-energy (MJ/kg)	R-carbon $CO_{2.eq}$ (kg/kg)
Ferrous metals	Cast iron	69	8.5	0.67
	High carbon steel	42	8.5	0.67
	Medium carbon steel	42	8.5	0.67
	Low carbon steel	42	8.1	0.64
	Low alloy steel	42	9	0.65
	Stainless steel	37	16	1.2
Non-ferrous metals	Aluminium alloys	43	33	2.6
	Copper alloys	43	13	1.1
	Lead alloys	72	8.6	0.67
	Magnesium alloys	39	49	3.8
	Nickel alloys	24	37	2.9
	Silver	6	150	12
	Tin	6	38	2.9
	Titanium alloys	23	81	6.4
	Tungsten alloys	36	75	5.9
	Zinc alloys	22	12	0.97

Ceramics		Recycle fraction (%)	R-Energy (MJ/kg)	R-Carbon $CO_{2.eq}$ (kg/kg)
Glasses	Borosilicate glass (Pyrex)	24	24	1.2
	Silica glass	25	32	1.6
	Soda-lime glass	24	8.7	0.53
Technical ceramics	Alumina	0.7	36	2
	Aluminium nitride	0.1	-	-
	Silicon	0.71	-	-
	Silicon carbide	0.1	-	-
	Tungsten carbide	31	-	-
	Zirconia	0.1		
Non-technical ceramics	Brick	17	-	-
	Cement	1.2	-	-
	Concrete	13	0.8	0.07
	Stone	1.4	-	-

Continued

Table A9   Recycle fraction, recycle energy, and recycle carbon footprint — cont'd

Polymers		Recycle fraction (%)	R-Energy (MJ/kg)	R-Carbon $CO_{2,eq}$ (kg/kg)
Elastomers	Butyl rubber (IIR)	2.8	-	-
	Ethylene vinyl acetate (EVA)	0.1	27	0.72
	Natural rubber (NR)	0.1	-	-
	Polychloroprene (Neoprene, CR)	1.4	-	-
	Polyisoprene (Isoprene, IR)	0.1	-	-
	Polyurethane elastomers (el-PU)	0.1	-	-
	Silicone elastomers (SI)	0.1	-	-
Thermoplastics	Acrylonitrile butadiene styrene (ABS)	4	32	1.2
	Cellulose polymers (CA)	0.7	30	1.2
	Ionomer (I)	0.1	38	1.9
	Polyamides (nylons, PA)	0.71	43	3
	Polycarbonate (PC)	0.7	37	2.4
	Polyetheretherketone (PEEK)	1.4	100	5.7
	Polyethylene (PE)	8.4	27	0.94
	Polyethylene terephthalate (PET)	21	28	1.5
	Polylactide (PLA)	0.33	18	0.95
	Polymethyl methacrylate (PMMA)	0.7	38	2.2
	Polyoxymethylene (Acetal, POM)	0.1	36	1.8
	Polypropylene (PP)	5.5	23	0.99
	Polystyrene (PS)	6	29	1.3
	Polytetrafluoroethylene (Teflon, PTFE)	0.7	39	1.9
	Polyurethane (tp-PU)	0.1	28	1.1
	Polyvinylchloride (PVC)	1.5	23	1.1
	Thermoplastic starch (TPS)	0.1	8.6	0.37
Thermosets	Epoxies (EP)	0.7	-	-
	Phenolics (PHEN)	0.7	-	-
	Polyester (PEST)	0.1	-	-
Foams	Flexible polymer foam	8.4	49	3.9
	Rigid polymer foam	0.1	-	-

Composites		Recycle fraction (%)	R-energy (MJ/kg)	R-carbon $CO_{2,eq}$ (kg/kg)
Isotropic	CFRP, epoxy matrix	0.1	-	-
	GFRP, epoxy matrix	0.1	-	-

Natural		Recycle fraction (%)	R-Energy (MJ/kg)	R-Carbon $CO_{2,eq}$ (kg/kg)
Wood-based	Paper and cardboard	72	22	1.2
	Softwood	9	-	-
	Hardwood	9	-	-

Table A10   Durability: preferred materials and coatings for given environments

Experience in balancing the conflicting factors that influence corrosion has led to a set of 'preferred' materials for use in each environment. They are listed here for 23 commonly encountered environments.

Environment	Metals	Polymers, composites, natural materials	Ceramics, glasses
Water and aqueous solutions			
Water (fresh)	Aluminium alloys, stainless steel, galvanised steel, copper alloys	PET, HDPE, PA, GFRP. All polymers are corrosion-free in fresh water, though some absorb up to 5%, causing swelling.	Glass, concrete, brick, porcelain
Water (salt)	Copper, bronze, stainless steel, galvanised steel, lead, platinum, gold, silver	PET, HDPE, PA, GFRP. All polymers are corrosion-free in salt water, though some absorb up to 5%, causing swelling.	Glass, concrete, brick
Soil	Steel, often coated or with galvanic protection	HDPE, PP, PVC, most polymers (except PHB, PLA, and those that are bio-degradable) and woods corrode only slowly in soil.	Brick, pottery, glass, concrete
Body fluids	316L stainless steel Cobalt-chromium alloys Titanium and its alloys Nickel-titanium alloys (Nitinols) Precious metals: gold, platinum, silver, amalgam	Acrylic (PMMA) Silicones Ultra-high mol. wt. polyethylene (UHMWPE) Polyimides	Hydroxyapatite Alumina Zirconia Calcium phosphate Vitreous carbon Bio-glass
Acids and alkalis			
Acetic acid, CH_3COOH (10%)	Aluminium, stainless steel, nickel and nickel alloys, titanium, Monel	HDPE, PTFE	Glass, porcelain, graphite
Hydrochloric acid, HCl (10%)	Copper, nickel and nickel alloys, titanium, Monel, molybdenum, tantalum, zirconium, platinum, gold, silver	HDPE, PP, GFRP, Neoprene	Glass

Continued

Table A10   Durability: preferred materials and coatings for given environments — cont'd

Environment	Metals	Polymers, composites, natural materials	Ceramics, glasses
Hydrofluoric acid (HF; 40%)	Lead, copper, stainless steel, carbon steels, Monel, Hastelloy C, platinum, gold, silver	PTFE, PE, PP	Graphite
Nitric acid HNO_3 (10%)	Stainless steel, titanium, 14.5% silicon cast iron, alloy 20 (40Fe, 35Ni, 20Cr, 4Cu)	PTFE, PVC, phenolics, PE-CTFE	Glass, graphite
Sulphuric acid, H_2SO_4 (10%)	Stainless steel, copper, nickel-based alloys, lead, tungsten, 14.5% silicon, cast iron, zirconium, tantalum, alloy 20 (40Fe, 35Ni, 20Cr, 4Cu), platinum, gold, silver	PET, PTFE, PE-CTFE, phenolics	Glass, graphite
Sodium hydroxide, NaOH (10%)	Nickel and its alloys, stainless steel, magnesium	PVC, LDPE, HDPE, PTFE, PE-CTFE	Glass, graphite
Fuels, oils, and solvents			
Benzene, C_6H_6	Carbon steel, stainless steel, aluminium, brass	PTFE, Polyamide (nylon), PEEK, PVC	Glass, graphite
Carbon tetrachloride CCl_4	Carbon steel, stainless steel, aluminium, Hastelloy C, Monel, platinum, gold, silver	POM, PTFE, rubber	Glass, graphite
Crude oil	Carbon steel, aluminium, stainless steel	HDPE, PTFE, epoxies, polyamides	Glass, porcelain, enameled metal
Diesel oil	Carbon steel, stainless steel, brass, copper, aluminium, Monel	HDPE, PP, PTFE, buna (nitrile) rubber, PVC, GFRP	Glass
Kerosene (paraffin oil)	Carbon steel, stainless steel, aluminium, Monel	HDPE, PP, PTFE, GFRP	Glass
Lubricating oil	Stainless steel, carbon steel, aluminium	HDPE, PP, PTFE, rubber	Glass
Petroleum (gasoline)	Carbon steel, stainless steel, brass, aluminium, Hastelloy C, alloy 20 (40Fe, 35Ni, 20Cr, 4Cu)	PTFE, HDPE, PP, buna (nitrile) rubber, Viton, GFRP	Glass

Table A10   Durability: preferred materials and coatings for given environments — cont'd

Environment	Metals	Polymers, composites, natural materials	Ceramics, glasses
Silicone fluid, $[(CH_3)2SiO]_n$	Carbon steel, aluminium	HDPE, PP, PET	Glass
Vegetable oil	Aluminium, carbon steel, stainless steel	HDPE, PP, PET	Glass
Alcohols, aldehydes, ketones			
Acetone, CH_3COCH_3	Aluminium, carbon steel, stainless steel	PTFE, HDPE, PP copolymer (PPCO)	Glass, graphite
Ethyl alcohol, C_2H_5OH	Steel, stainless steel, copper	PTFE, HDPE, PP, GFRP	Glass
Methyl alcohol, CH_3OH	Steel, stainless steel, lead, Monel	PTFE, HDPE, PP	Glass
Formaldehyde, CH_2O	Stainless steel, Monel, Hastelloy C	PTFE, HDPE, PET, POM, nitrile or butyl rubber	Glass, GFRP
Built environments			
Rural atmosphere	Galvanised steel, aluminium alloys, bronze, copper	PVC, polyurethane, epoxy, polyester, PTFE, nylon, CFRP, GFRP, PU, PP, PE, woods	Glass, brick, concrete, granite, slate
Industrial atmosphere	Stainless steel, aluminium alloys, lead, bronze, cast iron	PVC, polyurethane, epoxy, polyester, PTFE, nylon, CFRP, GFRP, PU, PP, PE	Glass, brick, concrete, granite, slate
Marine atmosphere	Galvanised steel, stainless steel, aluminium alloys, bronze, copper	PVC, polyurethane, epoxy, polyester, PTFE, nylon, CFRP, GFRP, PU, PP, PE, woods	Glass, brick, concrete, granite, slate

Index

Note: Page numbers followed by "*f*" indicate figures, "*t*" indicate tables, and "*b*" indicate boxes.